# ADVANCES IN BIOREMEDIATION OF WASTEWATER AND POLLUTED SOIL

# Advances in Bioremediation of Wastewater and Polluted Soil

*Editor:*

**Sarika Garg, Ph.D.**

ArclerPress

www.arclerpress.com

**Advances in Bioremediation of Wastewater and Polluted Soil**
*Editor: Sarika Garg, Ph.D.*


**708 3rd Avenue, 6th Floor**
**New York**
**NY 10017**
**United States of America**
**www.arclerpress.com**

ISBN: 978-1-68094-470-9

Library of Congress Control Number: 2016951361

# About the Editor

## Sarika Garg

### Ph.D. Montreal, Canada

Sarika Garg procured her doctoral degree at the Max Planck Unit For Structural Molecular Biology, University of Hamburg, Germany in 2011. This was followed by Postdoctoral Scientist position at University of Saskatchewan, Canada. Later, she moved to CRCHUM, University of Montreal for her second Postdoc. She has diverse research interests and has contributed immensely in many research fields viz. Enzyme Technology, Environmental Science, and Neuroscience. She has numerous research papers and review articles to her credit. She has coauthored two books 'Enzymes & Enzyme Technology'; and 'Biochemical Tests: Principles and Protocols'. She has also contributed several chapters in various books including a chapter on Regulation of Gene Expression in Encyclopedia of Molecular Cell Biology and Molecular Medicine. She is a recipient of 2012 SHRF Postdoctoral Fellowship Award and 2012 Laura E. Chapman Research Award. She is currently working as a Career Counselor in her very own venture 'higherstudiescounseling.com', Montreal, Canada and as a freelancer science writer.

# Contents

# List of Contributors

**Baba Uqab**
Department of Environmental Sciences/Centre of Research for Development, University of Kashmir, Jammu and Kashmir, India

**Syeed Mudasir**
Department of Environmental Sciences/Centre of Research for Development, University of Kashmir, Jammu and Kashmir, India

**Ruqeya Nazir**
Department of Environmental Sciences/Centre of Research for Development, University of Kashmir, Jammu and Kashmir, India

**Arindam Mitra**
Department of Microbiology, Adamas University, Kolkata 700126, West Bengal, India

**Suman Mukhopadhyay**
Division of Microbiology and Infectious Diseases, NIAID/NIH/HHS, Bethesda, MD, USA

**Ruchita Dixit**
National Bureau of Agriculturally Important Microorganisms (NBAIM), Kushmaur, Maunath Bhanjan 275 101, India

**Wasiullah**
National Bureau of Agriculturally Important Microorganisms (NBAIM), Kushmaur, Maunath Bhanjan 275 101, India

**Deepti Malaviya**
National Bureau of Agriculturally Important Microorganisms (NBAIM), Kushmaur, Maunath Bhanjan 275 101, India

**Kuppusamy Pandiyan**
National Bureau of Agriculturally Important Microorganisms (NBAIM), Kushmaur, Maunath Bhanjan 275 101, India

**Udai B. Singh**
National Bureau of Agriculturally Important Microorganisms (NBAIM), Kushmaur, Maunath Bhanjan 275 101, India

**Asha Sahu**
Division of Soil Biology, Indian Institute of Soil Science, Nabibagh, Berasia Road, Bhopal 462 038, India

**Renu Shukla**
National Bureau of Agriculturally Important Microorganisms (NBAIM), Kushmaur, Maunath Bhanjan 275 101, India

**Bhanu P. Singh**
Udai Pratap Autonomous College, Varanasi 221 002, India

**Jai P. Rai**
Department of Mycology and Plant Pathology (Krishi Vigyan Kendra), Institute of Agricultural Sciences, Banaras Hindu University, Varanasi 221 005, India

**Pawan Kumar Sharma**
National Bureau of Agriculturally Important Microorganisms (NBAIM), Kushmaur, Maunath Bhanjan 275 101, India

**Harshad Lade**
Department of Environmental Engineering, Konkuk University, Seoul 143-701, Korea

**Diby Paul**
Department of Environmental Engineering, Konkuk University, Seoul 143-701, Korea

**J. Paz-Ferreiro**
Key Laboratory of Vegetation Restoration and Management of Degraded Ecosystems, South China Botanical Garden, Chinese Academy of Sciences, Guangzhou 510650, China
Departamento de Edafología, ETSI Agrónomos, Universidad Politécnica de Madrid, Avenida Complutense 3, Madrid 28050, Spain

**H. Lu**
Key Laboratory of Vegetation Restoration and Management of Degraded Ecosystems, South China Botanical Garden, Chinese Academy of Sciences, Guangzhou 510650, China
University of Chinese Academy of Sciences, Beijing 100049, China

**S. Fu**
Key Laboratory of Vegetation Restoration and Management of Degraded Ecosystems, South China Botanical Garden, Chinese Academy of Sciences, Guangzhou 510650, China

**A. Méndez**
Departamento de Edafología, ETSI Agrónomos, Universidad Politécnica de Madrid, Avenida Complutense 3, Madrid 28050, Spain

**G. Gascó**
Departamento de Edafología, ETSI Agrónomos, Universidad Politécnica de Madrid, Avenida Complutense 3, Madrid 28050, Spain

**Du RY**
Public Monitoring Center for Agro-Product, Guangdong Academy of Agricultural Sciences, Guangzhou 510640, China and Key Laboratory of Testing and Evaluation for Agro-product Safety and Quality, Ministry of Agriculture, Guangzhou 510640, China

**Wen D**
Public Monitoring Center for Agro-Product, Guangdong Academy of Agricultural Sciences, Guangzhou 510640, China and Key Laboratory of Testing and Evaluation for Agro-product Safety and Quality, Ministry of Agriculture, Guangzhou 510640, China

**Zhao PH**
Public Monitoring Center for Agro-Product, Guangdong Academy of Agricultural Sciences, Guangzhou 510640, China and Key Laboratory of Testing and Evaluation for Agro-product Safety and Quality, Ministry of Agriculture, Guangzhou 510640, China

**Chen Y**
Public Monitoring Center for Agro-Product, Guangdong Academy of Agricultural Sciences, Guangzhou 510640, China and Key Laboratory of Testing and Evaluation for Agro-product Safety and Quality, Ministry of Agriculture, Guangzhou 510640, China

**Wang FH**
Public Monitoring Center for Agro-Product, Guangdong Academy of Agricultural Sciences, Guangzhou 510640, China and Key Laboratory of Testing and Evaluation for Agro-product Safety and Quality, Ministry of Agriculture, Guangzhou 510640, China

**Jaysankar De**
National Institute of Oceanography, Dona Paula, Goa 403004, India
Division of Microbiology, German Research Centre for Biotechnology, Mascheroder Weg 1, 38124 Braunschweig,
UNESCO Chair-Life Sciences International Postgraduate Educational Center. Acharian 31, Yerevan 0040, Armenia

**Johannes Leonha¨user**
Division of Microbiology, German Research Centre for Biotechnology, Mascheroder Weg 1, 38124 Braunschweig,
Bayer Technology Services GmbH; PT-PT-Environmental & Solids Processing Building E 41; 51368 Leverkusen, Germany

**Lilit Vardanyan**
Vanevan Institute, 14 Yerevanyan Street, Martuni, Gegharkunik, Armenia
UNESCO Chair-Life Sciences International Postgraduate Educational Center. Acharian 31, Yerevan 0040, Armenia

**Dhara Shukla**
204 'A', Facility for Ecological and Analytical Testing IIT, Kanpur, India

**Padma S. Vankar**
204 'A', Facility for Ecological and Analytical Testing IIT, Kanpur, India

**Sarvesh Kumar Srivastava**
204 'A', Facility for Ecological and Analytical Testing IIT, Kanpur, India

**Nilanjana Das**
Environmental Biotechnology Division, School of Biosciences and Technology, VIT University, Vellore, Tamil Nadu 632014, India

**Preethy Chandran**
Environmental Biotechnology Division, School of Biosciences and Technology, VIT University, Vellore, Tamil Nadu 632014, India

**Hussein Al-Nasrawi**
Fulbright visiting scholar, Florida state university, USA

**Asghari Bano**
Department of Plant Scineces Quaid-i-azam University Islamabad, Pakistan

**Asim Shahzad**
Department of Plant Scineces Quaid-i-azam University Islamabad, Pakistan

**Samina Siddiqui**
Centre of Excellence in Geology University of Peshwar, Peshawar, Pakistan

**Maria Genovese**
Institute for Coastal Marine Environment, CNR, Messina, Italy

**Francesca Crisafi**
Institute for Coastal Marine Environment, CNR, Messina, Italy

**Renata Denaro**
Institute for Coastal Marine Environment, CNR, Messina, Italy

**Simone Cappello**
Institute for Coastal Marine Environment, CNR, Messina, Italy

**Daniela Russo**
Institute for Coastal Marine Environment, CNR, Messina, Italy
Department of Biological and Environmental Sciences, University of Messina, Messina, Italy

**Rosario Calogero**
Institute for Coastal Marine Environment, CNR, Messina, Italy

**Santina Santisi**
Institute for Coastal Marine Environment, CNR, Messina, Italy
Department of Biological and Environmental Sciences, University of Messina, Messina, Italy
**Maurizio Catalfamo**
Institute for Coastal Marine Environment, CNR, Messina, Italy

**Alfonso Modica**
Environmental Laboratory, Syndial SpA, Priolo Gargallo, Italy

**Francesco Smedile**
Institute for Coastal Marine Environment, CNR, Messina, Italy

**Lucrezia Genovese**
Institute for Coastal Marine Environment, CNR, Messina, Italy

**Peter N. Golyshin**
Environmental Genomics, School of Biological Sciences, Bangor University, Bangor, UK

**Laura Giuliano**
Institute for Coastal Marine Environment, CNR, Messina, Italy
Mediterranean Science Commission, Monaco, Monaco

**Michail M. Yakimov**
Institute for Coastal Marine Environment, CNR, Messina, Italy

**Agnieszka Nowak**
Department of Biochemistry, Faculty of Biology and Environmental Protection, University of Silesia, Jagiellońska 28, 40-032 Katowice, Poland

**Agnieszka Mrozik**
Department of Biochemistry, Faculty of Biology and Environmental Protection, University of Silesia, Jagiellońska 28, 40-032 Katowice, Poland

**Samuel E. Agarry**
Biochemical Engineering and Biotechnology Laboratory, Department of Chemical Engineering, Ladoke Akintola University of Technology, Ogbomoso, Nigeria.

**Mujidat O. Aremu**
Biochemical Engineering and Biotechnology Laboratory, Department of Chemical Engineering, Ladoke Akintola University of Technology, Ogbomoso, Nigeria.

**Oluwafunmilayo A. Aworanti**
Biochemical Engineering and Biotechnology Laboratory, Department of Chemical Engineering, Ladoke Akintola University of Technology, Ogbomoso, Nigeria.

**Gabriella Caruso**
Institute for Coastal Marine Environment (IAMC), Italian National Research Council (CNR), Spianata S. Raineri, 86, 98122 Messina, Italy

**Katarzyna Kurnik**
Chair of Plant Physiology and Biotechnology, Nicolaus Copernicus University, Lwowska 1, 87-100 Toruń, Poland

**Krzysztof Treder**
Laboratory of Molecular Diagnostic and Biochemistry, Department of Potato Protection and Seed Science in Bonin, Plant Breeding and Acclimatization Institute-National Research Institute, 76-009 Bonin, Poland

**Monika Skorupa-Kłaput**
Chair of Plant Physiology and Biotechnology, Nicolaus Copernicus University, Lwowska 1, 87-100 Toruń, Poland

**Andrzej Tretyn**
Chair of Plant Physiology and Biotechnology, Nicolaus Copernicus University, Lwowska 1, 87-100 Toruń, Poland
Centre for Modern Interdisciplinary Technologies, Nicolaus Copernicus University, Wileńska 4, 87-100 Toruń,Poland

**Jarosław Tyburski**
Chair of Plant Physiology and Biotechnology, Nicolaus Copernicus University, Lwowska 1, 87-100 Toruń, Poland
Centre for Modern Interdisciplinary Technologies, Nicolaus Copernicus University, Wileńska 4, 87-100 Toruń,Poland

**Dhan Prakash**
Institute of Microbial Technology (CSIR), Sector 39-A, Chandigarh -160036, India
Department of Environmental Hydrology & Microbiology, Ben-Gurion University of the Negev, Sede- Boqer Campus, 84990, Israel

**Prashant Gabani**
Division of Biological and Health Sciences, University of Pittsburgh, Bradford, PA 16701, USA.

**Anuj K Chandel**
Department of Biotechnology, School of Engineering of Lorena, University of Sao Paulo, Lorena, Brazil.

**Zeev Ronen**
Department of Environmental Hydrology & Microbiology, Ben-Gurion University of the Negev, Sede- Boqer Campus, 84990, Israel

**Om V Singh**
Division of Biological and Health Sciences, University of Pittsburgh, Bradford, PA 16701, USA.

**Maria Romero-González**
Department of Geography, Kroto Research Institute, The University of Sheffield, Broad Lane, Sheffield, UK

**Bennett C. Nwaobi**
Department of Life Sciences, Imperial College London, South Kensington Campus, London, UK

**Joseph M. Hufton**
Department of Geography, Kroto Research Institute, The University of Sheffield, Broad Lane, Sheffield, UK

**Daniel J. Gilmour**
Department of Molecular Biology and Biotechnology, University of Sheffield, Western Bank, Sheffield, UK

**Ariel Marchlewicz**
Department of Biochemistry, Faculty of Biology and Environmental Protection, University of Silesia in Katowice, Jagiellonska 28, 40-032 Katowice, Poland

**Urszula Guzik,**
Department of Biochemistry, Faculty of Biology and Environmental Protection, University of Silesia in Katowice, Jagiellonska 28, 40-032 Katowice, Poland

**Danuta Wojcieszyńska**
Department of Biochemistry, Faculty of Biology and Environmental Protection, University of Silesia in Katowice, Jagiellonska 28, 40-032 Katowice, Poland

**Ariel Marchlewicz**
Department of Biochemistry, Faculty of Biology and Environmental Protection, University of Silesia in Katowice, Jagiellońska 28, 40-032 Katowice, Poland

**Dorota Domaradzka**
Department of Biochemistry, Faculty of Biology and Environmental Protection, University of Silesia in Katowice, Jagiellońska 28, 40-032 Katowice, Poland

**Urszula Guzik**
Department of Biochemistry, Faculty of Biology and Environmental Protection, University of Silesia in Katowice, Jagiellońska 28, 40-032 Katowice, Poland

**Danuta Wojcieszyńska**
Department of Biochemistry, Faculty of Biology and Environmental Protection, University of Silesia in Katowice, Jagiellońska 28, 40-032 Katowice, Poland

**Dorota Domaradzka**
Department of Biochemistry, Faculty of Biology and Environmental Protection, University of Silesia in Katowice, Jagiellońska 28, 40-032 Katowice, Poland

**Urszula Guzik**
Department of Biochemistry, Faculty of Biology and Environmental Protection, University of Silesia in Katowice, Jagiellońska 28, 40-032 Katowice, Poland

**Katarzyna Hupert-Kocurek**
Department of Biochemistry, Faculty of Biology and Environmental Protection, University of Silesia in Katowice, Jagiellońska 28, 40-032 Katowice, Poland

**Danuta Wojcieszyńska**
Department of Biochemistry, Faculty of Biology and Environmental Protection, University of Silesia in Katowice, Jagiellońska 28, 40-032 Katowice, Poland

# Preface

The deterioration of an environment due to industrial effluents is a worldwide challenge. There are several different types of toxic inorganic and organic pollutants, namely pesticides, heavy metals, petroleum hydrocarbons, lubricants, chlorophenols, plastic, drugs, phenol, and radionuclides. These pollutants have deleterious effects on human beings and animals. The monumental benefits obtained from the use of pesticides, herbicides, and chemical fertilizers in agriculture and forestry sectors come with environmental costs. These man-made chemicals destroy soil fertility and drain into the waterways thus polluting rivers, ponds, lakes and coastal areas. Similarly, the industrial waste containing several hazardous chemicals is detrimental to an environment. Therefore, the remediation of such toxic pollutants is imperative for the benefits of human health and environment.

Bioremediation is the process of using organisms to nullify or degrade the environmental pollutants in contaminated soil and water. Microorganisms can be used to cure several types of pollution including wastewater problems, oil spillage, and soil pollution. Naturally occurring microbes eat and digest organic substances for nutrients and energy. However, microorganisms cannot remediate all types of contaminants with ease. Under such circumstances, phytoremediation- a process that uses plants to accumulate toxic substances is taken into consideration.

This book gathers the recent published works related to biodegradation and bioremediation of wastewater and polluted soil with the hope that it will help students in understanding the basics and applied effects of bioremediation. The first Chapter highlights the bioremediation of pesticides. The second Chapter features the beneficial use of microbial biofilms in bioremediation. In Chapters 3-7, discussion is on the bioremediation of heavy metals. Biodegradation of petroleum hydrocarbon is described in Chapter 8. Cleaning up oil spills through bioremediation is discussed in Chapters 9-11. In Chapters 12 and 13, degradation of chlorophenols is described. Managing waste plastic materials with no adverse impacts through bioremediation is discussed in Chapter 14. Removal of phenol, a toxic substance, from an industrial effluent is illustrated in Chapter 15. Remediation of radionuclides is described in Chapters 16 and 17. Exploiting bioremediation process for degrading anti-inflammatory drugs is discussed in the last three Chapters of the book.

I hope that this book will be advantageous for students, researchers, teachers, and industrial experts.

**Editor**

**Sarika Garg, Ph.D.**

**Montreal, Canada**

# 1
CHAPTER

# REVIEW ON BIOREMEDIATION OF PESTICIDES

Baba Uqab, Syeed Mudasir, and Ruqeya Nazir

Department of Environmental Sciences/Centre of Research for Development, University of Kashmir, Jammu and Kashmir, India

## ABSTRACT

**With the onset of industrial revolution and manufacture of various pesticides has no doubt enhanced yield in our agricultural products and also protect our majority of crops from the pests. As we cannot lose the majority of crops to these pests. Pesticides at present play an important role in enhancing the yield and provide an economical benefit to our farmers, but the use of pesticides in the agricultural fields is a major concern today. Increase in soil pollution has caused a serious concern. Large numbers of contaminants in which pesticide is one the major concern have pose a serious threat to human health as well as to the natural ecosystem. The available methods (physical or chemical) are either incomplete or costly. Bioremediation provides a novel tool or such process. Bioremediation provides an eco-friendly, economical and efficient method for detoxification of pesticides.**

## INTRODUCTION

Pesticides are the chemical substances use to kill or manage pests at tolerable levels. The suffix "cide" literally means to kill. Pesticides consist of different products with different functions. Table 1 but the designation is formed by combining the names of pest with suffix [1].

**Citation:** Baba Uqab, Syeed Mudasir, Ruqeya Nazir, Review on Bioremediation of Pesticides, doi.org/10.4172/2155-6199.1000343.

The rapid increase in population has resulted in accumulation of variety of chemicals in the environment. Thus the production of these xenobiotics has forced to implement new technologies to reduce or eliminate them from the environment. Earlier techniques or technologies which were used to eliminate them from environment were landfills, recycling, pyrolysis etc., but these also have adverse effects on the environment and leads to formation of toxic intermediates [2]. These methods proved to be expensive and difficult to execute especially in case of pesticides [3].

The promising technology which utilizes the ability of microorganisms to remove pollution from the environment and are eco-friendly, economical and versatile is Bioremediation [4]. The extensive use of pesticides has resulted in serious environmental as well as health problems besides has effected biodiversity as well [5,6].

The use of pesticides not only degrade the soil quality but also reaches the water table hence enters aquatic environment also, so it can be infer that fate of pesticides is often uncertain, thus decontamination off pesticide polluted areas is very complex process [5]. Low degree of biodegradability has classified them into persistent toxic substances [7].

## Contamination of Soil

Soil being important resource on the planet earth is being degraded from variety of sources. Heavy metals, pesticides, municipal garbage. Municipal garbage contains discarded materials from home and industry contains paper, plastic and organic matter [8]. Heavy metals in soil come from atmospheric deposition, sewage, irrigation, industry and use of pesticides and fertilizers [9]. Effect of contamination may result in the loss of biodiversity and functioning of soil like nutrient cycling etc. Heavy metals also inhibit microbial activity [10].

## Contamination of Water

Water is arguably the most important resource on the planet earth providing the unique habitat to variety of organisms. Pollution levels in the water ecosystems had resulted in loss of fresh water content on planet earth. Lack of access to toilet facilities among Indians has crossed 700 million mark and about 1000 Indians die of diarrhea every day. Lack of fresh water has acute scarcity in china and 500 million people lack access to safe drinking water Excess use of fertilizers, herbicides and pesticides cause serious damage to life present in waters. Excess Phosphorous results in the Eutrophication. Among the pesticides 98% were classified as acutely toxic for fishes and crustaceans [11].

## Pesticide Scenario of India and World

Pesticide consumption in world has reached to 2 million tones as per Abhilash and Nandita [12] and from these 2 million tones Europe utilizes 45% followed by USA 24% and rest 25% in rest of the world. Pesticide consumption in Asia is also alarming. China uses highest percentage followed by Korea, Japan and India. In India use of pesticide is about 0.5 kg/hectare and large contribution is from organochlorine pesticides.

The usage is because of warm humid climatic conditions [13]. The concept of green revolution has played a important role for utilization of variety of pesticides for high yield varieties. Presently India is largest producer of pesticides in Asia and ranks 12th in world. Pesticide residue in several crops has also affected the export of agricultural commodities in the last few years. In this context, pesticide safety, regulation of pesticide use, proper application technologies, and integrated pest management are some of the key strategies for minimizing human exposure to pesticides and to maintain the fertility of the soils for proper productivity. There is a dearth of studies related to these issues in India Uttar Pradesh is the largest consumer followed by Punjab, Haryana and Maharashtra. Regarding the pesticide share across agricultural crops, cotton account for 45% followed by rice (25%), chillies/vegetables/fruits (13-24%), plantations (7-8%), cereals/ millets/oil seeds (6-7%), sugarcane (2-3%) and other (1-2%) [12,14].

**Table 1:** Classification of pesticides according to their target

| Pesticide | Target | Pesticide | Target |
|---|---|---|---|
| Algicides | Algae | Molluscicides | Snails |
| Avicides | Birds | Nematicides | Nematode |
| Bactericides | Bacteria | Virucides | Viruses |
| Fungicides | Fungi | Rodenticides | Rodents |
| Insecticides | Insects | Miticides or Acaricides | Mites |

## Pesticide Classification

Pesticides encompass a variety of different types of chemicals including herbicides, insecticides, fungicides and rodenticides. Pesticides are usually classified on the basis of structure (Table 2). The structural classification include organochlorine, organophosphorus, carbamates, nitrogen based pesticides [14].

## Bioremediation history and use

Bioremediation from it root meaning means to use microorganisms to remediate/ destroy or to immobilize pollutant from environment [15]. Natural Bioremediation has been used by civilizations for the treatment of waste water but intentional use for reduction of hazardous waste is more recent development. Modern bioremediation and use of microbes to consume pollutants are credited in part to George Robinson He used microbes to consume an oil spill along the coast of Santa Barbara, Calfornia in the Tale 1960.

## Pesticide Concerns

Pesticides are not only toxic to humans but they pose a threat to safety of soil water and air quality [16]. The pesticide contamination of surface and ground water pose a serious threat to surrounding ecosystems. The organochlorine and organophosphates cause tumors, irritability and convulsions [14]. Besides this organochlorine pesticides cause serious environmental issues due to Biomagnifications (Figure 1; Table 3).

## Pesticide Bioremediation Methods

The level of toxicity caused by the pesticides leads to the great need for bioremediation. No doubt in some cases intrinsic bioremediation occurs because of microbes that are already present in polluted ecosystems, but it is also true that in some cases intrinsic bioremediation is not adequate. The requirements for the process of bioremediation of pesticides given by Ref. [14] are summarized in Table 4.

## Strategies for Pesticide Remediation

Pesticide pollution is a serious environmental problem and their remediation is necessary. Ideally treatment should result in destruction of the compounds without generation of intermediates (Table 5).

## Bacterial Degradation of Pesticides

Bacteria species that degrade the pesticides belongs to genera Flavobacterium, Arthobacter, Aztobacter, Burkholderia, and pseudomonas [17]. Recently Bacteruim raoultella sp is also found to degrade pesticides.

The complete biodegradation of the pesticide involves the oxidation of the parent compound resulting in to carbon dioxide and water, this provides energy to microbes. The soil where innate microbial population cannot be able to manage pesticides, the external addition of pesticide degrading micro flora is recommended. Degradation of pesticides by microbes not only depends on the enzyme system but also the conditions like temperature, pH and nutrients. Some of the pesticides are easily degraded however some are recalcitrant because of presence of anionic species in the compound. Besides organophosphorus compounds, the Neonicotionoids are degraded by the Pseudomonas species (Figure 2).

## Role of Fungi

The minor structural changes that fungi does to degrade pesticides and render them into nontoxic substances and release them into soil where it is susceptible to further degradation. The various fungi which have shown ability to degrade pesticides are given in Table 6.

## Role of Enzymes

Enzymes take part in key role in Biodegradation of any xenobiotics and are able to renovate pollutants to a noticeable rate and have prospective to restore polluted environment [18]. Enzymes are also involved in the degradation of pesticide compounds, both in the target organism, through intrinsic detoxification mechanisms and evolved metabolic resistance, and in the wider environment, via biodegradation by soil and water microorganisms. P. putida theoretical oxygen demand (TOD) enzyme is a representative of a much larger family of enzymes with application in the biocatalysis of environmentally relevant reactions. Fungal enzymes especially, oxidoreductases, laccase and peroxidases have prominent application in removal of polyaromatic hydrocarbons (PAHs) contaminants either in fresh, marine water or terrestrial [19]. The enzymes play a key role in the biodegradation of any xenobiotics compounds.

**Table 2:** Types of Pesticides and Examples from Ref. [14]

| Pesticide | Examples |
|---|---|
| Insecticide | |
| Organophosphorus | Diazinon, dichlorvos, dimethoate, malathion, parathion |
| Carbamate | Carbaryl, propoxur, Aldicarb methiocarb |
| Organochlorine | DDT, methoxychlor, toxaphene, mirex, Kepone |
| Cyclodienes | Aldrin, chlordane, dieldrin, endrin, endosulfan, heptachlor |
| Herbicides | Chlorophenoxy acids, hexachlorobenzene (HCB) |
| Nitrogen-based | Picloram, Atrazine, diquat, paraquat |
| Organophosphates | Glyphosate (Roundup) |
| Fungicide | |
| Nitrogen-containing | Triazines, dicarboximides, phthalimide |
| Wood preservatives | Creosote, hexachlorobenzene |
| Botanicals | Perethrin, permethrin |
| Antimicrobial | Chlorine, quaternary alcohols |

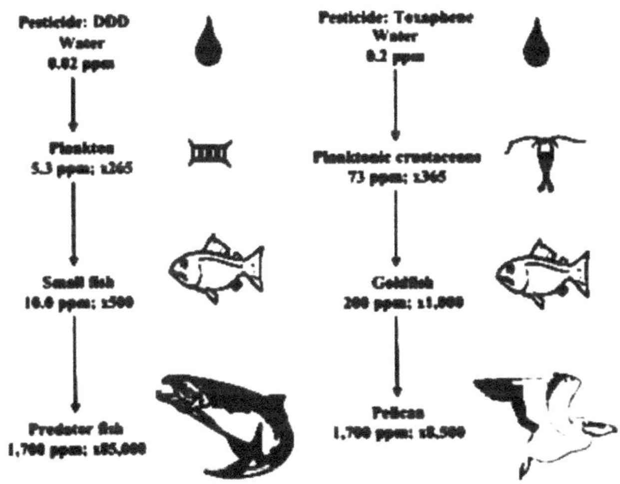


**Figure 1:** Biomagnification of Toxic Chemicals, from Ref. [14].

The organophosphorus compounds have been studied in detail and hence much of the literature is available describing the OP degrading enzymes. In 1973, the first bacterium to degrade OP compounds was isolated from a soil sample from the Philippines and was identified as Flavobacterium sp. ATCC 27551. Since then, several bacteria, a few fungi and cyanobacteria, have been isolated that can use OP compounds as a source of carbon, nitrogen or phosphorus (Tables 7 and 8).

**Table 3:** Health effects of common pesticides, from Ref. [23]

| Pesticide | Persistence (Half-life) | Health Effects |
|---|---|---|
| Aldrin | 20 days to 1 year | Nervous system effects. Probable carcinogen. **Large doses**: convulsions, death. **Moderate doses**: dizziness, headaches, vomiting, uncontrolled muscle movement |
| Dichlorodiphe nyltrichloroethane (DDT) | 2 to15 years | Nervous system effects (tremors, seizures); probable carcinogen |
| Chlordane | 4 years | Nervous system, digestive system, liver effects. Headaches, irritability, confusion, weakness, vision problems, vomiting, stomach cramps, diarrhea, and jaundice for lower doses. **Higher doses**: convulsions and death. |
| Dieldrin | Up to 7 years | Nervous system effects. Probable carcinogen. **Large doses:** Convulsions, death. **Moderate doses:** Dizziness, headaches, vomiting, uncontrolled muscle movement. |
| Heptachlor | 0.4 to 2 years | Nervous system damage, liver and adrenal gland damage, tremors |

**Table 4:** Requirements for the process of bioremediation of pesticides, from Ref. [15]

| Factor | Conditions required |
|---|---|
| Micro organisms | Aerobic or Anaerobic |
| Natural biological processes of micro organisms | Catabolism and Anabolism |
| Environmental factors | Oxygen content Temperature, pH, Electron acceptor/ donor |
| Nutrients | Carbon, Nitrogen, oxygen etc., |
| Soil moisture | 25-28 % of water holding capacity |
| Type of soil | Low clay or slit content |

**Table 5:** Technologies available for treatment of pesticide-contaminated sites

| Technology | Treatment time in months | Treatment media | Removal efficiency | References |
|---|---|---|---|---|
| Bioremediation | 3 (*ex-situ*) | Soil, sludge, ground water, sediments | Up to 99.8% | [24] |
| Phyto remediation | 3 (*ex-situ*) | Soil, sludge, ground water, sediments | Up to 80% | [29] |

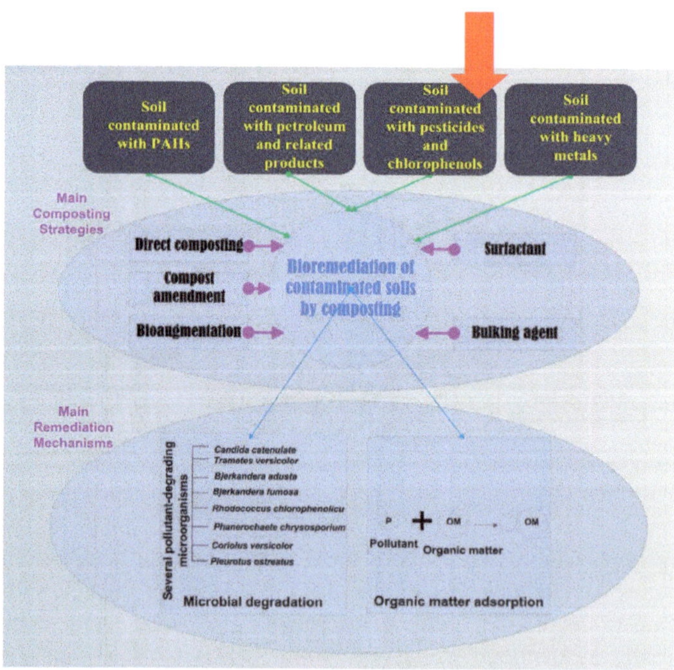


**Figure 2:** Bioavailability and degradation of phenanthrene in compost amended soils, from Ref. [25].

**Table 6:** Various fungi which have shown ability to degrade pesticides

| Species of fungi | Potential for degrading pesticide | Reference |
|---|---|---|
| *Flammulina velupites, Stereum hirsutum, Coriolus versicolor, Dichomitus squalens, Hypholoma fasciculare, Auricularia auricula, Pleurotus ostreatus,* *Avatha discolor* and *Agrocybe semiorbicularis* | triazine, phenylurea, dicarboximid, chlorinated organophosphorus compounds | [30] |
| White-rot fungi | Heptachlor atrazine, terbuthylazine, lindane, metalaxyl, chlordane mirex, gammahexachlorocyclohexane (g-HCH), dieldrin, diuron, aldrin, DDT, etc., | [30] |

**Table 7:** Sources of enzymes and there degradation property

| Enzyme | Source | Degradation |
|---|---|---|
| Arly aclyamidase | *Bacillus sphaericus* | Herbicide and fungicide |
| Organophosphorus hydrolase (OPH) | *B.diminuta* and *Flavobacterium* sp. | Xenobiotics compounds |
| Organophosphorus acid anhydrolase (OPAA) | *Alteromonas undina* and *Alteromonas haloplanktis* | Xenobiotics compounds |

**Table 8:** Role of different organisms in bioremediation

| Process | Source | Pollutant | Microbes/plants | References |
|---|---|---|---|---|
| Biodegradation | Garden, beach and mud | Saw | *Aspergillus* sp *Trichoderma* sp | [28] |
| Degradation | Cellulosic materials | Blue dye 2B | *Bacillus* sp | [21] |
| Phytoremedtion | Soil | Pb, Cd | *Vetiveria zizanioides* and *Eichornia crassipes* | [8] |
| Phytoremedation Adsorption | Sewage irrigated soils wastewater | Heavy metals | Flagellate sp. Of *Dunaliella* algae | [26] |
| Adsorption | Soil | Cu, Mn, Zn, Pb, Cr and Pd | Three herbaceous plants (*Plantago major*, L., *Taraxacum officinale* and *Urtica diocia* | [26] |

**Table 9:** Plant products used in pesticides to target pests

| Plant products used as pesticides | Target pests |
|---|---|
| Neem | Variety of sucking and chewing insects |
| Rotenone | Leaf feeding insects e.g., aphids, certain beetles and caterpillars as well as fleas and lice on animals |
| Limonene | Fleas, aphids, mites, ants and house cricket. |
| Linalool | Fleas, aphids, mites, ants and house cricket. |
| Pyrethrum | Ants, aphids, roaches and ticks |
| Ryania | Caterpillars and thrips |
| Sabadilla | Squash bugs, harlequin bugs, thrips, caterpillars |

Ample work has been done to discover various plant products that can be used as biopesticides, hence made it easier to destroy target pests with these products a list of such plant products in given in table given below [20].

Variety of organisms has been involved to degrade pesticides and the result has been a success [21-25]. Toxicity class of pesticides given by WHO in which the pesticides have been classified into three classes which represent their hazardous potential have also been found degradable by the microorganisms. Hence, the process of bioremediation has been accelerated by the use of such organisms [26-30] (Table 9).

## CONCLUSION

No doubt the pesticides have caused serious impact on the soil fertility. Soils contaminated with pesticides have attracted high attention because it impacts human health and natural ecosystem. Bioremediations has a tremendous potential for remediation of the soils that are affected by pesticides. Microorganisms that are present in the soils can remove pesticides from the environment. Biopesticide enzymatic degradation of polluted environment represents most important strategy for pollutant removal and degradation of persistent chemical substances by enzymatic reactions have been found high bioremediation potential. Hence bioremediation is much promising approach to overcome the pesticide pollution that can surely solve the problem of pesticide pollution of soils. This technology has proved again and again its potential to degrade not only pesticides but also the various in organic compounds. So time is to utilize this eco-friendly technology for better and safe future.

## REFERENCES


1. Norris A (2011) Past Alaska Master Gardener Manual.
2. Debarati P, Gunjan P, Janmejay P, RakeshVJK (2005) Accessing Microbial Diversity for Bioremediation and Environmental Restoration. Trends in Biotechnology 23: 135-142
3. Jain RK, Kapur M, Labana S, Lal B, Sarma PM, et al. (2005) Microbial Diversity: Application of Microorganisms for the Biodegradation of Xenobiotics. Current Science89: 101-112.
4. Finley SD, Broadbelt LJ, Hatzimanikatis V (2010) In silico feasibility of novel biodegradation pathways for 1,2,4-trichlorobenzene. MC Syst Biol 4: 7.
5. Gavrilescu M (2005) Fate of Pesticides in the Environment and its Bioremediation. Engineer in Life Science 5: 497-526.
6. Hussain S, Siddique T, Arshad M, Saleem M (2009) Bioremediation and Phytoremediation of Pesticides: Recent Advances. Criti Rev in Env Scieand Tech 39: 843-907.
7. Tayade S, Patel ZP, Mutkule DS, Kakde AM (2013) Pesticide contamination in food: A review. IOSR J Agri Vet Sci 6: 7-11.
8. Burken JG, Schnoor JL (2003) Uptake and metabolism of atrazine by poplar trees. Environ Sci Technol 31:1399-1406.
9. Zhang W, Jiang F, Ou J (2011) Global pesticide consumption and pollution: with China as a focus. Proc Int Acad Ecol Environ Sci 1: 125-144.
10. Su C, Jiang L, Zhang W (2014) A review on heavy metal contamination in the soil worldwide:Situation, impact and remediation techniques. Environ Skeptics Critics 3: 24-38.
11. de la Cruz E, Bravo-Durán V, Ramírez F, Castillo LE (2014) Environmental hazards associated with pesticide import into Costa Rica, 1977-2009. J Environ Biol 35: 43-55.
12. Abhilash PC, Singh N (2009) Pesticide use and application: an Indian scenario. J Hazard Mater 165: 1-12.
13. Bhat D, Padmaja P (2014) Assessment of organic pesticides in Ground and surface water in Bhopal India. IOSR journal of Environ Scie Toxicol and Food Technology 8: 51-52.

14. Vaccari A, Strom F, Alleman E (2006) Environmental Biology for Engineers and Scientists.

15. Shanahan P (2004) Bioremediation. Waste Containment and Remediation Technology. Spring.

16. Yao ZT, Ji XS, Sarker PK, Tang JH, Ge LQ, et al. (2015) A comprehensive review on the applications of coal fly ash. EarthScience Reviews 141: 105-121

17. Glazer AN, Nikaido H (2007) Microbial biotechnology: Fundamentals of applied Microbiology. Cambridge University Press, Cambridge, New York pp: 510-528.

18. Rao MA, Scelza Scotti R, Gianfreda L (2010) Role of enzymes in the remediation of polluted environments. J Soil Sci Plant Nutr 10: 333-353.

19. Balaji V, Arulazhagan P, Ebenezer P (2014) Enzymatic bioremediation of polyaromatic hydrocarbons by fungal consortia enriched from petroleum contaminated soil and oil seeds. J Environ Biol 35: 521-529.

20. Kandpal V (2014) Biopesticides. Int J Environ Res Dev 4: 191-196.

21. Bhoosreddy GL(2014) Decolorization and biodegradation of direct blue 2B by Mix Consortia of Bacillus. IOSR J Pharm Biol Sci 9: 34-40.

22. Gupta PK (2004) Pesticide exposure-Indian scene. Toxicology 198: 83-90.

23. CynthiaG, Ana H (2004) Phytoremediation Field Studies Database for Chlorinated Solvents. Pesticides, Explosives, and Metals.

24. Niu GL, Zhang JJ, Zhao S, Liu H, Boon N, et al. (2009) Bioaugmentation of a 4-chloronitrobenzene contaminated soil with Pseudomonas putida ZWL73. Environ Pollut 157: 763-771.

25. Puglisi E, Cappa F, Fragoulis G, Trevisan M, Del Re AA (2007) Bioavailability and degradation of phenanthrene in compost amended soils. Chemosphere 67: 548-556.

26. Qari HA, Hassan IA (2014) Removal of pollutants from waste water using Dunaliella algae. Biomed Pharmacol J 7: 147-151.

27. Rani K, Dhania G (2014) Bioremediation and Biodegradation of Pesticide from Contaminated Soil and Water - A Noval Approach. Inter j of cur micro and app Sci 3:23-33.

28. Sivakumaran S (2014) Isolation of Cellulolytic Fungi and their Degradation on Cellulosic Agricultural Wastes. Journal of Academia and Industrial Research (JAIR) 2: 458-463.

29. Vazquez S, Agha A, Granado A, Sarro M, Esteban E, et al. (2006) Use of white Lupin plant for phytostabilizaion of Cd and As polluted acid soil. Air and Soil pollution 177:349-365.

30. Watanabe K, Kodoma Y, Stutsubo K, Harayama S (2001) Molecular characterization of bacterial populations in petroleum contaminated groundwater discharge from undergoing crude oil storage cavities. Applied and Environmental Microbiology 66: 4803-4809.

# 2
CHAPTER

# BIOFILM MEDIATED DECONTAMINATION OF POLLUTANTS FROM THE ENVIRONMENT

Arindam Mitra[1], Suman Mukhopadhyay[2]

[1]Department of Microbiology, Adamas University, Kolkata 700126, West Bengal, India
[2]Division of Microbiology and Infectious Diseases, NIAID/NIH/HHS, Bethesda, MD, USA

## ABSTRACT

**In this review, we highlight beneficial use of microbial biofilms in remediation of environmental pollutants by bioremediation. Bioremediation is an environment friendly, cost effective, sustainable technology that utilizes microbes to decontaminate and degrade a wide variety of pollutants into less harmful products. Relative to free-floating planktonic cells, microbes existing in biofilm mode are advantageous for bioremediation because of greater tolerance to pollutants, environmental stress and ability to degrade varied harsh pollutants via diverse catabolic pathways. In biofilm mode, microbes are immobilized in a self-synthesized matrix which offers protection from stress, contaminants and predatory protozoa. Contaminants ranging from heavy metals, petroleum, explosives, pesticides have been remediated using microbial consortia of biofilms. In the industry, biofilm based bioremediation is used to decontaminate polluted soil and groundwater. Here we discuss conventional and newer strategies utilizing biofilms in environmental remediation.**

## BIOREMEDIATION

Bioremediation refers to the environment friendly process of detoxification of harmful pollutants from soil, water and air using microorganisms[1,2]. In bioremediation, harsh chemicals are not used and as such, damaging effect on the environment is reduced [3]. Bioremediation offers several advantages over other methods including no or minimum disruption of land or wildlife surrounding the treated area, reduction of noise and dust

**Citation:** Arindam Mitra and Suman Mukhopadhyay, Biofilm mediated decontamination of pollutants from the environment, doi: 10.3934/bioeng.2016.1.44.

during treatment as well as avoidance of harsh chemicals. Bioremediation is economical when applied to large areas compared to conventional decontamination methods. The process of bioremediation employs diverse microbes for degradation or treatment of xenobiotic compounds, aromatic hydrocarbons, volatile organic compounds, pesticides, herbicides, heavy metals, radionuclides, crude oil, jet fuels, petroleum products and explosives [4]. Even though bioremediation has been known for several decades, it's effectiveness has been the subject of public attention following a major oil spill by Exxon Oil in the shorelines of Prince William Sound, Alaska in the year 1989 [5,6]. This incident is followed by an era of development in the field of bioremediation across the world, particularly in the US, making the field one of the fastest growing sector in the hazardous waste industry [7,8].

Microbes particularly bacteria can be easily grown and genetically manipulated which makes them suitable for bioremediation. Besides bacteria, fungi are also used for pollution removal [9,10,11]. Importantly, microbes are extremely efficient in degrading natural organic compounds or waste pollutants via diverse catabolic pathways as microbes adapt to persist in diverse environments [12,13]. Some microbes used in bioremediation are extremophiles that can withstand acidic or heavy metal-contaminated or radioactive environment. Success of bioremediation relies on maintenance of conditions or factors that expedite biodegradation of pollutants by microbes [14]. The process of bioremediation depends on enzymatic activities of microbes for transformation and degradation of environmental pollutants or wastes into less toxic or harmless constituents such as carbon dioxide and water [15]. The process of metabolic pathways requires transfer of electrons from electron donors to electron acceptors. The electron donors act as food for microbes which is usually limited in a non-contaminated site. However, in a contaminated site release of an organic electron donor may stimulate microbes to compete for available acceptors to restore the balance of the system. Microorganisms can degrade pollutants without and with oxygen in anaerobic and aerobic mode of degradation respectively. In aerobic degradation, microbes use oxygen as final electron acceptor to convert organic and inorganic pollutant into harmless products, often carbon dioxide and water. In anaerobic degradation, oxygen is not present or limited; microbes use other electron acceptor such as nitrate, iron, manganese, sulphate, etc. to break down organic compounds often into carbon dioxide and methane. Generally, aerobic microbes are capable of faster contaminant degradation than anaerobic ones. Sometimes some microbes may break contaminants by fermentation. Pollutants that are electron donors may be readily degraded in presence of oxygen by aerobic microbes, whereas those contaminants that are poor electron donors may degrade pollutants under anaerobic conditions. Depending on the pollutant, both electron donor and acceptor may be supplied to facilitate the degradation process. Many redox reactions also immobilize trace elements found in the contaminated sites. A change in oxidation potential of metals is also associated with change in toxicity or solubility as evident in the case of uranium and chromium [16,17,18]. In case of heavy metals, the conversion of sulphate to sulphide changes solubility and facilitate immobilization and removal of sulphate from wastewater by sulphur reducing bacteria [19].

Bioremediation is primarily of two types depending on location of pollutant treatment. In case of *in situ* bioremediation, the contaminated sample is treated in the original location whereas in *ex-situ* remediation the sample is treated typically off-site [20,21]. *In situ* bioremediation is an attractive choice as movement of contaminants from the site is not required and as a consequence it minimizes transportation cost and site disruption. A general approach for bioremediation is optimization of physical and chemical conditions (such as pH, aeration, moisture) for the bacteria in the contaminated site to degrade pollutant faster along with supplemented nutrients. An alternative strategy is to choose microbial strains that can serve as efficient degraders or use genetically modified microorganisms that can modify pathways for efficient pollutant degradation [22,23,24]. Importantly, a wide variety of microorganisms present in a community mode of "biofilms" are utilized to degrade diverse pollutants in the natural environment and in engineered systems.

## ROLE OF BIOFILMS IN BIOREMEDIATION

A biofilm may be defined as an assemblage of microorganisms comprising of microbial species attached to a biological or inert surface and encased in a self-synthesized matrix comprising of water, proteins, carbohydrates and extracellular DNA [25]. It may be anticipated that different microbial species present in consortia of biofilms each with different metabolic degradation pathway are capable of degrading several pollutants either individually or collectively [26,27]. Biofilm forming bacteria are adapted to survive and suited for bioremediation as they compete with nutrients and oxygen and observations of tolerance of biofilms towards harsh environment found way in the process of bioremediation. Biofilm mediated remediation is environment friendly and cost effective option for cleaning up environmental pollutants. Use of biofilms is efficient for bioremediation as biofilms absorb, immobilize and degrade various environmental pollutants. Bacterial biofilms exist within indigenous populations near the heavily contaminated sites to better persist, survive and manage the harsh environment. Expressions of genes vary within the biofilms and are distinctive relative to free floating planktonic cells. Differential gene expressions within biofilms are owing to variable local concentration of nutrients and oxygen within biofilm matrix and division of labour among microbes. Such variable gene expression may be important for degradation of varied pollutants by numerous metabolic pathways. An important consideration for biofilm formation in microorganisms is chemotaxis and flagellar dependent motility [28]. Responses such as swimming, swarming, twitching motility, chemotaxis, quorum sensing in presence of xenobiotics commonly present in soil and water assist microbes to coordinate movement towards pollutant and improved biodegradation [29].

Under natural environmental conditions, most bacteria persist in biofilm mode encased in an extracellular polymeric substance (EPS) matrix which also provides a beneficial structure to biofilm forming microbes in bioremediation [30,31,32]. The EPS is made of both bound and secreted form of polysaccharides, proteins, lipids, nucleic acids, humic substances and water [33]. The composition of EPS varies between species and is dependent on growth conditions, surface on which biofilms are formed and environmental stress among others. Bacterial biofilms and EPS production can undergo changes both in structure and content depending on the environmental conditions in which the microbes are found [34,35,36]. Biofilms appear filamentous and mushroom-like shapes in fast moving and static water respectively [37,38]. In presence of predatory protozoa, bacteria often adapt to form biofilms in form of large inedible microcolonies that enable survival and persistence under harsh environment [39]. The biofilm matrix offers greater resistance than planktonic cells to microbes from environmental stress, shear stress, acid stress, antimicrobial agents, UV damage, desiccation, predation, biocides, solvents, high concentration of toxic chemicals and pollutants[40,41]. In contrast, free floating planktonic cells decontaminate environmental pollutants by metabolic activity but such cells are not stationery and not adapted to persist under mechanical and environmental stress. Biofilm forming microbes, in contrast, are particularly adept in bioremediation as they are immobilized in an EPS which also immobilizes pollutants during degradation [42]. Biofilm microbes also acquire relatively limited nutrients and oxygen as compared to planktonic cells because of diffusional mode of transport instead of convectional transport. The three-dimensional structure of EPS with reduced oxygen concentrations towards the centre brings in close proximity of aerobes and anaerobes, heterotrophs with nitrifiers, and sulphate reducers with sulphate oxidizers which promotes faster degradation of varied pollutants in natural and engineered systems [43]. EPS from cyanobacteria act as biosorbent and removes heavy metals from the aqueous phase [44,45,46]. EPS containing surfactants may also aid solubilisation of hydrophobic or other refractory substrates which would otherwise be inaccessible to microorganisms [47]. Presence of EPS also makes nutrient exchange and removal of by-products possible in biofilm community. Extracellular enzymes within EPS of biofilms decontaminate pollutants such as heavy metals and organic compounds [30]. EPS serve as traps for metal and metalloids due to presence of many negatively charged functional groups which enable formation of complexes with heavy metals and organic contaminants and their subsequent removal [48,49]. EPS is known to bind a wide variety of metals including lead, copper, manganese, magnesium, zinc, cadmium, iron and nickel [50]. Nutrient limitation can also increase EPS synthesis which can further absorb metals and pollutants from the environment. EPS of biofilms comprising of phosphorous-accumulating microorganisms act as reservoir and help removal and recovery of phosphorous from wastewater [51,52].

In nature, usually diverse mixed species biofilms are formed which may promote enhanced biofilm formation and greater capacity to withstand environmental stress [53]. Coexistence of multiple microbial species within biofilms in close proximity promotes interaction among its members. Biofilms can be found in natural and engineered systems. Common place natural environmental biofilms exist in soil, aquatic plants, sediments, covering rocks in streams and plants, lakes and rivers and in wetlands. Natural biofilms in the environment can degrade and remove pollutants from soil and river. Biofilms formed on the surface of water consists of bacteria, protozoa, fungi and algae. Naturally developed biofilms by algae such as *Nitzschia* in organic sediments can aid recovery of sediments by production of oxygen which in turn facilitates the biodegradation of aerobic bacteria present in the biofilm. Natural biofilms are an important part of food chain where predatory protozoa or grazers may feed on biofilm material. The protozoa in turn may be engulfed by small fish and subsequently small fish by larger fish and terrestrial animals. Microbial aggregates in aquatic systems such as microbial mats consisting of phototrophic bacteria and algae are formed on sediment surface under extreme conditions to survive predation from grazers and stress. Fragile structures called flocs are formed during bloom periods. Treatment of municipal wastewater is based on the floc activity in activated sludge plants. Slow-sand filters utilize biofilms formed on the surface of sand to remove organic compounds and metals from lakes, rivers and reservoir water [54]. Planctomycetes present in biofilms of seaweed in marine waters have the potential to remove nitrogen from wastewater because of its capacity to perform anammox reactions in which ammonia is anaerobically oxidized to dinitrogen [55,56].

Biofilms are increasingly used as early warning systems or an indicator for monitoring and evaluating heavy metal contamination in rivers and streams as structure and physiological alterations of biofilms occur rapidly in presence of toxicant. The choice of biofilms as an indicator system is beneficial as biofilms can adsorb pollutants from the environment, develops rapidly and offer easy sampling methods [57]. Biofilms are the first to interact with nutrients and pollutants in aquatic systems and as such biofilms can be used as early environmental monitor in aqueous bodies[58]. Various indicator properties of biofilm may be assessed for monitoring environmental pollution including change in biomass, species composition, pigment production, photosynthesis and enzymatic activity. Species present in river biofilms varies seasonally and on the level of pollution [59]. Heavy environmental contamination such as Zinc and Cadmium can influence the species diversity within biofilms and as such species diversity estimation of microbial biofilms can also indicate environmental pollution [60]. Apart from convention techniques, modern molecular biology methods such as denaturing gradient gel electrophoresis, ribosomal spacer analysis and terminal restriction fragment length polymorphism has been used to estimate species diversity within biofilms [61,62,63,64]. Biofilm sampling has been shown to provide an improved indicator of heavy metal contamination of aquatic microbial community [65]. Biomass change of natural algal biofilm community upon exposure to heavy metals and herbicides has also been assessed and may be used as an environmental pollution indicator [66]. Furthermore, as pigment composition among members of biofilm community can alter following exposure of toxic chemicals, analysis and patterns of pigment composition can serve as a biomarker [67,68]. Often, short term toxicity tests on multiple parameters affecting biofilm structure and function are assessed including photosynthetic activity and pollution induced community tolerance [69].

## STRATEGIES FOR USE OF BIOFILMS IN REMEDIATION

Biofilm mediated remediation *in situ* may be performed in various ways. The process of natural attenuation relies on natural processes without the use of engineered steps or intervention including addition of specific strains for bioremediation. For example, microbial biofilm community present in the soil can biotransform certain pollutants into less harmful components. Natural attenuation is based on premise that under favourable conditions, certain contaminants can be degraded, transformed, immobilized and detoxified without any human intervention [70]. This passive remediation process requires resident profile of microbes which may be present in biofilm mode capable of degrading pollutants and requires long time. Extra nutrients such as carbon and phosphorus compounds, air to improve oxygen availability and additives may be added to increase growth and faster degradation of the pollutant in a process is called biostimulation. Natural attenuation may be monitored at defined times [20,71]. Natural attenuation strategy is typically used when the level of contaminant is relatively low and has been quite widely used in remediation of petroleum hydrocarbon sites [72]. This process is also known as monitored natural attenuation.

Bioaugmentation or bioenhancement, on the other hand, relies on inoculation of specific competent microbes or consortia of microbes in contaminated sites to perform degradation [73]. It is also likely that indigenous microbes capable of degrading pollutants may be present but requires substrates and nutrients to be added to assist the process. Microbial populations formed near the contaminated site can often degrade the pollutants found in the contaminated sites. Such microbial consortium can be stored under laboratory conditions and subsequently added to contaminated sites for cleaning the pollutant. Bioaugmentation is preferred method for freshly contaminated sites where indigenous population of microbes might lack or exhibits reduced capacity for efficient degradation. Such environment also facilitates development of microbial biofilms which increases the efficiency of degradation. The process of bioaugmentation may be monitored using biomarkers based on *gfp* or *luc* to track the efficacy of the inoculated microbes [74]. Bioaugmentation may be improved by incorporation of conventional genetic engineering techniques or by methods to increase the nutrient concentrations or persistence of microbes or by airventing and biostimulation methods [75]. In biostimulation, stimulus such as nutrients, growth substrates, electron donors and acceptors are provided to enhance the activity of microbes that are present near the site to better biodegrade the pollutant from the site [76]. A study on comparative bioremediation strategies showed that use of bioaugmentation and biostimulation were more efficient in remediation for petroleum hydrocarbon oil [77]. In another study, bioaugmentation and biostimulation have been shown to enhance nitrification performance [78]. In air venting, air is pumped into the site of contamination present below the soil surface to enrich aerobic microbial community and establish biofilms.

Contaminated sites lacking suitable endogenous microbial degrading population or favourable conditions for degradation that accelerate degradation of a pollutant in question may be subjected to *ex situ* remediation frequently in a reactor. In engineered systems, biofilms are used in a bioreactor in an inert support. The biofilm bioreactors are used for sorption and biochemical conversion of pollutants particularly for heavy metals, hydrocarbons and industrial and municipal wastewater [79]. Bioreactors based on biofilms particularly have been used commercially for cleaning up industrial wastewaters for decades [80,81]. Biofilm reactors offer many advantages over conventional treatment processes; high concentration and retention of biomass for long periods of time, enhanced metabolic activity, increased process flow rates, greater tolerance to harsh pollutants, large mass transfer area, enhanced volumetric biodegradation capacity, coexistence of anoxic and aerobic metabolic activity and reduced interruption in the bioreactor. In industrial set up, biofilm reactors are used in situations where free floating microorganisms in suspension do not produce adequate biomass or where biomass may not be retained long enough for efficient volumetric conversion. This can happen when microorganisms grow very slowly in suspensions or when a diluted feed streams are used in bioreactors. In a typical biofilm reactor, a support medium is needed for adhesion of microbes and development of biofilms. Biofilm bioreactors are of different types including batch, continuous stirred tank, trickle bed, air-lift reactors, upflow anaerobic sludge blanket, fluidized bed, expanded granular sludge blanket, biofilm airlift suspension batch reactors [80,82,83]. Biofilm reactors can be used off-site or used near the site of contamination. A schematic diagram of important types of bioreactors is shown in Figure 1.

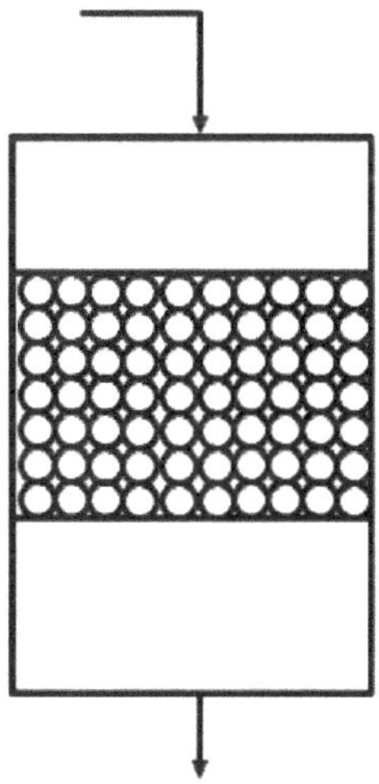

A. Fixed-bed reactor

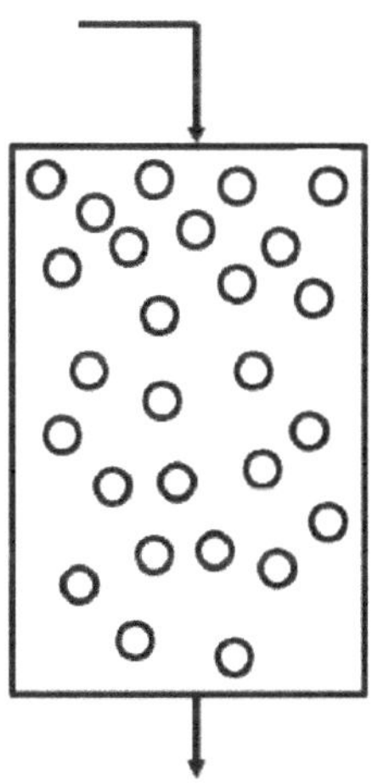

B. Fluidized-bed reactor

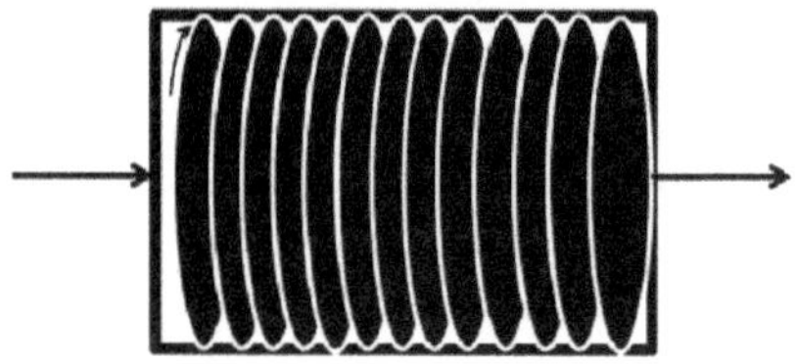

C. Rotating-disc contactor

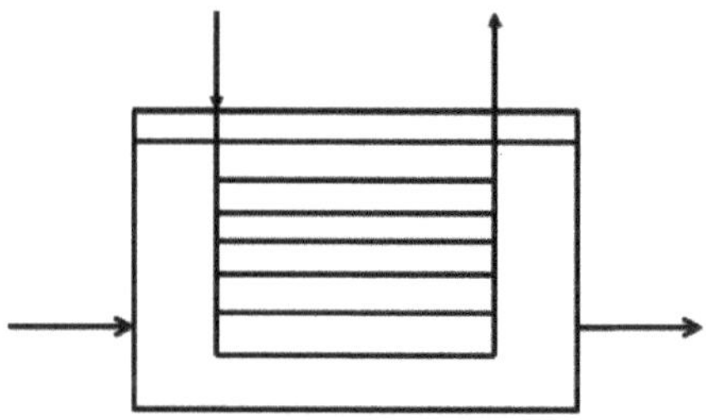

D. Membrane biofilm reactor

**Figure 1:** Various types of biofilm reactors. A. Fixed-bed reactor B. Fluidized-bed reactor C. Rotating-disc contactor D. Membrane biofilm reactor.

Fixed-bed reactor, also known as packed bed reactor (Fig.1A) is a common biofilm based reactor in which solid supports (media) are packed tightly where biofilms are colonised and provide high interface between the biofilm mass and the liquid. A packed bed reactor containing a biofilm of mercury resistant strains has been successfully employed in bioremediation of mercury [84]. However, fixed-bed reactors achieve high biofilm mass, which at times can clog the fixed-bed. Trickle-bed biofilm bioreactor (TBR), a special type of fixed-bed reactor, is one of the oldest types of biofilm reactor which has been extensively used for treatment of wastewater. Media used in TBR are usually plastic, rock, ceramics and other materials where biofilm develops. In TBR, wastewater trickles downward from the top via distribution system over the biofilm surface held on a fixed media. Pollutants in the water get metabolized as it diffuses through the biofilms. Oxygen may be supplied upward or downward which diffuses through the water to reach the biofilm. A net production of suspended solids in TBR requires a liquid-solid separation via a clarifier. The biofilm in the reactors may not have enough feed in certain areas and may cause reduced productivity.

Fluidized-bed reactor (Fig.1B) employs a column of biofilm coated beads in which polluted water is slowly pumped upward and keeps biofilm beads suspended during treatment of contaminated water [85]. This is in sharp contrast to fixed-bed reactor, where the media is not suspended. Solids are suspended either by flow of liquid or gas at certain velocity. Fluidization enables biofilms to develop on a large surface area and produce high biomass. Aeration is done either via oxygenator or supplied from the bottom of the reactor. Fluidized-bed reactor has been used for the treatment of streams contaminated with organic and inorganic compounds [85,86,87]. In Rotating Biological

Contactors (RBC) or modified forms of RBC (Fig.1C) has been used globally for wastewater treatment by reducing chemical oxygen demand and biochemical oxygen demand and also for nitrification and denitrification process [88,89,90]. RBC uses a thin biofilms of aerobic microbes which are grown on a rotating cylinder or biodiscs. Typically, the discs are partly submerged in effluent and slowly rotated such that biofilm microbes are alternately exposed to the effluent and air during which the attached biofilm on the disc degrades the pollutant. Excess biomass may slough off the RBC media and may be removed by clarifiers. Rotating biological contactors are economical in functioning and do not require much space or land. Some of the parameters that affect pollutant removal are rotational speed of the discs, disc submergence and hydraulic retention time. RBCs may be used for treatment of water contaminated with heavy metals, degradation of dyes, volatile organic compounds and PAH[88,91,92,93]. Membrane biofilm reactor (MBfR) (Fig.1D) delivers pressurized air or oxygen via gas permeable membranes to the attached biofilms formed on the membrane exterior. Such bubble-free, high transfer of oxygen prevents stripping of volatile organic compounds and greenhouse gases and foaming when surfactant is being used. MBfR is particularly suited for treatment of high oxygen demanding wastewater. The membrane may also act as support media for development of biofilms. In some case, hydrogen-based MBfR, hydrogen may be delivered to the a biofilm comprising of autotrophic bacteria, which then oxidizes hydrogen and use electron donor to various pollutants such as chlorate and nitrate [94,95]. A methane fed membrane biofilm reactor has been used to remove nitrate and pesticides from contaminated water [96].

Various new biofilm bioreactors have been designed for treatment of recalcitrant pollutants. In sequential biofilm reactors, different bioconversion processes may be performed in separate stages. For example, a sequential aerobic-anaerobic two-stage biofilm reactor has been used to degrade polychlorinated hydrocarbons [97]. Simultaneous nitrification and denitrification was made possible due to presence of aerobic and anoxic biofilms in a novel air-lift internal loop biofilm bioreactor [98]. An intensified biofilm-electrode reactor combining autotrophic and heterotrophic denitrification has been demonstrated to remove nitrate from contaminated groundwater [99]. Biofilm reactors using sulfate-reducing bacteria entrap or precipitate metals such as copper and zinc at the interface of biofilms [100,101,102]. Studies are performed to better understand and optimize the processing conditions and parameters for biofilm based bioremediation in bioreactors [103,104,105]. Modeling and simulation studies are conducted to better design and optimize the biodegradation processes [106,107,108].

## TYPES OF POLLUTANTS REMEDIATED BY BIOFILMS

Microbial biofilm mediated bioremediation are being increasingly used in removal of different types of pollutants including persistent organic pollutants, oil spills, heavy metals pesticides and xenobiotics. Biofilm remediation has been particularly useful in treatment of heavy metal contaminated samples from groundwater and soil for frequently encountered heavy metals such as cadmium, copper, uranium and chromium[109,110]. Phosphatase enzyme in presence of biofilm matrix facilitates metal precipitation both for aerobic bacteria and for anaerobic conditions [111]. In certain cases, biofilm formation is induced by addition of carbon sources in contaminated ground water to create a barrier or reduce flow of pollutants away from the site of contamination minimizing its spread.

Persistent organic pollutants (POP) are the most persistent of the pollutants with long half-lives due to hydrophobicity and can be found in air, water and sediments. Examples of POPs include polycyclic aromatic hydrocarbons (PAH), polychlorinated biphenyls, polychlorinated ethenes, all of which are listed as toxic or harmful pollutants in the United States Environmental Protection Agency [112,113,114,115]. Many bacterial biofilm formers have been isolated from the environments that degrade POP which may be further modified and used in bioremediation [113]. POPs are hydrophobic and can be toxic when moved to the food chain [116]. Bacterial biofilms have been engineered for detoxification of POP. Studies have shown that biofilm formation and cometabolism in biofilms are key factors in remediation of polycyclic aromatic hydrocarbons[117,118,119].

Increasing use of petroleum in the industry is associated with rise in its accumulation in the water environment causing toxicity to marine life. Oil spills can itself shift the production of hydrocarbon degrading microbes in the soil [120]. Oil spills in the marine environment can be also decontaminated using hydrocarbon degradation bacteria exhibiting speedy growth [5,121]. In this context, several microorganisms including*Pseudomonas*, *Arthrobacter*, *Rhodococcus*, *Bacillus*, *Alcanivorax* and *Cycloclasticus* spp. of gamma proteobacteria have been employed [122]. Biofilm formation has been stimulated by turning off unused oilfields. New microbial strains capable of biofilm mediated oil degradation are being continuously screened [123]. In case of petroleum degradation, a microbial consortia comprising of *Bacillus subtilis* and *Acinetobacter radioresistens* with a surfactant producing strain has been shown to better degrader than microbial consortia consisting of degraders alone[124]. Use of water-insoluble fertilizers such as uric acid can provide the required nitrogen source for hydrocarbon degraders to facilitate the biodegradation of oil in open environment [125].

Heavy metals such as copper, zinc, nickel, cadmium, cobalt have been remediated using diverse biofilm reactors. Biofilm forming sulphate reducing bacteria (SRB) are particularly useful in mines for scavenging metals from metal contaminated water into precipitates of metal sulphides [102,126]. Recent studies suggest that certain bacterial strains are capable of forming electroactive films or electrochemically active biofilms (EAB) which directly exchange electrons with a conductive solid surface [127]. EAB are increasingly explored in the field of bioremediation and used for reduction of heavy metals from contaminated groundwater and soil [128,129].

## CONCLUSIONS

Biofilm based bioremediation have certain limitations. Bioremediation is relatively slow as compared to the chemical treatment for pollutant degradation. A major limitation of bioremediation is reliability which limits its application in certain situations particularly during heavy contamination. Not all chemicals are amenable to biodegradation, particularly the man-made unnatural recalcitrant compounds such as plastics and certain halogenated aromatic compounds. Another limitation of bioremediation is that some metabolic toxic products could be generated post microbial degradation. The strategy of bioremediation is ideally useful in situations when the level of pollution is relatively low or does not require immediate restoration or where chemical treatment is not ideal. Bioremediation may be slow or ineffective when essential nutrients for supporting microbial growth are limiting, particularly when the level of pollution is high. Bioavailability of pollutant to microbes also determines the efficiency of bioremediation and pollutants that are not enclosed by other materials such as clay are more amenable to biodegradation. Adequate technical expertise and interdisciplinary approach from different field such as environmental microbiology, civil engineering soil science is also required for successful performance of biofilm mediated bioremediation. Biofilm reactors may not be used with rapid growing microorganisms where the reactor capacity is dependent on oxygen diffusion. Because of these limitations, bioremediation may be assessed for effectiveness relative to other methods available for decontamination of environmental pollutants [130].

While biodegradation by microbes has been taking place since the beginning of life, the field of bioremediation is relatively new. Novel approaches are constantly being tested and employed in the field of bioremediation. Improved strains are continuously being designed that energetically favour pollutants as preferred substrates over other available compounds for effective bioremediation [109]. A novel approach for improving bioremediation uses natural transformation process within biofilms utilizing uptake of DNA harbouring catabolic genes that facilitate biodegradation of selected pollutants [131]. Genetically modified microorganisms (GEM) have been constructed with the capacity to degrade diverse pollutants including halogenated aromatic compounds [132,133]. Horizontal transfer of genes capable of biodegradation from GEM to members of biofilm populations can further enhance biodegradation process. Cloning of genes for biosurfactant synthesis and chemotactic ability of GEM can further enhance the biodegradative capability of modified microbes. Nevertheless, the release and use of GEM in the nature and transmission is under much debate and controversial. However, the majority of organisms usually have other disabling mutations that will not permit the microbes to grow outside a given environment. Reengineering of secreted proteins in biofilm matrix is also an area for further development in the field of bioremediation [134]. Cell-cell interaction among members of biofilm community can be explored further to improve the bioremediation process. Combined processes are increasingly used such as trickling filters with RBC or with activated sludge to maximize the strengths and minimize the weaknesses of these individual processes [135,136]. In certain cases of recalcitrant pollutants, biofilm mediated bioremediation can be used in combination with phytoremediation or with chemical treatments. Still in other cases, microbial consortia of bacteria-fungi may be used in combination to degrade xenobiotic compounds. Overall biofilm mediated bioremediation remains an attractive choice in mitigating environmental pollution despite of its limitations.

## REFERENCES

1. Alexander M, Loehr RC (1992) Bioremediation review. *Science* 258: 874.

2. Prasad MN, Prasad R (2012) Nature›s cure for cleanup of contaminated environment—a review of bioremediation strategies. *Rev Environ Health* 27: 181-189.
3. Vidali M (2001) Bioremediation. *an overview. Pure and Applied Chemistry* 73: 1163-1172.
4. Gaur N, Flora G, Yadav M, et al. (2014) A review with recent advancements on bioremediation-based abolition of heavy metals. *Environ Sci Process Impacts* 16: 180-193.
5. Atlas RM, Hazen TC (2011) Oil biodegradation and bioremediation: a tale of the two worst spills in U. *S. history. Environ Sci Technol* 45: 6709-6715.
6. Bragg JR, Prince RC, Harner EJ, et al. (1994) Effectiveness of bioremediation for the Exxon Valdez oil spill. *Nature* 368: 413-418.
7. Day SM (1993) US environmental regulations and policies--their impact on the commercial development of bioremediation. *Trends Biotechnol* 11: 324-328.
8. Caplan JA (1993) The worldwide bioremediation industry: prospects for profit. *Trends Biotechnol* 11: 320-323.
9. Mishra A, Malik A (2014) Novel fungal consortium for bioremediation of metals and dyes from mixed waste stream. *Bioresour Technol* 171: 217-226.
10. Cerniglia CE (1997) Fungal metabolism of polycyclic aromatic hydrocarbons: past, present and future applications in bioremediation. *J Ind Microbiol Biotechnol* 19: 324-333.
11. Balaji V, Arulazhagan P, Ebenezer P (2014) Enzymatic bioremediation of polyaromatic hydrocarbons by fungal consortia enriched from petroleum contaminated soil and oil seeds. *J Environ Biol* 35: 521-529.
12. Bouwer EJ, Zehnder AJ (1993) Bioremediation of organic compounds--putting microbial metabolism to work. *Trends Biotechnol* 11: 360-367.
13. Bruins MR, Kapil S, Oehme FW (2000) Microbial resistance to metals in the environment. *Ecotoxicol Environ Saf* 45: 198-207.
14. Prince RC (2000) Bioremediation. *Kirk-Othmer Encyclopedia of Chemical Technology*: John Wiley & Sons, Inc.
15. Das S, Dash HR (2014) 1 - Microbial Bioremediation: A Potential Tool for Restoration of Contaminated Areas. In: Das S, editor. *Microbial Biodegradation and Bioremediation*. Oxford: Elsevier. pp. 1–21.
16. Joutey NT, Sayel H, Bahafid W, et al. (2015) Mechanisms of hexavalent chromium resistance and removal by microorganisms. *Rev Environ Contam Toxicol* 233: 45-69.
17. Kumar R, Singh S, Singh OV (2007) Bioremediation of radionuclides: emerging technologies. *OMICS* 11: 295-304.
18. Wall JD, Krumholz LR (2006) Uranium reduction. *Annu Rev Microbiol* 60: 149-166.
19. Beyenal H, Sani RK, Peyton BM, et al. (2004) Uranium immobilization by

sulfate-reducing biofilms. *Environ Sci Technol* 38: 2067-2074.

20. Vogt C, Richnow HH (2014) Bioremediation via in situ microbial degradation of organic pollutants. *Adv Biochem Eng Biotechnol* 142: 123-146.
21. Jorgensen KS (2007) In situ bioremediation. *Adv Appl Microbiol* 61: 285-305.
22. Singh JS, Abhilash PC, Singh HB, et al. (2011) Genetically engineered bacteria: an emerging tool for environmental remediation and future research perspectives. *Gene* 480: 1-9.
23. Hedlund BP, Staley JT (2001) Vibrio cyclotrophicus sp. *nov., a polycyclic aromatic hydrocarbon (PAH)-degrading marine bacterium. Int J Syst Evol Microbiol* 51: 61-66.
24. Nakajima-Kambe T, Ichihashi F, Matsuzoe R, et al. (2009) Degradation of aliphatic–aromatic copolyesters by bacteria that can degrade aliphatic polyesters. *Polym Degrad Stab* 94: 1901-1905.
25. Costerton JW, Cheng KJ, Geesey GG, et al. (1987) Bacterial biofilms in nature and disease. *Annu Rev Microbiol* 41: 435-464.
26. Gieg LM, Fowler SJ, Berdugo-Clavijo C (2014) Syntrophic biodegradation of hydrocarbon contaminants. *Curr Opin Biotechnol* 27: 21-29.
27. Horemans B, Breugelmans P, Hofkens J, et al. (2013) Environmental dissolved organic matter governs biofilm formation and subsequent linuron degradation activity of a linuron-degrading bacterial consortium. *Appl Environ Microbiol* 79: 4534-4542.
28. Pratt LA, Kolter R (1999) Genetic analyses of bacterial biofilm formation. *Curr Opin Microbiol* 2: 598-603.
29. Lacal J, Reyes-Darias JA, García-Fontana C, et al. (2013) Tactic responses to pollutants and their potential to increase biodegradation efficiency. *J Appl Microbiol* 114: 923-933.
30. Flemming HC, Wingender J (2010) The biofilm matrix. *Nat Rev Microbiol* 8: 623-633.
31. More TT, Yadav JSS, Yan S, et al. (2014) Extracellular polymeric substances of bacteria and their potential environmental applications. *J Environ Manage* 144: 1-25.
32. Flemming HC, Wingender J (2001) Relevance of microbial extracellular polymeric substances (EPSs)--Part II: Technical aspects. *Water Sci Technol* 43: 9-16.
33. Branda SS, Vik S, Friedman L, et al. (2005) Biofilms: the matrix revisited. *Trends Microbiol* 13: 20-26.
34. Jung JH, Choi NY, Lee SY (2013) Biofilm formation and exopolysaccharide (EPS) production by Cronobacter sakazakii depending on environmental conditions. *Food Microbiol* 34: 70-80.
35. Kreft JU, Wimpenny JW (2001) Effect of EPS on biofilm structure and function as revealed by an individual-based model of biofilm growth. *Water Sci Technol*

43: 135-141.

36. Miqueleto AP, Dolosic CC, Pozzi E, et al. (2010) Influence of carbon sources and C/N ratio on EPS production in anaerobic sequencing batch biofilm reactors for wastewater treatment. *Bioresour Technol* 101: 1324-1330.
37. Reysenbach AL, Cady SL (2001) Microbiology of ancient and modern hydrothermal systems. *Trends Microbiol* 9: 79-86.
38. Edwards KJ, Bond PL, Gihring TM, et al. (2000) An archaeal iron-oxidizing extreme acidophile important in acid mine drainage. *Science* 287: 1796-1799.
39. Matz C, Kjelleberg S (2005) Off the hook--how bacteria survive protozoan grazing. *Trends Microbiol* 13: 302-307.
40. Davey ME, O›Toole G A (2000) Microbial biofilms: from ecology to molecular genetics. *Microbiol Mol Biol Rev* 64: 847-867.
41. Mah TF, O›Toole GA (2001) Mechanisms of biofilm resistance to antimicrobial agents. *Trends Microbiol* 9: 34-39.
42. Sutherland IW (2001) The biofilm matrix--an immobilized but dynamic microbial environment. *Trends Microbiol* 9: 222-227.
43. Field JA, Stams AJ, Kato M, et al. (1995) Enhanced biodegradation of aromatic pollutants in cocultures of anaerobic and aerobic bacterial consortia. *Antonie Van Leeuwenhoek* 67: 47-77.
44. De Philippis R, Colica G, Micheletti E (2011) Exopolysaccharide-producing cyanobacteria in heavy metal removal from water: molecular basis and practical applicability of the biosorption process. *Appl Microbiol Biotechnol* 92: 697-708.
45. De Philippis R, Paperi R, Sili C (2007) Heavy metal sorption by released polysaccharides and whole cultures of two exopolysaccharide-producing cyanobacteria. *Biodegradation* 18: 181-187.
46. Micheletti E, Colica G, Viti C, et al. (2008) Selectivity in the heavy metal removal by exopolysaccharide-producing cyanobacteria. *J Appl Microbiol* 105: 88-94.
47. Iwabuchi N, Sunairi M, Urai M, et al. (2002) Extracellular polysaccharides of Rhodococcus rhodochrous S-2 stimulate the degradation of aromatic components in crude oil by indigenous marine bacteria. *Appl Environ Microbiol* 68: 2337-2343.
48. Li W-W, Yu H-Q (2014) Insight into the roles of microbial extracellular polymer substances in metal biosorption. *Bioresource Technology* 160: 15-23.
49. Pal A, Paul AK (2008) Microbial extracellular polymeric substances: central elements in heavy metal bioremediation. *Indian J Microbiol* 48: 49-64.
50. Ferris FG, Schultze S, Witten TC, et al. (1989) Metal Interactions with Microbial Biofilms in Acidic and Neutral pH Environments. *Appl Environ Microbiol* 55: 1249-1257.
51. Zhang HL, Fang W, Wang YP, et al. (2013) Phosphorus removal in an enhanced biological phosphorus removal process: roles of extracellular polymeric substances. *Environ Sci Technol* 47: 11482-11489.

52. Yuan Z, Pratt S, Batstone DJ (2012) Phosphorus recovery from wastewater through microbial processes. *Curr Opin Biotechnol* 23: 878-883.

53. Burmolle M, Webb JS, Rao D, et al. (2006) Enhanced biofilm formation and increased resistance to antimicrobial agents and bacterial invasion are caused by synergistic interactions in multispecies biofilms. *Appl Environ Microbiol* 72: 3916-3923.

54. Logsdon GS, Kohne R, Abel S, et al. (2002) Slow sand filtration for small water systems. *J Environ Eng Sci* 1: 339-348.

55. Kartal B, Kuenen JG, van Loosdrecht MC (2010) Engineering. *Sewage treatment with anammox. Science* 328: 702-703.

56. Bengtsson MM, Ovreas L (2010) Planctomycetes dominate biofilms on surfaces of the kelp Laminaria hyperborea. *BMC Microbiol* 10: 261.

57. Fuchs S, Haritopoulou T, Schäfer M, et al. (1997) Heavy metals in freshwater ecosystems introduced by urban rainwater runoff—Monitoring of suspended solids, river sediments and biofilms. *Water Sci Technol* 36: 277-282.

58. Peacock AD, Chang YJ, Istok JD, et al. (2004) Utilization of microbial biofilms as monitors of bioremediation. *Microb Ecol* 47: 284-292.

59. Brummer IH, Fehr W, Wagner-Dobler I (2000) Biofilm community structure in polluted rivers: abundance of dominant phylogenetic groups over a complete annual cycle. *Appl Environ Microbiol* 66: 3078-3082.

60. Arini A, Feurtet–Mazel A, Maury-Brachet R, et al. (2012) Recovery potential of periphytic biofilms translocated in artificial streams after industrial contamination (Cd and Zn). *Ecotoxicology* 21: 1403-1414.

61. SzabÓ KÉ, Makk J, Kiss KT, et al. (2008) Sequential colonization by river periphyton analysed by microscopy and molecular fingerprinting. *Freshwater Biology* 53: 1359-1371.

62. Dorigo U, Bérard A, Humbert JF (2002) Comparison of Eukaryotic Phytobenthic Community Composition in a Polluted River by Partial 18S rRNA Gene Cloning and Sequencing. *Microbial Ecology* 44: 372-380.

63. Kostanjsek R, Lapanje A, Drobne D, et al. (2005) Bacterial community structure analyses to assess pollution of water and sediments in the Lake Shkodra/Skadar, Balkan Peninsula. *Environ Sci Pollut Res Int* 12: 361-368.

64. Bricheux G, Le Moal G, Hennequin C, et al. (2013) Characterization and evolution of natural aquatic biofilm communities exposed in vitro to herbicides. *Ecotoxicol Environ Saf* 88: 126-134.

65. Ancion PY, Lear G, Dopheide A, et al. (2013) Metal concentrations in stream biofilm and sediments and their potential to explain biofilm microbial community structure. *Environ Pollut* 173: 117-124.

66. Navarro E, Guasch H, Sabater S (2002) Use of microbenthic algal communities in ecotoxicological tests for the assessment of water quality: the Ter river case study. *J Appl Phycol* 14: 41-48.

67. Dorigo U, Bourrain X, Bérard A, et al. (2004) Seasonal changes in the sensitivity

of river microalgae to atrazine and isoproturon along a contamination gradient. *Sci Total Environ* 318: 101-114.

68. Dewez D, Didur O, Vincent-Héroux J, et al. (2008) Validation of photosynthetic-fluorescence parameters as biomarkers for isoproturon toxic effect on alga Scenedesmus obliquus. *Environ Pollution* 151: 93-100.
69. Schmitt-Jansen M, Altenburger R (2008) Community-level microalgal toxicity assessment by multiwavelength-excitation PAM fluorometry. *Aquat Toxicol* 86: 49-58.
70. Sayler GS, Layton A, Lajoie C, et al. (1995) Molecular site assessment and process monitoring in bioremediation and natural attenuation. *off. Appl Biochem Biotechnol* 54: 277-290.
71. Jorgensen KS, Salminen JM, Bjorklof K (2010) Monitored natural attenuation. *Methods Mol Biol* 599: 217-233.
72. Rittmann BE (2004) Definition, objectives, and evaluation of natural attenuation. *Biodegradation* 15: 349-357.
73. Tyagi M, da Fonseca MM, de Carvalho CC (2011) Bioaugmentation and biostimulation strategies to improve the effectiveness of bioremediation processes. *Biodegradation* 22: 231-241.
74. Jansson JK, Bjorklof K, Elvang AM, et al. (2000) Biomarkers for monitoring efficacy of bioremediation by microbial inoculants. *Environ Pollut* 107: 217-223.
75. Gentry T, Rensing C, Pepper IAN (2004) New Approaches for Bioaugmentation as a Remediation Technology. *Crit Rev Environ Sci Technol* 34: 447-494.
76. Morgan P, Watkinson RJ (1989) Hydrocarbon degradation in soils and methods for soil biotreatment. *Crit Rev Biotechnol* 8: 305-333.
77. Grace Liu P-W, Chang TC, Whang L-M, et al. (2011) Bioremediation of petroleum hydrocarbon contaminated soil: Effects of strategies and microbial community shift. *Int Biodeterior Biodegradation* 65: 1119-1127.
78. Abeysinghe DH, De Silva DG, Stahl DA, et al. (2002) The effectiveness of bioaugmentation in nitrifying systems stressed by a washout condition and cold temperature. *Water Environ Res* 74: 187-199.
79. Boon N, De Gelder L, Lievens H, et al. (2002) Bioaugmenting bioreactors for the continuous removal of 3-chloroaniline by a slow release approach. *Environ Sci Technol* 36: 4698-4704.
80. Qureshi N, Annous BA, Ezeji TC, et al. (2005) Biofilm reactors for industrial bioconversion processes: employing potential of enhanced reaction rates. *Microb Cell Fact* 4: 24.
81. Bryers JD (1993) Bacterial biofilms. *Curr Opin Biotechnol* 4: 197-204.
82. Rosche B, Li XZ, Hauer B, et al. (2009) Microbial biofilms: a concept for industrial catalysis. *Trends Biotechnol* 27: 636-643.
83. Singh R, Paul D, Jain RK (2006) Biofilms: implications in bioremediation. *Trends Microbiol* 14: 389-397.

84. Wagner-Dobler I (2003) Pilot plant for bioremediation of mercury-containing industrial wastewater. *Appl Microbiol Biotechnol* 62: 124-133.

85. Shieh W, Keenan J (1986) Fluidized bed biofilm reactor for wastewater treatment. *Bioproducts*: Springer Berlin Heidelberg. pp. 131–169.

86. Denac M, Dunn IJ (1988) Packed- and fluidized-bed biofilm reactor performance for anaerobic wastewater treatment. *Biotechnol Bioeng* 32: 159-173.

87. Kumar TA, Saravanan S (2009) Treatability studies of textile wastewater on an aerobic fluidized bed biofilm reactor (FABR): a case study. *Water Sci Technol* 59: 1817-1821.

88. Costley SC, Wallis FM (2001) Bioremediation of heavy metals in a synthetic wastewater using a rotating biological contactor. *Water Res* 35: 3715-3723.

89. Eker S, Kargi F (2008) Biological treatment of 2,4-dichlorophenol containing synthetic wastewater using a rotating brush biofilm reactor. *Bioresour Technol* 99: 2319-2325.

90. Eker S, Kargi F (2010) COD, para-chlorophenol and toxicity removal from synthetic wastewater using rotating tubes biofilm reactor (RTBR). *Bioresour Technol* 101: 9020-9024.

91. Abraham TE, Senan RC, Shaffiqu TS, et al. (2003) Bioremediation of textile azo dyes by an aerobic bacterial consortium using a rotating biological contactor. *Biotechnol Prog* 19: 1372-1376.

92. Jeswani H, Mukherji S (2012) Degradation of phenolics, nitrogen-heterocyclics and polynuclear aromatic hydrocarbons in a rotating biological contactor. *Bioresour Technol* 111: 12-20.

93. Sarayu K, Sandhya S (2012) Rotating biological contactor reactor with biofilm promoting mats for treatment of benzene and xylene containing wastewater. *Appl Biochem Biotechnol* 168: 1928-1937.

94. Rittmann BE (2006) The membrane biofilm reactor: the natural partnership of membranes and biofilm. *Water Sci Technol* 53: 219-225.

95. Nerenberg R, Rittmann BE (2004) Hydrogen-based, hollow-fiber membrane biofilm reactor for reduction of perchlorate and other oxidized contaminants. *Water Sci Technol* 49: 223-230.

96. Modin O, Fukushi K, Yamamoto K (2008) Simultaneous removal of nitrate and pesticides from groundwater using a methane-fed membrane biofilm reactor. *Water Sci Technol* 58: 1273-1279.

97. Fathepure BZ, Vogel TM (1991) Complete degradation of polychlorinated hydrocarbons by a two-stage biofilm reactor. *Appl Environ Microbiol* 57: 3418-3422.

98. Zhang C, Wang L, Yan N, et al. (2013) Air-lift internal loop biofilm reactor for realized simultaneous nitrification and denitrification. *Bioprocess Biosyst Eng* 36: 597-602.

99. Zhao Y, Feng C, Wang Q, et al. (2011) Nitrate removal from groundwater by cooperating heterotrophic with autotrophic denitrification in a biofilm–electrode

reactor. *J Hazard Mater* 192: 1033-1039.

100. White C, Gadd GM (1998) Accumulation and effects of cadmium on sulphate-reducing bacterial biofilms. *Microbiology* 144: 1407-1415.
101. White C, Gadd GM (2000) Copper accumulation by sulfate-reducing bacterial biofilms. *FEMS Microbiology Letters* 183: 313-318.
102. Smith WL, Gadd GM (2000) Reduction and precipitation of chromate by mixed culture sulphate-reducing bacterial biofilms. *J Appl Microbiol* 88: 983-991.
103. Hosseini Koupaie E, Alavi Moghaddam MR, Hashemi SH (2013) Evaluation of integrated anaerobic/aerobic fixed-bed sequencing batch biofilm reactor for decolorization and biodegradation of azo dye acid red 18: comparison of using two types of packing media. *Bioresour Technol* 127: 415-421.
104. Lin YH, Hsien TY (2009) Kinetics of biodegradation of phenolic wastewater in a biofilm reactor. *Water Sci Technol* 59: 1703-1711.
105. Moreno-Andrade I, Buitron G, Vargas A (2009) Effect of starvation and shock loads on the biodegradation of 4-chlorophenol in a discontinuous moving bed biofilm reactor. *Appl Biochem Biotechnol* 158: 222-230.
106. Coelhoso I, Boaventura R, Rodrigues A (1992) Biofilm reactors: an experimental and modeling study of wastewater denitrification in fluidized-bed reactors of activated carbon particles. *Biotechnol Bioeng* 40: 625-633.
107. Masic A, Eberl HJ (2014) A modeling and simulation study of the role of suspended microbial populations in nitrification in a biofilm reactor. *Bull Math Biol* 76: 27-58.
108. Martin KJ, Picioreanu C, Nerenberg R (2015) Assessing microbial competition in a hydrogen-based membrane biofilm reactor (MBfR) using multidimensional modeling. *Biotechnol Bioeng*.
109. Valls M, de Lorenzo Vc (2002) Exploiting the genetic and biochemical capacities of bacteria for the remediation of heavy metal pollution. *FEMS Microbiol Rev* 26: 327-338.
110. Diels L, De Smet M, Hooyberghs L, et al. (1999) Heavy metals bioremediation of soil. *Molecular Biotechnology* 12: 149-158.
111. Macaskie LE, Yong P, Doyle TC, et al. (1997) Bioremediation of uranium-bearing wastewater: biochemical and chemical factors influencing bioprocess application. *Biotechnol Bioeng* 53: 100-109.
112. Shukla SK, Mangwani N, Rao TS, et al. (2014) 8 - Biofilm-Mediated Bioremediation of Polycyclic Aromatic Hydrocarbons. In: Das S, editor.*Microbial Biodegradation and Bioremediation*. Oxford: Elsevier. pp. 203–232.
113. Chen M, Xu P, Zeng G, et al. (2015) Bioremediation of soils contaminated with polycyclic aromatic hydrocarbons, petroleum, pesticides, chlorophenols and heavy metals by composting: Applications, microbes and future research needs. *Biotechnol Adv* 33: 745-755.
114. Cerniglia C (1993) Biodegradation of polycyclic aromatic hydrocarbons. In: Rosenberg E, editor. *Microorganisms to Combat Pollution*: Springer Netherlands.

pp. 227–244.

115. Jones KC, de Voogt P (1999) Persistent organic pollutants (POPs): state of the science. *Environ Pollut* 100: 209-221.

116. Bonefeld-Jorgensen EC, Hjelmborg PS, Reinert TS, et al. (2006) Xenoestrogenic activity in blood of European and Inuit populations. *Environ Health* 5: 12.

117. Johnsen AR, Karlson U (2004) Evaluation of bacterial strategies to promote the bioavailability of polycyclic aromatic hydrocarbons. *Appl Microbiol Biotechnol* 63: 452-459.

118. Rodriguez S, Bishop P (2008) Enhancing the Biodegradation of Polycyclic Aromatic Hydrocarbons: Effects of Nonionic Surfactant Addition on Biofilm Function and Structure. *J Environ Eng* 134: 505-512.

119. Plosz BG, Vogelsang C, Macrae K, et al. (2010) The BIOZO process--a biofilm system combined with ozonation: occurrence of xenobiotic organic micro-pollutants in and removal of polycyclic aromatic hydrocarbons and nitrogen from landfill leachate. *Water Sci Technol* 61: 3188-3197.

120. Song HG, Bartha R (1990) Effects of jet fuel spills on the microbial community of soil. *Appl Environ Microbiol* 56: 646-651.

121. Ron EZ, Rosenberg E (2014) Enhanced bioremediation of oil spills in the sea. *Curr Opin Biotechnol* 27: 191-194.

122. Harayama S, Kasai Y, Hara A (2004) Microbial communities in oil-contaminated seawater. *Curr Opin Biotechnol* 15: 205-214.

123. Dasgupta D, Ghosh R, Sengupta TK (2013) Biofilm-mediated enhanced crude oil degradation by newly isolated pseudomonas species. *ISRN Biotechnol* 2013: 250749.

124. Mnif I, Mnif S, Sahnoun R, et al. (2015) Biodegradation of diesel oil by a novel microbial consortium: comparison between co-inoculation with biosurfactant-producing strain and exogenously added biosurfactants. *Environ Sci Pollut Res Int* 22: 14852-14861.

125. Koren O, Knezevic V, Ron EZ, et al. (2003) Petroleum pollution bioremediation using water-insoluble uric acid as the nitrogen source. *Appl Environ Microbiol* 69: 6337-6339.

126. Muyzer G, Stams AJ (2008) The ecology and biotechnology of sulphate-reducing bacteria. *Nat Rev Microbiol* 6: 441-454.

127. Erable B, Duţeanu NM, Ghangrekar MM, et al. (2009) Application of electro-active biofilms. *Biofouling* 26: 57-71.

128. Li Z, Zhang X, Lei L (2008) Electricity production during the treatment of real electroplating wastewater containing Cr6+ using microbial fuel cell. *Process Biochemistry* 43: 1352-1358.

129. Cong Y, Xu Q, Feng H, et al. (2013) Efficient electrochemically active biofilm denitrification and bacteria consortium analysis. *Bioresour Technol* 132: 24-27.

130. Heitzer A, Sayler GS (1993) Monitoring the efficacy of bioremediation. *Trends*

*Biotechnol* 11: 334-343.

131. Perumbakkam S, Hess TF, Crawford RL (2006) A bioremediation approach using natural transformation in pure-culture and mixed-population biofilms. *Biodegradation* 17: 545-557.
132. Urgun-Demirtas M, Stark B, Pagilla K (2006) Use of Genetically Engineered Microorganisms (GEMs) for the Bioremediation of Contaminants. *Crit Rev Biotechnol* 26: 145-164.
133. Cases I, de Lorenzo V (2005) Genetically modified organisms for the environment: stories of success and failure and what we have learned from them. *Int Microbiol* 8: 213-222.
134. Absalon C, Ymele-Leki P, Watnick PI (2012) The bacterial biofilm matrix as a platform for protein delivery. *MBio* 3: e00127-00112.
135. Safa M, Alemzadeh I, Vossoughi M (2014) Biodegradability of oily wastewater using rotating biological contactor combined with an external membrane. *J Environ Health Sci Eng* 12: 117.
136. Kaindl N (2010) Upgrading of an activated sludge wastewater treatment plant by adding a moving bed biofilm reactor as pre-treatment and ozonation followed by biofiltration for enhanced COD reduction: design and operation experience. *Water Sci Technol* 62: 2710-2719.

# 3
CHAPTER

# BIOREMEDIATION OF HEAVY METALS FROM SOIL AND AQUATIC ENVIRONMENT: AN OVERVIEW OF PRINCIPLES AND CRITERIA OF FUNDAMENTAL PROCESSES

Ruchita Dixit [1], Wasiullah [1], Deepti Malaviya [1], Kuppusamy Pandiyan [1], Udai B. Singh [1], Asha Sahu [2], Renu Shukla [1], Bhanu P. Singh [3], Jai P. Rai [4], Pawan Kumar Sharma [1], Harshad Lade [5], and Diby Paul [5]

[1]National Bureau of Agriculturally Important Microorganisms (NBAIM), Kushmaur, Maunath Bhanjan 275 101, India

[2]Division of Soil Biology, Indian Institute of Soil Science, Nabibagh, Berasia Road, Bhopal 462 038, India

[3]Udai Pratap Autonomous College, Varanasi 221 002, India

[4]Department of Mycology and Plant Pathology (Krishi Vigyan Kendra), Institute of Agricultural Sciences, Banaras Hindu University, Varanasi 221 005, India

[5]Department of Environmental Engineering, Konkuk University, Seoul 143-701, Korea

## ABSTRACT

**Heavy metals are natural constituents of the environment, but indiscriminate use for human purposes has altered their geochemical cycles and biochemical balance. This results in excess release of heavy metals such as cadmium, copper, lead, nickel, zinc *etc.* into natural resources like the soil and aquatic environments. Prolonged exposure and higher accumulation of such heavy metals can have deleterious health effects on human life and aquatic biota. The role of microorganisms and plants in biotransformation of heavy metals into nontoxic forms is well-documented, and understanding the molecular mechanism of metal accumulation has numerous biotechnological implications for bioremediation of metal-contaminated sites. In**

**Citation:** Ruchita Dixit, Wasiullah, Deepti Malaviya, Kuppusamy Pandiyan, Udai B. Singh, Asha Sahu, Renu Shukla, Bhanu P. Singh , Jai P. Rai , Pawan Kumar Sharma, Harshad Lade, and Diby Paul, Bioremediation of Heavy Metals from Soil and Aquatic Environment: An Overview of Principles and Criteria of Fundamental Processes, doi:10.3390/su7022189.

**view of this, the present review investigates the abilities of microorganisms and plants in terms of tolerance and degradation of heavy metals. Also, advances in bioremediation technologies and strategies to explore these immense and valuable biological resources for bioremediation are discussed. An assessment of the current status of technology deployment and suggestions for future bioremediation research has also been included. Finally, there is a discussion of the genetic and molecular basis of metal tolerance in microbes, with special reference to the genomics of heavy metal accumulator plants and the identification of functional genes involved in tolerance and detoxification.**

## INTRODUCTION

With the growth of industry, there has been a considerable increase in the discharge of industrial waste to the environment, chiefly soil and water, which has led to the accumulation of heavy metals, especially in urban areas. Slow depletion of heavy metals also takes place through leaching, plant uptake, erosion and deflation. The indiscriminate release of heavy metals into the soil and waters is a major health concern worldwide, as they cannot be broken down to non-toxic forms and therefore have long-lasting effects on the ecosystem. Many of them are toxic even at very low concentrations; arsenic, cadmium, chromium, copper, lead, mercury, nickel, selenium, silver, zinc *etc.* are not only cytotoxic but also carcinogenic and mutagenic in nature [1]. Some metals are required by plants in very small amounts for their growth and optimum performance. However, the increasing concentration of several metals in soil and waters due to industrial revolution has created an alarming situation for human life and aquatic biota. This is evident from various reports citing harmful effects of heavy metals on human health (Table 1).

In order to make the environment healthier for human beings, contaminated water bodies and land need to be rectified to make them free from heavy metals and trace elements. There are several techniques to remove these heavy metals, including chemical precipitation, oxidation or reduction, filtration, ion-exchange, reverse osmosis, membrane technology, evaporation and electrochemical treatment. But most of these techniques become ineffective when the concentrations of heavy metals are less than 100 mg/L [2]. Most heavy metal salts are water-soluble and get dissolved in wastewater, which means they cannot be separated by physical separation methods [3]. Additionally, physico-chemical methods are ineffective or expensive when the concentration of heavy metals is very low. Alternately, biological methods like biosorption and/or bioaccumulation for removal of heavy metals may be an attractive alternative to physico-chemical methods [4]. Use of microorganisms and plants for remediation purposes is thus a possible solution for heavy metal pollution since it includes sustainable remediation technologies to rectify and re-establish the natural condition of soil. However, introduction of heavy metals into the soil causes considerable modification of the microbial community, despite their vital importance for the growth of microorganisms at relatively low concentrations [5]. The modification of the microbial make up is mainly brought about by exerting an inhibitory action through blockage of essential functional groups, displacement of essential metal ions or modification of active conformations of biological molecules [6,7]. The response of microbial communities to heavy metals depends on the concentration and availability of heavy metals and is a complex process which is controlled by multiple factors, such as type of metal, the nature of the medium, and microbial species [8].

**Table 1:** Toxic effect of some heavy metals on human health

| Heavy Metal | EPA Regulatory Limit (ppm) [9] | Toxic Effects | Ref. |
|---|---|---|---|
| Ag | 0.10 | Exposure may cause skin and other body tissues to turn gray or blue-gray, breathing problems, lung and throat irritation and stomach pain. | [10] |
| As | 0.01 | Affects essential cellular processes such asoxidative phosphorylation and ATP synthesis | [11] |
| Ba | 2.0 | Cause cardiac arrhythmias, respiratory failure, gastrointestinal dysfunction, muscle twitching and elevated blood pressure | [12] |
| Cd | 5.0 | Carcinogenic, mutagenic, endocrine disruptor, lung damage and fragile bones, affects calcium regulation in biological systems | [1,13] |
| Cr | 0.1 | Hair loss | [1] |
| Cu | 1.3 | Brain and kidney damage, elevated levels result in liver cirrhosis and chronic anemia, stomach and intestine irritation | [1,14] |
| Hg | 2.0 | Autoimmune diseases, depression, drowsiness, fatigue, hair loss, insomnia, loss of memory, restlessness, disturbance of vision, tremors, temper outbursts, brain damage, lung and kidney failure | [15–17] |
| Ni | 0.2 (WHO permissible limit) | Allergic skin diseases such as itching, cancer of the lungs, nose, sinuses, throat through continuous inhalation, immunotoxic, neurotoxic, genotoxic, affects fertility, hair loss | [1,18–20] |
| Pb | 15 | Excess exposure in children causes impaired development, reduced intelligence, short-term memory loss, disabilities in learning and coordination problems, risk of cardiovascular disease | [1,14,21] |
| Se | 50 | Dietary exposure of around 300 μg/day affects endocrine function, impairment of natural killer cells activity, hepatotoxicity and gastrointestinal disturbaces | [22] |
| Zn | 0.5 | Dizziness, fatigue *etc.* | [23] |

Bioremediation is an innovative and promising technology available for removal of heavy metals and recovery of the heavy metals in polluted water and lands. Since microorganisms have developed various strategies for their survival in heavy metal-polluted habitats, these organisms are known to develop and adopt different detoxifying mechanisms such as biosorption, bioaccumulation, biotransformation and biomineralization, which can be exploited for bioremediation either *ex situ* or *in situ* [24,25,26,27]. A global survey to examine the use of bioremediation technologies for addressing the environmental problems was carried out by Elekwachi *et al.* [28]. They found that despite aspirations from respondents to apply bioremediation techniques, it should not become the current practice. Developed economies made higher use of low-cost *in situ* bioremediation technologies such as monitored natural attenuation, while their developing counterparts appeared to focus on occasionally more expensive *ex situ* technologies. Despite significant investment and widespread availability of online resources, their limited use underlines the need to explore improved training and development of more user-friendly resources. There are many reports about biodegradation and bioremediation strategies being utilized by bacteria or plant species [29,30,31,32], but so far none of these investigations suggest possible drivers in the global use of the said techniques [28]. Among the preferred methods for treatment of contaminated areas, 51% of the respondents preferred environment friendly approaches, including microbial remediation (35%) and phytoremediation (16%) [33,34].

Microorganisms uptake heavy metals actively (bioaccumulation) and/or passively (adsorption) [35]. The microbial cell walls, which mainly consist of polysaccharides, lipids and proteins, offer many functional groups that can bind heavy metal ions, and these include carboxylate, hydroxyl, amino and phosphate groups [36]. Among various microbe-mediated methods, the biosorption process seems to be more feasible for large scale application compared to the bioaccumulation process, because microbes will require addition of nutrients for their active uptake of heavy metals, which increases the biological oxygen demand or chemical oxygen demand in the waste. Further, it is very difficult to maintain a healthy population of microorganisms due to heavy metal toxicity and other environmental factors [37,38]. Fungi of the genera*Penicillium*, *Aspergillus* and *Rhizopus* have been studied extensively as potential microbial agents for the removal of heavy metals from aqueous solutions [39,40]. Xiao *et al.* reported a novel technology for obtaining highly efficient biosorbents from endophytes, a hyperaccumulator, which is more convenient than the traditional method of obtaining biosorbents [41]. Sun *et al.* evaluated the genetic diversity of endophytic bacteria from the copper-tolerant species of *Elshotzia apliendens* and*Commelina communis*, reporting increased dry weights of roots and aboveground tissues compared to uninoculated plants [42]. Further, they also reported significant amounts of (ranging from 63% to 125%) Cu content in inoculated plants compared to uninoculated ones.

In view of such reports on the use of microorganisms and plants for removal of heavy metals from contaminated sites, the present review focuses on recent developments in bioremediation techniques. Additionally, new approaches such as the designer plant approach and rhizosphere modification to achieve the goal of bioremediation in a cheaper and safer way are also discussed.

## SOURCES OF HEAVY METAL IN THE ENVIRONMENT

Heavy metals occur naturally in the environment from pedogenetic processes of weathering of parent materials and also through anthropogenic sources (Figure 1). The most significant natural sources are weathering of minerals, erosion and volcanic activity, while the anthropogenic sources depend upon human activities such as mining, smelting, electroplating, use of pesticides and phosphate fertilizer discharge, as well biosolids (e.g., livestock manures, composts, and municipal sewage sludge), atmospheric deposition, *etc.* [14,43,44,45,46,47]. The disturbance of nature's slowly occurring geochemical cycle of metals by man results in accumulation of one or more of heavy metals in the soil and waters, and above defined levels, this is enough to cause risk to human health, plants, animals and aquatic biota [48]. The heavy metals essentially become contaminants in the soil and water environment because of their excess generation by natural and man-made activities, transfer from mines to other locations where higher exposure to humans occurs, discharge of high concentration of metal waste through industries, and greater bioavailability.

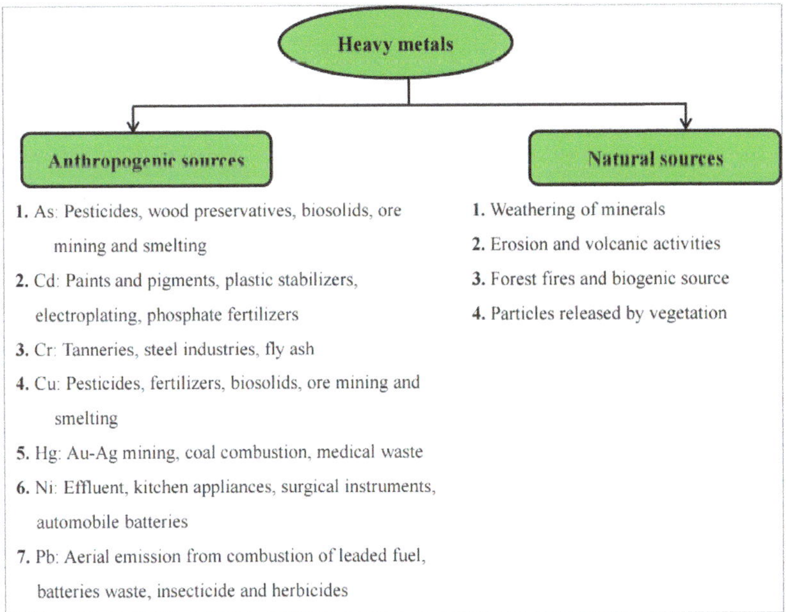


**Figure 1:** Sources of heavy metals in the environment.

A mass balance of heavy metals in the soil environment can be expressed using the formula [49]:

$$M_{total} = (M_p + M_a + M_f + M_{ag} + M_{ow} + M_{ip}) - (M_{cr} + M_l) \quad (1)$$

where M is the heavy metal, *p* is the parent material, *a* is atmospheric deposition, *f* is fertilizer source, *ag* is agrochemical source, *ow* is organic waste source, *ip* is inorganic pollutant, *cr* is crop removal and *l* is losses by leaching, volatilization and other processes. It is estimated that emission of several heavy metals in atmosphere from anthropogenic sources is one to three orders higher than natural sources [50].

## BIOREMEDIATION: INTRODUCING MICROBE BASED CLEAN UP SYSTEM

Remediation of environment niches such as soil, sediments and water amended with heavy metals can be achieved through biologically encoded changes in the oxidation state. Bioremediation is the microbe-mediated process for clearance or immobilization of the contaminants, including all possible toxins like hydrocarbons, agrochemicals and other organic toxicants. But for inorganic toxic compounds such as heavy metals, microbes are unable to simplify them into harmless compounds, and they should be used according to their specialization for the type of contaminants. Thus the bioremediation strategy for heavy metals depends on the active metabolizing capabilities of microorganisms. Several microorganisms are known to require varying amounts of heavy metals as essential micronutrients for growth and development. For example, $Fe^{3+}$ is essentially required by all bacteria while $Fe^{2+}$ is important for anaerobic bacteria [51]. However, the adsorption capacity depends on microbial total biomass and geochemistry of the system. Some oxyanions of metals do not interact with microbes, and their bioremediation is based on their catalyzed redox conversion to insoluble forms. These reduction or oxidation reactions take place due to enzymatic activity and biomass concentration of microbes. Microorganisms have a great deal of undiscovered and unexplored potential for remediation of soil pollutants and increasing the production of agricultural crops with low input. Selection of rhizospheric microbes should be done based on an understanding of mechanisms involved in the adsorption and mobilization of heavy metals and trace elements in the soil to restore soil health. Microorganisms as metal accumulators possess an inherent novel remediation property for toxic metals in the soil. The study of genetics of such metal accumulator microorganisms can help us to transfer the traits in the microbes that are missing through the development of microarrays, which result in differentially expressed microbe genes.

Detoxification and rehabilitation of contaminated soil with the use of microbes has emerged as the most safe, easy and effective technology. Native soil microorganisms have been explored and harnessed for their ability to remove or detoxify toxic products released due to human activities in the environment *viz.* mining of ores, oil and gas extraction, pesticides, pigments, plastic, organic solvents, fuel and industrial processes [52]. But the lack of information on the cellular responses of microbes towards utilization and interaction with trace element and heavy metal pollutants restricts their successful execution. Studies have been carried out to promote the use of modified microbes designed especially to increase sensitivity towards toxic metals. The biochemical route for the redistribution of the organic pollutant in the soil starts from various physical, chemical and biological processes resulting in adsorption by soil particles and root tissues, volatilization, transport through water and air, microbial degradation and leaching, *etc.*

## Mechanisms of Bioremediation

Microorganisms are omnipresent that dominate in heavy metal-contaminated soil and can easily convert heavy metals into non-toxic forms. In bioremediation processes, microorganisms mineralize the organic contaminants to end-products such as carbon dioxide and water, or to metabolic intermediates which are used as primary substrates for cell growth. Microorganisms are capable of two-way defense *viz.* production of degradative enzymes for the target pollutants as well as resistance to relevant heavy metals. Different mechanisms of bioremediation are known, including biosorption, metal-microbe interactions, bioaccumulation, biomineralisation, biotransformation and bioleaching. Microorganisms remove the heavy metals from soil by using chemicals for their growth and development. They are capable of dissolving metals and reducing or oxidizing transition metals. Different methods by which microbes restore the environment are oxidizing, binding, immobilizing, volatizing and transformation of heavy metals. Bioremediation can be made successful in a particular location by the designer microbe approach, and by understanding the mechanism controlling growth and activity of microorganisms in the contaminated sites, their metabolic capabilities and their response to environmental changes. Many contaminants are organic solvents which disrupt membranes, but cells may develop defense mechanisms including formation of outer cell-membrane-protective material, often hydrophobic or solvent efflux pumps [53]. For instance, plasmid-encoded and energy-dependent metal efflux systems involving ATPases and chemiosmotic ion/proton pumps are reported for As, Cr and Cd resistance in many bacteria [54].

## Bioremediation by Adsorption

Heavy metals can be biosorbed by microbes at binding sites present in cellular structure without the involvement of energy. Among the various reactive compounds associated with bacterial cell walls, the extracellular polymeric substances are of particular importance and are well known to have significant effects on acid-base properties and metal adsorption [55]. Studies on the metal binding behavior of extracellular polymeric substances (EPS) revealed a great ability to complex heavy metals through various mechanisms, which include proton exchange and micro-precipitation of metals [56,57]. Recent studies have characterized and quantified the proton and adsorbed metals on bacterial cells and EPS free cells in order to determine the relative importance of EPS molecules in metal removal [58]. Bioremediation research and practice are still hampered in the current scenario due to an incomplete understanding of genetics and genome level characteristics of the organisms used in metal adsorption, the metabolic pathway and their kinetics. This results in an inability to model and predict the process behavior and develop a natural bioremediation process in the field [29,30,32,59,60].

## Bioremediation by Physio-Bio-Chemical Mechanism

Biosorption is the process which involves higher affinity of a biosorbent towards sorbate (metal ions), continued until equilibrium is established between the two components [19]. *Saccharomyces cerevisiae* acts as a biosorbent for the removal of Zn (II) and Cd (II) through the ion exchange mechanism [61,62]. *Cunninghamella elegans* emerged as a promising sorbent against heavy metals released by textile wastewater [63]. Heavy metal degradation involves energy for the cell metabolic cycle. The combined active and passive modes of toxic metal bioremediation can be called bioaccumulation [64]. Fungi have emerged as potential biocatalysts to access heavy metals and transform them into less toxic compounds [65]. Some fungi such as *Klebsiella oxytoca, Allescheriella* sp., *Stachybotrys* sp., *Phlebia* sp. *Pleurotus pulmonarius*, *Botryosphaeria rhodina* have metal binding potential [66]. Pb (II) contaminated soils can be biodegraded by fungal species like *Aspergillus parasitica* and*Cephalosporium aphidicola* with biosorption process [67,68]. Hg resistant fungi (*Hymenoscyphus ericae, Neocosmospora vasinfecta*and *Verticillum terrestre*) were able to biotransform a Hg (II) state to a nontoxic state [69]. Many of the contaminants are hydrophobic, and these substances appear to be taken up by microbes through the secretion of some biosurfactant and direct cell-contaminant association. Biosurfactants form stronger ionic bonds with metals and form complexes before being desorbed from soil matrix to water phase due to low interfacial tension [70].

Bioremediation may also involve aerobic or anaerobic microbial activities. Aerobic degradation often involves introduction of oxygen atoms into the reactions mediated by monooxygenases, dioxygenases, hydroxylases, oxidative dehalogenases, or chemically reactive oxygen atoms generated by enzymes such as ligninases or peroxidases. Anaerobic degradations of contaminants involve initial activation reactions followed by oxidative catabolism mediated by anoxic electron acceptors. The technique used to reduce the mobilization of heavy metals from contaminated sites by changing the physical or chemical state of the toxic metals is called immobilization. Solidification treatment involves mixing of chemical agents at the contaminated sites or precipitation of hydroxides [71]. Microbes mobilize the heavy metals from the contaminated sites by leaching, chelation, methylation and redox tansformation of toxic metals. Heavy metals can never be destroyed completely, but the process transforms their oxidation state or organic complex, so that they become water-soluble, less toxic and precipitated [52]. Microorganisms use heavy metals and trace elements as terminal electron acceptors or reduce them through the detoxification mechanism, used for the removal of metals from the contaminated environment. Microorganisms remove heavy metals through the mechanisms which they employ to derive energy from metals redox reactions, to deal with toxic metal through enzymatic and non-enzymatic processes.

Two main mechanisms for development of resistance in bacteria are detoxification (transformation of the toxic metal state and making it unavailable) and active efflux pumping of the toxic metal from cells [72]. The basic redox (oxidation and reduction) reaction takes place in the soil between toxic metals and microorganisms; microorganisms act as an oxidizing agent for heavy metals and cause them to lose electrons, which are accepted by alternative electron acceptors (nitrate, sulphate and ferric oxides). In aerobic conditions, oxygen acts as an electron acceptor, while in anaerobic conditions microbes oxidize organic contaminants by reducing electron acceptors. The microorganism takes energy for growth by oxidizing the organic compound with Fe (III) or Mn (IV) as an electron acceptor [73]. Anaerobic degradation of organic contamination is stimulated with the higher availability of Fe (III) for microbial reduction [74,75]. Metals being used as terminal electron acceptors is called dissimilatory metal reduction [76]. Biodegradation of chlorines from contaminants takes place through reductive dechlorination, where contaminants as chlorinated solvents acts as an electron acceptors in respiration. Microorganisms reduce the state of metals and change their solubility, like the *Geobaccter* species, and reduce the Uranium soluble state ($U^{6+}$) to insoluble state ($U^{4+}$) [77].

Different defense systems (exclusion, compartmentalization, complex formation and synthesis of binding protein and peptides) reduce the stress developed by toxic metals [78]. Heavy metal accumulation by microorganisms can be studied by the expression of metal binding protein and peptides (phytochelatins and metallothionein) [79]. These metal binding protein transcription factors are known to mediate in hormone and redox signaling process in the context of toxic metal (Cd, Zn, Hg, Cu, Au, Ag, Co, Ni and Bi) exposure [80]. *Synechococcus* sp. (cynobacterial strains) has been reported with the expression of the smtA gene and production of metal-binding protein [81]. *Ralstonia eutropha* has been genetically modified to express mouse metallothionein on the cell surface and decrease the toxic effect of the Cd (II) in the contaminated sites [82].

Expression of different proteins and peptides by the *Escherichia coli* regulates the range of accumulation of cadmium [83]. Co-expression of precursor glutathione (GSH) along with phytochelatins (PC) resulted in the 10 fold increase in PC that finally increased cadmium accumulation twofold [84]. Natural resistant pathways for heavy metals (Hg and Ar) in microorganisms have been regulated by metalloregulatory protein [85].

### Molecular Mechanisms Involved in Bioremediation Process

Various mechanisms involved in the removal of heavy metals by microorganisms are known. In a genetically engineered bacterium *Deinococcus geothemalis*, Hg reduction has been reported at high temperatures due to the expression of mer operon from *E. coli* coded for $Hg^{2+}$ reduction [86]. Mercury resistant bacteria *Cupriavidus metallidurans* strain MSR33 was modified genetically by introducing a pTP6 plasmid that provided genes (merB and merG) regulating Hg biodegradation along with the synthesis of organomercurial lyase protein (MerB) and mercuric reductase (MerA) [87]. Modification of *Pseudomonas*strain with the pMR68 plasmid with novel genes (*mer*) made that strain resistant to mercury [88]. Two different mechanisms for Hg degradation by bacteria (*Klebsiella pneumonia* M426) are mercury volatilization by reduction of Hg (II) to Hg (0) and mercury precipitation as insoluble Hg due to volatile thiol ($H_2S$) [89]. Genetic engineering of *Deinococcus radiodurans*(radiation resistant bacterium) which naturally reduces Cr (IV) to Cr (III) has been done for complete toluene (fuel hydrocarbon) degradation by cloned genes of *tod* and *xyl* operons of *Pseudomonas putida* [90]. Microbial metabolites like metal bound coenzymes and siderophores mainly involved the degradation pathway [91].

## PHYTOREMEDIATION

Phytoremediation basically refers to the use of plants and associated microorganisms to partially or completely remediate selected contaminants from soil, sludge, sediments, wastewater and ground water. It can be used for removal of radionuclides, organic pollutants as well as heavy metals [92]. Phytoremediation utilizes a variety of plant processes and the physical characteristics of plants to aid in remediation of contaminated sites. Over the recent years, a special emphasis has been placed on phytoremediation since this property can be exploited for remediation of heavy metal polluted soils [93,94]. It is a cost-effective, efficient and eco-friendly *in situ* remediation technology driven by solar energy. The technique of phytoremediation includes a number of different processes such as phytoextraction, phytofiltration, phytostabilization, phytovolatilization and phytodegradation [95]. A summary of various processes involved in the phytoremediation of heavy metals is shown in Figure 2. The initial step of phytoremediation is phytoextraction, the uptake of contaminants from soil or water by plant roots and their translocation to and accumulation in biomass, *i.e.*, shoots [96]. Translocation of metals to shoots is an important biochemical process and is desirable in an effective phytoextraction. The next important process of phytoremediation is phytofiltration, which includes rhizofiltration (use of plant roots), blastofiltration (use of seedlings) or caulofiltration (use of excised plant shoots) [97]. In this, the metals are absorbed or adsorbed and thus their movement in underground water is minimized.

In addition to the above process, phytostabilization or phytoimmobilization occurs, which reduces the mobility and bioavailability of metals in the environment and thus prevents their migration into groundwater or the food chain [98]. Plants perform the immobilization of heavy metals in soils by sorption through roots, precipitation, complex formation or metal valence reduction in the rhizosphere [99]. Organic pollutants taken by plants are metabolized by enzymes such as dehalogenase and oxygenase, which are not dependent on rhizospheric microorganisms [100]. However, several heavy metals absorbed by plants get converted into volatile forms and subsequently released into the atmosphere by the process called phytovolatilization. This process has been used for removal of some volatile heavy metals like Hg and Se from polluted soils [101]. However, this is limited by the fact that it does not remove the metals completely but rather transfers them from one medium (soil or water) to another (atmosphere) from which they can reenter soil and water.

Recently, removal of heavy metals through phytoremediation, especially hyperaccumulators to degrade and detoxify contaminants, has received wide attention due to its efficacy and cost efficiency [51]. The criteria used for hyperaccumulation varies according the metal, ranging from 100 $mg \cdot kg^{-1}$ dry mass for Cd to 1000 $mg \cdot kg^{-1}$ for Cu, Co, Cr and Pb. These values exhibit a shoot-to-soil ratio of metal concentration and the factor for bioaccumulation is higher than 1 [102].

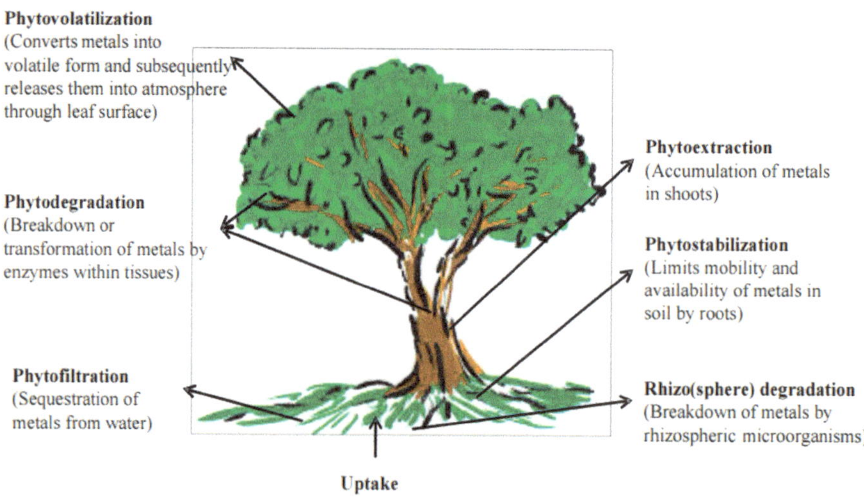


**Figure 2:** Various processes involved in the phytoremediation of heavy metals.

Hyperaccumulators have been found to exhibit higher heavy metal tolerance and accumulating abilities compared to other plants [103]. Many such plants like *Arabidopsis halleri* [104] and *Solanum nigrum* L. [105] have been utilized for phytoremediation of cadmium. Table 2 summarizes the list of different plants reported for remediation of heavy metals. However, the disadvantages that limit the use of hyperaccumulators include difficulty in finding heavy metal hyperaccumulators, slow growth and lower biomass yield. This makes the process quite time-consuming and therefore not feasible for rapidly contaminated sites or sewage treatments [41]. However, different rhizospheric microorganisms that may play important roles in plant growth and/or metal tolerance via different mechanisms are known, and these can be beneficial for the design of a phytoremediation plan to select appropriate multifunctional microbial combinations from the rhizosphere, which may include arbuscular mycorrhizal fungi and plant growth-promoting rhizobacteria. It is suggested that the remediation role of rhizosphere is the main part of phytoremediation and one of the main basic theories for removing contaminants by the combined activity of plants and microorganisms [106]. The main reason for the enhanced removal of metals in the rhizosphere is likely the increase in the number and metabolic activities of microorganisms. In the rhizospheric degradation process, the metal toxicity to plants can be reduced by the use of plant growth-promoting bacteria, free-living soil microorganisms that exert beneficial effects on plant growth. In this process, plants can stimulate microbial activity about 10–100 times by the secretion of exudates which contain carbohydrates, amino acids, flavonoids *etc.* [107]. In return, the rhizosphere bacteria that contain ACC deaminase may act to insure that the ethylene level does not impair root development and to facilitate the generation of larger roots which enhance seedling survival [108]. It is reported that nickel-resistant soil bacterium *Kluyvera ascorbata* SUD 165 promoted the growth of *Brassica campestris* in the presence of high concentration of nickel due to its ability to lower the level of ethylene stress in the seedlings [109].

**Table 2:** List of selected plants reported for phytoremediation of heavy metals

| Heavy Metal | Plant Species | Ref. |
| --- | --- | --- |
| Cd, Cu, Pb, Zn | *Salix* spp. (*Salix viminalis, Salix fragilis*) | [110–112] |
| Cd | Castor (*Ricinus communis*) | [113] |
| Cd, Pb, Zn | Corn (*Zea mays*) | [114] |
| Cd, Cu, Pb, Zn | *Populus* spp. (*Populus deltoides, Populus nigra, Populus trichocarpa*) | [112] |
| Cd, Cu, Ni, Pb | Jatropha (*Jatropha curcas* L.) | [115,116] |
| Hg | *Populus deltoides* | [117] |
| Se | *Brassica juncea, Astragalus bisulcatus* | [118] |
| Zn | *Populus canescens* | [119] |

## BIOTECHNOLOGICAL INTERVENTION/GENETIC ENGINEERING IN BIOREMEDIATION PROCESSES

### "Designer" Microbes Approach

Genetically engineered microorganisms (GEM) are organisms whose genetic material has been altered using recombinant DNA technology to generate a character-specific efficient strain for bioremediation of soil, water and activated sludge by exhibiting enhanced degrading capabilities against a wide range of chemical contaminants [120]. It offers the advantage of constructing microbial strains which can withstand adverse stressful situations and can be used as a bioremediators under various and complex environmental conditions. A list of selected genetically engineered bacteria for removal of heavy metals is presented in Table 3. Genetic engineering has led to the development of "microbial biosensors" to measure the degree of contamination in contaminated sites quickly and accurately. Various biosensors have been designed to evaluate heavy metal concentrations like mercury (Hg), cadmium (Cd), nickel (Ni), copper (Cu) and arsenic (As) [121,122]. Genetic engineering of endophytes and rhizospheric bacteria for plant-associated degradation of pollutants in soil is considered to be one of the most promising new technologies for remediation of metal contaminated sites [123]. Bacteria like *Escherichia coli* and *Moreaxella* sp. expressing phytochelatin 20 on the cell surface have been shown to accumulate 25 times more Cd or Hg than the wild-type strains [124,125]. However, one major obstacle for utilizing these GEMs in hostile field conditions is sustaining the recombinant bacteria population in soil, with various environmental conditions and competition from native bacterial populations [126]. Further, the molecular approaches have been applied to only limited bacterial strains like*Escherichia coli, Pseudomonas putida, Bacillus subtilis etc.* This means other microorganisms need to be explored for their application in heavy metal bioremediation through molecular intervention.

**Table 3:** Genetically engineered bacteria for remediation of heavy metals

| Heavy Metal | Initial Conc. (ppm) | Removal Efficiency (%) | Genetically Engineered Bacteria | Expressed Gene | Ref. |
|---|---|---|---|---|---|
| As | 0.05 | 100 | *E. coli* strain | Metalloregulatory protein ArsR | [127] |
| $Cd^{2+}$ | - | - | *E .coli* strain | SpPCS | [84] |
| $Cr^{6+}$ | 1.4–1000 | 100 | *Methylococcus capsulatus* | CrR | [128] |
| Cr | - | - | *P. putida* strain | Chromate reductase (ChrR) | [129] |
| $Cd^{2+}$, Hg | - | - | *Ralstonia eutropha* CH34, *Deinococcus radiodurans* | merA | [82,130] |
| Hg | - | - | *E. coli* strain | Organomurcurial lyase | [131] |
| Hg | 7.4 | 96 | *E. coli* JM109 | $Hg^{2+}$ transporter | [132] |
| Hg | - | - | *Pseudomonas* K-62 | Organomercurial lyase | [133] |
| Hg | - | - | *Achromobacter* sp AO22 | mer | [134] |
| Ni | 145 | 80 | *P. fluorescens* 4F39 | Phytochelatin synthase (PCS) | [135,136] |

## Designer Plant Approach

Recent advances in omics technologies such as genomics, proteomics, transcriptomics and metabolomics play important roles in identifying traits that maximize the benefits of field remediation technologies. The main constraint of phytoremediation technology is the accumulation of pollutants or their metabolites in plant tissues, which shortens plant life and releases contaminants into the atmosphere via volatilization. This problem can be minimized by manipulation of metal tolerance, accumulation and degradation potential of plants against various inorganic pollutants. Using the designer plant approach, the bacterial genes responsible for metal degradation can be introduced in plant tissues to allow degradation of metals within the plant tissues [137]. Application of genetically engineered (GE) plant-based bioremediation for various heavy metals pollutants is in the forefront due to its eco-friendliness and reduced health hazards compared to physico-chemical based strategies, which are less eco-friendly and more hazardous to human health [138]. Various microbial genes can be harnessed in the transgenic plant for detoxification and accumulation of inorganic contaminants [139]. The metal-detoxifying chelators such as metallothineins and phytochetains can confer resistance to the plant by enhancing uptake, transport and accumulation of various heavy metals [140]. Similarly, transgenic plants carrying bacterial reductase can enhance the volatilization of Hg and Se while accumulating the arsenic in plant shoots [118,139]. Fast-growing as well as high-biomass-yielding plants like poplar, willow and *Jatropa* could be used for both phytoremediation and energy production. However, transfer of the metals from soil or water does not solve the problem, and burning metal-contaminated plant material for energy production will release the metals into the atmosphere, moving the problem from soil or water to the air. Thus such metal-accumulating biomasses should be stored or disposed of appropriately so that they do not pose any hazard to the environment.

Among the fast-growing and high-biomass-yielding plants, poplar is the most commonly studied because of its rapid growth rate and potential to produce high biomass within a short period of time (5–8 years). Many of the poplar hybrid varieties have been genetically modified with microbial catabolic genes and specific transporters for increased remediation [141,142]. For example, mercuric reductase and γ-glutamylsysteine synthetase genes showed increased resistance to Hg and Cd and Cu, respectively, through accumulation of higher concentrations of these metals [119,142]. As mentioned earlier, manipulation of desired plant species with multiple genes will facilitate complete degradation of pollutants to ensure that the harvested biomass can be utilized completely for additional benefits [143].

## Rhizosphere Engineering

Engineered bioremediation strategies involve either the addition of growth stimulators (electron acceptors/donors) to the rhizosphere for reduction of heavy metals or addition of nutrients to the contaminated soil for enhancement of microbial growth and bioremediation properties of microorganisms or genetically modified plants [144]. Many engineered bacteria with heavy metal reduction capacity through the expression of improved enzymes like chromate and uranyl reductase were applied in a specific rhizosphere to perform a specific function. Similarly, genetically modified plants are also known to produce specific compounds which may support the rhizospheric transformation of heavy metals. Recently, several researchers have applied these approaches to remediate heavy metal-contaminated soils using the rhizosphere ecosystem.

## Manipulation of Plant-Microbe Symbiosis

The main drawbacks of phytoremediation technology are storage and accumulation of pollutants in the plant materials and the remediation process slowing down and often becoming inadequate when the contaminated site has multiple pollutants [145]. The appropriate solution to these problems is to combine the microbe-plant symbiosis within the plant rhizosphere [126] or to introduce microbes as endophytes to allow degradation of pollutants within the plant tissues [137]. The microbial population in the rhizosphere is much higher than present in vegetation-less soil, and this is due to the facilitation provided by the plants through release of substances that are nutrients for microorganisms. This approach has been evaluated under laboratory conditions, and if it succeeds in field conditions, this technology could facilitate accelerated removal of pollutants, which in turn will support high plant biomass production for bioenergy [146]. The major strategies for implementing bioremediation processes include biostimulation and bioaugmentation approaches guided by specific microbes in combination with plants. Biostimulation involves adding supplements to a contaminated site with the objective of stimulating growth of the microbial population already present there, which may be capable of degrading the contaminants. Bioaugmentation refers to addition of selected and acclimated microbial inocula to the environment that do not contain microbes capable of degrading the contaminants.

## Application of Nano-Biotechnology

Apart from the above discussed strategies, the remediation of heavy metals and trace elements can be achieved by an emerging technology *i.e.*, nanotechnology. Various nanoparticles or nanomaterials have been found to be very effective for the removal of a wide range of toxic metals from the environment as compared to conventional methods. Nanoparticles enhancing microbial activity to remove toxic pollutants is called "nanobioremediation." Nano-based technologies not only reduce the costs of cleaning up contaminated sites at a large scale, but also reduce the process time as well. "Bionanotechnology" or "nanotechnology through biotechnology" is the bio-fabrication of nano-objects or bifunctional macromolecules used as tools to construct or manipulate nano-objects. Wide physiological diversity, small size, genetic manipulability and controlled culturability make microbial cells ideal producers of nanostructures ranging from natural products, such as polymers and magnetosomes, to engineered proteins or protein constructs, such as virus-like proteins (VLP) and tailored metal particles [147]. *Deinococcus radiodurans*, a radioactive-resistant organism, has the ability to withstand radiation well beyond the naturally occurring levels, thus its application in radioactive waste clean-up initiatives funded by US Department of Energy (DOE) [130,148]. Metal chelating polymers require toxic solvents for their synthesis and ultrafiltration for their separation, and this can be solved by developing metal binding materials that can be recovered by changing the environment surround them like pH, temperature *etc.* One such material is nanoscale modified biopolymers, which are manufactured by genetic and protein engineering of microorganisms, and their size can be controlled at the molecular level [149]. This innovative technique would be a promising tool to address the escalating problem of heavy metal as well as organic contaminants in the environment.

## Application of Genomics

Genomics has been explored and used mainly in microbial genetics and in agriculture, e.g., GMO crops, but it is now a new tool for bioremediation. Knowledge of genomics in of the context of understanding microbe-mediated remediation provides a view of genes related to the sensitivity of microbes towards toxic metals in the soil. Application of genomics in bioremediation makes it possible to analyze the microorganism on the basis of not only biochemical parameters but also molecular levels related to mechanism.

## FUTURE PROSPECTS

Rapid industrialization and technology development have adverse side effects like soil contamination and degrading soil health. Due to the complexity involved in the conventional methods for remediation of soil, the use of microbes has arisen as a time-saver for bioremediation. However, bioremediation technology has limitations; several microorganisms cannot break toxic metals into harmless metabolites, and these have inhibitory effects on microbial activity. Modification in the outer membrane proteins of bacteria with potential bioremediation properties for improving metal binding abilities is the likely way to enhance their capacity for biotransformation of toxic metals. Future studies should focus on the factors involved in improving *in situ* bioremediation strategies using genetically engineered microorganisms (GEM) and also theapplicability and adaptability of these GEMs in all the possible adverse/stress conditions and multiple-heavy-metal-polluted conditions. The reluctance among the public to accept GEM for bioremediation also needs to be considered in future studies, and they must proved non-toxic to the environment.

## ACKNOWLEDGMENTS

The authors would like to thank all the three anonymous referees for their constructive comments and suggestions.

## AUTHOR CONTRIBUTIONS

Ruchita Dixit, Wasiullah, Deepti Malaviya and Kuppusamy Pandiyan mostly contributed equally at designing of manuscript. Renu Shukla, Asha Sahu, Bhanu P. Singh and Jai P. Rai carried out data collection and processing. Pawan Kumar Sharma was involved in the statistical analysis. Udai B. Singh, Harshad Lade and Diby Paul wrote and revise the paper. All authors read and approved the final manuscript.

## REFERENCES

1. Salem, H.M.; Eweida, E.A.; Farag, A. Heavy metals in drinking water and their environmental impact on human health. In *ICEHM 2000*; Cairo University: Giza, Egypt, 2000; pp. 542–556. [Google Scholar]
2. Ahluwalia, S.S.; Goyal, D. Microbial and plant derived biomass for removal of heavy metals from wastewater.*Bioresour. Technol.* 2007, *98*, 2243–2257.
3. Hussein, H.; Farag, S.; Moawad, H. Isolation and characterization of Pseudomonas resistant to heavy metals contaminants. *Arab. J. Biotechnol.* 2004, *7*, 13–22.

4. Kapoor, A.; Viraraghvan, T. Fungal biosorption—An alternative treatment option for heavy metal bearing wastewater: A review. *Bioresour. Technol.* 1995, *53*, 195–206.

5. Doelman, P.; Jansen, E.; Michels, M.; van Til, M. Effects of heavy metals in soil on microbial diversity and activity as shown by the sensitivity-resistance index, an ecologically relevant parameter. *Biol. Fertil. Soils* 1994, *17*, 177–1784.

6. Wood, J.M.; Wang, H.K. Microbiol resistance to heavy metals. *Environ. Sci. Technol.* 1983, *17*, 582–590.

7. Li, F.; Tan, T.C. Monitoring BOD in the presence of heavy metal ions using a poly (4-vinylpyr-idine) coated microbial sensor. *Biosen. Bioelectron.* 1994, *9*, 445–455.

8. Goblenz, A.; Wolf, K.; Bauda, P. The role of glutathione biosynthesis in heavy metal resistance in the fission yeast*Schizosaccharomyces pombe*. *FEMS Microbiol. Rev.* 1994, *14*, 303–308.

9. US EPA. *Drinking Water Contaminants*; United States Environmental Protection Agency (EPA): Washington, DC, USA, 2009.

10. Agency for Toxic Substances and Disease Registry (ATSDR). *Toxicological Profile for Silver*; U.S. Department of Health and Human Services, Public Health Service: Atlanta, GA, USA, 1990.

11. Tripathi, R.D.; Srivastava, S.; Mishra, S.; Singh, N.; Tuli, R.; Gupta, D.K.; Maathuis, F.J.M. Arsenic hazards: Strategies for tolerance and remediation by plants. *Trends Biotech.* 2007, *25*, 158–165.

12. Acobs, I.A.; Taddeo, J.; Kelly, K.; Valenziano, C. Poisoning as a result of barium styphnate explosion. *Am. J. Ind. Med.*2002, *41*, 285–288.

13. Degraeve, N. Carcinogenic, teratogenic and mutagenic effects of cadmium. *Mutat. Res.* 1981, *86*, 115–135.

14. Wuana, R.A.; Okieimen, F.E. Heavy metals in contaminated soils: A review of sources, chemistry, risks and best available strategies for remediation. *ISRN Ecol.* 2011, *2011*. Article 20.

15. Neustadt, J.; Pieczenik, S. Toxic-metal contamination: Mercury. *Integr. Med.* 2007, *6*, 36–37.

16. Ainza, C.; Trevors, J.; Saier, M. Environmental mercury rising. *Water Air Soil Poll.* 2010, *205*, 47–48.

17. Gulati, K.; Banerjee, B.; Bala Lall, S.; Ray, A. Effects of diesel exhaust, heavy metals and pesticides on various organ systems: Possible mechanisms and strategies for prevention and treatment. *Indian J. Exp. Biol.* 2010, *48*, 710–721. [PubMed]

18. Khan, M.A.; Ahmad, I.; Rahman, I. Effect of environmental pollution on heavy metals content of *Withania somnifera*. *J. Chin. Chem. Soc.* 2007, *54*, 339–343.

19. Das, N.; Vimala, R.; Karthika, P. Biosorption of heavy metals—An overview. *Indian J. Biotechnol.* 2008, *7*, 159–169.

20. Duda-Chodak, A.; Baszczyk, U. The impact of nickel on human health. *J. Elementol.* 2008, *13*, 685–696.

21. Padmavathiamma, P.K.; Li, L.Y. Phytoremediation technology: Hyperaccumulation metals in plants. *Water Air Soil Poll.* 2007, *184*, 105–126.

22. Vinceti, M.; Wei, E.T.; Malagoli, C.; Bergomi, M.; Vivoli, G. Adverse health effects of selenium in humans. *Rev. Environ. Health* 2001, *16*, 233–251.

23. Hess, R.; Schmid, B. Zinc supplement overdose can have toxic effects. *J. Pediat. Hematol. Onc.* 2002, *24*, 582–584.

24. Gadd, G.M. Bioremedial potential of microbial mechanisms of metal mobilization and immobilization. *Curr. Opin. Biotechnol.* 2000, *11*, 271–279.

25. Lim, P.E.; Mak, K.Y.; Mohamed, N.; Noor, A.M. Removal and speciation of heavy metals along the treatment path of wastewater in subsurface-flow constructed wetlands. *Water Sci. Technol.* 2003, *48*, 307–313.

26. Malik, A. Metal bioremediation through growing cells. *Environ. Int.* 2004, *30*, 261–278.

27. Lin, C.C.; Lin, H.L. Remediation of soil contaminated with the heavy metal ($Cd^{2+}$). *J. Hazard. Mater.* 2005, *122*, 7–15.

28. Elekwachi, C.O.; Andresen, J.; Hodgman, T.C. Global use of bioremediation technologies for decontamination of ecosystems. *J. Bioremed. Biodeg.* 2014, *5*, 1–9.

29. Gan, S.; Lau, E.V.; Ng, H.K. Remediation of soils contaminated with polycyclic aromatic hydrocarbons (PAHs). *J. Hazard. Mater.* 2009, *172*, 532–549.

30. Haritash, A.K.; Kaushik, C.P. Biodegradation aspects of polycyclic aromatic hydrocarbons (PAHs): A review. *J. Hazard. Mater.* 2009, *169*, 1–15.

31. Satinder, K.B.; Verma, M.; Surampalli, R.Y.; Misra, K.; Tyagi, R.D.; Meunier, N. Bioremediation of hazardous wastes-A review. *Pract. Period. Hazard. Toxic Radioact. Waste Manag.* 2006, *10*, 59–72.

32. Onwubuya, K.; Cundy, A.; Puschenreiter, M.; Kumpiene, J.; Bone, B. Developing decision support tools for the selection of "gentle" remediation approaches. *Sci. Total Environ.* 2009, *407*, 6132–6142.

33. Environmental Agency (EA). Reporting the Evidence: Dealing with Contaminated Land in England and Wales. A Review of Progress from 2000–2007 with Part 2A of the Environmental Protection Act. Available online: https://www.gov.uk/government/uploads/system/uploads/attachment_data/file/313964/geho0109bpha-e-e.pdf (accessed on 1 January 2015).

34. US EPA. *Treatment Technologies for Site Cleanup: Annual Status Report*; United States Environmental Protection Agency (EPA): Washington, DC, USA, 2007.

35. Hussein, H.; Krull, R.; Abou El-Ela, S.I.; Hempel, D.C. Interaction of the different heavy metal ions with immobilized bacterial culture degrading xenobiotic wastewater compounds. In Proceedings of the Second International Water Association World Water Conference, Berlin, Germany, 15–19 October 2001; pp. 15–19.

36. Scott, J.A.; Karanjkar, A.M. Repeated cadmium biosorption by regenerated *Enterobacter aerogenes* biofilm attached to activated carbon. *Biotechnol. Lett.* 1992, *14*, 737–740.

37. Ajmal, M.; Rafaqat, A.K.; Bilquees, A.S. Studies on removal and recovery of Cr (VI) from electroplating wastes. *Water Res.* 1996, *30*, 1478–1482.

38. Dilek, F.B.; Gokcay, C.F.; Yetis, U. Combined effects of Ni(II) and Cr(VI) on activated sludge. *Water Res.* 1998, *32*, 303–312.

39. Volesky, B.; Holan, Z.R. Biosorption of heavy metals. *Biotechnol. Prog.* 1995, *11*, 235–250.

40. Huang, C.; Huang, C.P. Application of *Aspergillus oryzae* and *Rhizopus oryzae* for Cu (II) removal. *Water Res.* 1996, *30*, 1985–1990.

41. Xiao, X.; Luo, S.; Zeng, G.; Wei, W.; Wan, Y.; Chen, L.; Guo, H.; Cao, Z.; Yang, L.; Chen, J.; *et al.* Biosorpiton of cadmium by endophytic fungus (EF) *Microsphaeropsis* sp. LSE10 isolated from cadmium hyperaccumulator *Solanum nigrum* L. *Bioresour. Technol.* 2010, *101*, 1668–1674.

42. Sun, L.N.; Zhang, Y.F.; He, L.Y.; Chen, Z.J.; Wang, Q.Y.; Qian, M.; Sheng, X.F. Genetic diversity and characterization of heavy metal-resistant-endophytic bacteria from two copper-tolerant plant species on copper mine wasteland. *Bioresour. Technol.* 2010, *101*, 501–509.

43. Modaihsh, A.; Al-Swailem, M.; Mahjoub, M. Heavy metal contents of commercial inorganic fertilizer used in the Kingdom of Saudi Arabia. *Agric. Mar. Sci.* 2004, *9*, 21–25.

44. Chehregani, A.; Malayeri, B.E. Removal of heavy metals by native accumulator plants. *Int. J. Agric. Biol.* 2007, *9*, 462–465.

45. Fulekar, M.; Singh, A.; Bhaduri, A.M. Genetic engineering strategies for enhancing phytoremediation of heavy metals.*Afr. J. Biotechnol.* 2009, *8*, 529–535.

46. Sabiha-Javied; Mehmood, T.; Chaudhry, M.M.; Tufai, M.; Irfan, N. Heavy metal pollution from phosphate rock used for the production of fertilizer in Pakistan. *Microchem. J.* 2009, *91*, 94–99.

47. Sumner, M.E. Beneficial use of effluents, wastes, and biosolids. *Commun. Soil Sci. Plant Anal.* 2000, *31*, 1701–1715.

48. D'Amore, J.J.; Al-Abed, S.R.; Scheckel, K.G.; Ryan, J.A. Methods for speciation of metals in soils: A review. *J. Environ. Qual.* 2005, *34*, 1707–1745.

49. Lombi, E.; Gerzabek, M.H. Determination of mobile heavy metal fraction in soil: Results of a pot experiment with sewage sludge. *Commun. Soil Sci. Plant Anal.* 1998, *29*, 2545–2556.

50. Sposito, G.; Page, A.L. Cycling of metal ions in the soil environment. In *Metal Ions in Biological Systems, Circulation of Metals in the Environment*; Sigel, H., Ed.; Marcel Dekker, Inc.: New York, NY, USA, 1984; pp. 287–332.

51. Ahemad, M. Remediation of metalliferous soils through the heavy metal resistant plant growth promoting bacteria: Paradigms and prospects. *Arab. J. Chem.* 2014.

52. Garbisu, C.; Alkorta, I. Phytoextraction: A cost-effective plant-based technology for the removal of metals from the environment. *Bioresour. Technol.* 2001, *77*, 229–236.

53. Sikkema, J.; de Bont, J.A.; Poolman, B. Mechanisms of membrane toxicity of hydrocarbons. *Microbiol. Rev.* 1995, *59*, 201–222.

54. Roane, T.M.; Pepper, I.L. Microorganisms and metal pollution. In *Environmental Microbiology*; Maier, I.L., Pepper, C.B., Eds.; Gerba, Academic Press: London, UK, 2000; p. 55.

55. Guiné, V.; Spadini, L.; Sarret, G.; Muris, M.; Delolme, C.; Gaudet, J.P.; Martins, J.M. Zinc sorption to three gram-negative bacteria: Combined titration, modeling and EXAFS study. *Environ. Sci. Technol.* 2006, *40*, 1806–1813.

56. Comte, S.; Guibaud, G.; Baudu, M. Biosorption properties of extracellular polymeric substances (EPS) towards Cd, Cu and Pb for different pH values. *J. Hazard. Mater.* 2008, *151*, 185–193.

57. Fang, L.C.; Huang, Q.Y.; Wei, X.; Liang, W.; Rong, X.M.; Chen, W.L.; Cai, P. Microcalorimetric and potentiometric titration studies on the adsorption of copper by extracellular polymeric substances (EPS), minerals and their composites. *Bioresour. Technol.* 2010, *101*, 5774–5779.

58. Fang, L.; Wei, X.; Cai, P.; Huang, Q.; Chen, H.; Liang, W.; Rong, X. Role of extracellular polymeric substances in Cu(II) adsorption on *Bacillus subtilis* and *Pseudomonas putida*. *Bioresour. Technol.* 2011, *102*, 1137–1141.

59. Carter, P.; Cole, H.; Burton, J. Bioremediation: Successes and Shortfalls. In Proceedings of Key International Conference and Exhibition for Spill Prevention, Preparedness, Response and Restoration (Interspill), London, UK, 23 March 2006.

60. Kinya, K.; Kimberly, L.D. Current use of bioremediation for TCE cleanup: Results of a survey. *Remediat. J.* 1996, *6*, 1–14.

61. Chen, C.; Wang, J.L. Characteristics of $Zn^{2+}$ biosorption by *Saccharomyces cerevisiae*. *Biomed. Environ. Sci.* 2007, *20*, 478–482.

62. Talos, K.; Pager, C.; Tonk, S.; Majdik, C.; Kocsis, B.; Kilar, F.; Pernyeszi, T. Cadmium biosorption on native*Saccharomyces cerevisiae* cells in aqueous suspension. *Acta Univ. Sapientiae Agric. Environ.* 2009, *1*, 20–30.

63. Tigini, V.; Prigione, V.; Giansanti, P.; Mangiavillano, A.; Pannocchia, A.; Varese, G.C. Fungal biosorption, an innovative treatment for the decolourisation and detoxification of textile effluents. *Water* 2010, *2*, 550–565.

64. Brierley, C.L. Bioremediation of metal-contaminated surface and groundwater. *Geomicrobiol. J.* 1990, *8*, 201–223.

65. Pinedo-Rivilla, C.; Aleu, J.; Collado, I.G. Pollutants biodegradation by fungi. *Curr. Org. Chem.* 2009, *13*, 1194–1214.

66. D'Annibale, A.; Leonardi, V.; Federici, E.; Baldi, F.; Zecchini, F.; Petruccioli, M. Leaching and microbial treatment of a soil contaminated by sulphide ore ashes and aromatic hydrocarbons. *Appl. Microbiol. Biotechnol.* 2007, *74*, 1135–1144.

67. Tunali, S.; Akar, T.; Oezcan, A.S.; Kiran, I.; Oezcan, A. Equilibrium and kinetics of biosorption of lead(II) from aqueous solutions by *Cephalosporium aphidicola*. *Sep. Purif. Technol.* 2006, *47*, 105–112.

68. Akar, T.; Tunali, S.; Cabuk, A. Study on the characterization of lead (II) biosorption by fungus *Aspergillus parasiticus*.*Appl. Biochem. Biotech.* 2007, *136*, 389–406.

69. Kelly, D.J.A.; Budd, K.; Lefebvre, D.D. The biotransformation of mercury in pH-stat cultures of microfungi. *Can. J. Bot.*2006, *84*, 254–260.

70. Thavasi, R. Microbial biosurfactants: From an environment application point of view. *J. Bioremed. Biodegrad.* 2011, *2*. Article 104e.

71. Evanko, C.R.; Dzombak, D.A. Remediation of metals-contaminated soil and groundwater. *Environ. Sci.* 1997, *412*, 1–45.

72. Silver, S. Bacterial heavy metal resistance: New surprises. *Annu. Rev. Microbiol.* 1996, *50*, 753–789.

73. Lovley, D.R.; Phillips, E.J.P. Novel mode of microbial energy metabolism: Organic carbon oxidation to dissimilatory reduction of iron or manganese. *Appl. Environ. Microbiol.* 1988, *54*, 1472–1480.

74. Lovley, D.R.; Coates, J.D.; Blunt-Harris, E.L.; Philips, E.J.P.; Woodward, J.C. Humic substances as electron acceptors for microbial respiration. *Nature* 1996, *382*, 445–448.

75. Spormann, A.M.; Widdel, F. Metabolism of alkylbenzenes, alkanes, and other hydrocarbons in anaerobic bacteria.*Biodegradation* 2000, *11*, 85–105.

76. Lovely, D.R. Dissimilatory metal rduction: From early life to bioremediation. *ASM News* 2002, *68*, 231–237.

77. Lovley, D.R.; Philips, E.J.; Gorby, Y.A.; Landa, E.R. Microbial reduction of uranium. *Nature* 1991, *350*, 413–416.

78. Gómez Jiménez-T, R.; Moliternib, E.; Rodríguezb, L.; Fernándezc, F.J.; Villaseñorc, J. Feasibility of mixed enzymatic complexes to enhanced soil bioremediation processes. *Procedia Environ. Sci.* 2011, *9*, 54–59.

79. Cobbett, C.; Goldsbrough, P. Phytochelatins and metallothioneins: Role in heavy metals detoxification and homeostatis. *Annu. Rev. Plant Biol.* 2002, *53*, 159–182.

80. Kagi, J.H.R.; Schaffer, A. Biochemistry of metallothionein. *Biochemistry* 1988, *27*, 8509–8515.

81. Huckle, J.W.; Morby, A.P.; Turner, J.S.; Robinson, N.J. Isolation of a prokaryotic metallothionein locus and analysis of transcriptional control by trace metal ions. *Mol. Microbiol.* 1993, *7*, 177–187.

82. Valls, M.; Atrian, S.; de Lorenzo, V.; La, F. Engineering a mouse metallothionein on the cell surface of *Ralstonia eutropha* CH34 for immobilization of heavy metals in soil. *Nat. Biotechnol.* 2000, *18*, 661–665.

83. Mejare, M.; Bulow, L. Metal binding proteins and peptides in bioremediation and phytoremediation of heavy metals.*Trends Biotechnol.* 2001, *19*, 67–73.

84. Kang, S.H.; Singh, S.; Kim, J.Y.; Lee, W.; Mulchandani, A.; Chen, W. Bacteria

metabolically engineered for enhanced phtochelatin production and cadmium accumulation. *App. Environ. Microbiol.* 2007, *73*, 6317–6320.

85. Singh, S.; Kang, S.H.; Mulchandani, A.; Chen, W. Bioremediation: environmental cleanup through pathway engineering. *Curr. Opin. Biotechnol.* 2008, *19*, 437–444.

86. Brim, H.; Venkateshwaran, A.; Kostandarithes, H.M.; Fredrickson, J.K.; Daly, M.J. Engineering *Deinococcus geothermalis*for bioremediation of high temperature radioactive waste environments. *App. Environ. Microbiol.* 2003, *69*, 4575–4582.

87. Rojas, L.A.; Yanez, C.; Gonzalez, M.; Lobos, S.; Smalla, K.; Seeger, M. Characterization of the metabolically modified heavy metal-resistant *Cupriavidus metallidurans* strain MSR33 gener-ated for mercury bioremediation. *PLoS One 20*2011, *6*, e17555.

88. Sone, Y.; Mochizuki, Y.; Koizawa, K.; Nakamura, R.; Pan-Hou, H.; Itoh, T.; Kiyono, M. Mercurial resistance determinants in *Pseudomonas* strain K-62 plasmid pMR68. *AMB Express* 2013, *3*. Article 41.

89. Essa, A.M.M.; Macaskie, L.E.; Brown, N.L. Mechanisms of mercury bioremediation. *Biochem. Soc. Trans.* 2002, *30*, 672–674.

90. Brim, H.; Osborne, J.P.; Kostandarithes, H.M.; Fredrickson, J.K.; Wackett, L.P.; Daly, M.J. *Deinococcus radiodurans*engineered for complete toluene degradation facilities Cr(IV) reduction. *Microbiology* 2006, *152*, 2469–2477.

91. Penny, C.; Vuilleumier, S.; Bringel, F. Microbial degradation of tetrachloromethane: Mechanisms and perspectives for bioremediation. *FEMS Microbiol. Ecol.* 2010, *74*, 257–275.

92. Ali, H.; Khan, E.; Sajad, M.A. Phytoremediation of heavy metals-Concepts and applications. *Chemosphere* 2013, *91*, 869–881.

93. Robinson, B.H.; Brooks, R.R.; Howes, A.W.; Kirkman, J.H.; Gregg, P.E.H. The potential of the high biomass nickel hyperaccumulator *Berkheya coddii* for phytoremediation and phytomining. *J. Geochem. Explor.* 1997, *60*, 115–126.

94. Martinez, M.; Bernal, P.; Almela, C.; Vélez, D.; García-Agustín, P.; Serrano, R.; Navarro-Aviñó, J. An engineered plant that accumulates higher levels of heavy metals than *Thlaspi caerulescens*, with yields of 100 times more biomass in mine soils. *Chemosphere* 2006, *64*, 478–485.

95. Alkorta, I.; Hernández-Allica, J.; Becerril, J.M.; Amezaga, I.; Albizu, I.; Garbisu, C. Recent findings on the phytoremediation of soils contaminated with environmentally toxic heavy metals and metalloids such as Zinc, Cadmium, Lead, and Arsenic. *Rev. Environ. Sci. Biotechnol.* 2004, *3*, 71–90.

96. Sekara, A.; Poniedzialeek, M.; Ciura, J.; Jedrszczyk, E. Cadmium and lead accumulation and distribution in the organs of nine crops: implications for phytoremediation. *Pol. J. Environ. Stud.* 2005, *14*, 509–516.

97. Mesjasz-Przybylowicz, J.; Nakonieczny, M.; Migula, P.; Augustyniak, M.; Tarnawska, M.; Reimold, W.U.; Koeberl, C.; Przybylowicz, W.; Glowacka, E.

Uptake of cadmium, lead, nickel and zinc from soil and water solutions by the nickel hyperaccumulator *Berkheya coddii*. *Acta Biol. Cracov. Bot.* 2004, *46*, 75–85.

98. Erakhrumen, A.A. Phytoremediation: An environmentally sound technology for pollution prevention, control and remediation in developing countries. *Educ. Res. Rev.* 2007, *2*, 151–156.
99. Barceló, J.; Poschenrieder, C. Phytoremediation: principles and perspectives. *Contrib. Sci.* 2003, *2*, 333–344.
100. Vishnoi, S.R.; Srivastava, P.N. Phytoremediation-green for environmental clean. In Proceeding of Taal 2007: The 12th World Lake Conference, Jaipur, India, 29 October–2 November 2008.
101. Karami, A.; Shamsuddin, Z.H. Phytoremediation of heavy metals with several efficiency enhancer methods. *Afr. J. Biotechnol.* 2010, *9*, 3689–3698.
102. Baker, A.J.M.; McGrath, S.P.; Sidoli, C.M.D.; Reeves, R.D. The possibility of *in situ* heavy metal decontamination of polluted soils using crops of metal accumulating plants. *Resour. Conserv. Recycl.* 1994, *11*, 42–49.
103. Prasad, M.N.V.; Freitas, H.M.D.O. Metal hyperaccumulation in plants—Biodiversity prospecting for phytoremediation technology. *Electro. J. Biotechnol.* 2003, *6*, 285–321.
104. Dahmani-Muller, H.; Van-Oort, F.; Gelie, B.; Balabane, M. Strategies of heavy metal uptake by three plant species growing near a metal smelter. *Environ. Pollut.* 2000, *109*, 231–238.
105. Wei, S.H.; Zhou, Q.X.; Wang, X. Cadmium-hyperaccumulator *Solanum nigrum* L. and its accumulating characteristics.*Environ. Sci.* 2005, *26*, 167–171.
106. Wei, S.; Zhou, Q.; Zhang, K.; Liang, J. Roles of rhizosphere in remediation of contaminated soils and its mechanisms.*Ying Yong Sheng Tai Xue Bao* 2003, *14*, 143–147.
107. Kuiper, I.; Lagendijk, E.L.; Bloemberg, G.V.; Lugtenberg, B.J.J. Rhizoremediation: A beneficial plant-microbe interaction. *Mol. Plant-Microbe Interact.* 2004, *17*, 6–15.
108. Glick, B.R.; Penrose, D.M.; Li, J. A model for the lowering of plant ethylene concentrations by plant growth-promoting bacteria. *J. Theor. Biol.* 1998, *190*, 63–68.
109. Burd, G.I.; Dixon, D.G.; Glick, B.R. A plant growth-promoting bacterium that decreases nickel toxicity in seedlings.*Appl. Environ. Microbiol.* 1998, *64*, 3663–3668.
110. Pulford, I.; Watson, C. Phytoremediation of heavy metal-contaminated land by trees-a review. *Environ. Int.* 2003, *29*, 529–540.
111. Volk, T.A.; Abrahamson, L.P.; Nowak, C.A.; Smart, L.B.; Tharakan, P.J.; White, E.H. The development of short-rotation willow in the northeastern United States for bioenergy and bioproducts, agroforestry and phytoremediation.*Biomass Bioener.* 2006, *30*, 715–727.

112. Ruttens, A.; Boulet, J.; Weyens, N.; Smeets, K.; Adriaensen, K.; Meers, E.; van Slycken, S.; Tack, F.; Meiresonne, L.; Thewys, T.; *et al*. Short rotation coppice culture of willows and poplars as energy crops on metal contaminated agricultural soils. *Int. J. Phytorem.* 2011, *13*, 194–207.

113. Huang, H.; Yu, N.; Wang, L.; Gupta, D.K.; He, Z.; Wang, K.; Zhu, Z.; Yan, X.; Li, T.; Yang, X.E. The phytoremediation potential of bioenergy crop *Ricinus communis* for DDTs and cadmium co-contaminated soil. *Bioresour. Technol.* 2011,*102*, 11034–11038.

114. Meers, E.; van Slycken, S.; Adriaensen, K.; Ruttens, A.; Vangronsveld, J.; Du Laing, G.; Witters, N.; Thewys, T.; Tack, F.M. The use of bio-energy crops (Zea mays) for "phytoattenuation" of heavy metals on moderately contaminated soils: A field experiment. *Chemosphere* 2010, *78*, 35–41.

115. Abhilash, P.C.; Jamil, S.; Singh, N. Transgenic plants for enhanced biodegradation and phytoremediation of organic xenobiotics. *Biotechnol. Adv.* 2009, *27*, 474–488.

116. Jamil, S.; Abhilash, P.C.; Singh, N.; Sharma, P.N. *Jatropha curcas*: A potential crop for phytoremediation of coal fly ash.*J. Hazard. Mater.* 2009, *172*, 269–275.

117. Che, D.; Meagher, R.B.; Heaton, A.C.; Lima, A.; Rugh, C.L.; Merkle, S.A. Expression of mercuric ion reductase in Eastern cottonwood (*Populus deltoides*) confers mercuric ion reduction and resistance. *Plant Biotechnol. J.* 2003, *1*, 311–319.

118. Bitther, O.P.; Pilon-Smits, E.A.H.; Meagher, R.B.; Doty, S. Biotechnological approaches for phytoremediation. In *Plant Biotechnology and Agriculture*; Arie Altman, A., Hasegawa, P.M., Eds.; Academic Press: Oxford, UK, 2012; pp. 309–328.

119. Bittsanszkya, A.; Kömives, T.; Gullner, G.; Gyulai, G.; Kiss, J.; Heszky, L.; Radimszky, L.; Rennenberg, H. Ability of transgenic poplars with elevated glutathione content to tolerate zinc(2+) stress. *Environ. Int.* 2005, *31*, 251–254.

120. Sayler, G.S.; Ripp, S. Field applications of genetically engineered microorganisms for bioremediation process. *Curr. Opin. Biotechnol.* 2000, *11*, 286–289.

121. Verma, N.; Singh, M. Biosensors for heavy metals. *J. Biometals* 2005, *18*, 121–129.

122. Bruschi, M.; Goulhen, F. New bioremediation technologies to remove heavy metals and radionuclides using Fe(III)-sulfate- and sulfur reducing bacteria. In *Environmental Bioremediation Technologies*; Singh, S.N., Tripathi, R.D., Eds.; Springer Publication: NY, USA, 2006; pp. 35–55.

123. Divya, B.; Deepak Kumar, M. Plant-Microbe interaction with enhanced bioremediation. *Res. J. BioTechnol.* 2011, *6*, 72–79.

124. Bae, W.; Mehra, R.K.; Mulchandani, A.; Chen, W. Genetic engineering of *Escherichia coli* for enhanced uptake and bioaccumulation of mercury. *Appl. Environ. Microbiol.* 2001, *67*, 5335–5338.

125. Bae, W.; Wu, C.H.; Kostal, J.; Mulchandani, A.; Chen, W. Enhanced mercury biosorption by bacterial cells with surface-displayed MerR. *App. Environ. Microbiol.* 2003, *69*, 3176–3180.

126. Wu, C.H.; Wood, T.K.; Mulchandani, A.; Chen, W. Engineering plant-microbe symbiosis for rhizoremediation of heavy metals. *Appl. Environ. Microbiol.* 2006, *72*, 1129–1134.

127. Kostal, J.R.Y.; Wu, C.H.; Mulchandani, A.; Chen, W. Enhanced arsenic accumulation in engineered bacterial cells expressing ArsR. *Appl. Environ. Microbiol.* 2004, *70*, 4582–4587.

128. Hasin, A.A.; Gurman, S.J.; Murphy, L.M.; Perry, A.; Smith, T.J.; Gardiner, P.E. Remediation of chromium (VI) by a methane-oxidizing bacterium. *Environ. Sci. Technol.* 2010, *44*, 400–405.

129. Ackerley, D.F.; Gonzalez, C.F.; Keyhan, M.; Blake, R.; Matin, A. Mechanism of chromate reduction by the *Escherichia coli* protein, NfsA, and the role of different chromate reductases in minimizing oxidative stress during chromate reduction. *Environ. Microbiol.* 2004, *6*, 851–860.

130. Brim, H.; McFarlan, S.C.; Fredrickson, J.K.; Minton, K.W.; Zhai, M.; Wackett, L.P.; Daly, M.J. Engineering *Deinococcus radiodurans* for metal remediation in radioactive mixed waste environments. *Nat. Biotechnol.* 2000, *18*, 85–90.

131. Murtaza, I.; Dutt, A.; Ali, A. Biomolecular engineering of *Escherichia coli* organomercurial lyase gene and its expression. *Indian J. Biotech.* 2002, *1*, 117–120.

132. Zhao, X.W.; Zhou, M.H.; Li, Q.B.; Lu, Y.H.; He, N.; Sun, D.H.; Deng, X. Simultaneous mercury bioaccumulation and cell propagation by genetically engineered *Escherichia coli. Process Biochem.* 2005, *40*, 1611–1616.

133. Kiyono, M.; Pan-Hou, H. Genetic engineering of bacteria for environmental remediation of mercury. *J. Health Sci.*2006, *52*, 199–204.

134. Ng, S.P.; Davis, B.; Polombo, E.A.; Bhave, M. A Tn5051-like mer-containing transposon identified in a heavy metal tolerant strain *Achromobacter* sp. AO22. *BMC Res. Notes* 2009, *7*, 2–38.

135. Lopez, A.; Lazaro, N.; Morales, S.; Margues, A.M. Nickel biosorption by free and immobilized cells of *Pseudomonas fluorescens* 4F39: A comparative study. *Water Air Soil Pollut.* 2002, *135*, 157–172.

136. Sriprang, R.; Hayashi, M.; Ono, H.; Takagi, M.; Hirata, K.; Murooka, Y. Enhanced accumulation of $Cd^{2+}$ by a*Mesorhizobium* sp. transformed with a gene from *Arabidopsis thaliana* coding for phytochelatin synthase. *Appl. Environ. Microbiol.* 2003, *69*, 79–796.

137. Van Aken, B.; Tehrani, R.; Schnoor, J. Endophyte-assisted phytoremediation of explosives in poplar trees by*Methylobacterium populi* BJ001T. In *Endophytes of Forest Trees: Biology and Applications, Forestry Sciences*; Pirttilä, A.M., Frank, A.C., Eds.; Springer: Heidelberg, The Netherlands, 2011; Volume 80, pp. 217–

234.

138. Singh, B.K. Emerging and genomic approaches in bioremediation. In Proceedings of the 4th International Contaminated Site Remediation Conference, Adelaide, Australia, 11–15 September 2011.

139. Maestri, E.; Marmiroli, M. Genetic and molecular aspects of metal tolerance and hyperaccumualtion. In *Metal Toxicity in Plants: Perception, Signalling and Remediation*; Gupta, D.K., Sandalio, L.M., Eds.; Springer: Berlin, Germany, 2011; pp. 41–61.

140. Ruis, O.N.; Daniell, H. Genetic engineering to enhance mercury phytoremediation. *Curr. Opin. Biotechnol.* 2009, *20*, 213–219.

141. Doty, S.L. Enhanced metabolism of halogenated hydrocarbons in transgenic plants contain mammalian P450 2E1.*Proc. Natl. Acad. Sci. USA* 2007, *97*, 6287–6291.

142. Gullner, G. Enhanced tolerance of transgenic poplar plants overexpressing gamma-glutamylcysteine synthetase towards chloroacetanilide herbicides. *J. Exp. Bot.* 2001, *52*, 971–979.

143. Abhilash, P.C.; Powell, J.R.; Singh, H.B.; Singh, B.K. Plant–microbe interactions: Novel applications for exploitation in multipurpose remediation technologies. *Trends Biotechnol.* 2012, *30*, 416–420.

144. Lovley, D.R. Cleaning up with genomics: Applying molecular biology to bioremediation. *Nat. Rev. Microbiol.* 2003, *1*, 35–44.

145. Ma, Y.; Prasad, M.N.; Rajkumar, M.; Freitas, H. Plant growth promoting rhizobacteria and endophytes accelerate phytoremediation of metalliferous soils. *Biotechnol. Adv.* 2011, *29*, 248–258.

146. Lehmann, J.; Rillig, M.C.; Thies, J.; Massielo, C.A.; Hockaday, W.C.; Crowley, D. Biochar effects on soil biota—A review. *Soil Biol. Biochem.* 2011, *43*, 1812–1836.

147. Sarikaya, M.; Tamerler, C.; Jen, A.K.; Schulten, K.; Baneyx, F. Molecular biomimetics: Nanotechnology through biology. *Nat. Mater.* 2003, *2*, 577–585.

148. Smith, M.D.; Lennon, E.; McNeil, L.B.; Minton, K.W. Duplication insertion of drug resistance determinants in the radioresistant bacterium *Deinococcus radiodurans*. *J. Bacteriol.* 1998, *170*, 2126–2135.

149. Vishwanathan, B. *Nanomaterials*; Narosa Publishing House Pvt Ltd.: New Delhi, India, 2009.

# 4

CHAPTER

# USE OF PHYTOREMEDIATION AND BIOCHAR TO REMEDIATE HEAVY METAL POLLUTED SOILS: A REVIEW

J. Paz-Ferreiro[1,2], H. Lu[1,3], S. Fu[1] , A. Méndez[2], and G. Gascó[2]

[1]Key Laboratory of Vegetation Restoration and Management of Degraded Ecosystems, South China Botanical Garden, Chinese Academy of Sciences, Guangzhou 510650, China

[2]Departamento de Edafología, ETSI Agrónomos, Universidad Politécnica de Madrid, Avenida Complutense 3, Madrid 28050, Spain

[3]University of Chinese Academy of Sciences, Beijing 100049, China

## ABSTRACT

**Anthropogenic activities are resulting in an increase of the use and extraction of heavy metals. Heavy metals cannot be degraded and hence accumulate in the environment, having the potential to contaminate the food chain. This pollution threatens soil quality, plant survival and human health. The remediation of heavy metals deserves attention, but it is impaired by the cost of these processes. Phytoremediation and biochar are two sound environmental technologies which could be at the forefront to mitigate soil pollution. This review provides an overview of the state of the art of the scientific research on phytoremediation and biochar application to remediate heavy-metalcontaminated soils. Research to date has attempted only in a limited number of occasions to combine both techniques, however we discuss the potential advantages of combining both, and the potential mechanisms involved in the interaction between phytoremediators and biochar. We identified specific research needs to ensure a sustainable use of phytoremediation and biochar as remediation tools.**

**Citation:** J. Paz-Ferreiro, H. Lu, S. Fu1 , A. Méndez, and G. Gascó, Use of phytoremediation and biochar to remediate heavy metal polluted soils: a review, doi:10.5194/se-5-65-2014.

## INTRODUCTION

Industrialisation and technical advances have led to an increase in the use of heavy metals and heavy metal pollution. Contrary to organic substances, heavy metals are nondegradable and accumulate in the environment. While some soils can have a high background level of heavy metals due to volcanic activity or weathering of parent materials, in other soils anthropogenic activities, including smelting, mining, use of pesticides, fertilisers and sludges are responsible for these high levels of heavy metals. Soil heavy metal pollution has a pernicious effect on soil microbial properties (Yang et al., 2012) and on the taxonomic and functional diversity of soils (Vacca et al., 2012). Soil heavy metal pollution poses a risk to the environment and to human health (Roy and McDonald, 2014) due to biomagnifi- cation (increases in metal concentration as the element passes from lower to higher trophic levels). Some of these elements can be essential for living organisms while some others are non-essential. Even concentrations of essential elements beyond a certain threshold will have pernicious health effects, as they interfere with the normal metabolism of living systems. It is not the purpose of this article to review the adverse effects of heavy metals on human or plant health. KabataPendias and Pendias (2001) provide a list of toxic effects of heavy metals on plants and the mechanism involved, while a summary of adverse effects of heavy metals on human health was provided by Ali et al. (2013). We would like to remind the reader that studies on heavy metal pollution are focused on As, Cd, Cr, Hg and Pb as they are toxic, non-essential heavy metals, and on Cu, Ni and Zn which, although essential, can cause health problems in humans or can result in phytotoxicity at high concentrations.

With an increasing amount of literature on heavy metal remediation, we aim to summarise the state of art of two of these techniques situated at the forefront of remediation practices: phytoremediation, with a focus on phytoextraction, and biochar soil amendment, and to discuss their mechanism and how we could combine them to improve remediation efforts.

## BIOCHAR

Biochar is a porous, carbonaceous product obtained from the pyrolysis of organic materials. Numerous materials can be used as feedstocks, including sludges, plant materials and manures. Although the use of charcoal (wood biochar) has been common since preterit times, the idea of using other feedstocks for biochar production is new and relatively unexplored. Typically biochars have high cation exchange capacity and are alkaline. Biochar has many potential benefits on soil properties as an increase in soil biological activity (Lehmann et al., 2011; Paz-Ferreiro et al., 2014), diminishing soil greenhouse gas emissions from agricultural sources and thus enhancing soil carbon sequestration due to its elevated content of recalcitrant forms of carbon (Gascó et al., 2012). The changes brought about by biochar addition to the soil will cause alterations in soil quality (Paz-Ferreiro and Fu, 2014) with the potential to increase agricultural yields (Jeffery et al., 2011; Liu et al., 2013). The multiple benefits of biochar for soil have been compiled recently in the book by Lehmann and Joseph (2009). However, little information was available in this book about the effect of biochar on soil heavy metals. Although there is a recent review on the role of biochar to remediate polluted soils, with a particular interest in the metalloid arsenic (Beesley et al., 2011), our article has a more focused scope on the combination of phytoremediation and biochar with respect to heavy metal remediation. Moreover, in the last years there have been an increasing number of articles devoted to understanding the interaction between heavy metals, vegetation and biochar.

## MECHANISM OF INTERACTION BETWEEN BIOCHAR AND HEAVY METALS

Biochar characteristics are a function of several factors, including the type of feedstock, the particle size of the feedstock and temperature and conditions of pyrolysis. The wide range of characteristics that biochar might posses makes some particular materials more suitable than others to remediate different heavy metals. Therefore, when selecting a biochar for remediation purposes, scientists should be aware not only of soil type and characteristics but also on biochar properties. Moreover, it should also be considered that key biochar properties such as surface area, pH, ash and carbon contents can be affected by post-treatments and thus enhance biochars' ability to immobilise heavy metals (Lima et al., 2014).

Before reviewing the mechanisms implied in the interaction between biochar and heavy metal it is necessary to note that biochars act on the bioavailable fraction of soil heavy metals and that they can reduce also their leachability.

One of the characteristics of biochars is possessing large surface areas, which implies a high capacity for complex heavy metals on their surface. Surface sorption of heavy metals on biochar has been demonstrated on multiple occasions using scanning electron microscopy (Beesley and Marmiroli, 2011; Lu et al., 2012). This sorption can be due to complexation of the heavy metals with different functional groups present in the biochar, due to the exchange of heavy metals with cations associated with biochar, such as $Ca^{+2}$ and $Mg^{+2}$ (Lu et al., 2012), $K^{+}$, $Na^{+}$ and S (Uchimiya et al., 2011c), or due to physical adsorption (Lu et al., 2012). Also oxygen functional groups are known to stabilise heavy metals in the biochar surface, particularly (Uchimiya et al., 2011c) for softer acids like $Pb^{+2}$ and $Cu^{+2}$ . In addition, Méndez et al. (2009) observed that $Cu^{+2}$ sorption was related to the elevated oxygenated surface groups and also with high average pore diameter, elevated superficial charge density and $Ca^{+2}$ and $Mg^{+2}$ exchange content of biochar. Possibly, sorption mechanisms are highly dependent on soil type and the cations present in both biochar and soil. Some other compounds present in the ash, such as carbonates, phosphates or sulphates (Cao et al., 2009; Karimi et al., 2011; Park et al., 2013) can also help to stabilise heavy metals by precipitation of these compounds with the pollutants.

Alkalinity of biochar can also be partially responsible for the lower concentrations of available heavy metals found in biochar-amended soils. Higher pH values after biochar addition can result in heavy metal precipitation in soils. Biochar pH value increases with pyrolysis temperature (Wu et al., 2012), which has been associated with a higher proportion of ash content (Cantrell et al., 2012).

Biochar can also reduce the mobility of heavy metals, altering their redox state of those (Choppala et al., 2012). As an example, biochar addition could lead to the transformation of $Cr^{+6}$ to the less mobile $Cr^{+3}$ (Choppala et al., 2012).

The relative contribution of the different mechanisms to heavy metal immobilisation by different biochar remains unknown, although some authors like Houben et al. (2013a) postulate that it is mostly a pH effect.

## STUDIES ON THE EFFECT OF BIOCHAR ON SOIL HEAVY METALS

Table 1 shows a brief summary of the latest papers about the effect of biochar on soil heavy metals. Fellet et al. (2011) tried to use biochar to remediate a multicontaminated mine soil. Biochar addition did not result in the decrease of the total heavy metal content of the soil, however, biochar addition reduced the bioavailability of Cd, Pb and Zn and the mobility (measured using a leaching experiment) of Cd, Cr and Pb.

Park et al. (2011) studied the effect of two biochars in a heavy-metal-spiked soil and a naturally strongly polluted soil. They performed a sequential extraction of some heavy metals. They found chicken manure biochar effective reducing extractable concentrations of Cd and Pb, but not Cu concentration, while green waste biochar was more effective in diminishing all of the heavy metals studied. Heavy metal fractions that were bonded to organic matter increased after biochar addition. Both biochars also decreased Cd and Pb presence in soil pore water.

Uchimiya et al. (2012a) analysed the effects on soil heavy meals concentrations of 10 biochars prepared from 5 feedstocks at 2 different temperatures. They observed that manures with a high or low proportion of ash or P were less effective to immobilise heavy metals. In contrast, biochars prepared at 700 ◦C were more effective, which could be attributed to transformations in the material, including the removal of nitrogen containing heteroaromatic and leachable aliphatic functional groups. They found Cu and Pb relatively easy to stabilise in soil, while Cd and Ni response depended strongly on the type of biochar added to the soil.

Beesley and Marmiroli (2011) detected a retention of As, Cd and Zn on biochar surfaces. These authors proved that sorption of the metal was produced at the biochar surface and that this process was not immediately reversible. Leachate concentrations of Cd and Zn were reduced 300- and 45-fold, respectively. However, leachate concentrations of As did not diminish.

Namgay et al. (2010) reported that the concentrations of Cd, As and Pb in maize shoots decreased after biochar application. Beesley et al. (2013) reported interesting results, finding that As can increase in soil pore water after biochar addition, but transfer to the plant be reduced. This would imply that at least some biochars could pose no risk of increasing heavy metals in plants and hence are safe in terms of food chain transfer, but leaching of As to nearby waters must be considered. Karami et al. (2011) added biochar to a mine soil polluted with Pb and Cu.

**Table 1:** Studies considering the effect of biochar application on soil heavy metals. Blank indicates not specified in the article

| Feedstock (temperature) | Soil type | Heavy metals | Reference |
|---|---|---|---|
| Sewage sludge | Haplic Cambisol | Cu, Ni, Zn, Cd, Pb | Méndez et al. (2012) |
| Rice husk, rice straw and rice bran (400 °C) | Technosol | As, Cd, Pb, Zn | Zheng et al. (2012) |
| Wastewater sludge (550 °C) | Chromosol (Australian system) | As, Cd, Cr, Cu, Pb, Ni, Se, Zn, Sb, B, Ag, Ba, Be, Co, Sn, Sr | Hossain et al. (2010) |
| | | Cu, Pb, Zn | Sızmur et al. (2011) |
| Broiler litter (350 and 700 °C), pecan shells (450 °C) | Abruptic Durixeralfs | Cu, Cd, Ni | Uchimiya et al. (2010) |
| Pecan shell (450 °C), broiler litter samples (700 °C) | Typic Kandiudult and Abruptic Durixeralfs | Cu | Uchimiya et al. (2011a) |
| Chicken manure (550 °C), green waste (550 °C) | | Cd, Cu, Pb | Park et al. (2011) |
| Forest green waste (600–800 °C) | Peat | Cu | Buss et al. (2012) |
| Dairy manure (350 and 700 °C), paved feedlot manure (350 and 700 °C), poultry litter (350 and 700 °C), turkey litter (350 and 700 °C), separated swine solids (350 and 700 °C) | Typic Kandiudult | Pb, Cu, Ni, Cd | Uchimiya et al. (2012a) |
| Mix of hardwoods (400 °C) | | As, Cd, Zn | Beesley and Marmiroli (2011) |
| Mix of hardwoods (400 °C) | Technosol | Pb, Cu | Karami et al. (2011) |
| Orchard prune residue (500 °C) | Technosol | Cd, Cr, Cu, Ni, Pb, Tl, Zn | Fellet et al. (2011) |
| Eucalyptus | | As, Cd, Cu, Pb, Zn | Namgay et al. (2010) |
| Wheat straw (350–550 °C) | Technosol | Cd | Cui et al. (2011) |
| Wheat straw (350–550 °C) | Technosol | Cd | Cui et al. (2012) |
| Rice straw | Ultisol | Cu, Cd, Pb | Jiang et al. (2012) |
| Orchard prune residues (500 °C) | Technosol | As | Beesley et al. (2013) |
| Miscanthus (600 °C) | | Cd, Zn, Pb | Houben et al. (2013a) |
| Chicken manure (550 °C), green waste (550 °C) | | Cd, Pb | Park et al. (2013) |
| De-inking paper sludge (300 and 500 °C) | Vertisol | Ni | Méndez et al. (2014) |

They found that biochar addition reduced pore water Pb concentrations to half their values in the mine soil. When biochar was combined with greenwaste compost the levels of Pb concentrations in the pore water were 20 times lower than in the control. Jiang et al. (2012) found that the acid-soluble fractions of $Pb^{+2}$ and $Cu^{+2}$ diminished by 18.8–77.0 % and 19.7–100.0 %, respectively, depending on biochar concentration.

However, only 5.6–14.1 % of acid-soluble $Cd^{+2}$ was immobilised. Park et al. (2013) compared the sorption capacity of two biochars, made from chicken manure and from green waste. They found chicken manure biochar more effective to immobilise Cd and Pb compared to green waste biochar. Both biochars presented a higher sorption capacity for Pb, possibly as a consequence of precipitation and complexation of Pb with carbonate, sulphate and phosphate present in the biochar.

Hydrochars could also be used for soil heavy metal immobilisation, however there is a lack of studies on the topic. Hydrochars are produced after pyrolysis of organic-matterrich materials in the presence of subcritical liquid water. This technique can be applied to obtain pyrolysed products from wet feedstocks. In principle, the adsorption capacity of hydrochars seems to be reduced compared to biochars or other adsorbents due to the fewer functional groups containing oxygen present on hydrochar surfaces. However, Xue et al. (2012) have demonstrated experiments that the use of activated hydrochars could overcome these problems. They performed a series of batch and columns experiments to show how this type of hydrochar could reduce Pb on water. The potential applicability of hydrochar to address soil heavy metal pollution remains untested. However, hydrochars tend to be acidic and could possess phytotoxic or genotoxic risks (Busch et al., 2013), which would deem them unsuitable in restoration projects.

There is a lack of studies concerning how pyrolysis conditions affect biochar properties as heavy metal sorbent. To fill this gap, Uchimiya et al. (2011b) performed an experiment using wood and grass biochars prepared at five different temperatures and another one (Uchimiya et al., 2012b) used poultry litter prepared at four different temperatures to study lead retention. From the first experiment they suggested using biochars prepared at high temperature (650 °C to 800 °C) for remediation purposes. In addition they recommended performing acid or other oxidant post-treatment in order to increase oxygen-containing surface functional groups (carboxyl, carbonyl and hydroxyl) which have a great importance in relation to heavy metal sorption into biochar. In the case of the chicken litter biochar, they found that lower production temperatures were more suitable than higher ones due to the stabilising effect. Higher rates of amendment were necessary in their experiments for chicken manure biochar to get the same remediation effect as plant-derived biochars.

It is expected that as biochar is in contact with soil for a prolonged period of time, oxidation, both biotic and abiotic, would result in the alteration of biochar, a process known as aging. This process, which would result in the formation of carboxylic, phenolic, carbonyl, quinones and hydroxyl functional groups and which can be emulated under laboratory conditions was studied by Uchimiya et al. (2010). These authors found that the immobilisation of heavy metals by biochar was related to the metal lability, this means that heavy metal immobilisation followed the order $Cu^{+2} > Cd^{+2} > Ni^{+2}$ . Heavy metal immobilisation was not affected by biochar aging, except for a small increase in Ni observed in soils with aged biochar.

Earthworms can be added to soil at some stages of ecological restoration due to their well-established positive effects on soil properties as organic matter content, soil formation, soil aeration and nutrient cycling. Sizmur et al. (2011) tested a polluted soil collected in the vicinity of a Cu mine using biochar in combination with compost and earthworms (Lumbricus terrestris). They found all treatments (biochar alone, biochar + compost, and biochar + compost + earthworms) to reduce the amount of heavy metals compared to the control soils. A limiting aspect when using earthworms with remediation purposes is that their addition to soil could lead to the mobilisation of heavy metals and hence to an increase of plant heavy metal concentrations. Interestingly, Sizmur et al. (2011) found that the treatments containing biochar and earthworms did not result in higher heavy metal mobility or plant availability.

As a consequence of heavy metal immobilisation, biochars can reduce the phytotoxicity of polluted soils, resulting in increases in the percentage of germinated seeds and root length (Ahmad et al., 2012).

All of the above experiments have been conducted under laboratory conditions. We would urge scientists to design experiments to help to demonstrate the benefits of biochar against heavy metal pollution under field conditions, as done by Zheng et al. (2012) and Cui et al. (2011, 2012). Zheng et al. (2012) studied the effect of three biochars on different heavy metals (see Table 2) using a multi-polluted soil planted with rice. They found Cd, Pb and Zn to be reduced on rice shoots, in particular when using straw-derived biochar. However, as in rice shoots was increased by biochar addition. More importantly, we believe that this is one of the first studies considering the effects of biochar particle size on plant heavy metals. The authors found that decreases in particle size resulted in less Cd, Zn and Pb accumulating in the rice plants. Similarly, Cui et al. (2011) and Cui et al. (2012) found reduced Cd uptake in paddy fields and in a soil cropped with wheat, respectively. Both studies consisted of two annual measurements, so the need to reapply biochar after more extended periods of time remains to be explored.

## PHYTOREMEDIATION

Phytoremediation is an umbrella term for a series of techniques that combine the disciplines of soil microbiology and chemistry and plant physiology (Cunningham and Ow, 1996). Currently the most extended practice for soil heavy metal remediation does not address the problem of contamination, as it consists of encapsulation or digging and dumping. Immobilisation or extraction can be expensive and, as a consequence, phytoremediation can be considered comparatively attractive as it can be used at a relatively low cost to restore or partially decontaminate a site compared to other options, as the cost is 5 % that of other alternative methods (Prasad, 2003). Other advantages would include its good perception as a remediation technique among the general public and being more environmentally friendly than other options, as the introduction of vegetation in the polluted area can also help to prevent erosion or contaminant leaching. Phytoremediation consists in the use of plants to remove contaminants from the environment or to transform them into less harmful forms (see Table 3 for a summary of phytoremediation techniques). Phytoremediation is a relatively new technology, as research studies have been mostly conducted from 1990 onwards.

Phytoextraction is the main and most promising technique to remove soil heavy metals. It is based on the use of hyperaccumulators which uptake heavy metals and then translocate them to aboveground tissues (Table 1). One common way of defining a hyperaccumulator is as a plant that can store heavy metals at a level 100-fold greater than common plants without yield reduction (Chaney et al., 2007). On other occasions, these types of plants are defined on their basis to accumulate more than 100 mg $kg^{-1}$ dry weight of Cd, more than 1000 mg $kg^{-1}$ of Cu, Co, Cr, Ni or Pb, or more than 10 000 mg $kg^{-1}$ of Mn or Zn (Baker and Brooks, 1989). Some other authors have mentioned that these values are conservative and propose these criteria to be lowered (van der Ent et al., 2013). Species used for phytoextraction must not only accumulate high amounts of the target element but also have a high growth rate, tolerate the toxic effects of the heavy metals, be adapted to local environment and climate, be resistant to pathogen and pests, be easy to cultivate and repulse herbivores to avoid food chain contamination (Ali et al., 2013).

To date, more than 400 species have been identified as hyperaccumulators, including more than 300 Ni hyperaccumulators (Li et al., 2003). In contrast with Ni, only a few plant species have demonstrated the potential to accumulate Cd, Cu, Pb, and Zn (Brooks, 1998). Many phytoremediators belong to the taxonomical order of Brassicales and phytoremediators are also abundant in Asterales, Solanales, Poales, Malpighiales, Fabales, Caryophyllales and Rosales (Shao et al., 2011). The amount of metal extracted from the soil depends not only on the plant species utilised but also on the type of soil and climate of the region (Shao et al., 2011).

The mechanism and reasons of phytoaccumulation remain unknown. Metal concentrations are higher in the shoots compared to the roots, suggesting that there could be an ecological role, leading to protection against insect, herbivore or fungal attack, by making the leaves toxic or unpalatable.

Phytoextraction has three main purposes: firstly, to remove the contaminant from the soil or contain it, secondly phytoextraction of elements that have market value and finally gradually improving soil quality to cultivate crops with higher market value (Vangrosveld et al., 2009).

There are a number of problems associated with the effectiveness of this remediation technique. Phytoremediation might not be suitable in areas were the heavy metal concentration is too elevated, as plants could show symptoms of phytotoxicity. In addition, most of the phytoaccumulators have slow growth rate or produce few biomass, limiting the amount of metal uptaken.

Manipulation of soil pH, soil nutrient content or soil organic matter can also be undertaken to improve metal hyperaccumulation. In this sense, these additional agronomic practices can be carried out when heavy metal concentrations in the soil are too elevated to reduce plant stress (Adriano et al., 2004; Gabos et al., 2011; de Abreu et al., 2012). Thus, liming can allow the decrease of the heavy metal available fraction, therefore enabling vegetative growth, while fertilisers can improve phytoextractor growth. On the other hand, both liming and fertiliser addition can alter the mobility and speciation of soil heavy metals. As an example, Li et al. (2012) found that Cd removal from soil was enhanced by the phytoaccumulator Amaranthus hypocondriacus after NPK or NP fertilisation due to an increase on plant biomass. However, they found that N alone did not increase plant biomass and led to a limited increment in phytoextraction. Other studies (Huang et al., 2013) have found that P fertilisers can decrease soil pH, enhancing the mobility of Cd and leading to increased phytoextraction by Sedum alfredii. When adding a phosphate fertiliser to promote phytoremediation, the choice of amendment should be carefully chosen as cations ($K^+$, $Na^+$, $Ca^{+2}$or $NH_4^+$) associated with the phosphate could affect the mobility of heavy metals (Bolan et al., 2003; Huang et al., 2013). Indeed, plant growth (Oo et al., 2014) and the mobility of different elements in the soil (Ahmad et al., 2013) can be related to soil salinity. For example, Stevens et al. (2003) observed that $Zn^{+2}$ and $Pb^{+2}$ mobility increased with the increment of electrical conductivity. Differences in soil pH caused by the addition of different phosphate fertilisers can also lead to differences in phytoextraction (Mandal et al., 2012). Urea has also been used to alleviate plant stress and improve B phytoextraction by the plant species Brassica juncea (Giansoldati et al., 2012).

**Table 2:** Studies considering the effect of biochar application on soil heavy metals in combination with phytoremediators. Blank indicates not specified in the article

| Feedstock (temperature) | Soil type | Pollutants | Plant species | Reference |
|---|---|---|---|---|
| Mix of hardwoods (400 °C) | 3 soils | As | *Miscanthus x giganteus* | Hartley et al. (2009) |
| Miscanthus (600 °C) | | Cd, Pb, Zn | *Brassica napus* L. | Houben et al. (2013b) |
| Pruning residues from orchards (550 °C), fir tree pellets (350–400 °C) and manure pellets mixed with fir tree pellets (350–400 °C) | Technosol | Cd, Cr, Cu, Fe, Ni, Pb, Tl and Zn | *Anthyllis vulneraria* subsp. *polyphylla* (Dc.) Nyman, *Noccaea rotundifolium* (L.) Moench subsp. *cepaeifolium* and *Poa alpina* L. subsp. *alpina* | Fellet et al. (2014) |

**Table 3:** Summary of the different techniques of phytoremediation

| Technique | Description |
|---|---|
| Phytoextraction | Plants accumulate contaminants in harvestable biomass i.e., shoots |
| Phytofiltration | Sequestration of pollutants from contaminated waters by plants |
| Phytostabilisation | Limiting the mobility and bioavailability of polluting substances by prevention of migration or immobilisation |
| Phytovolatilisation | Conversion of pollutants to volatile form followed by their release to the atmosphere |
| Phytodegradation | Degradation of organic xenobiotics by plant enzymes within plant tissues |
| Rhizodegradation | Degradation of organic xenobiotics in the rhizosphere by rhizospheric microorganisms |
| Phytodesalination | Removal of excess salts from saline soils by halophytes |

Organic amendments such as chicken manure have also been shown to increase growth of the species Rorippa globosa (Wei et al., 2011). Chicken manure addition resulted in a decrease in soil extractable Cd and thus, the concentration of Cd in the shoots was lower in soils amended with chicken manure than in soils amended with urea or in the controls (soil + phytoremediator). However, the total concentration of metal extracted in the shoots was in both cases higher than in the control. Other materials such as pig manure vermicompost can also be used to improve plant yield and assist phytoremediation, as demonstrated by Wang et al. (2012) in an experiment using Cd as target heavy metal and Sedum alfredii as phytoremediator. Indeed, the use of organic amendments has numerous applications, for example, Siebielec and Chaney (2012) have demonstrated the effectiveness of biosolids compost in the rapid stabilisation of Pb and Zn and revegetation of military range contaminated soils increasing tall fescue growth by more than 200 %, while Clemente et al. (2012) recovered land contaminated by mining activity with Cd, Cu, Pb, and Zn by a combination of the halophytic shrub Atriplex halimus L. with pig slurry.

The use of chelators such as citric acid or EDTA has also been sometimes advised to assist phytoremediation, with the aim of increasing the mobility of soil heavy metals and thus plant extraction (Zhou et al., 2007; Freitas et al., 2013). However, we should bear in mind that the use of chelators can originate other environmental problems, including toxicity for plants and metal leaching (Zhou et al., 2007).

In addition, experiments should be done to account for the potential impact of climate change on the capability of phytoextractors to accumulate heavy metals, which at the moment is uncertain (Rajkumar et al., 2013).

Finally, we would like to remark that pot experiments are a good first approach to evaluate the potential of a phytoextractor, but they cannot substitute field experiments as the uptake of heavy metals is higher in pots than for the same soil in the field (see for example, Marschner, 1986). This can be due to differences in soil moisture or microclimate and to the fact that field-grown plants can reach down to less polluted soil.

Phytostabilisation is another phytoremediation technique and has been used mostly in relation with the stabilisation and containment of mine tailings (Conesa et al., 2007; Méndez et al., 2007). Thus, the vegetative cover diminishes eolian dispersion while roots prevent water erosion and leaching and contributes to the immobilisation of heavy metals. Mechanisms involved in phytostabilisation include precipitation, root sorption, complexation or metal valence reduction. Phytostabilisation, contrary to phytoextraction, primarily focuses on heavy metal sequestration within the rhizosphere but not in plant tissues.

## COMBINING BIOCHAR AND PHYTOREMEDIATION

There is an abundance of reports in the literature about amendments, such as lime and compost being used to reduce the bioavailability of heavy metals (Komárek et al., 2013) and thus having the potential to be combined with phytoremediators (de Abreu et al., 2012). Biochar, as reviewed before, can also stabilise heavy metals in soils and thus reduce plant uptake. However, until recently there was a lack of experiments trying to combine both approaches to soil remediation.

Biochar is commonly reported in the literature to increase plant growth, hence there is a potential of biochar to increase the yield of phytoremediators. This increase in plant productivity is highly heterogeneous and has overall been quanti- fied as 10 % (Jeffery et al., 2011; Liu et al., 2013). However, there are several factors that limit the accuracy of the figure provided by Jeffery et al. (2011) and Liu et al. (2013) and that could skew the data. To date, most of the field experiments have been conducted in the short term, being limited to a period of 1–2 yr and there are a high relative number of laboratory mesocosm incubations (with a duration of 1– 2 months) included in the data set. Also, the data set in this review comprises a higher number of experiments in tropical latitudes compared to temperate ones. Finally, we should bear in mind that a high heterogeneity in the response was detected, depending on the type of soil and plant utilised.

Improvements in plant yield after biochar addition are often attributed to increased water and nutrient retention, improved biological properties and CEC, effects on nutrient cycling and turnover and improvements in soil pH. Many of these effects are interrelated and potentially they could act synergistically. In general, acid soils with a coarse texture or a medium texture are more prone to produce increases in crop productivity (Jeffery et al., 2011; Liu et al., 2013). In the last years the scientific community has also raised awareness over the improvement of plant responses to disease as an additional benefit of biochar soil amendment (Graber et al., 2010). As said before, biochar can alter soil microbial community, possibly including an increase in beneficial organisms that produce antibiotics and can protect plants against pathogens. Another mechanism could be compounds included in biochar such as 2-phenoxyethanol, benzoic acid, hydroxy-propionic and butyric acids, ethylene glycol and quinones suppressing some of the pathogens present in the microbiota (Graber et al., 2010; Elad et al., 2011).

In principle, biochar prepared from any material would have the potential to increase plant yield and thus be used in combination with phytoremediation. However, the use of sewage sludge biochar would be unadvised due to its generally negative effect on crop performance (Jeffery et al., 2011). Caution should also be taken with the presence of heavy metals in sewage sludge biochars, although some studies (Méndez et al., 2012; Hossain et al., 2010) show that the metals present in the biochar are not in mobile forms.

It is also worth mentioning that for a long time, phytoextractors were considered to be non-mycorrhizal. However, in the last year it has been demonstrated that hyperaccumulators can form symbiosis with arbuscular mycorrhizal fungi (AMF) and these enhance plant growth and lead to higher contents of metal extracted (Al Agely et al., 2005; Orlowska et al., 2011). Positive effects of biochar have usually been found in arbuscular mycorrhizal fungi, although exceptions can be found in nutrient-rich soils (Lehmann et al., 2011).

There has been a recent interest about the possibility of combining phytoremediation with other potential plant uses, such as using plants that can be used to obtain bioenergy (de Abreu et al., 2012). While heavy-metal-contaminated areas are not suitable for food production, planting biocrops could promote soil organic matter stocks and reduce soil pollutants (Hartley et al., 2009). Willow and poplar have been commonly used as biocrops and they can be utilised for phytoremediation purposes due to their high uptake of heavy metals and fast growing rates (Baum et al., 2009). Recently, Hartley et al. (2009) observed no increase on As transfer to plants in three soils planted with Miscanthus and amended with hardwood biochar. They warned, however, that alkalyne biochars could mobilise As. It is a well-known fact that As behaves differently to other metals with respect to pH, as As mobility is reduced in acid soils due to adsorption on iron oxide surfaces. The results of Hartley et al. (2009) show that biochar can be used in combination with Miscanthus for phytostabilisation. More recent research has proved that biochar can have an added environmental benefit, improving the greenhouse gas balance of other bioenergy crops such as Miscanthus (Case et al., 2014).

Biochar and phytoremediation techniques have been used recently (see Table 3) to target at Cd-polluted soils (Houben et al., 2013b) using Brassica napus L. as Cd and Zn phytoextractor in combination with Miscanthus biochar and for the case of multicontaminated soils using different biochars and plant species (Fellet et al., 2014). Houben et al. (2013b) observed that phytoextraction of Cd and Zn by Brassica napus was impeded by biochar and, due to the lower BCF achieved in pots with biochar, suggested using biochar and Brassica napus as a phytostabilisation alternative. While Fellet et al. (2014) used three biochars, produced from pruning residues from orchards, fir tree pellets and fir tree pellets mixed with manure at two different doses. Fellet et al. (2014) observed higher concentrations of Pb in plants grown with the fir tree pellets biochar. However, no increase in yield was obtained with this treatment, and the value of the translocation index, although significantly higher than in the control, was insufficient for the purposes of phytoextraction. Overall, they found the manure biochar to immobilise more heavy metals and also to produce the most noticeable increase in plant biomass, thus, making manure biochar more suitable for phytostabilisation purposes.

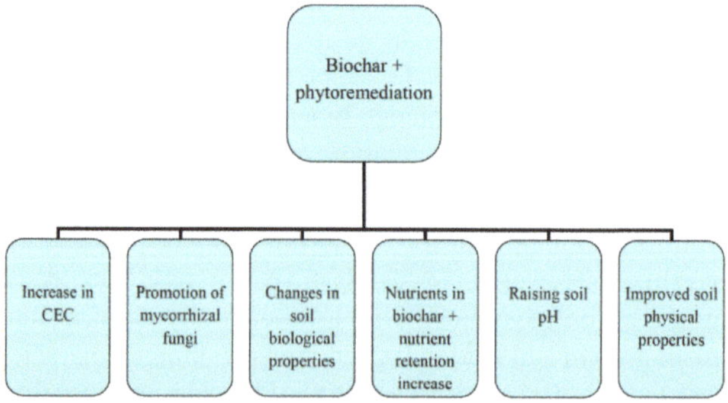


**Figure 1:** An overview of the potential positive effects attained by combining phytoremediation and biochar in heavy metal pollution remediation.

It seems plausible that one of the best approaches to combine biochar and phytoextractors would be in multicontaminated soils, where both can target at different elements. Biochar could also be used as a soil conditioner prior to plant colonisation in acidic, polluted mine tailings. However, these two approaches in relation with phytoextraction remains to be tested.

## CONCLUSIONS AND RESEARCH NEEDS

Biochar and phytoremediation techniques have the potential to be combined in the remediation on heavy metal polluted soils (see Fig. 1). Biochar can reduce the bioavailability and leachability of heavy metals in the soil. On the other hand phytoextractors can reduce the amount of soil heavy metals in polluted areas.

We anticipate that in the next years there will be a growing interest to study the interaction between phytoremediators and biochars and we identify the next areas as the ones warranting research.

Biochars have highly heterogeneous properties, which should be understood as maximising the efficacy of soil remediation. We should comprehend, firstly, how these properties are relevant for heavy metal adsorption and how they contribute to the different mechanism of heavy metal immobilisation, and secondly how to optimise the choice of pyrolysis conditions and feedstocks in order to produce the desired products.

Most experiments utilising biochar or phytoremediators alone and not in combination have been carried out under laboratory conditions. In the case of phytoremediators this can result in an overestimation of heavy metal extraction.

For biochar most of the experiments (both in field and under laboratory conditions) have been conducted in the short term, which poses an interrogation on the long-term fate of these heavy metals. In fact it could be expected that, due to aging processes, the ability of biochar to sequester heavy metals decreases with time. More research will be needed to understand the aging process in biochar.

Thus, well-designed, large-scale and long-term field trials will be essential to evaluate the feasibility on the approach proposed in this article. The economics of these new remediation processes should be assessed against other options.

## ACKNOWLEDGEMENTS

J. Paz-Ferreiro thanks the Chinese Academy of Sciences for financial support (fellowship for young international scientists number 2012Y1SA0002).

## REFERENCES

1. Adriano, D. C., Wenzel, W. W., Vangrosveld, J., and Nolam, N. S.: Role of assisted natural remediation in environmental clean-up, Geoderma, 122, 121–142, 2004.
2. Ahmad, M., Lee, S. S., Yang, J. E., Ro, H. M., Lee, Y. H., and Ok, Y. S.: Effects of soil dilution and amendments (mussel shell, cow bone and biochar) on Pb availability and phytotoxicity in military shooting range soil, Ecotox. Environ. Safe., 79, 225–231, 2012.
3. Ahmad, S., Ghafoor, A., Akhtar, M. E., and Khan, M. Z.: Ionic displacement and reclamation of saline-sodic soils using chemical amendments and crop rotation, Land Degrad. Dev., 24, 170–178, 2013.

4. Al Agely, A., Sylvia, D. M., and Ma, L. Q.: Mycorrhizae increase arsenic uptake by the hyperaccumulator Chinese brake fern (Pteris vittata L.), J. Environ. Qual., 6, 2181–2186, 2005.
5. Ali, H., Khan, E., and Sajad, M. A.: Phytoremediation of heavy metals – Concepts and applications, Chemosphere, 91, 869–881, 2013.
6. Baker, A. J. M. and Brooks, R. R.: Terrestrial higher plants which hyperaccumulate metallic elements. A review of their distribution, ecology and phytochemistry, Biorecovery, 1, 81–126, 1989.
7. Baum, C., Leinweber, P., Weih, M., Lamersdorf, N., and Dimitriou, I.: Effects of short rotation coppice with willows and poplar on soil ecology, Agric. Forestry Res., 3, 183–196, 2009.
8. Beesley, L. and Marmiroli, M.: The immobilisation and retention of soluble arsenic, cadmium and zinc by biochar, Environ. Pollut., 159, 474–480, 2011.
9. Beesley, L., Moreno-Jiménez, E., Gómez-Eyles, J.L., Harris, E., Robinson, B., and Sizmur, T.: A review of biochars' potential role in the remediation, revegetation and restoration of contaminated soils, Environ. Pollut., 159, 3269–3282, 2011.
10. Beesley, L., Marmiroli, M., Pagano, L., Pigoni, V., Fellet, G., Fresno, T., Vamerali, T., Bandiera, M., and Marmiroli, N.: Biochar addition to an arsenic contaminated soil increases arsenic concentrations in the pore water but reduces uptake to tomato plants (Solanum lycopersicum L.), Sci. Total Environ., 454–455, 598–603, 2013.
11. Bolan, N. S., Adriano, D. C., and Naidu, R.: Role of phosphorus in (im) mobilization and bioavailability of heavy metals in the soil– plant system, Rev. Environ. Contam. T., 177, 1–44, 2003.
12. Brooks, R. R.: Geobotany and hyperaccumulator, in: Plants that hyperaccumulate heavy metals, their role in phytoremediation, microbiology, archaeology, mineral exploration and phytomining, edited by: Brooks, R. R., Wallingford, UK: CAB, International, 55–94, 1998.
13. Busch, D., Stark, A., Kammann, C. I., and Glaser, B.: Genotoxic and phytotoxic risk assessment of fresh and treated hydrochar from hydrothermal carbonization compared to biochar from pyrolysis, Ecotox. Environ. Safe., 97, 59–66, 2013.
14. Buss, W., Kammann, C., and Koyro, H. W.: Biochar reduces copper toxicity in Chenopodium quinoa Willd. in a sandy soil, J. Environ. Qual., 41, 1157–1165, 2012.
15. Cantrell, K. B., Hunt, P. G., Uchimiya, M., Novak, J. M., and Ro, K. S.: Impact of pyrolysis temperature and manure source on physicochemical characteristics of biochar, Bioresource Technol., 107, 419–428, 2012.
16. Cao, X. D., Ma, L. N., Gao, B., and Harris, W.: Dairy-manure derived biochar effectively sorbs lead and atrazine, Environ. Sci. Technol., 43, 3285–3291, 2009.
17. Case, S. D. C., McNamara, N. P., Reay, D. S., and Whitaker, J.: Can biochar reduce soil greenhouse gas emissions from a Miscanthus bioenergy crop?, GCB Bioenergy, 6, 76–89, 2014.

18. Chaney, R. L., Angle, J. S., Broadhurst, C. L., Peters, C. A., Tappero, R. V., and Sparks, D. L.: Improved understanding of hyperaccumulation yields commercial phytoextraction and phytomining technologies, J. Environ. Qual., 36, 1429–1443, 2007.

19. Choppala, G. K., Bolan, N. S., Megharaj, M., Chen, Z., and Naidu, R.: The influence of biochar and black carbon on reduction and bioavailability of chromate in soils, J. Environ. Qual., 41, 1175– 1184, 2012.

20. Clemente, R., Walker, D. J., Pardo, T., Martínez-Fernández, D., and Bernal, M. P.: The use of a halophytic plant species and organic amendments for the remediation of a trace elementscontaminated soil under semi-arid conditions, J. Hazard. Mater., 223–224, 63–71, 2012.

21. Conesa, H. M., Faz, A., and Arnaldos, R.: Initial studies for the phytostabilization of a mine tailing from the Cartagena-La Unión mining district (SE Spain), Chemosphere, 66, 38–44, 2007.

22. Cui, L., Li, L., Zhang, A., Pan, G., Bao, D., and Chang, A.: Biochar amendment greatly reduces rice Cd uptake in a contaminated paddy soil: A two-year field experiment, Bioresources, 6, 2605– 2618, 2011.

23. Cui, L., Pan, G., Li, L., Yan, J., Zhang, A., Bian, R., and Chang, A.: The reduction of wheat Cd uptake in contaminated soil via biochar amendment: A two-year field experiment, Bioresources, 7, 5666–5676, 2012.

24. Cunningham, S. D. and Ow, D. W.: Promises and prospects of root zone of crops phytoremediation, Plant Physiol., 110, 715–719, 1996.

25. de Abreu, C. A., Coscione, A. R., Pires, A. M., and Paz-Ferreiro, J.: Phytoremediation of a soil contaminated by heavy metals and boron using castor oil plants and organic matter amendments. J. Geochem. Explor., 123, 3–7, 2012.

26. Elad, Y., Cytryn, E., Meller Harel, Y., Lew, B., and Graber, E. R.: The Biochar effect: Plant resistance to biotic stresses, Phytopathol. Mediterr., 50, 335–349, 2011.

27. Fellet, G., Marchiol, L., Delle Vedove, G., and Peressotti, A.: Application of biochar on mine tailings: effects and perspectives for land reclamation, Chemosphere, 83, 1262–1297, 2011.

28. Fellet, G., Marmiroli, M., and Marchiol, L.: Elements uptake by metal accumulator species grown on mine tailings amended with three types of biochar, Sci. Total Environ., 468–469, 598–608, 2014.

29. Freitas, E. V., Nascimento, C. W., Souza, A., and Silva, F. B.: Citric acid-assisted phytoextraction of lead: A field experiment, Chemosphere, 92, 213–217, 2013.

30. Gabos, M. B., Casagrande, G., Abreu, C. A., and Paz-Ferreiro, J.: Use of organic matter to mitigate multicontaminated soil and sunflower plants as phytoextractor, R. Bras. Eng. Agric. Ambiental, 15, 1298–1306, 2011.

31. Gascó, G., Paz-Ferreiro, J. and Méndez, A.: Thermal analysis of soil amended with sewage sludge and biochar from sewage sludge pyrolysis, J. Therm. Anal.

Calorim., 108, 769–775, 2012.

32. Giansoldati, V., Tassi, E., Morelli, E., Gabellieri, E., Pedron, F., and Barbafieri, M.: Nitrogen fertilizer improves boron phytoextraction by Brassica juncea grown in contaminated sediments and alleviates plant stress, Chemosphere, 87, 1119–1125, 2012.
33. Graber E. R., Meller-Harel, Y., Kolton, M., Cytryn, E., Silber, A., Rav David, D., Tsechansky, L., Borenshtein M., and Elad, Y.: Biochar impact on development and productivity of pepper and tomato grown in fertigated soilless media, Plant Soil, 337, 481– 496, 2010.
34. Hartley, W., Dickinson, N. M., Riby, P., and Lepp, N. W.: Arsenic mobility in brownfield soils amended with green waste compost or biochar and planted with Miscanthus, Environ. Pollut., 157, 2654–2662, 2009.
35. Hossain, M. K., Strezov, V., Chan, K. Y., and Nelson, P. F.: Agronomic properties of wastewater sludge biochar and bioavailability of metals in production of cherry tomato (Lycopersicon esculentum), Chemosphere, 78, 1167–1171, 2010.
36. Houben, D., Evrard, L., and Sonnet, P.: Mobility, bioavailability and pH-dependent leaching of cadmium,zinc and lead in a contaminated soil amended with biochar, Chemosphere, 92, 1450–1457, 2013a.
37. Houben, D., Evrard, L., and Sonnet, P.: Beneficial effects of biochar application to contaminated soils on the bioavailability of Cd, Pb and Zn and the biomass production of rapeseed (Brassica napus L.), Biomass Bioenerg., 57, 196–204, 2013b.
38. Huang, H., Wang, K., Zhu, Z., Li, Y., He, Z., Yang, X. E., and Gupta, D. K.: Moderate phosphorus application enhances Zn mobility and uptake in hyperaccumulator Sedum alfredii, Environ. Sci. Pollut. R., 20, 2844–2853, 2013.
39. Jeffery, S., Verheijen, F. G. A., van der Velde, M., and Bastos, A. C.: A quantitative review of the effects of biochar application to soils
40. on crop productivity using meta-analysis, Agr. Ecosyst. Environ., 144, 175–187, 2011.
41. Jiang, J., Xu, R. K., Jiang, T. Y., and Li, Z.: Immobilization of Cu(II), Pb(II) and Cd(II) by the addition of rice straw derived biochar to a simulated polluted Ultisol, J. Hazard. Mater., 229– 230, 145–150, 2012.
42. Kabata-Pendias, A. and Pendias, H.: Trace elements in soil and plants, 3rd Edn., CRC press, 403 pp., 2001.
43. Karami, N., Clemente, R., Moreno-Jiménez, E., Lepp, N., and Beesley, L.: Efficiency of green waste compost and biochar soil amendments for reducing lead and copper mobility and uptake to ryegrass (Lolium perenne), J. Hazard. Mater., 191, 41–48, 2011.
44. Komárek, M., Vanek, A., and Ettler, V.: Chemical stabilization of ˇ metals and arsenic in contaminated soils using oxides-a review, Environ. Pollut., 172, 9–22, 2013.
45. Lehmann, J. and Joseph, S.: Biochar for environmental management: science and

technology. Earthscan, London and Sterling, VA USA, 2009.

46. Lehmann, J., Rillig, M. C., Thies, J., Masiello, C. A., Hockaday, W. C., and Crowley, D.: Biochar effects on soil biota. A review, Soil Biol. Biochem., 43, 1812–1836, 2011.
47. Li, N. Y., Fu, Q. L., Zhuang, P., Guo, B., Zou, B., and Li, Z. A.: Effect of fertilizers on Cd uptake of Amaranthus Hypochondriacus, a high biomass, fast growing and easily cultivated potential Cd hyperaccumulator, Int. J. Phytoremediat., 14, 162–173, 2012.
48. Li, Y. M., Chaney, R., Brewer, E., Roseberg, R., Angle, J. S., Baker, A., Reeves, R., and Nelkin, J.: Development of a technology for commercial phytoextraction of nickel: economic and technical considerations, Plant Soil, 249, 107–115, 2003.
49. Lima, I. M., Boykin, D. L., Klasson, K. T., and Uchimiya, M.: In- fluence of post-treatment strategies on the properties of activated chars from broiler manure, Chemosphere, 95, 96–104, 2014.
50. Liu, X., Zhang, A., Ji, C., Joseph, S., Bian, R., Li, L., Pan, G., and Paz-Ferreiro, J.: Biochar's effect on crop productivity and the dependence on experimental conditions- a meta-analysis of literature data, Plant Soil, 373, 583–594, 2013.
51. Lu, H., Zhang, Y. Y., Huang, X., Wang, S., and Qiu, R.: Relative distribution of Pb2+ sorption mechanisms by sludge-derived biochar, Water Res., 46, 854–862, 2012.
52. Mandal, A., Purakayastha, T. J., Patra, A. K., and Sanyal, S. K.: Phytoremediation of arsenic contaminated soils by Pteris Vittata L. I. Influence of phosphatic fertilizers and repeated harvests, Int. J. Phytoremediat., 14, 978–995, 2012.
53. Marschner, H.: Mineral nutrition in higher plants, Academic, London, 1986.
54. Méndez, A., Barriga, S., Fidalgo, J. M., and Gascó, G.: Adsorbent materials from paper industry waste materials and their use in Cu(II) removal from water, J. Hazard. Mater., 165, 736–743, 2009.
55. Méndez, A., Gómez, A., Paz-Ferreiro, J., and Gascó, G.: Effects of biochar from sewage sludge pyrolysis on Mediterranean agricultural soils, Chemosphere, 89, 1354–1359, 2012.
56. Méndez, A., Paz-Ferreiro, J., Araujo, F., and Gasco, G.: Biochar from pyrolysis of de-inking paper sludge and its use in the treatment of a nickel polluted soil, J. Anal. Appl. Pyrol., doi:10.1016/j.jaap.2014.02.001, in press, 2014.
57. Méndez, M. O., Glenn, E. P., and Maier, R. M.: Phytostablization potential of quailbush for mine tailings, J. Environ. Qual., 36, 245–253, 2007.
58. Namgay, T., Singh, B., and Singh, B. P.: Influence of biochar application to soil on the availability of As, Cd, Cu, Pb, and Zn to maize (Zea mays L.), J. Aust. Soil Res., 48, 638–647, 2010.
59. Oo, A. N., Iwai, C. B., and Saenjan, P.: Soil properties and maize growth in saline and nonsaline soils using cassava-industrial waste compost and vermicompost with or without earthworms, Land Degrad. Dev., doi:10.1002/ldr.2208, in press, 2014.

60. Orlowska, E., Przybylowicz, W., Orlowski, D., Turnau, K., and Mesjasz-Przybylowicz, J.: The effect of mycorrhiza on the growth and elemental composition of Ni hyperaccumulating plant Berkheya coddii Roessler, Environ. Pollut., 159, 3730– 3738, 2011.

61. Park, J. H., Choppala, G. H., Bolan, N. S., Chung, J. W., and Chuasavathi, T.: Biochar reduces the bioavailability and phytotoxicity of heavy metals, Plant Soil, 348, 439–451, 2011.

62. Park, J. H., Choppala, G. H., Lee, S. J., Bolan, N., Chung, J. W., and Edraki, M.: Comparative sorption of Pb and Cd by biochars and its implication for metal immobilization in soil, Water Air Soil Poll., 224, 1711, doi:10.1007/s11270-013-1711-1, 2013.

63. Paz-Ferreiro, J. and Fu, S.: Biological indices for soil quality evaluation: perspectives and limitations, Land Degrad. Dev., doi:10.1002/ldr.2262, in press, 2014.

64. Paz-Ferreiro, J., Fu, S., Méndez, A., and Gasco, G.: Interactive effects of biochar and the earhworm Pontoscolex corethrurus on plant productivity and soil enzymes activities, J. Soils Sediments, doi:10.1007/s11368-013-0806-z, in press, 2014.

65. Prasad, M. N. V.: Phytoremediation of metal-polluted ecosystems: hype for commercialization, Russ. J. Plant Physiol., 50, 686–700, 2003.

66. Rajkumar, M., Prasad, M. N. V., Swaminathan, S., and Freitas, H.: Climate change driven plant–metal–microbe interactions, Environ. Int., 53, 74–86, 2013.

67. Roy, M. and McDonald, L. M.: Metal uptake in plants and health risk assessments in metal-contaminated smelter soils, Land Degrad. Dev., doi:10.1002/ldr.2237, in press, 2014.

68. Shao, H., Chu, L. Y., Xu, G., Yan, K., Zhang, L. H., and Sun, J. N.: Progress in phytoremediating heavy-metal contaminated soils, in: Detoxification of Heavy Metals, edited by: Sherameti, I. and Varma, A., Springer, 73–90, 2011.

69. Siebielic, G. and Chaney, R. L.: Testing amendments for remediation of military range contaminated soil, J. Environ. Manage., 108, 8–13, 2012.

70. Sizmur, T., Wingate, J., Hutchings, T., and Hodson, M. E.: Lumbricus terrestris L. does not impact on the remediation efficiency of compost and biochar amendments, Pedobiologia, 54, S211– S216, 2011.

71. Stevens, D. P., McLaughlin, M. J., and Heinrich, T.: Determining toxicity of lead and zinc runoff in soils: salinity effects on metal partitioning and on phytotoxicity, Environ. Toxicol. Chem., 22, 3017–3024, 2003.

72. Uchimiya, M., Lima, I. M., Klasson, K. T., and Wartelle, L. H.: Contaminant immobilization and nutrient release by biochar soil amendment: Roles of natural organic matter, Chemosphere, 80, 935–940, 2010.

73. Uchimiya, M., Klasson, K. T., Wartelle, L. H., and Lima, I. M.: In- fluence of soil properties on heavy metal sequestration by biochar amendment: 1. Copper sorption isotherms and the release of cations, Chemosphere, 82, 1431–1437, 2011a.

74. Uchimiya, M., Wartelle, L. H., Klasson, K. T., Fortier, C. A., and Lima, I. M.: Influence of pyrolysis temperature on biochar property and function as a heavy metal sorbent in soil, J. Agr. Food Chem., 59, 2501–2510, 2011b.

75. Uchimiya, M., Chang, S. C., and Klasson, K. T.: Screening biochars for heavy metal retention in soil: Role of oxygen functional groups, J. Hazard. Mater., 190, 432–444, 2011c.

76. Uchimiya, M., Cantrell, K. B., Hunt, P. G., Novak, J. M., and Chang, S. C.: Retention of heavy metals in a Typic Kandiudult amended with different manure-based biochars, J. Environ. Qual., 41, 1138–1149, 2012a.

77. Uchimiya, M., Bannon, D. I., Wartelle, L. H., Lima, I. M., and Klasson, K. T.: Lead retention by broiler litter biochars in small arms range soil: Impact of pyrolysis temperature, J. Agr. Food Chem., 60, 5035–5044, 2012b.

78. Vacca, A., Bianco, M. R., Murolo, M., and Violante, P.: Heavy metals in contaminated soils of the Rio Sitzerri floodplain (Sardinia, Italy): Characterization and impact on pedodiversity, Land Degrad. Dev., 23, 250–364, 2012.

79. van der Ent, A., Baker, A. J. M., Reeves, R. D., Pollard, A. J., and Schat, H.: Hyperaccumulators of metal and metalloid trace elements: facts and fiction, Plant Soil, 362, 319–334, 2013.

80. Vangronsveld, J., Herzig, R., Weyens, N., Boulet, J., Adriaensen, K., Ruttens, A., Thewys, T., Vassilev, A., Meers, E., Nehnevajova, E., Van der Lelie, D., and Mench, M.: Phytoremediation of contaminated soils and groundwater: lessons from the field, Environ. Sci. Pollut. R., 16, 765–794, 2009.

81. Wang, K., Zhang, J., Zhu, Z., Huang, H., Li, T., He, Z., Yang, X., and Alva, A.: Pig manure vermicompost (PMVC) can improve phytoremediation of Cd and PAHs co-contaminated soil by Sedum alfredii, J. Soils Sediments, 12, 1089–1099, 2012.

82. Wei, S., Zhu, J., Zhou, Q.X., and Zhan, J.: Fertilizer amendment for improving the phytoextraction of cadmium by a hyperaccumulator Rorippa globosa (Turcz.) Thell. J. Soils Sediments, 11, 915-922, 2011.

83. Wu, W., Yang, M., Feng, Q., McGrouther, K., Wang, H., Lu, H., and Chen, Y.: Chemical characterization of rice straw-derived biochar for soil amendment, Biomass Bioenerg., 47, 268–276, 2012.

84. Xue, Y. W., Gao, B., Yao, Y., Inyang, M., Zhang, M., Zimmerman, A. R., and Ro, K. S.: Hydrogen peroxide modification enhances the ability of biochar (hydrochar) produced from hydrothermal carbonization of peanut hull to remove aqueous heavy metals: Batch and column tests, Chem. Eng. J., 200–202, 673–680, 2012.

85. Yang, D., Zeng, D. H., Li, L. J., and Mao, R.: Chemical and microbial properties in contaminated soils around a magnesite mine in Northeast China, Land Degrad. Dev., 23, 256–262, 2012.

86. Zheng, R. L., Cai, C., Liang, J. H., Huang, Q., Chen, Z., Huang, Y. Z., Arp, H. P. H, and Sun, G. X.: The effects of biochars from rice residue on the formation of iron plaque and the accumulation of Cd, Zn, Pb, As in rice (Oryza sativa L.) seedlings, Chemosphere, 89, 856–863, 2012.

87. Zhou, Q. X., Cui, S., Wei, S. H., Zhang, W., Cao, L., and Ren, L. P.: Effects of exogenous chelators on phytoavailability and toxicity of Pb in Zinnia elegans Jacq, J. Hazard. Mater., 146, 341–346, 2007.

# 5
CHAPTER

# EFFECT OF BACTERIAL APPLICATION ON METAL AVAILABILITY AND PLANT GROWTH IN FARMLAND-CONTAMINATED SOILS

Du RY, Wen D, Zhao PH, Chen Y and Wang FH

Public Monitoring Center for Agro-Product, Guangdong Academy of Agricultural Sciences, Guangzhou 510640, China and Key Laboratory of Testing and Evaluation for Agro-product Safety and Quality, Ministry of Agriculture, Guangzhou 510640, China

## ABSTRACT

**Conventional approaches for phytoremediation of heavy metal-contaminated soils face challenges in farmland werlands. The combined use of functional microorganisms is a promising option to enhance phytoremediation. Two heavy metal-resistant model strains, designated DBM1 (rhizosphere Arthrobacter) and DBM2 (non-rhizosphere Bacillus), were isolated from polluted farmland soils downstream of a multi-metal mine. The effects of these two bacterial strains on soil metal availability and plant (rice and red ramie) growth were determined under greenhouse conditions. Both microbes (especially DBM2) significantly reduced metal availability and increased pH level in lowcontaminated soil, but the passivation ability of DBM2 disappeared in high-contaminated soil. Soil metal availability was markedly increased by DBM1 after addition of carbon source, while it was reduced by DBM1 or DBM2 with pH increase after addition of nitrogen, phosphorous, or modified fly-ash amendment. Both microbes significantly increased plant biomass and reduced total metal contents of low-metal accumulating rice. For high-metal accumulating red ramie, DBM1 significantly increased plant biomass while DBM2 increased total metal contents. Proper selection of microbial species to match plants for phytoremediation is the key to improve microbe-assisted phytoremediation efficiency in mining-contaminated soils.**

**Citation:** Du RY, Wen D, Zhao PH, Chen Y and Wang FH, Effect of Bacterial Application on Metal Availability and Plant Growth in Farmland-Contaminated Soils, doi: 10.4172/2155-6199.1000341.

## INTRODUCTION

Ecological problems caused by mine exploitation have become global [1-3]. An example is from the Dabaoshan Multi-metal Mine in Guangdong Province, China, where mineral resources are characterized by complex paragenetic associations with a variety of metal elements such as Pb, Cd, Cu, Zn, and As [4]. As a consequence, either the heavily contaminated mining wasteland or the moderately contaminated downstream agricultural soil is commonly associated with multiple toxic metals [5].

Dynamic interactions between concomitant heavy metals can lead to changes in the ecotoxicity of mining-contaminated soils, as well as affect the form and bioavailability of metal contaminants [4]. These problems bring enormous difficulty in applying conventional remediation technologies to remediation of heavy metal-contaminated soils. Compared with conventional chemical and physical treatments, phytoremediation is an advantageous option that is cost-effective, environment-friendly, capable of maintaining soil productivity, and free of secondary contamination [6]. However, simple application of plant remediation techniques has failed in ecological remediation of combined contamination, Because the heavy metal contaminated soil lead pH, nutrient matter content, microorganisms activity were 51 lower and the known hyperaccumulators plant often have relatively low biomass or limited accumulation of heavy metals in contaminated mining areas. This situation urges the use of chemical enhancers or fertilizers to improve the efficiency of phytoremediation, thus increasing the cost of remediation and bringing the risk of secondary pollution [7-9].

Soil microorganisms are mediators in the biogeochemical cycling of nutrient elements in terrestrial ecosystems. A diversity of bacteria is involved in the migration and transformation of heavy metals in the soil [10]. There is increasing evidence that microbial components play an important role in phytoremediation of mining-contaminated soils. More than 20 species of rhizosphere bacteria (e.g., Pseudomonas, Bacillus, Rhizobium, and Serratia) are known to have a positive role in plant establishment and colonization on mining wastelands, mainly by increasing plant nutrient uptake and improving the soil structure [11]. Soil microorganisms, especially the rhizosphere microflora, utilize the nutrients secreted by plant roots and in turn provide metabolites to facilitate plant growth. Moreover, microbial active substances such as extracellular enzymes, organic acids and chelates convert soil heavy metals from the unavailable to bioavailable forms, promoting their uptake and accumulation by plants. Compared to chemical enhancement, the biological enhancement of phytoremediation is obviously associated with lower cost and environmental risk the soil environmental quality will be affect plant growth. Therefore, microbeassisted phytoremediation has become a hot topic for research on the remediation of heavy mental-contaminated soils [7,9,12,13].

Heavy metal-tolerant microbes are frequently isolated from mining-contaminated soils at low pH and high concentrations of multiple heavy metals. Although such environments are lethal for ordinary microbes to live in, specific genes in the tolerant microbes can be activated under extreme conditions, inducing unique biological metabolic pathways for survival. In practical applications, functional microbes can be isolated from mining-contaminated soil and then applied back into the original or adjacent soil via expanded propagation. This microbe-assisted method is reported to enhance the remediation of heavy metal-contaminated soil compared with non microbial treatment [14]. However, the effect of microbial function on soil metal availability and plant growth in heavy metal-contaminated soils has not been fully elucidated. The influence mechanism will be change with the different functional microorganisms combined the different plants are different.

Isolated two functional bacterial strains from rice and red ramie root plant in polluted farmland soils downstream of a multi-metal mine in preliminary studies. Found that two strains can still grow normally in the high concentration of heavy metal treatment. So through research the bacterial functional in the different levels of soil contaminations, and combined with rice and red ramie. Analysis their effects on soil metal availability and plant growth were examined under greenhouse conditions. The results will provide reference data for microbe-assisted phytoremediation of mining-contaminated soils.

## MATERIALS AND METHODS

### Study Site

The Dabaoshan Mine is a large multi-metal mine in Shaoguan City, Guangdong Province, China. Its main orebody consists of limonite ore in the upper part (20 million ton reserves) and large copper ore in the lower part (28 million ton reserves). Associated rare and noble metals include tungsten, bismuth, molybdenum, gold, and silver. The primary finished products are iron ore, copper concentrate ore and grade 1 electrolytic copper [15,16]. In the process of mining, the $S^{2-}$ with the copper ore will be oxidized into $H_2SO_4$. Accompanied by the efflux of mineral water lead the surroundings farmland soil pH value is very low.

Since 1970, the Dabaoshan Mine has experienced large-scale exploitation. Over the past four decades, due to lack of environmental awareness and unplanned, disorganized exploitation of individual miners, local vegetation has largely been destroyed while substantial heavy metals and high concentrations of sulfates are produced from oxidation of metal sulfides. The mining waste products caused serious contamination in the vicinity, especially the eco-environment of Hengshi River Basin. Heavy metal contamination also brought disastrous consequences to local agricultural production and people's health [17].

### Soil Sampling and Physicochemical Analysis

Along the drainage channel, mining wastewater often outflows during the rainy season, resulting in downstream farmland contaminating in the vicinity. Soil sampling was conducted at two points of different distances from the drainage channel: 20 m as the high-contaminated sampling point and 200 m as the low-contaminated sampling point.

Soil samples (1000 g each) were taken from the surface down to a depth of 20 cm using a wooden spatula. Each sample was divided into two portions, one stored at 4°C for bacterial isolation, and the other air-dried for physicochemical analysis. Plant residuals and stones were removed by hand, and soil samples were passed through a 1-mm sieve prior to soil physicochemical analysis.

The air-dried soil was sieved and digested with HCl-HNO3-HClO4 (3:1:2, V/V/V). Total soil concentrations of Cd, Pb, Cu, Mn and Zn were determined by inductively coupled plasma-optical spectrophotometry (ICP-MS, Agilent 7700X, USA). Soil pH and organic matter content were determined using standard methods for soil agro-chemical analysis [18]. The major physicochemical properties of the test soils are shown in Table 1.

**Table 1:** The physicochemical properties of mining-contaminated farmland soil samples

| Soil | pH | Organic matter (%) | Total N (mg·kg$^{-1}$) | Total amount of heavy metals (mg·kg$^{-1}$) | | | | |
|---|---|---|---|---|---|---|---|---|
| | | | | Cd | Cu | Mn | Pb | Zn |
| Low-contaminated | 4.34 ± 0.07 | 2.11 ± 0.03 | 1439.5 ± 123.58 | 3.07 ± 1.28 | 323 ± 5.38 | 139.62 ± 5.23 | 392 ± 12.13 | 320 ± 8.52 |
| High-contaminated | 3.61 ± 0.02 | 1.20 ± 0.03 | 671.19 ± 12.45 | 10.37 ± 0.26 | 615 ± 20.27 | 158.08 ± 4.79 | 1644 ± 61.12 | 470 ± 14.25 |

## Bacterial Isolation and Identification

For bacterial isolation, soil samples were collected from rhizosphere and adjacent non-rhizosphere soils of *Jatropha curcas* L and *Hibiscus cannabinus* L. Up to 10.00 g of mixed fresh soil (highand low-contaminated) was added into a triangular flask containing 90 mL sterile water and ten 3-5-mm glass beads. Cd, Pb, Cu and Zn are the main pollutants in farmland soil. The soil suspension was shaken at 160 rpm·min$^{-1}$for 30 min at 28°C and then inoculated to a solid beef-extract-peptone medium containing $Pb^{2+}$ (1000 mg·kg$^{-1}$), $Cd^{2+}$ (20 mg·kg$^{-1}$), $Cu^{2+}$ (800 mg·kg$^{-1}$), $Zn^{2+}$ (1000 mg·kg$^{-1}$), and other heavy metals. $Pb^{2+}$, $Cu^{2+}$, $Zn^{2+}$ and $Cd^{2+}$ concentrations in the medium were designed based on total metal concentrations in farmland soils surrounding the Dabaoshan Mine area according to our research group investigate.

The inoculated plates were incubated at 30°C for 3-7 days. Isolated single colonies were chosen and fully grown to obtain a pure culture using the streak plate method. After incubation (three passages) with the heavy metal-containing medium, two purified rod-shaped bacteria resistant to high concentrations of Cd, Zn, Cu, and Pb were obtained [11]. The main purpose is to screen functional bacteria. Cd, Pb, Cu and Zn are the main pollutants in farmland soil. So add those into culture medium for screen stronger functional bacteria. Then the bacteria were cultured and purified. These two isolates were identified based on 16S rDNA sequences by the Microbiology Institute of Guangdong Province, Guangzhou, China. One was *Arthrobacter sp.* DBM, referred to as DBM1, and the other was *Bacillus sp.* DBM, referred to as DBM2.

## Bacterial Growth Experiments

To eliminate the impact of broth medium on the experimental results, the culture broths of DBM1 and DBM2 were diluted with sterile water to obtain bacterial cell suspensions with an optical density at 600 nm ($OD_{600}$) of 0.5. The suspensions (1 mL) were dispensed into 50- mL centrifuge tubes containing 1.00 g of high- or low-contaminated soil sample each. The mixtures were incubated at 30°C for 7 days by shaking at 200 rpm. Three replicates were prepared for each treatment. After the growth experiment, the concentrations of available heavy metals and the level of pH in the soils were determined as described above.

For parallel experiments, sterilized (121°C-heated, 2 h) highcontaminated soil weighing 1.00 g was added into 50 mL centrifuge tubes. Then, filter-sterilized glucose solution was added as carbon source, urea as nitrogen source, and $KH_2PO_4$ as phosphorus source (N0.08 g·kg$^{-1}$, $P_2O_5$ 0.08 g·kg$^{-1}$, K2O 0.16 k·kg$^{-1}$; 1 mL each). Three replicates for each treatment were prepared. Additionally, the high-contaminated soil was mixed with fly ash (Zhujiang Cement Plant, Guangzhou, China) to equilibrium for one month and then sterilized as a modified treatment. Approximately 1 mL of bacterial suspensions of DBM1 and DBM2 ($OD_{600}$=0.5) was separately added to each treatment. The mixtures were incubated at 30°C for 7 days by shaking at 200 rpm. The concentrations of available heavy metals and pH level in the soil were then determined as described above.

## Plant Growth Test

Rice and red ramie seeds were purchased from Agro-technical Station, Wengyuan County, Guangdong Province. Five rice (or red ramie) seeds were sown in 200-mL plastic containers containing 150 g of steam-sterilized soil. Each container was inoculated with 2 mL of bacterial suspension (DBM1 or DBM2, $OD_{600}$=0.5). The control was inoculated with an equal volume of distilled water. Three replicates for each treatment were made. The plants were incubated in a greenhouse under natural light conditions. Up to 1 mL of bacterial suspensions were inoculated into each cup every other week, along with 1 mL of deionized water. After 30 days, plant biomass (fresh and dry weights of shoots and roots) and total contents of available heavy metals were determined.

## Data Analysis

Data were expressed as the mean ± standard deviation ($n$=3). Data comparison between treatments was performed using the analysis of variance. Multiple comparisons were performed using Duncan's multiple range test. The relevance between soil pH and heavy mental availability was examined using Pearson correlation analysis. P values less than 0.05 were considered statistically significant. SPSS 11.0 (SPSS Inc., Chicago, IL, USA) was used for data analysis.

# RESULTS

## Metal Availability and pH level in Mining-Contaminated Soils

In the low-contaminated soil, the availability of Al, Cu, Fe, Pb, Zn, and Mn was significantly reduced due to the presence of DBM1 and DBM2 compared with that in the control. The reducing effect of microbes on heavy metal availability was most significant for Al and Pb, with approximately 6.5- and 3.5-fold differences, respectively. The capability of DBM1 to passivate heavy metals (except Cu) was slightly stronger than that of DBM2, but the differences were not statistically significant (Figure 1a).

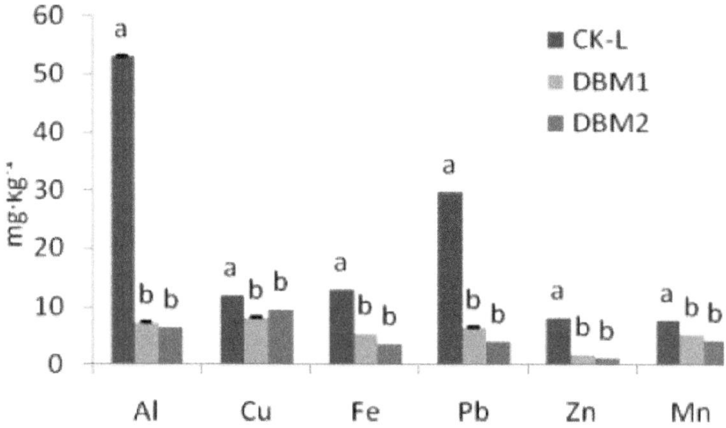


**Figure 1a:** Metal availability in mining-contaminated soil as influenced by Arthrobacter sp. DBM1 and Bacillus sp. DBM2 at Low-contaminated soil.

In the high-contaminated soil, DBM1 significantly reduced the availability of heavy metals tested, except for Mn. The passivation capability of DBM2 to these heavy metals was weaker than that of DBM1, which only significantly affected Al and Fe availability (Figure 1b).

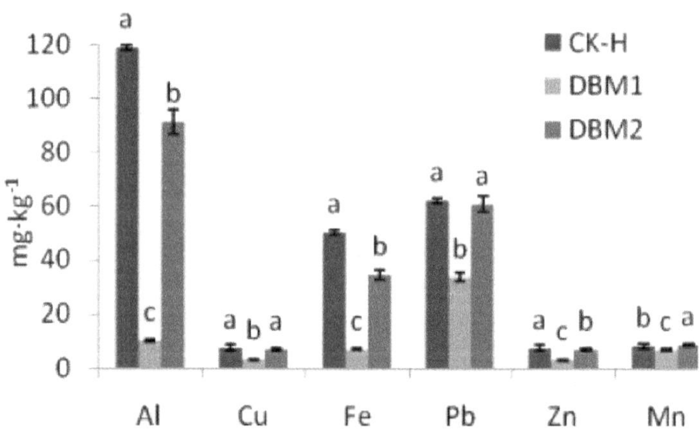


**Figure 1b:** Metal availability in mining-contaminated soil as influenced by Arthrobacter sp. DBM1 and Bacillus sp. DBM2 at High-contaminated soil. CKL, low-contaminated soil control; CK-H, high-contaminated soil control.

DBM1 and DBM2 raised soil pH to different levels, with greater pH increase in the low-contaminated soil ($P<0.05$ for both microbes) than in the high-contaminated soil ($P<0.05$ for DBM1 only, Figure 2). The capability of DBM1 and DBM2 to raise soil pH was almost identical in the low-contaminated soil ($P<0.05$). No significant pH increase was observed in the high-contaminated soil in the presence of DBM2 compared the control ($P>0.05$, Figure 2).

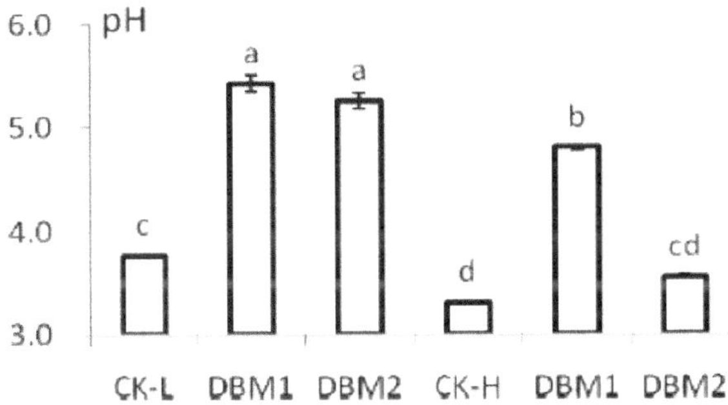


**Figure 2:** The pH of mining-contaminate soil as influenced by *Arthrobacter*sp. DBM1 and *Bacillus* sp. DBM2. CK-L, low-contaminated soil control; CK-H, high-contaminated soil control.

The relevance analysis showed that soil pH was significantly negatively correlated with metal availability ($P<0.01$). By contrast, there were significant positive correlations between the availability of different metals ($P<0.01$ or 0.05; Table 2).

**Table 2:** Relevance analysis between the pH and metal availability in miningcontaminated soil samples

| Factor | Al | Cu | Fe | Pb | Zn | Mn |
|---|---|---|---|---|---|---|
| pH | -0.943** | -0.912** | -0.862** | -0.961** | -0.961** | -0.906 ** |
| Al | | 0.811** | 0.966** | 0.984** | 0.986** | 0.920** |
| Cu | | | 0.672* | 0.825** | 0.824** | 0.722* |
| Fe | | | | 0.936** | 0.939** | 0.892** |
| Pb | | | | | 0.999** | 0.956** |
| Zn | | | | | | 0.960** |

*$P<0.05$ and ** $P<0.01$

## Metal Availability and ph Level In Mining-Contaminated Soils after Addition of Nutrients and Modified Amendment

The interaction between functional microbes and multiple metals changed after addition of C, N, and P sources and modified amendment (Table 3). With extra C source, DBM1 significantly enhanced the availability of most metals ($P<0.05$) except for Pb and Zn, whereas DBM2 significantly reduced the availability of most metals ($P<0.05$) except for Mn. After the addition of N source, P source, or modified amendment, both DBM1 and DBM2 significantly reduced the concentrations of available metals by 1.5- to 590-fold compared with the control ($P<0.05$). The extent of metal availability reduction by microbes generally decreased as N>P>modified amendment, except for Cu (modified amendment>N>P).

After the addition of N source, P source, or modified amendment, both DBM1 and DBM2 significantly raised soil pH. However, no significant change was observed in soil pH after the addition of C source (Table 4). The relevance analysis showed that soil pH was significantly negatively correlated with available metal concentrations, with the correlation coefficients of -0.943, -0.912, -0.862, -0.961, -0.961, and -0.906 ($P<0.01$).

### Plant Biomass and Metal Absorption

DBM1 and DBM2 had different effects on plant biomass of rice and red ramie grown in contaminated soils (Table 5). For rice, DBM2 increased plant biomass significantly more than DBM1 ($P<0.05$). For red ramie, DBM1 increased plant biomass ($P<0.05$) while DBM2 reduced plant biomass to different extents compared with the control ($P<0.05$ or$>0.05$).

DBM1 and DBM2 showed different effects on metal absorption capacity of plants. DBM1 increased Cu, Zn, Fe, and Al contents but decreased Pb and Mn contents in rice roots, whereas DBM2 generally reduced metal contents in rice shoots and roots (Figure 3). Red ramie gained increased Al, Fe, and Pb contents after inoculated with DBM1 or DBM2. Additionally, DBM2 increased Cu, Mn, and Zn contents, whereas DBM1 showed opposite effects on metal contents in red ramie (Figure 4).

## DISCUSSION

Slow growth, low biomass, and even failure of growth are the major issues encountered in phytoremediation around mining wastelands due to poor soil fertility and heavy metal toxicity [19]. In order to survive in heavy metal-contaminated soils, functional microbes [20] have to alter their metabolic activities with the ambient environment [8,13]. In this study, DBM1 and DBM2 exhibited strong ability to reduce heavy metal availability in mining-contaminated soils. DBM1 as rhizobacteria strongly reduced heavy metal availability and increased the pH in both high- and low-contaminated soils. However, the metal passivation ability of non-rhizobacteria DBM2 disappeared in the high-contaminated soil although it still increased soil pH (Figure 1). The increase in soil pH by functional bacteria could be attributed to their secretion of alkaloids [21], which inevitably led to decrease in metal availability. Although it remains unclear why DBM2 failed to reduce soil metal availability under high contamination, the results indicate that rhizosphere microorganisms play a more important role in the process of phytoremediation of heavy metal-contaminated soils [12,22,23].

Microbes can affect the availability of heavy metals through direct or indirect metabolic activities, which play a decisive role in maintaining soil function. Alvareng (2009) reported that microbial activities were negatively correlated with Cu, Pb, and Zn availability. In the present study, we observed that following the addition of C source, soil metal availability was strongly increased by DBM1 in the contaminated soil. Especially, the increase of available Fe concentration (Table 3) could be attributed to secretion of siderophores by the rhizosphere bacteria [24,25]. In a microbial habitat, high C/N ratio often leads to gradual decrease in the pH, while low C/N ratio is associated with increases in the pH [26]. In the present study, glucose was added as the C source which did not help increase soil pH value (Table 4). The lower pH possibly induced specific metabolic pathways and triggered secretions of special metabolites to increase metal availability in the soil [27], thereby facilitating metal solution through directly and/or indirectly.

In contrast, the addition of N source, P source, or modified fly-ash amendment raised soil pH to different extents (Table 4), thereby reduces the toxicity of contaminated soil to microbes [14]. Additionally, the N and P sources increased their availability in a suitable range of soil pH (Table 4), which could benefit microbial growth and reproduction. Moreover, N and P sources might stimulate specific metabolic pathways of soil functional bacteria, further increasing the pH of soil solution. These hypotheses explained the decrease in soil metal availability upon the addition of N or P source in the presence of functional bacteria (Table 3).

**Table 3:** Metal availability in mining-contaminated soil after addition of nutrient sources and modified amendment ($mg{\cdot}kg^{-1}$)

| | | Al | Cu | Fe | Pb | Zn | Mn |
|---|---|---|---|---|---|---|---|
| C | CK-C | 168.45 ± 8.47b | 9.16 ± 0.26b | 88.45 ± 3.14b | 19.77 ± 0.17b | 90.00 ± 3.69a | 20.57 ± 0.48b |
| | DBM1[a] | 329.25 ± 9.01a | 13.18 ± 0.37a | 149.97 ± 3.61a | 20.64 ± 0.59ab | 97.81 ± 4.00a | 24.60 ± 0.77a |
| | DBM2[b] | 123.84 ± 2.74c | 7.42 ± 0.21c | 23.66 ± 0.72c | 21.42 ± 0.31a | 70.07 ± 1.73b | 18.35 ± 0.71b |
| N | CK-N | 179.47 ± 7.23a | 10.69 ± 0.07a | 76.78 ± 3.53a | 21.61 ± 0.39a | 94.93 ± 5.41a | 24.37 ± 0.00a |
| | DBM1 | 1.39 ± 0.26b | 3.15 ± 0.09b | 2.38 ± 0.48b | 0.51 ± 0.03b | 0.34 ± 0.07b | 0.07 ± 0.02c |
| | DBM2 | 1.63 ± 0.23b | 2.94 ± 0.39b | 2.57 ± 0.19b | 0.86 ± 0.07b | 0.16 ± 0.02b | 0.29 ± 0.09b |
| P | CK-P | 101.75 ± 4.49a | 11.01 ± 0.44a | 28.85 ± 0.26a | 37.18 ± 0.69a | 51.79 ± 1.30a | 27.32 ± 0.91a |
| | DBM1 | 1.74 ± 0.20b | 6.17 ± 0.35b | 3.89 ± 0.30b | 2.97 ± 0.08b | 0.30 ± 0.06b | 0.31 ± 0.08b |
| | DBM2 | 3.17 ± 2.48b | 7.67 ± 0.33b | 6.23 ± 3.71b | 2.47 ± 0.20b | 0.29 ± 0.05b | 0.51 ± 0.09b |
| Amendments | CK-G | 84.69 ± 2.97a | 7.25 ± 0.41a | 24.73 ± 0.90a | 40.06 ± 0.87a | 60.86 ± 1.24a | 15.46 ± 0.68a |
| | DBM1 | 1.12 ± 0.04b | 1.71 ± 0.10b | 1.62 ± 0.03b | 26.58 ± 3.50b | 3.47 ± 0.17b | 3.24 ± 0.17b |
| | DBM2 | 1.21 ± 0.21b | 1.45 ± 0.06b | 2.18 ± 0.35b | 30.81 ± 4.36ab | 2.85 ± 0.87b | 2.67 ± 0.74b |

[a]DBM1, *Arthrobacter* sp.; [b]DBM2, *Bacillus* sp.; Data are presented as the mean ± SD ($n$=3). In each treatment, values with different lowercase letters in the same column are significantly different ($P$ <0.05).

**Table 4:** Soil pH after addition of nutrient sources and modified amendment

| | C | N | P | Modified fly-ash amendment |
|---|---|---|---|---|
| Ctrl | 3.65 ± 0.03a | 3.67 ± 0.03b | 4.81 ± 0.70b | 3.92 ± 0.01b |
| DBM1[a] | 3.39 ± 0.01a | 6.83 ± 0.02a | 6.50 ± 0.04a | 5.06 ± 0.02a |
| DBM2[b] | 3.86 ± 0.02a | 6.72 ± 0.11a | 6.42 ± 0.05a | 5.18 ± 0.06a |

[a]DBM1, *Arthrobacter* sp.; [b]DBM2, *Bacillus* sp.; Data are presented as the mean ± SD ($n$=3). Values with different lowercase letters in the same column are significantly different ($P$ <0.05).

An increasing attention is paid to the use of growth-promoting rhizosphere microbes to improve the efficiency of remediating heavy metal polluted soils through the interaction between microbes and plants. Rice is the grain crop most widely planted in heavy metalpolluted farmland in Guangdong province and other areas of China, which can easily accumulate Cd from contaminated soils. There is a need to control heavy metal transfer from the soil into the plant body of rice. Red ramie is an inedible economic crop, for which mainly plant biomass is of interest when grown in contaminated soil. Thus, proper selection of functional microbes to match plants used for phytoremediation is the key to improve microbe-assisted phytoremediation efficiency in mining-contaminated soils. Different functional microbes should be selected properly for phytoremediation of contaminated soils. Under greenhouse conditions, DBM1 significantly increased the biomass of both rice and red ramie, while DBM2 more significantly increased the biomass of rice grown in contaminated soil (Table 5). Moreover, DBM2 reduced all metal contents in rice shoots but increased metal contents in red ramie (Figures 3 and 4). When DBM1 was inoculated into rice and red ramie, total Cu, Pb, and Zn contents showed opposite trends in the two plants (Figures 3 and 4). The results demonstrate that a combined use of rhizosphere bacteria and functional plants is an effective approach to promote plant growth and attenuate metal hazard.

**Table 5:** Plant biomass of rice and red ramie grown in mining-contaminated soil under greenhouse conditions

| Plant | | Fresh weight ($g \cdot plant^{-1}$) | | Dry weight ($g \cdot plant^{-1}$) | | Root/shoot ratio |
|---|---|---|---|---|---|---|
| | | Shoot | Root | Shoot | Root | |
| Rice | CK | 3.52c | 3.77c | 0.75c | 0.70b | 0.93b |
| | DBM1[a] | 4.88b | 4.86b | 0.94b | 0.96a | 1.03a |
| | DBM2[b] | 5.97a | 5.20a | 1.09a | 1.00a | 0.92b |
| Red ramie | CK | 0.39b | 0.05b | 0.04b | 0.01b | 0.20b |
| | DBM1 | 0.52a | 0.09a | 0.06a | 0.02a | 0.28b |
| | DBM2 | 0.33b | 0.01c | 0.02c | 0.01b | 0.56a |

[a]DBM1, *Arthrobacter* sp.; [b]DBM2, *Bacillus* sp.

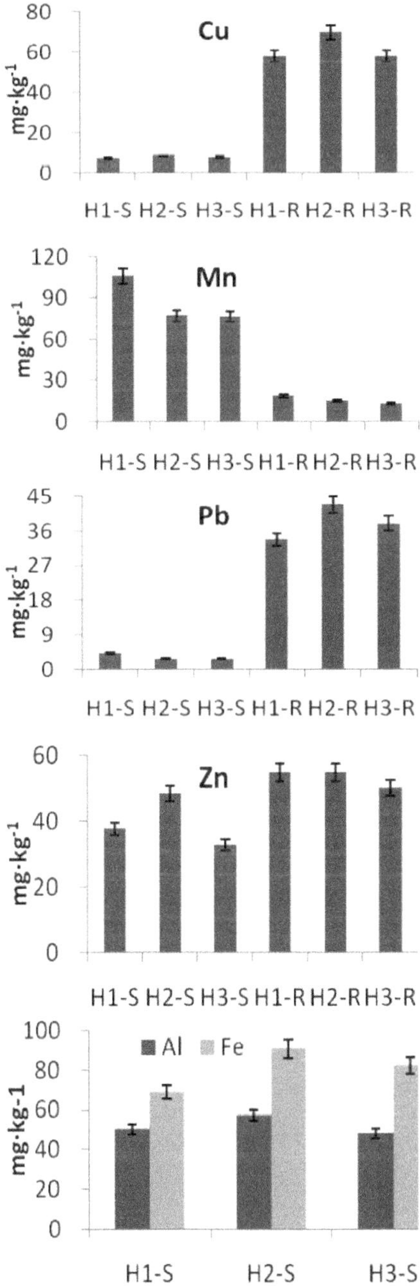


**Figure 3:** Metal contents of rice plants grown in mining-contaminated soil as influenced by Arthrobacter sp. DBM1 and Bacillus sp. DBM2 under greenhouse conditions. H1, control; H2, DBM1; H3, DBM2; S, shoots; and R, roots.

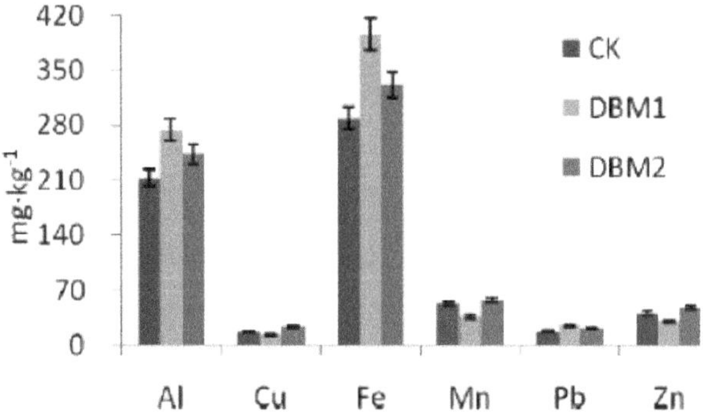


**Figure 4:** Metal contents of red ramie plants grown in mining-contaminated soil as influenced by Arthrobacter sp. DBM1 and Bacillus sp. DBM2 under greenhouse conditions.

## CONCLUSIONS

Two bacterial model strains representing Arthrobacter sp. and Bacillus sp. were screened out from heavy metal-contaminated farmland soil downstream of a multi-metal mine. When inoculated into the contaminated soil, both bacteria reduced metal availability and increased the pH in low-contaminated soil, but metal passivation ability of Bacillus sp. disappeared in high-contaminated soil. Additionally, Arthrobacter sp. increased plant biomass and available Fe in rice and red ramie, whereas Bacillus sp. reduced metal contents in rice shoots but increased metal contents in red ramie. Both kinds of bacteria can be used as functional microbes for microbe-assisted phytoremediation of mining-contaminated soils. In practice, the degree of soil contamination must be taken into account for choosing suitable microbes, phytoremediation plants and their combinations. In the optimal combination of plants, microbes, and chemicals, not only suitable species but also the most reasonable combination among the three factors should be considered for rational remediation of miningcontaminated soils.

## ACKNOWLEDGEMENT

This study was supported by the National Natural Science Foundation of China (No. 41401367).

## REFERENCES

1. Aguilar J, Dorronsoro C, Fernandez E, Fernandez J, Garcia I, et al. (2004) Soil pollution by a pyrite mine spill in Spain: evolution in time. Environ Pollut 132: 395-401.
2. Lim TT, Chui PC, Goh KH (2005) Process evaluation for optimization of EDTA use and recovery for heavy metal removal from a contaminated soil. Chemosphere 58: 1031-1040.

3. Tang ZL, Li XH, Run HQ (2005) Problems and countermeasures of mine geologic environment. Chinese Journal of Earth Sciences and Environment 27:l-4.

4. Li YT, Becquer T, Dai J, Quantin C, Benedetti MF (2009) Ion activity and distribution of heavy metals in acid mine drainage polluted subtropical soils. Environ Pollut 157: 1249-1257.

5. Luo YM, Teng Y, Guo Y (2005) Soil remediation - a new branch discipline of soil science. Soils 37: 230-235.

6. Meers E, Ruttens A, Hopgood M, Lesage E, Tack FM (2005) Potential of Brassicrapa, Cannabis sativa, Helianthus annuus and Zea mays for phytoextraction of heavy metals from calcareous dredged sediment derived soils. Chemosphere 61: 561-572.

7. Sinegani AAS, Khalilikhab F (2010) Effects of EDTA, sheep manure extract, and their application time on Cd uptake by Helianthus annuus from a Calcareous mine soil. Soil&Sediment Contamination 19:378-390.

8. Turgut C, Bahahan I, Atatanir L, Cutright TJ (2010) Assessment of two new ligands for increasing the uptake of Cd, Cr and Ni in Helianthus annuus grown in a Sandy-Loam soil. Water Air and Soil Pollution 210:289-295.

9. Bianchi V, Masciandam G, Ceceanti B (2010) Phytoremediation and biophysical conditioning of dredged marine sediments for their re-use in the environment. Water Air and Soil Pollution 210:187-195.

10. Xie XD, Zhang GS (2001) Environmental significance of the intimation between minerals and microbes. Acta Petrol Mineral 20:382-386.

11. Giasson P, Jaouich A, Gagn S (2005) Phytoremediation of zinc and cadmium: A study of arbuscularmycorrhizal hyphae. Remediation Journal 15:113-122.

12. Sheng X, Sun L, Huang Z, He L, Zhang W, et al. (2012) Promotion of growth and Cu accumulation of bio-energy crop (Zea mays) by bacteria: implications for energy plant biomass production and phytoremediation. J Environ Manage 103: 58-64.

13. Kumar KV, Srivastava S, Singh N, Behl HM (2009) Role of metal resistant plant growth promoting bacteria in ameliorating fly ash to the growth of Brassica juncea. J Hazard Mater 170: 51-57.

14. Kumpiene J, Ore S, Lagerkvist A, Maurice C (2007) Stabilization of Pb- and Cu-contaminated soil using coal fly ash and peat. Environ Pollut 145: 365-373.

15. Lin C, Tong X, Lu W, Yan L, Wu Y, et al. (2005) Ecological degradation in an area affected by surface mining southern China. Land DegradDev 16:463-474.

16. Lin C, Wu Y, Lu W, Chen A, Liu Y (2007) Water chemistry and ecotoxicity of an acid mine drainage-affected stream in subtropical China during a major flood event. J Hazard Mater 142: 199-207.

17. Liu Y, Lin C, Ma Y, Lu W, Wu Y, et al. (2009) Toxic effects of two acid sulfate soils from the Dabaoshan Mine on Corymbiacitriodoravar.variegata and Daphnia carinata. J Hazard Mater 166: 1162-1168.

18. Bao SD (2000) Agricultural Soil Analysis. 3rd edn. China Agricultural Published, Beijing, China.

19. Wong MH (2003) Ecological restoration of mine degraded soils, with emphasis on metal contaminated soils. Chemosphere 50: 775-780.

20. Drake HL, Liu SJ (2008) Current and future perspectives on the environmental importance of microorganisms. In: Liu SJ, Drake HL (eds.), Microbes and the Environment: Perspectives and Challenges, Science Press, Beijing.

21. Li JZ (1981) Research progress and prospects on microbially produced alkaloids. Microbiology 1:30-36.

22. Glick BR (2010) Using soil bacteria to facilitate phytoremediation. BiotechnolAdv 28: 367-374.

23. Ma Y Prasad MN, Rajkumar M, Freitas H (2011) Plant growth promoting rhizobacteria and endophytes accelerate phytoremediation of metalliferous soils. BiotechnolAdv 29: 248-258.

24. Rajkumar M, Ae N, Prasad MN, Freitas H (2010) Potential of siderophore-producing bacteria for improving heavy metal phytoextraction. Trends Biotechnol 28: 142-149.

25. Hardoim PR, van Overbeek LS, Elsas JD (2008) Properties of bacterial endophytes and their proposed role in plant growth. Trends Microbiol 16: 463-471.

26. Alvarenga P, Palma P, Gonçalves AP, Fernandes RM, de Varennes A, et al. (2009) Organic residues as immobilizing agents in aided phytostabilization: (II) effects on soil biochemical and ecotoxicological characteristics. Chemosphere 74: 1301-1308.

27. Wang X, Zhou QX (2004) Interaction between heavy metals and soil microorganisms and remediation of contaminated soils. Techniques and Equipment for Environmental Pollution Control 5:1-5 (In Chinese).

# 6
CHAPTER

# REMOVAL OF MERCURY IN FIXED-BED CONTINUOUS UPFLOW REACTORS BY MERCURY-RESISTANT BACTERIA AND EFFECT OF SODIUM CHLORIDE ON THEIR PERFORMANCE

Jaysankar De[1,2,4], Johannes Leonha¨user[2,5], Lilit Vardanyan[3,4]

[1]National Institute of Oceanography, Dona Paula, Goa 403004, India

[2]Division of Microbiology, German Research Centre for Biotechnology, Mascheroder Weg 1, 38124 Braunschweig, Germany

[3]Vanevan Institute, 14 Yerevanyan Street, Martuni, Gegharkunik, Armenia

[4]UNESCO Chair-Life Sciences International Postgraduate Educational Center. Acharian 31, Yerevan 0040, Armenia

[5]Bayer Technology Services GmbH; PT-PT-Environmental & Solids Processing Building E 41; 51368 Leverkusen, Germany

## ABSTRACT

**Urgent need to reduce the amount of toxic mercury compounds in the wastewater of industries and subsequent reuse of metal ions, has led to an increasing interest in microbial bioremediation. Two Pseudomonas aeruginosa strains, namely, isolate $CH0_7$ and isolate Bro12, and a genetically engineered strain P. putida ($KT_{2442}$ $_{mer::73}$) were used to study the kinetics of mercury removal from liquid M9 medium, considering the potential of the bacteria in volatilizing ionic mercury to its gaseous form. The P. aeruginosa strains were further used to remove toxic mercury from synthetic wastewater in fixed-bed, continuous upflow reactors and thereafter to recover the toxic metal from the reactor beds. We also studied the effect of sodium chloride on the kinetics of mercury removal by the isolate CH07 from marine sediment, as well as the other two non-marine bacteria. After a successful run of over a month, the bioreactors were able to retain the toxic metal, which resulted in a recovery of approximately 64% of the influent mercury. No major alteration in the retention capacity of the bioreactors occurred during drastic changes in concentration of inflowing metals or salt concentration.**

**Citation:** Jaysankar De, Johannes Leonha¨user, Lilit Vardanyan, Removal of mercury in fixed-bed continuous upflow reactors by mercury-resistant bacteria and effect of sodium chloride on their performance, doi.org/ 10.5339/connect.2014.17.

## INTRODUCTION

There are several practical reasons for selectively separating all types of heavy metal ions from aqueous media. A few examples include the remediation of hazardous or radioactive wastes, the remediation of contaminated groundwater, and recovery of precious and/or toxic metals from industrial processing solutions. In modern applications of available techniques, recovery and re-use of the extracted material is becoming more important. This is being driven by tougher environmental regulations, high initial costs of new, more effective and selective extractants, and the need to minimize the volume of waste destined for permanent disposal. Mercury (Hg), deriving from atmospheric deposition, erosion, urban discharges, agricultural materials, mining, and combustion/industrial discharges, including chloralkali, paper and pulp, oil refinery, paint, pharmaceutical and battery manufacturing and industrial wastewater discharges etc., present an ongoing and serious threat to human health and natural water quality.1–3

There are huge variations in the level of toxic mercury in the biosphere and almost no physiological roles for this metal have been documented. Several bacteria have adapted to this highly toxic metal and have developed systems for metabolizing and utilizing the built-in energy of mercuric compounds to drive their own biosynthetic processes.4,5 Mechanisms of microbial metal resistance include the precipitation of metals as phosphates, carbonates and sulfides; metal volatilization by methyl or ethyl group addition; physical exclusion by electronegative components in membranes and extracellular polymeric substances; energy-dependent metal efflux systems; and intracellular sequestration with low molecular weight, cysteine-rich proteins.6–8 Comparatively lower costs and higher efficiency at low metal concentrations make biotechnological processes very attractive in comparison with physicochemical methods for heavy metal removal.9 Initially interest focused on technologies that could be applied to achieve in situ immobilization of metals, later on these shifted towards the actual metal removal, as it is difficult to guarantee metals will remain immobilized indefinitely.10 The need to reduce mercury ions in wastewater streams of industries and its subsequent reuse, has led to an increasing interest in microbial remediation.3,6,7 Enzymatic reduction of $Hg^{2+}$ to $Hg^{o}$ has been identified as the major bacterial mechanism to deal with Hg toxicity.11 It has been realized that the mer operon that confers both resistance and detoxification capabilities to its possessor, is almost universally distributed in resistant bacterial populations.5,12 High concentrations of sodium chloride (NaCl), causing less bioavailability of metal usually inhibit the volatilization of mercury by bacteria. It has been shown that NaCl interferes with the activity of mercuric reductase, probably due to the formation of mercuro-chloro complexes.13 Therefore, it was necessary to determine the efficiency of microbial mercury removal in the presence of different salt concentrations. This was important as the isolate CH07 is from a marine environment, whereas the other bacteria, being non-marine, would prefer lower salt concentrations.

Mercury removal processes utilize physical and chemical approaches that involve either trapping or collecting mercury from the contaminated sites, or the chemical precipitation of mercuric compounds. Such processes are costly and often leave behind hazardous by-products. Remediation technologies based on mercury volatilization have barely been explored beyond laboratory scale as collecting this toxic metal is technically tedious and expensive.14 Removal of mercury in a laboratory test reactor using mercury-resistant bacteria was first reported nearly three decades ago.15 However, serious consideration towards the bioremediation of this metal started in the 1990s.16 Several researchers17,18 demonstrated that the elemental mercury formed could be retained in a packed bed bioreactor, consisting of inert porous carrier material that was covered by a biofilm of Mercury-Resistant Bacteria (MRB). The aim of this study was to investigate the reduction potential of three mercury-resistant strains of Pseudomonas putida (one isolate) and Pseudomonas aeruginosa (two isolates) to high $Hg^{2+}$ loads. These strains of bacteria were tested for their capacity in remediation and retention of mercury from artificial wastewater in fixed-bed continuous upflow bioreactors. We also studied the effect of NaCl on mercury removal rate by the mercury-resistant bacteria.

## MATERIALS AND METHODS

### Bacterial Cultures

Two P. aeruginosa isolates namely, isolate $CH0_7$ and isolate Bro12, and a genetically engineered strain P. putida $KT_{2442::mer-73}$ were used in the experiments. The P. aeruginosa $CHO_7$ was isolated from marine sediment and has been deposited with accession number: NRRL B-30604.3 The other strain P. aeruginosa Bro12 was isolated from chloralkali wastewater and has been described elsewhere.19 The P. putida $KT_{2442::mer73}$ was a genetically engineered strain, which had a high and constitutive expression of the mer genes.20 The bacteria were grown overnight (16 hrs) in modified M9 minimal medium (10x salts containing $Na_2HPO_4$.7H2O, 70 g; $KH_2PO_4$, 30 g; NaCl, 25 g; $NH_4Cl$, 10.0 g; distilled water 1 litre) amended with trace elements (MgO, 10.75 g; $CaCO_3$, 2.0 g; $ZnSO_4$, 1.44 g; $MnSO_4$, 1.12 g; $CuSO_4$, 0.25 g; $CaSO_4$, 0.28 g; $H_3BO_3$, 0.06 g; MgSO4, 120 g; FeSO4.7H2O, 10 g; conc. HCl, $_{51.3}$ ml; distilled water, 1 litre), glucose (4 g/l) and 1 ppm (mg/l) $HgCl_2$ at 308C, on a rotary shaker. The "preculture" was made by adding this overnight grown culture to fresh medium and allowing the bacteria to grow to a desired cell density in the exponential phase. This was examined by frequent measurements of $OD_{600}$. For reliable comparison between the experimental runs, care was taken to transfer the preculture at a specific time to ensure that the bacterial cells always came from the growth phase.

### Kinetics of Mercury Removal

The set up for an online system to measure the kinetics of mercury transformation by bacteria is shown in Figure 1. It comprised a Cold Vapour Atomic Absorption Spectrophotometer (CVAAS), a three-reaction vessels-assembly, one magnetic stirrer, a water bath (with temperature control device), and a motor with timer and channels for carrying air and mercury vapour to the AAS. Gaseous mercury ($Hg^{o}$ ) was generated by enzymatic transformation caused by the MRB and the accumulated gaseous mercury was blown out into the AAS by airflow through the reaction vessel. A magnetic stirrer was used to mix the liquid bacterial culture and M9 medium amended with mercury.

The reaction vessels had a volume of 50 ml and had a screw cap lid. Each lid had a plastic tube for inflow and outflow of air and a needle to couple the syringe. Direction of airflow was always from the reaction vessels directly into the AAS. Air tubes had unidirectional valves above the reaction vessels so that there was no back-flow into the vessels. Therefore, the reaction vessel was airproof when the inflow valve was closed. The whole experiment was carried out at room temperature of approximately 25°C.

## *Mercury Biotransformation Assay*

For this assay to biotransform mercury, 20 ml of M9 minimal medium (including 4g/l glucose as carbon source) was prepared with defined mercury concentrations (for example 1, 2, 5, 8, 10, 15, 20 ppm). Mercury transformation was allowed to take place in 3 parallel reaction vessels, where 600ml of culture containing approximately the same cell numbers was added to 5.4 ml M9 medium, amended with defined concentrations of mercury. A magnetic stirrer in each reaction vessel allowed continuous mixing of these solutions. Effects of NaCl concentration on mercury removal were checked with all three isolates in the experiment at a fixed mercury concentration of 8 ppm. Varying NaCl concentrations in a series of 5 g/l to 30 g/l were used for this experiment. A standard solution of 1000 ppm $Hg(NO_3)_2$ was used to prepare different concentrations to derive a calibration curve for mercury throughout the experiment.

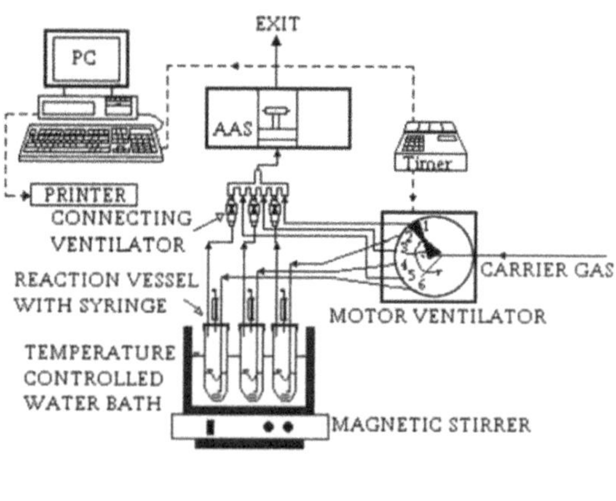


**Figure 1:** Apparatus for measuring mercury removal kinetics.

## *Calculation*

Transformed gaseous mercury ($Hg^o$) was blown out to the CVAAS, which was detected as a distinct output peak corresponding to the amount of Hg0 . The highest peak area resulting from each vessel was taken into consideration and the background level was subtracted from the last cycle of the preblowouts to obtain the abstract values. This value was calculated based on the calibration curve into the total amount of mercury in nanograms (ng) per cycle. The kinetics of mercury biotransformation (n) was expressed as $Hg^{2+}$/cell/min.

$\nu = Hg^{2+}(ng)/0.6\ ml\ culture * cells\ (no./ml) * 3.2\ minutes = Hg^{2+}(ng)/cell/minute.$

## Remediation of Mercury by Mercury-Resistant Bacteria in Bioreactors

### *Reactor Instillation*

The apparatus consisted of six reactors (numbered 1–6) arranged in parallel, with pairs of the reactor vessels containing similar bacterial inocula (reactors 1 & 2 with strain $CH0_7$, 3 & 4 with strain $Bro_{12}$, and 5 & 6 with mixed cultures of both strains). Each reactor unit consisted of a reactor bed filled with pumice stones (80 ml measured by water displacement), one side-arm glass tube containing provisions for inlet and outlet, along with a bubble trap, and one glass tube for inoculating the reactor with liquid bacterial culture. The complete set up (photograph 1) was steam sterilized by autoclaving and was placed inside a fume hood with provisions for disposal of waste effluent.

### Reactor Function

The sterilized reactors were operated to measure the efficiency of the bacteria, either singly or in combination, for removal of mercury. The reactors were all connected to individual medium reservoirs and one common wastewater reservoir via separate routes. Filter (0.2 m) sterilized solution of mercury salt was added to the wastewater under sterile conditions. The mercury-containing waste (with different concentrations of mercury viz., 1, 2, 3.32, 4, 5.08, 6 and 8 ppm containing different salt concentrations ranging from 8 g/l to 16 g/l) and duly sterilized medium (modified M9 medium supplemented with 4 g/l glucose) were pumped into the reactors in an up-flow mode at a controlled speed. Mechanical disturbances, such as the blockage of pipes or the occurrence of bubbles inside the bioreactors causing disturbances to the functioning of the reactors, were corrected periodically. Mercury in the inflow and outflow was measured using CVAAS.

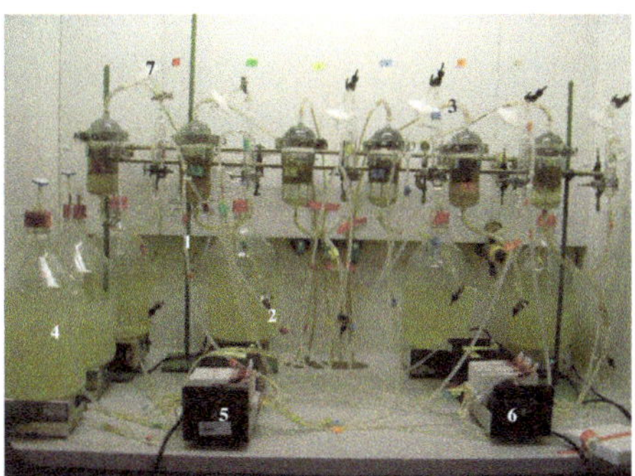


**Photograph 1:** Bioreactor set up showing different reactor vessels and connections to influent and effluent Hg flows and other accessories. 1, channel for medium; 2, channel for wastewater inflow; 3, channel for wastewater outflow; 4, medium; 5, pump for medium; 6, pump for wastewater; 7, bubble trap.

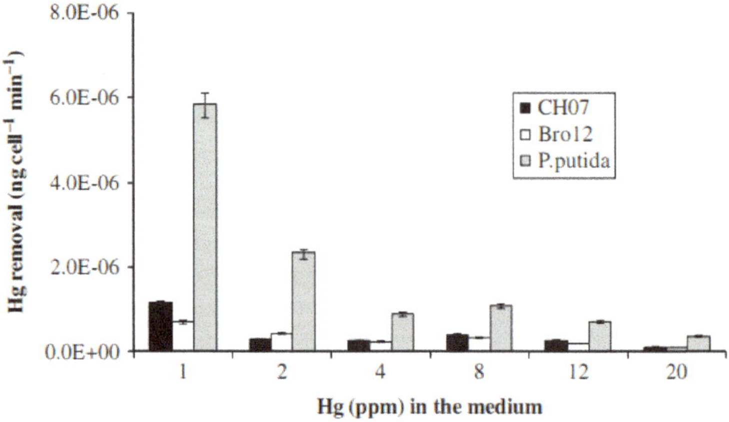


**Figure 2:** Rates of mercury removal by P. aeruginosa $CH_{07}$, P. putida $KT_{2442\ mer::73}$, and P. aeruginosa $Bro_{12}$.

### *Mercury Recovery*

After the completed run the bioreactors were disassembled and pumice stones collected to recover the transformed metallic and thus, trapped, mercury. The pumice stones from each reactor were separately immersed in a solution of concentrated $HNO_3$ and $H_2SO_4$ (1:1) for 48 hrs. Stones were washed with deionized water 5 times, and the washings were collected and pooled together. Concentration of mercury from the washings was measured by CVAAS.

## RESULTS

### Kinetics of Mercury Removal

All three isolates were able to remove mercury by means of volatilization from the M9 assay medium containing up to 20 ppm mercury. The mercury removal rate was the highest in the reaction with 1 ppm $Hg^{2+}$, though the efficiency was relatively good up to 8 ppm $Hg^{2+}$, in the normal M9 medium (Figure 2). However, with all mercury concentrations tested, the P. putida strain achieved the highest removal rate when compared to the other two strains, possibly due to its constitutive expression of the mer genes and higher enzyme production. The removal rate at 1 ppm mercury was $1.16 \times 10^{-6}$ ng/cell/min for $CH0_7$, whereas the genetically engineered strain removed mercury at a rate four times higher, $5.84 \times 10^{-6}$ ng/cell/min. The trend in removal magnitude changed with increasing concentrations of NaCl (Figure 3). The marine pseudomonas isolate CH07 was most effective at the salt concentration of 30 g/l. The $Bro_{12}$ strain was most efficient at the salt concentration of 20 g/l NaCl, whereas the P. putida had its best removal potential at moderate salt concentration peaking at 12 g/l NaCl.

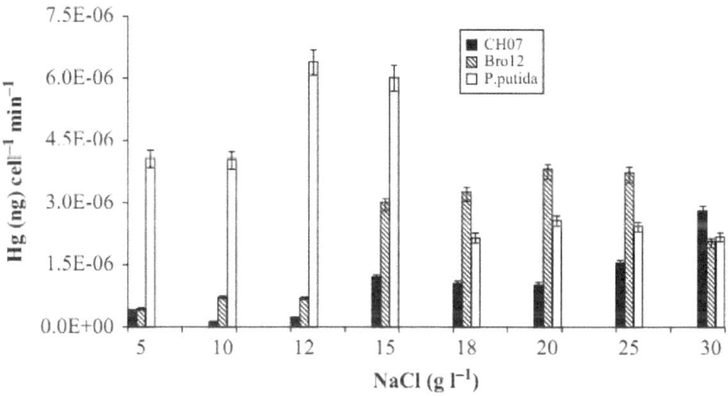


**Figure 3:** Effect of NaCl on mercury removal rates by different mercury-resistant bacteria.

## Remediation of Mercury by Mercury-Resistant Bacteria in Bioreactors

Both P. aeruginosa strains (isolate CH07 and Bro12) reduced mercury concentrations in the bioreactors relatively efficiently over a month (Figure 4). An average of 204.33 mg mercury was passed through each reactor, totaling 1225.99 mg mercury during this operation (Table 1). A total of 786.74 mg was recovered from the pumice stones. The retention and recovery of mercury in one of the replicates (reactor 2) was very high (>95%) with isolate CH07, whereas the corresponding data with isolate Bro12 were moderate, with a recovery of 56 and 60% metallic mercury in two respective vessels. The two reactors run with mixed cultures of isolates CH07 and Bro12, resulted in a slightly higher recovery of mercury, amounting to 60 and 66% mercury, respectively from reactors 5 and 6 (Table 2). Mechanical disturbances, such as pipe blockage, bubbles in the reactor, differential flow rate, effected mercury retention in the bioreactor by mercury resistant bacteria, however the two strains were able to retain 42 to >.95% (totaling 64%) of the influent mercury. Within the concentration ranges studied, the functioning of the reactors was not markedly affected by the change in mercury or NaCl levels. There were some large deviations in the outflow mercury concentrations at times probably due to bubbles in the pipes that may have blocked the flow of medium, as well as wastewater.

## DISCUSSION

Technologies for treating mercury-polluted environments have been a major concern over the last couple of decades.6,18,8 One of the initial efforts to retain mercury in bioreactors using bacterial strains was made by Brunke et al.20 Chang and Law21 developed a detoxification process using P. aeruginosa PU21 in batch, fed-batch and continuous bioreactor systems. Reduction of $Hg^{2+}$ by mercury-resistant bacteria, one of the best mechanisms for its removal from chloralkali waste, has been demonstrated by Canstein et al.19,22 using several P. putida strains in which nearly 97% of the mercury was recovered from waste containing 3– 10 ppm mercury. In such bioreactors, outflow concentrations of mercury did not depend on inflow mercury concentrations.14 In the bioreactors run with isolates CH07 and Bro12, more than 60% influent $Hg^{2+}$ rapidly accumulated. Though a consortium of different bacterial strains have been reported to be far more effective,19 combining CH07 and Bro12 in this study did not make a significant difference. This might imply that the compatibility of the consortium-candidates is of relevance. Both isolates belonging to the same species might have competed for similar substrate and/or niche in the reactors, instead of sharing the microenvironment.

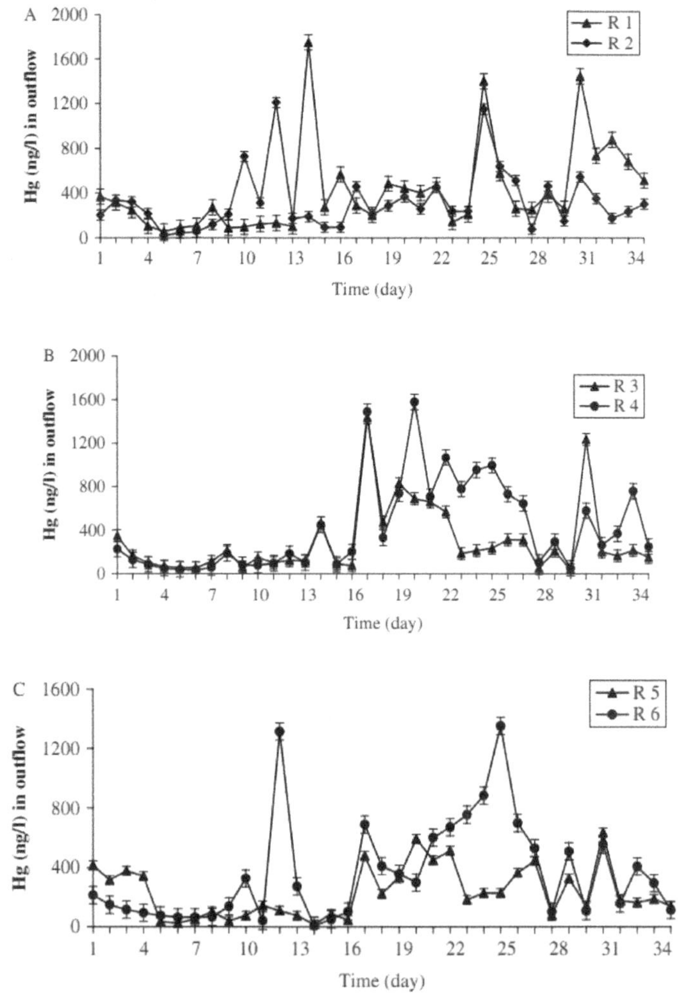


**Figure 4:** Mercury concentration in waste from reactors using isolate $CH0_7$ (A; $R_1$ & $R_2$), and isolate $Bro_{12}$ (B; $R_3$ & $R_4$), and mixed cultures of both isolates (C; $R_5$ & $R_6$).

**Table 1:** Mercury retention in the reactors

| Days of operation | Hg inflow | | $^a$Hg retention rate in the reactors (μg/hr) | | | | | | |
|---|---|---|---|---|---|---|---|---|---|
| | Conc. (ppb) | $R^b$ (μg/hr) | $r^c$-1 | r-2 | r-3 | r-4 | r-5 | r-6 | NaCl (g/l) |
| 1–7 | 1000 | 80 | 65.35 | 66.47 | 69.78 | 72.98 | 62.17 | 71.15 | 8 |
| 8–11 | 2000 | 160 | 148.48 | 132.72 | 149.68 | 151.28 | 152.72 | 148.56 | 8 |
| 12–16 | 4000 | 320 | 274.82 | 291.78 | 306.05 | 303.65 | 314.62 | 292 | 8 |
| 17–22 | 3320 | 265.6 | 235.06 | 238.53 | 203.63 | 186.8 | 231.01 | 225.2 | 8 |
| 23–26 | 4000 | 320 | 273.4 | 274.72 | 301.16 | 250.96 | 299.92 | 246.12 | 8 |
| 27–28 | 6000 | 480 | 459.68 | 456.48 | 465.68 | 450.08 | 458.96 | 454.72 | 8 |
| 29–30 | 8000 | 640 | 614.32 | 615.6 | 630.24 | 626.24 | 621.28 | 615.36 | 8 |
| 31–32 | 8000 | 640 | 552.8 | 604.24 | 582.48 | 606.4 | 607.52 | 611.44 | 4 |
| 33 | 4000 | 320 | 249.76 | 249.76 | 249.76 | 249.76 | 249.76 | 249.76 | 8 |
| 34–35 | 4000 | 320 | 272.24 | 298.64 | 305.44 | 279.68 | 306.88 | 303.72 | 16 |

[a]Reactors 1 & 2 were inoculated with *Pseudomonas aeruginosa* CH07, 3 & 4 with *Pseudomonas aeruginosa* Br012, and 5 & 6, with the mix culture of CH07 and Br012. [b]R, rate; [c]r, reactor.

Biosorption is another biological method, which involves adsorption of metals into the biomass, such as algal or bacterial (either dead or alive), has been inexpensive and promising.23,24 Biosorption using natural25,8 or recombinant microbial biomass26 has been experimentally successfully. Essa et al.27 reported three mechanisms of mercury detoxification of wastewater in Klebsiella pneuomoniae M426. These mechanisms are enzyme-mediated reduction, aerobic precipitation of ionic $Hg^{2+}$ as insoluble HgS, as a result of H2S production, and biomineralization of $Hg^{2+}$ as an insoluble mercury-sulfur complex other than HgS. However, under aerobic condition HgS is methylated to the considerably toxic methylmercury, for this reason anaerobic treatment of mercury-contaminated matrices require extreme safety measures. Recently, it has been claimed that the process "In Situ Mercury Stabilization" (ISMS) can treat and remove mercury contamination from the ground in a cost-effective manner.28

Kiyono et al.29 successfully used bioaccumulation as a measure of bioremediation of Hg using mer-ppk fusion plasmid. They were able to remove mercury at low concentrations (2–4 ppm), but the efficiency was not as good at higher mercury concentrations i.e., 8– 16 ppm, due to inactivation of viable cells inside the alginate beads. Chen and Wilson30 constructed a genetically engineered E. coli strain for simultaneously expressing an $Hg^{2+}$ transport protein, as well as metallothionein, to improve the limitation of trans-membrane mercury transport. This was effective at low concentrations of mercury as its capacity for $Hg^{2+}$ accumulation was limited by the number of cellular metallothionein. The meroperon has been expressed in the radio-resistant Deinococcus radiodurans, which tolerates extremely high dosages of radiation and could be useful for cleaning up sites contaminated with radioactive wastes.31 Phytoremediation expressing merA and merB genes in plants (especially rhizospheres), have been investigated for the cleanup of contaminated soils.32,33 A genetically engineered rice plant34 has been used for the bioremediation of mercury. However, the volatile mercury ($Hg^0$) produced this way is released into the atmosphere directly, which is again a matter of concern.

**Table 2:** Total mercury recovered from the reactors inoculated with isolates Pseudomonas aeruginosa $CH0_7$ (in reactors 1 & 2), Pseudomonas aeruginosa $Bro_{12}$ (in reactors 3 & 4) and of both isolates (in reactors 5 & 6)

| Reactor | Total Hg (mg) inflow | Total Hg (mg) recovered | % Recovered |
|---|---|---|---|
| 1 | 192.65 | 81.54 | 42.32 |
| 2 | 209.62 | 201.12 | 95.94 |
| 3 | 211.03 | 119.72 | 56.73 |
| 4 | 201.72 | 122.5 | 60.73 |
| 5 | 214.95 | 129.62 | 60.31 |
| 6 | 196.02 | 132.24 | 66.44 |

Results from the present study clearly demonstrate the possibilities of employing mercury-resistant bacteria for mercury detoxification and recovery of this costly metal under special circumstances. Mercury bioremediation using the resistant bacteria stated here, can principally be applied to bioremediation of most kinds of mercury contamination. Further, leads from this study can be explored to adapt these bacteria for bioremediation under various other situations.

## ACKNOWLEDGEMENTS

De acknowledges CSIR-SRF grants 31/26/75/2002 EMR-I, and UGC-DAAD short-term scholarship.

## REFERENCES

1. Nascimento AMA, Chartone-Souza E. Operon mer: bacterial resistance to mercury and potential for bioremediation of contaminated environments. Genet Mol Res. 2003;2:92– 101.
2. Clarkson TW, Laszlo M. The toxicology of mercury and its chemical compounds. Crit Rev Toxicol. 2006;36(8):609–662.
3. De J, Ramaiah N, Bhosle NB, Garg A, Vardanyan L, Nagle VL, Fukami K. Potential of mercury-resistant marine bacteria for detoxification of chemicals of environmental concern. Microbes Environ. 2007;22(4):336–345.
4. Wang Q, Kim D, Dionysiou DD, Sorial GA, Timberlake D. Sources and remediation for mercury contamination in aquatic systems-a literature review. Environ Pollut. 2004;131(2):323–336.
5. Mindlin S, Minakhin L, Petrova M, Kholodii G, Minakhina S, Gorlenko Z, Nikiforov V. Present-day mercury resistance transposons are common in bacteria preserved in permafrost grounds since the Upper Pleistocene. Res Microbiol. 2005;156(10):994–1004.
6. Barkay T, Wagner-Do¨bler I. Microbial transformations of mercury: potentials, challenges, and achievements in controlling mercury toxicity in the environment. Adv Appl Microbiol. 2005;57:1–52.
7. De J, Ramaiah N, Vardanyan L. Detoxification of toxic heavy metals by marine bacteria highly resistant to mercury. Mar Biotechnol (NY). 2008;10(4):471–477.

8. De J, Dash HR, Das S. Mercury pollution and bioremediation–a case study on biosorption by a mercury-resistant marine bacterium. Microbial Biodegradation and Bioremediation. Elsevier Inc.; 2014:137–166. DOI: 10.1016/B978-0- 12-800021-2.00006-6.
9. Gadd GM, White C. Microbial treatment of metal pollution–a working biotechnology? Trends Biotechnol. 1993;11(8):353–359.
10. De J, Ramaiah N. Characterization of marine bacteria highly resistant to mercury exhibiting multiple resistance to toxic chemicals. Ecol Indic. 2007;7(3):511–520.
11. Summers AO, Silver S. Mercury resistance in a plasmid-bearing strain of Escherichia coli. J Bacteriol. 1972;112(3):1228–1236.
12. Kannan SK, Krishnamoorthy R. Isolation of mercury resistant bacteria and influence of abiotic factors on bioavailability of mercury – a case study in Pulicat Lake North of Chennai, South East India. Sci Total Environ. 2006;367:341–353.
13. Barkay T, Gillman M, Turner RR. Effects of dissolved carbon and salinity on bioavailability of mercury. Appl Environ Microbiol. 1997;63(11):4267–4271.
14. Wagner-Do¨bler I. Pilot plant for bioremediation of mercury-containing industrial wastewater. Appl Microbiol Biotechnol. 2003;62(2-3):124– 133.
15. Williams JW, Silver S. Bacterial resistance and detoxification of heavy metals. Enzyme Microb Technol. 1984;6(12):530–537.
16. Summers AO. The hard stuff: metals in bioremediation. Curr Opin Biotech. 1992;3(3):271–276.
17. Deckwer WD, Becker FU, Ledakowicz S, Wagner-Do¨bler I. Microbial removal of ionic mercury in a three-phase fluidized bioreactor. Environ Sci Technol. 2004;38(6):1858– 1865.
18. Głuszcz P, Zakrzewska K, Wagner-Do¨bler I, Ledakowicz S. Bioreduction of ionic mercury from wastewater in a fixed-bed bioreactor with activated carbon. Chem Pap. 2008;62(3):232–238.
19. Canstein VH, Li Y, Timmis KN, Deckwer WD, Wagner-Do¨bler I. Removal of mercury from chloralkali electrolysis wastewater by a mercury-resistant Pseudomonas putida strain. Appl Environ Microbiol. 1999;65(12):5279–5284.
20. Brunke M, Deckwer WD, Fritschmuth JM, Horn H, Lunsdorf M, Rhode M, Rohricht M, Timmis KN, Weppen P. Microbial retention of mercury from waste systems in a laboratory column containing merA gene bacteria. FEMS Microbiol Rev. 1993;11:45–52.
21. Chang JS, Law WS. Development of microbial mercury detoxification processes using mercury-hyperresistant strain of Pseudomonas aeruginoasa PU21. Biotechnol Bioeng. 1998;57(4):462–470.
22. Canstein VH, Li Y, Wagner-Do¨bler I. Long-term performance of bioreactors cleaning mercury-contaminated wastewater and their response to temperature and mercury stress and mechanical perturbation. Biotechnol Bioeng. 2001;74(3):212–219.

23. Chang JS, Hong J. Biosorption of mercury by the inactivated cells of Pseudomonas aeruginosa PU21 (Rip64). Biotechnol Bioeng. 1994;44(8):999– 1006.

24. Mulligan CN, Yong RN, Gibbs BF. Remediation technologies for metal-contaminated soils and groundwater: an evaluation. Eng Geol. 2001;60(1-4):193–207.

25. Volesky B, Holan ZR. Biosorption of heavy metals. Biotechnol Progr. 1995;11(3):235–250.

26. Pazirandeh M, Chrisey LA, Mauro JM, Campbell JR, Gaber BP. Expression of the Neurospora crassa metallothionein gene in Escherichia coli and its effect on heavy-metal uptake. Appl Microbiol Biotech. 1995;43(6):1112– 1117.

27. Essa AM, Macaskie LE, Brown NL. Mechanisms of mercury bioremediation. Biochem Soc Trans. 2002;30(4):672–674.

28. Fuhrmann M, Heiser J, Kalb PD. Brookhaven Science Associates, Llc. Mercury contamination extraction. Patent US7589248 B2. 2009. Sep.

29. Kiyono M, Omura H, Omura T, Murata S, Hou-Pan H. Removal of inorganic and organic mercurials by immobilized bacteria having mer-ppk fusion plasmids. Appl Microbiol Biotech. 2003;62(2-3):274–278.

30. Chen S, Wilson DB. Genetic engineering of bacteria and their potential for Hg2þ bioremediation. Biodegradation. 1997;8(2):97–103.

31. Brim H, McFarlan SC, Fredrickson JK, Minton K, Zhai M, Wackett LP, Daly MJ. Engineering Deinococcus radiodurans for metal remediation in radioactive mixed waste environments. Nat Biotechnol. 2000;18:85–90.

32. Bizily SP, Rugh CL, Summers AO, Meagher RB. Phytoremediation of methylmercury pollution: merB expression in Arabidopsis thaliana confers resistance to organomercurials. Proc Nat Acad Sci USA. 1999;96(12):6808–6813.

33. Rugh CL, Wilde HD, Stack NM, Thompson DM, Summers AO, Meagher RB. Mercuric ion reduction and the resistance in transgenic Arabidopsis thaliana plants expressing a modified bacterial mer A gene. Proc Nat Acad Sci USA. 1996;93(8):3182–3187.

34. Heaton AC, Rugh CL, Kim T, Wang NJ, Meagher RB. Toward detoxifying mercury-polluted aquatic sediments with rice genetically engineered for mercury resistance. Environ Toxicol Chem. 2003;22(12):2940–2947.

# 7
CHAPTER

# BIOREMEDIATION OF HEXAVALENT CHROMIUM BY A CYANOBACTERIAL MAT

Dhara Shukla, Padma S. Vankar, Sarvesh Kumar Srivastava

204 'A', Facility for Ecological and Analytical Testing IIT, Kanpur, India

## ABSTRACT

**The study comprises the use of cyanobacterial mat (collected from tannery effluent site) to remove hexavalent chromium. This mat was consortium of cyanobacteria/ blue-green algae such as *Chlorella* sp., *Phormidium* sp. and *Oscillatoria* sp. The adsorption experiments were carried out in batches using chromium concentrations 2–10, 15–30 and 300 ppm at pH 5.5–6.2. The adsorption started within 15 min; however, 96 % reduction in metal concentration was observed within 210 min. The adsorption phenomenon was confirmed by Fourier transform–infrared spectroscopy and energy dispersive X-ray analysis. This biosorption fitted Freundlich adsorption isotherm very well. It was observed that the best adsorption was at 4 ppm, and at 25 ppm in the chosen concentration ranges. Scanning electron micrograph showed the physiology of mat, indicating sites for metal uptake. The main focus was collection of the cyanobacterial mat from local environments and its chromium removal potential at pH 5.5–6.2.**

## INTRODUCTION

The discharge of chromium (VI) into aquatic ecosystems has become a matter of concern in all the tannery areas over the last few decades. This pollutant is introduced into the aquatic systems significantly from the chrome tanning effluents of leather processing units. In this process about 60–70 % of chromium reacts with the hides and about 30–40 % of chromium remains in the solid and liquid wastes (especially as spent tanning solutions). Chromium (VI) has been designated as the priority pollutant by United States Environment Protection Act (Srinath et al. 2002; Laxman and More 2002).

**Citation:** Dhara Shukla, Padma S. Vankar, Sarvesh Kumar Srivastava, Bioremediation of hexavalent chromium by a cyanobacterial mat, DOI 10.1007/s13201-012-0044-3.

Conventionally, chromium (VI) containing industrial effluent is treated by physico-chemical methods such as reduction, precipitation, ion exchange, reverse osmosis and electrodialysis. However, it has been observed that these processes are costly and unreliable (Camargo et al. 2005).

Presently, biotechnological methods are becoming attractive alternatives to remove toxicants from effluents. The ability of some microorganisms especially cyanobacteria to tolerate and interact with metal/chromium ions makes them attractive substrate for environmental biotechnology (Cervantes et al. 2001). Cyanobacteria-based bioremediation technologies have received recent attention as strategies to clean up contaminated soil and water. The microbial mats containing several microbial species are complex systems, but require few external inputs to be exploited for bioremediation. Several workers have shown the use of different biomaterials such as non-living bacteria, microalgae, fungi, seaweed and even agricultural byproducts, which can retain relatively high quantities of metal ions by passive sorption and/or by complexation mechanism.

It has been shown that the short-term uptake/accumulation of chromium (VI) by interactions with cyanobacteria, *Anabaena variabilis and Synechococcus* PCC 6301, using their ability to sequester chromium (Garnham and Green 1995).

Aksu and Balibek (2007) showed the biosorption of chromium (VI) by dried *Rhizopus arrhizus* and its effect on addition of salt (NaCl). Arica et al. (2005) demonstrated biosorption of chromium (VI) from aqueous solutions using free and immobilized biomass prepared from *Lentinus sajor-caju* and studied the kinetic characterization. Thus, fungal species are also efficient in removing chromium metal. Biosorption of the chromium ion chromium (VI) onto the cell surface of non-pathogenic species of *Trichoderma* fungal species in aerobic condition at pH 5.5 gave 97.39 % reduction (Vankar and Srivastava2008). Arica and Bayramoglu (2005) also showed the utilization of native, heat and acid-treated microalgae *Chlamydomonas reinhardtii* for biosorption of chromium (VI) ions. Deepa et al. (2006) have shown the sorption of chromium (VI) from dilute solutions and wastewater by live and pretreated biomass of *Aspergillus flavus*. Similarly, Zhou et al. (2007) have demonstrated the kinetic and equilibrium studies of chromium (VI) biosorption by dead *Bacillus licheniformis* biomass.

In the present study, the work has been carried out using a cyanobacterial mat (collected from tannery effluent disposal site) to remove chromium (VI) from tannery effluent at relatively higher pH, i.e., 5.5–6.2. The consortium of bacteria present was *Chlorella* sp.,*Phormidium* sp. and *Oscillatoria* sp. (cyanobacteria/blue-green algae). The main focus was hexavalent chromium removal potential of cyanobacterial mat collected from the local environment at relatively higher pH 5.5–6.2.

This is the first report where chromium (VI) removal has been carried out at relatively higher pH so efficiently, although it is known that at low pH (0–2) the biosorption of chromium (VI) is more predominant than bioreduction of chromium (VI) to (III). On the other hand, at higher pH such as 3.0 and above, less of chromium (III) species were known to be produced due to the sharp decrease in both biosorption and bioreduction processes (Han et al. 2008). Some researchers have reported that the optimal pH for total chromium biosorption was around 2–3. The optimal pH for *Sphagnum*-moss peat, leaf, mould and coconut-husk fiber were reported to be 1.5, 2.0, 2.05 and 2.0, respectively while 2–2.5 was the optimal pH reported for *Sargassum* (Kratochvil et al. 1998; Cabatingan et al. 2001).

Seven exopolysaccharide-producing cyanobacteria were tested with regard to their capability to remove Cr(VI) from the wastewater of a plating industry. The cyanobacterium which showed, under lab conditions, the most promising features with regard to both Cr(VI) removal (about 12 mg of Cr(VI) removed per gram of dry biomass) and growth characteristics (highest growth rate and simplest culture medium) was Nostoc PCC7936 (Colica et al. 2010).

Very recently a study exploring the utilization of waste cyanobacterial biomass of *Nostoc linckia* from a lab-scale hydrogen fermentor for the biosorption of Cr(VI) from aqueous solution has been reported (Sharma et al. 2011). The biomass immobilized in alginate beads was used for removal of the metal in batch mode optimizing the process conditions adopting response surface methodology (RSM). Kinetic studies were done to get useful information on the rate of chromium adsorption onto the cyanobacterial biomass, which was found to follow pseudo second-order model.

## EXPERIMENTAL

### Cyanobacterial Mat

The cyanobacterial mat was obtained from a disposal site near tannery sludge in Jajmau tannery area in Kanpur. The Tannery effluent was of strongly acidic nature (pH 3–4) with TDS ranging from 850 to 1,000 and salinity toward 600–700 PPT. The mat contained three species of bacteria, i.e., *Chlorella sp. Phormidium* sp. and *Oscillatoria* sp. as identified by Dr R.D. Tripathi, Taxonomer, National Botanical Research Institute, Lucknow.

### Reagents

Metal stock solution (1,000 ppm) was prepared by potassium dichromate salt of Merck AR grade and deionized water (DI). It was then diluted to required concentration 2–10, 15–30 and 300 ppm by DI. The standard calibration was also carried out.

### Biosorption Methodology

In this study, batch biosorption experiments were carried out in triplicates for determining the biosorption capacity of cyanobacterial mat under aerobic conditions. Before the biosorption experiments, the cyanobacterial mat was desorbed for 30 min in 0.1 N HCl. An acclimatization period of 30 min was given for every batch study. Aqueous chromium (VI) solution with initial concentrations (2–10, 15–30 and 300 ppm) was prepared and 40 ml of each solution was taken and the pH was maintained between 5.5 and 6.2 to which the weighed amount (1.5 gm) of wet cyanobacterial mat was poured. The batch was then kept at room temperature, samples were withdrawn at different time intervals from it for colorimetric determination of chromium (VI) by Thermo Heλios α model spectrophotometer and percentage reduction in chromium (VI) concentration was calculated.

## Treatment of Cyanobacterial Mat with EDTA

A new set of biosorption study was set up by pretreatment of cyanobacterial mat. The cyanobacterial mat was treated with N/56 EDTA for 15 min. Biosorption study was carried out by treated mat to compare the results.

Estimation of chromium VI by DPC method

### *Principle*

Diphenyl carbazide reagent combines with hexavalent chromium to form a complex which gives a beautiful pink/magenta color in acidic medium as described in the "Method" (ISO Wq 1994).

### *Reagents*

1. Standard solution of potassium dichromate (2–10, 15–30 and 300 ppm).
2. Diphenyl carbazide solution (0.1 % w/v) in acetone.
3. 3.1 N Hydrochloric acid solution.

### *Method*

All the samples were filtered first through common filter paper then by Whatman No. 42 and finally by membrane filter (Millipore, 0.45 µm). The pH of the filtrate was maintained between 1 and 1.3, and to this solution 1 ml of DPC solution was added and the solution was shaken, readings were taken after 10 min on spectrophotometer at the requisite wavelength 540 nm.

## Scanning Electron Microscopy and EDX

The control and adsorbed samples of cyanobacterial mat of size 1 × 1 mm were used for SEM and EDX. Scanning electron micrographs were taken on FEI Quanta 200 and EDX was recorded on Genesis Spectrum Ver. 3.6 (EDAX Inc.).

## Adsorption Isotherm

The sorption data have been subjected to different sorption isotherms, namely the Freundlich and Langmuir. The data fitted well to Freundlich isotherm well. The Freundlich model stipulates that the ratio of solute adsorbed to the solute concentration is a function of the solution. The empirical model was shown to be consistent with exponential distribution of active centers characteristic of heterogeneous surfaces. The Freundlich adsorption isotherm was tested in the following linearized form:

$$\log Qe=\log Kf+1/n\log Ce$$

by plotting log Qe versus log Ce. Ce is the equilibrium concentration of metal ions and Qe is metal ion adsorbed per unit mass of adsorbent (mg/g). Both Kf and *n* are constants, being indicative of the extent of adsorption and the degree of non-linearity between solution and concentration, respectively. The result is represented in Tables 2 and 3. From the slope and intercept of straight portion of the plot, the values of Freundlich parameters, i.e., $1/n$ and Kf are computed, the adsorption isotherm experimental data obtained was fitted to Freundlich isotherm. The magnitude of Kf and *n* shows easy separation of heavy metal ions from wastewater and high adsorption capacity. The value of *n*, which is related to the distribution of bonded ions on the sorbent surface, represents beneficial adsorption if it is between 1 and 10. The *n* value for the biosorbent used was found (Kadirvelu and Namasivayam 2000) to be >1, indicating that adsorption is favorable. Tables 2 and 3 give the isotherm parameters for Freundlich isotherms. From linear correlation coefficients of the adsorption isotherm, it can be said that sorption data is obeyed well, which is indicative that multilayer sorption takes place on the surface of cyanobacterial mat.

## RESULTS AND DISCUSSION

In recent years, there has been an increasing research interest in microorganisms that are able to transform the highly toxic and water-soluble chromium (VI) to the less toxic and insoluble chromium (III) or remove chromium (VI) by adsorption. Considering the versatility of cyanobacteria and their ability to survive under diverse environmental condition, strains originating from different habitats must be studied with respect to their applicability (Hameed and Hasnain 2005). Many microorganisms are known to reduce chromium (VI) to chromium (III). Mats sequester or precipitate metals by surface adsorption or by conditioning the surrounding chemical environment, thus bio concentrating the metal in a small volume.

Cyanobacteria are the member of morphologically diverse group of prokaryotes. The oxygenic photosynthetic pathway adopted by such prokaryote is the presence of two photosystems (PSII and PSI), and the use of $H_2O$ as the photoreductant in photosynthesis (Tamagnini et al. 2002; Boone et al. 2001) is well known. Mats display wide variety of mechanisms for removal of metals and metalloids. These mechanisms take place at the cellular level of the constituent microorganism and at the community level of entire consortium.

The biosorption experiments were carried out in batches with standard chromium concentration (2–10, 15–30 and 300 ppm) and pH of 5.5–6.2. The percentage of reduction in the chromium (VI) concentration by cyanobacterial mat at different time periods are shown in Table 1. A totally new phenomenon was observed in the present case. The substantial biosorption that occurred at initial pH 5.5–6.2, could be attributed to the fact that the mat itself was releasing $H^+$ during the process and thus the pH had been observed to have lowered with time duration.

**Table 1**: Reduction of chromium VI (%) by cyanobacterial mat in standard dichromate solution

| Concentration | 15 min | 30 min | 45 min | 60 min | 90 min | 210 min | Overnight |
|---|---|---|---|---|---|---|---|
| 2 ppm | 22.80 | 35.10 | 59.80 | 51.15 | 65.35 | 91.45 | 100.0 |
| 4 ppm | 36.20 | 56.30 | 72.22 | 87.00 | 100.0 | – | – |
| 6 ppm | 25.41 | 52.85 | 65.01 | 66.32 | 69.47 | 100.0 | – |
| 8 ppm | 18.96 | 31.52 | 47.80 | 62.67 | 79.995 | 96.85 | 100.0 |
| 10 ppm | 3.890 | 33.82 | 48.43 | 58.01 | 53.39 | 95.16 | 100.0 |

In this study, the pathway adapted for biosorption by the cyanobacterial mat seems to be chromate reduction. The probable mechanism of reduction could be envisaged as following: (1) the participation of photosynthesis by the *Chlorella* taking place in the upper cyanobacteria stratum which provided a continuous supply of electron for the chromium (VI) to (III) reduction process, (2) the consortium was able to partially adsorb chromium as well as to partially reduce chromium (VI) to chromium (III), and (3) at very high chromium (VI) concentration (300 ppm) the mat showed very efficient removal. This indicated that chromium (VI) was first getting adsorbed and then the adsorbed metal species were getting reduced by the bioreductant released by the mat.

In most of the cases the mat showed 100 % chromium (VI) removal in overnight, except in the cases of 4 and 6 ppm, where it showed total chromium (VI) removal in 90 and 210 min, respectively. In other cases adsorption of this metal ion was above 90 %, i.e., 91.45, 96.85 and 95.16 % for 2, 8 and 10 ppm, respectively. The short-term uptake of chromate (biosorption) was concentration dependent and the data were found to conform to the Freundlich adsorption isotherm as shown in Tables 2, 3.

**Table 2**: Freundlich isotherm parameters for Cr(VI) adsorption by cyanobacterial mat

| Concentration | Adsorbent (g/100 ml) | $K_f$ | $1/n$ | $R^2$ |
|---|---|---|---|---|
| 2 ppm | 3.75 | 2.393 | 0.536 | 0.734 |
| 4 ppm | 3.75 | 0.742 | 1.107 | 0.705 |
| 6 ppm | 3.75 | 2.030 | −1.091 | 0.978 |
| 8 ppm | 3.75 | 1.332 | −0.370 | 0.604 |
| 10 ppm | 3.75 | 1.535 | −0.648 | 0.456 |

*Concentration* initial concentration of solute

**Table 3**: Freundlich isotherm parameters for Cr(VI) adsorption by cyanobacterial mat at higher concentration

| Concentration | Adsorbent (g/100 ml) | $K_f$ | $1/n$ | $R^2$ |
|---|---|---|---|---|
| 15 ppm | 3.0 | 1.850 | −2.010 | 0.495 |
| 20 ppm | 3.0 | 1.713 | −2.779 | 0.627 |

| | | | | |
|---|---|---|---|---|
| 25 ppm | 3.0 | 2.913 | −8.65 | 0.774 |
| 30 ppm | 3.0 | −3.271 | 26.11 | 0.440 |

*Concentration* initial concentration of solute

Another set of chromium biosorption with higher concentration of chromium, i.e., 15, 20, 25 and 30 ppm was also carried out as shown in (Fig. 1).

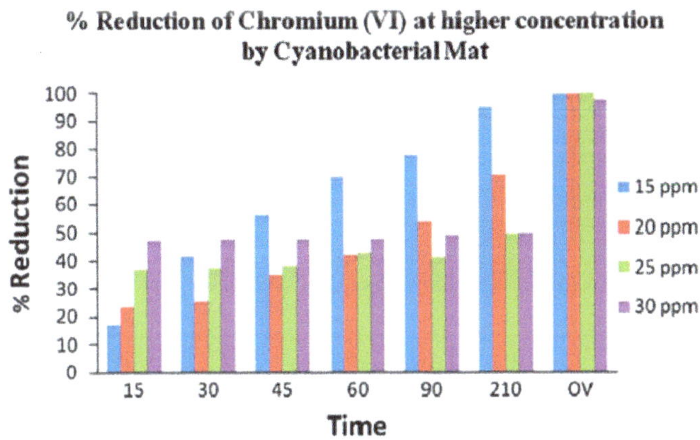


**Figure 1**: Reduction of chromium (VI) (%) by mat at high concentration.

In this set an average removal of chromium (VI) has been observed except in the case of 15 ppm where the removal was 94.76 % in 210 min. However, after overnight the mat showed 100 % chromium (VI) removal in every batch. It was observed that best adsorption among the lower concentration range was at 6 ppm while for higher concentration good adsorption was observed at 25 ppm. This could be attributed to low acclimatization for higher concentration of chromium (VI) ion for the mat.

In another study, cyanobacterial mat was treated with EDTA to check the effect of treatment on metal removal by the mat. The mat was suspended in EDTA to increase cellular apertures so that cell surface could be increased. These experiments were carried out with 2 and 10 ppm standard dichromate solution. EDTA treatment did not show any positive effect on the removal/reduction of chromium (VI) as compared to control set. The results did not show the expected trend of higher reduction for the EDTA treated mat. The untreated mat seemed to be more efficient in metal removal.

This bacterial consortium was also tested against 300 ppm of chromium (VI) concentration and it showed efficient removal of chromium by 1.5 g of the mat. This indicates that surface area is not a limiting factor and both biosorption and bioreduction are occurring simultaneously.

Scanning electron micrograph, shown in Figs. 2 and 3, respectively, explains the morphology of mat before and after biosorption.

**Figure 2**: SEM of mat collected from effluent site (before biosorption).

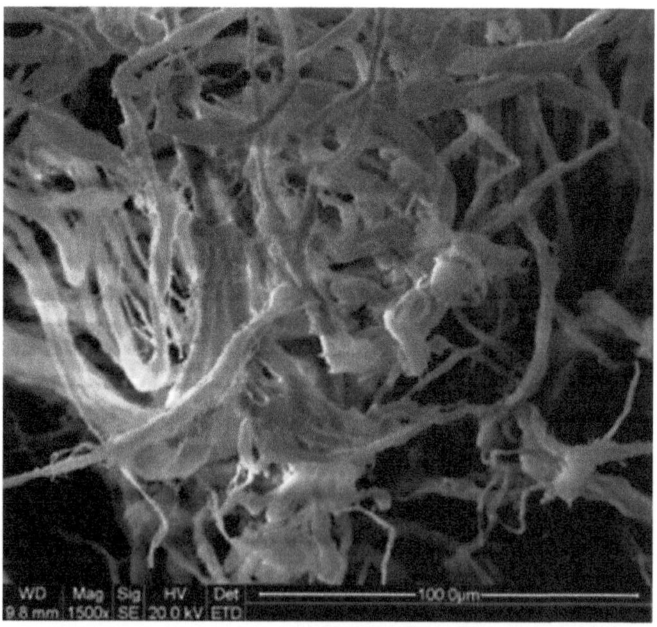


**Figure 3**: SEM of cyanobacterial mat after biosorption.

It was found to be a well-established mat containing *Oscillotoria* sp., *Phormidium* sp. and*Chlorella* sp. The entire network was a mesh containing *Chlorella* sp. on upper surface of the mat, while *Oscillotoria* sp. and *Phormidium* occupied the lower side. The cyanobacterial physiology did not get affected by the presence of high concentrations of chromium possibly due to the fact that the mat surface could be highly negatively charged due to the hydrolysis of the surface polysaccharides causing more number of hydroxyl ions to be present on the mat surface. The charges would be possibly distributed along the filamentous cyanobacteria which would have provided an enormous surface area for binding of positively charged chromium ions. Thus, it was the ability of the entire mat surface which would be eventually responsible for higher sequestering of chromium (VI) ion.

The actual mechanism of the initial chromium (VI) reduction by the mat is not known. Chromate reductase has been identified in other microbial groups which demonstrate both aerobic and anaerobic reduction of chromium (VI) as demonstrated by Tamagnini et al. (2002) and Castenzholz et al. (2001). However, the presence of chromate reductase has not been investigated in the cyanobacterial mat.

FT-IR spectra of the mat after chromium (VI) biosorption showed four apparent changes of the functional groups on the biomass. (1) The enhancement of the intensity at the region 3,200–3,500 $cm^{-1}$ indicates an increase of the free hydroxyl group on the biomass (Fig. 4). This could be due to hydrolysis of some polysaccharides on the cell wall to shorter saccharides such as oligosaccharides, dioses and monoses under acidic condition as observed (Lin et al. 2005). (2) The weakening of the peak at 1,470–1,450 $cm^{-1}$, which was typical of the complexation of the carboxylate functional group by coordination with chromium (III) metal formed from bioreduction of chromium (VI). (3) The third change was the shift of the peak at 1034–1028 $cm^{-1}$, which could be due to the involvement of the C–O bond of polysaccharides in chromium (III) biosorption. (4) The last change was the presence of a new peak at around 940 $cm^{-1}$ in the chromium (VI)-treated biomass, and it could be attributed to the presence of chromium (VI)–O bond as suggested (Holman et al.1999).

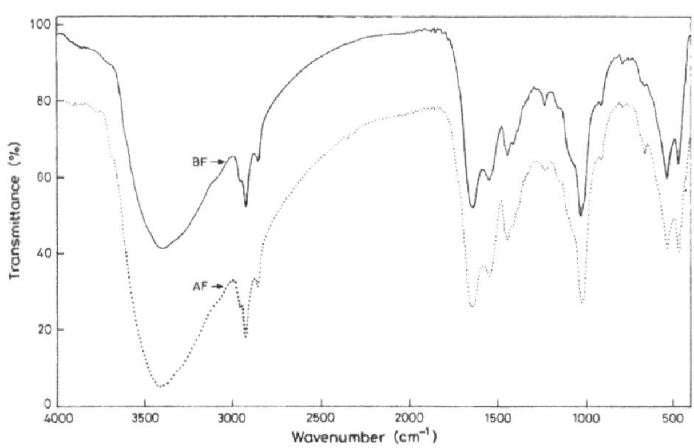


**Figure 4**: FT-IR of the cyanobacterial mat before (*BF*) and after (*AF*) the adsorption.

During the metal remediation study it was also observed that the mat secreted purple colored dye identified as phycoerythrin by UV–Visible spectrum having a peak at 550 nm. It is known that *Phormidium* possess phycocyanin such as phycoerythrin. *Phormidium* sp. cells containing phycoerythrin has been reported to show a single absorption peak at 565 nm in the visible wavelength region (Haxo 1960).

The EDX data of mat and biosorbed mat shown in Figs. 5 and 6 respectively clearly showed the enhancement of chromium content in the biosorbed mat. The mat initially contained 29.68 % of chromium along with other metal ions as it was collected from effluent disposal site and the mat after biosorption contained 47.25 %. A net increase of 17.57 % in chromium content was observed.

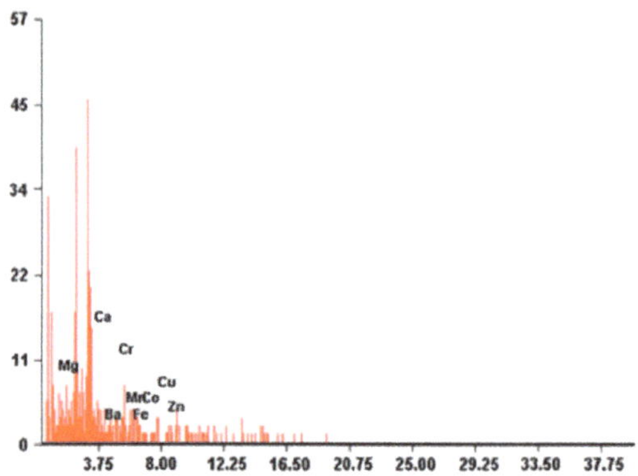


| *Element* | *Wt %* | *At %* |
|---|---|---|
| *N K* | 08.60 | 13.23 |
| *O K* | 51.93 | 69.93 |
| *MgK* | 03.65 | 03.24 |
| *BaL* | 04.24 | 00.67 |
| *CrK* | **29.68** | **12.30** |
| *CuK* | 00.43 | 00.14 |
| *ZnK* | 01.47 | 00.49 |

**Figure 5**: EDX of cyanobacterial mat collected from tannery site (before biosorption).

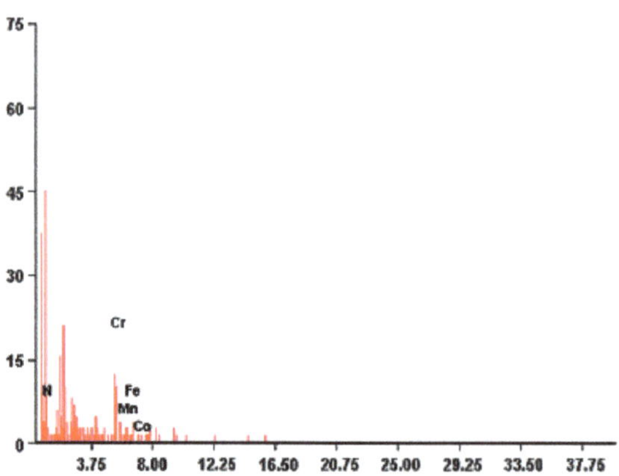


| *Element* | *Wt %* | *At %* |
|---|---|---|
| *MgK* | 17.26 | 31.89 |
| *CrK* | **47.25** | **40.81** |
| *MnK* | 14.52 | 11.87 |
| *FeK* | 06.34 | 05.10 |
| *CoK* | 00.00 | 00.00 |
| *CuK* | 14.63 | 10.34 |

**Figure 6**: EDX of cyanobacterial mat after Biosorption.

Based on the results, it is reasonable to conclude that the mechanism involved in the removal of chromium (VI) by cyanobacteria mat was found to biosorption–bioreduction. The sequence of reaction that may be occurring is probably (1) the protons released by the mat (from the photosynthetic pathway of *Chlorella*) would be creating protonated sites on the mat for biosorption of chromium (VI), (2) bioreduction of chromium (VI) to chromium (III), the chromium (VI) adsorbed on the biomass surface was bioreduced to chromium (III) by the reductant on the biomass such as polysaccharides and the electrons generated during the photosynthesis carried out by *Chlorella. Phormidium* and the other microbes produce hydrogen sulfide that leads to anoxic conditions. It has been observed that elemental sulfur (showed positive test for sulfate ion) was deposited which seems to have come from anaerobic environments with abundant hydrogen sulfide produced by*Phormidium* and in the presence of light, *Oscillatoria* are known to undergo an oxygenic photosynthesis to excrete elemental sulfur rather than oxygen gas. Thus, the entire consortium was found to be participated in the biosorption process.

## CONCLUSIONS

Although cyanobacterial mats occur in nature as stratified communities of cyanobacteria and some other bacteria, but they can be cultured on large scale and used for bioremediation. This study is an example of bioremediation where a cyanobacterial mat is used in form of biological treatment to clean up chromium (VI) contaminant in surface water at pH 5.5–6.2 for both low and high levels of contamination. Moreover, it offers permanent in situ remediation rather than simply moving the pollution to other site. One of the important facts about this study was, due to possible release of acid by the mat during the biosorption, the process could easily be carried out at higher initial pH 5.5–6.2. Thus, the cyanobacterial mat, which is genetically and ecologically modified species, seems to be potential savior for the chromium (VI) pollution caused by the tannery industry.

## ACKNOWLEDGMENTS

The authors express their sincere thanks to Ministry of Environment and Forest (MOEF), Govt. of India, New Delhi for financial support.

## REFERENCES

1. Aksu Z, Balibek E (2007) Chromium(VI) biosorption by dried Rhizopus arrhizus: Effect of salt (NaCl) concentration on equilibrium and kinetic parameters. J Hazard Mater 145(1–2):210–220. doi:10.1016/j.jhazmat.2006.11.011
2. Arica MY, Bayramoglu G (2005) Cr(VI) biosorption from aqueous solutions using free and immobilized biomass of Lentinus sajor-caju: preparation and kinetic characterization. Colloids Surf, A 253(1–3):203–211. doi:10.1016/j.colsurfa.2004.11.012
3. Arica MY, Tüzün I, Yalçin E, Ince Ö, Bayramoglu G (2005) Utilisation of native, heat and acid-treated microalgae *Chlamydomonas reinhardtii* preparations for biosorption of Cr(VI) ions. Process Biochem 40(7):2351–2358. doi:10.1016/j.procbio.2004.09.008
4. Boone DR, Castenholz RW, Garrity GM (eds) (2001) Bergey's manual of systematic bacteriology, vol 1, 2nd edn. Springer, New York, p 173

5. Cabatingan LK, Agapay RC, Rakels JLL, Ottens M, van der Wielen LAM (2001) Potential of biosorption for the recovery of chromate in industrial wastewaters. Ind Eng Chem Res 40(10):2302–2309. doi:10.1021/ie0008575

6. Camargo FAO, Okeke BC, Bento FM, Frankenberger WT (2005) Diversity of chromium-resistant bacteria isolated from soils contaminated with dichromate. Appl Soil Ecol 29(2):193–202. doi:10.1016/j.apsoil.2004.10.006

7. Cervantes C, Campos-Garcia J, Devars S, Gutiérrez-Corona F, Loza-Tavera H, Torres Guzmán JC, Moreno-Sánchez R (2001) Interactions of chromium with microorganisms and plants. FEMS Microbiol Rev 25(3):335–347. doi:10.1016/s0168-6445(01)00057-2

8. Colica G, Mecarozzi PC, Philippis RD (2010) Treatment of Cr(VI)-containing wastewaters with exopolysaccharide-producing cyanobacteria in pilot flow through and batch systems. Appl Microbiol Biotechnol 87:1953–1961

9. Deepa KK, Sathishkumar M, Binupriya AR, Murugesan GS, Swaminathan K, Yun SE (2006) Sorption of Cr(VI) from dilute solutions and wastewater by live and pretreated biomass of *Aspergillus flavus*. Chemosphere 62(5):833–840. doi:10.1016/j.chemosphere.2005.04.087

10. Garnham GW, Green M (1995) Chromate (VI) uptake by and interactions with cyanobacteria. J Ind Microbiol Biotechnol 14(3):247–251. doi:10.1007/bf01569935

11. Hameed A, Hasnain S (2005) Cultural characteristics of chromium resistant unicellular cyanobacteria isolated from local environment in Pakistan. Chin J Oceanol Limnol 23(4):433–441. doi:10.1007/bf02842688

12. Han X, Wong YS, Wong MH, Tam NFY (2008) Effects of anion species and concentration on the removal of Cr(VI) by a microalgal isolate Chlorella miniata. J Hazard Mater 158(2–3):615–620. doi:10.1016/j.jhazmat.2008.02.024

13. Haxo FT (1960) Comparative biochemistry of photoreactive systems. Academic Press, MB Allen New York, pp 339–360

14. Holman H-YN, Perry DL, Martin MC, Lamble GM, McKinney WR, Hunter-Cevera JC (1999) Real-time characterization of biogeochemical reduction of Cr(VI) on basalt surfaces by SR-FTIR imaging. Geomicrobiol J 16(4):307–324

15. ISO Wq (1994) Determination of chromium (VI)—spectrometric method using 1,5-diphenylcarbazide

16. Kadirvelu K, Namasivayam C (2000) Agricutural by-product as metal adsorbent: sorption of lead(II) from aqueous solution onto Coirpith carbon. Environ Technol 21(10):1091–1097

17. Kratochvil D, Pimentel P, Volesky B (1998) Removal of trivalent and hexavalent chromium by seaweed biosorbent. Environ Sci Technol 32(18):2693–2698. doi:10.1021/es971073u

18. Laxman RS, More S (2002) Reduction of hexavalent chromium by *Streptomyces griseus*. Miner Eng 15(11):831–837. doi:10.1016/s0892-6875(02)00128-0

19. Lin Z, Wu J, Xue R, Yang Y (2005) Spectroscopic characterization of Au3 + biosorption by waste biomass of *Saccharomyces cerevisiae*. Spectrochim Acta Part A Mol Biomol Spectrosc 61(4):761–765. doi:10.1016/j.saa.2004.03.029
20. Sharma M, Kaushik A, Kaushik CP (2011) Int Biodeterior Biodegradation 65(4):656–663
21. Srinath T, Garg SK, Ramteke PW (2002) Chromium (VI) accumulation by *Bacillus circulans*: effect of growth conditions. Indian J Microbiol 142(2):141–146
22. Tamagnini P, Axelsson R, Lindberg P, Oxelfelt F, Wunschiers R, Lindblad P (2002) Hydrogenases and hydrogen metabolism of cyanobacteria. Microbiol Mol Biol Rev 66(1):1–20. doi:10.1128/mmbr.66.1.1-20.2002
23. Vankar PS, Srivastava J (2008) Comparative study of total phenol, flavonoid contents and antioxidant activity in *Canna indica* and *Hibiscus rosa sinensis*: prospective natural food dyes. Int J Food Eng 4(3)
24. Zhou M, Liu Y, Zeng G, Li X, Xu W, Fan T (2007) Kinetic and equilibrium studies of Cr(VI) biosorption by dead. *Bacillus licheniformis* biomass. World J Microbiol Biotechnol 23(1):43–48. doi:10.1007/s11274-006-9191-8

# 8

CHAPTER

# MICROBIAL DEGRADATION OF PETROLEUM HYDROCARBON CONTAMINANTS: AN OVERVIEW

Nilanjana Das and Preethy Chandran

Environmental Biotechnology Division, School of Biosciences and Technology, VIT University, Vellore, Tamil Nadu 632014, India

*ABSTRACT*

**One of the major environmental problems today is hydrocarbon contamination resulting from the activities related to the petrochemical industry. Accidental releases of petroleum products are of particular concern in the environment. Hydrocarbon components have been known to belong to the family of carcinogens and neurotoxic organic pollutants. Currently accepted disposal methods of incineration or burial insecure landfills can become prohibitively expensive when amounts of contaminants are large. Mechanical and chemical methods generally used to remove hydrocarbons from contaminated sites have limited effectiveness and can be expensive. Bioremediation is the promising technology for the treatment of these contaminated sites since it is cost-effective and will lead to complete mineralization. Bioremediation functions basically on biodegradation, which may refer to complete mineralization of organic contaminants into carbon dioxide, water, inorganic compounds, and cell protein or transformation of complex organic contaminants to other simpler organic compounds by biological agents like microorganisms. Many indigenous microorganisms in water and soil are capable of degrading hydrocarbon contaminants. This paper presents an updated overview of petroleum hydrocarbon degradation by microorganisms under different ecosystems.**

## INTRODUCTION

Petroleum-based products are the major source of energy for industry and daily life. Leaks and accidental spills occur regularly during the exploration, production, refining,

**Citation:** Nilanjana Das and Preethy Chandran, "Microbial Degradation of Petroleum Hydrocarbon Contaminants: An Overview," Biotechnology Research International, vol. 2011, Article ID 941810, 13 pages, 2011. doi:10.4061/2011/941810.

transport, and storage of petroleum and petroleum products. The amount of natural crude oil seepage was estimated to be 600,000 metric tons per year with a range of uncertainty of 200,000 metric tons per year [1]. Release of hydrocarbons into the environment whether accidentally or due to human activities is a main cause of water and soil pollution [2]. Soil contamination with hydrocarbons causes extensive damage of local system since accumulation of pollutants in animals and plant tissue may cause death or mutations [3]. The technology commonly used for the soil remediation includes mechanical, burying, evaporation, dispersion, and washing. However, these technologies are expensive and can lead to incomplete decomposition of contaminants.

The process of bioremediation, defined as the use of microorganisms to detoxify or remove pollutants owing to their diverse metabolic capabilities is an evolving method for the removal and degradation of many environmental pollutants including the products of petroleum industry [4]. In addition, bioremediation technology is believed to be noninvasive and relatively cost-effective [5]. Biodegradation by natural populations of microorganisms represents one of the primary mechanisms by which petroleum and other hydrocarbon pollutants can be removed from the environment [6] and is cheaper than other remediation technologies [7].

The success of oil spill bioremediation depends on one's ability to establish and maintain conditions that favor enhanced oil biodegradation rates in the contaminated environment. Numerous scientific review articles have covered various factors that influence the rate of oil biodegradation [7–12]. One important requirement is the presence of microorganisms with the appropriate metabolic capabilities. If these microorganisms are present, then optimal rates of growth and hydrocarbon biodegradation can be sustained by ensuring that adequate concentrations of nutrients and oxygen are present and that the pH is between 6 and 9. The physical and chemical characteristics of the oil and oil surface area are also important determinants of bioremediation success. There are the two main approaches to oil spill bioremediation: (a) bioaugmentation, in which known oil-degrading bacteria are added to supplement the existing microbial population, and (b) biostimulation, in which the growth of indigenous oil degraders is stimulated by the addition of nutrients or other growth-limiting cosubstrates.

The success of bioremediation efforts in the cleanup of the oil tanker Exxon Valdez oil spill of 1989 [13] in Prince William Sound and the Gulf of Alaska created tremendous interest in the potential of biodegradation and bioremediation technology. Most existing studies have concentrated on evaluating the factors affecting oil bioremediation or testing favored products and methods through laboratory studies [14]. Only limited numbers of pilot scale and field trials have provided the most convincing demonstrations of this technology which have been reported in the peer-reviewed literature [15–18]. The scope of current understanding of oil bioremediation is also limited because the emphasis of most of these field studies and reviews has been given on the evaluation of bioremediation technology for dealing with large-scale oil spills on marine shorelines.

This paper provides an updated information on microbial degradation of petroleum hydrocarbon contaminants towards the better understanding in bioremediation challenges.

## MICROBIAL DEGRADATION OF PETROLEUM HYDROCARBONS

Biodegradation of petroleum hydrocarbons is a complex process that depends on the nature and on the amount of the hydrocarbons present. Petroleum hydrocarbons can be divided into four classes: the saturates, the aromatics, the asphaltenes (phenols, fatty acids, ketones, esters, and porphyrins), and the resins (pyridines, quinolines, carbazoles, sulfoxides, and amides) [19]. Different factors influencing hydrocarbon degradation have been reported by Cooney et al. [20]. One of the important factors that limit biodegradation of oil pollutants in the environment is their limited availability to microorganisms. Petroleum hydrocarbon compounds bind to soil components, and they are difficult to be removed or degraded [21]. Hydrocarbons differ in their susceptibility to microbial attack. The susceptibility of hydrocarbons to microbial degradation can be generally ranked as follows: linear alkanes > branched alkanes > small aromatics > cyclic alkanes [6, 22]. Some compounds, such as the high molecular weight polycyclic aromatic hydrocarbons (PAHs), may not be degraded at all [23].

Microbial degradation is the major and ultimate natural mechanism by which one can cleanup the petroleum hydrocarbon pollutants from the environment [24–26]. The recognition of biodegraded petroleum-derived aromatic hydrocarbons in marine sediments was reported by Jones et al. [27]. They studied the extensive biodegradation of alkyl aromatics in marine sediments which occurred prior to detectable biodegradation of n-alkane profile of the crude oil and the microorganisms, namely, Arthrobacter, Burkholderia, Mycobacterium, Pseudomonas, Sphingomonas, and Rhodococcus were found to be involved for alkylaromatic degradation. Microbial degradation of petroleum hydrocarbons in a polluted tropical stream in Lagos, Nigeria was reported by Adebusoye et al. [28]. Nine bacterial strains, namely, Pseudomonas fluorescens, P. aeruginosa, Bacillus subtilis, Bacillus sp., Alcaligenes sp., Acinetobacter lwoffi, Flavobacterium sp., Micrococcus roseus, and Corynebacteriumsp. were isolated from the polluted stream which could degrade crude oil.

Hydrocarbons in the environment are biodegraded primarily by bacteria, yeast, and fungi. The reported efficiency of biodegradation ranged from 6% [29] to 82% [30] for soil fungi, 0.13% [29] to 50% [30] for soil bacteria, and 0.003% [31] to 100% [32] for marine bacteria. Many scientists reported that mixed populations with overall broad enzymatic capacities are required to degrade complex mixtures of hydrocarbons such as crude oil in soil [33], fresh water [34], and marine environments [35, 36].

Bacteria are the most active agents in petroleum degradation, and they work as primary degraders of spilled oil in environment [37, 38]. Several bacteria are even known to feed exclusively on hydrocarbons [39]. Floodgate [36] listed 25 genera of hydrocarbon degrading bacteria and 25 genera of hydrocarbon degrading fungi which were isolated from marine environment. A similar compilation by Bartha and Bossert [33] included 22 genera of bacteria and 31 genera of fungi. In earlier days, the extent to which bacteria, yeast, and filamentous fungi participate in the biodegradation of petroleum hydrocarbons was the subject of limited study, but appeared to be a function of the ecosystem and local environmental conditions [7]. Crude petroleum oil from petroleum contaminated soil from North East India was reported by Das and Mukherjee [40]. Acinetobacter sp. was found to be capable of utilizing n-alkanes of chain length $C_{10}$–$C_{40}$ as a sole source of carbon [41]. Bacterial genera, namely, Gordonia, Brevibacterium, Aeromicrobium, Dietzia, Burkholderia, and Mycobacterium isolated from petroleum contaminated soil proved to be the potential organisms for hydrocarbon degradation [42]. The degradation of poly-aromatic hydrocarbons by Sphingomonas was reported by Daugulis and McCracken [43].

Fungal genera, namely, Amorphoteca, Neosartorya, Talaromyces, and Graphium and yeast genera, namely,Candida, Yarrowia, and Pichia were isolated from petroleum-contaminated soil and proved to be the potential organisms for hydrocarbon degradation [42]. Singh [44] also reported a group of terrestrial fungi, namely,Aspergillus, Cephalosporium, and Pencillium which were also found to be the potential degrader of crude oil hydrocarbons. The yeast species, namely, Candida lipolytica, Rhodotorula mucilaginosa, Geotrichum sp, and Trichosporon mucoides isolated from contaminated water were noted to degrade petroleum compounds [45].

Though algae and protozoa are the important members of the microbial community in both aquatic and terrestrial ecosystems, reports are scanty regarding their involvement in hydrocarbon biodegradation. Walker et al. [51] isolated an alga, Prototheca zopfi which was capable of utilizing crude oil and a mixed hydrocarbon substrate and exhibited extensive degradation of n-alkanes and isoalkanes as well as aromatic hydrocarbons. Cerniglia et al. [52] observed that nine cyanobacteria, five green algae, one red alga, one brown alga, and two diatoms could oxidize naphthalene. Protozoa, by contrast, had not been shown to utilize hydrocarbons.

## FACTORS INFLUENCING PETROLEUM HYDROCARBON DEGRADATION

A number of limiting factors have been recognized to affect the biodegradation of petroleum hydrocarbons, many of which have been discussed by Brusseau [53]. The composition and inherent biodegradability of the petroleum hydrocarbon pollutant is the first and foremost important consideration when the suitability of a remediation approach is to be assessed. Among physical factors, temperature plays an important role in biodegradation of hydrocarbons by directly affecting the chemistry of the pollutants as well as affecting the physiology and diversity of the microbial flora. Atlas [54] found that at low temperatures, the viscosity of the oil increased, while the volatility of the toxic low molecular weight hydrocarbons were reduced, delaying the onset of biodegradation.

Temperature also affects the solubility of hydrocarbons [62]. Although hydrocarbon biodegradation can occur over a wide range of temperatures, the rate of biodegradation generally decreases with the decreasing temperature. Figure 1 shows that highest degradation rates that generally occur in the range 30–40°C in soil environments, 20–30°C in some freshwater environments and 15–20°C in marine environments [33, 34]. Venosa and Zhu [63] reported that ambient temperature of the environment affected both the properties of spilled oil and the activity of the microorganisms. Significant biodegradation of hydrocarbons have been reported in psychrophilic environments in temperate regions [64, 65].

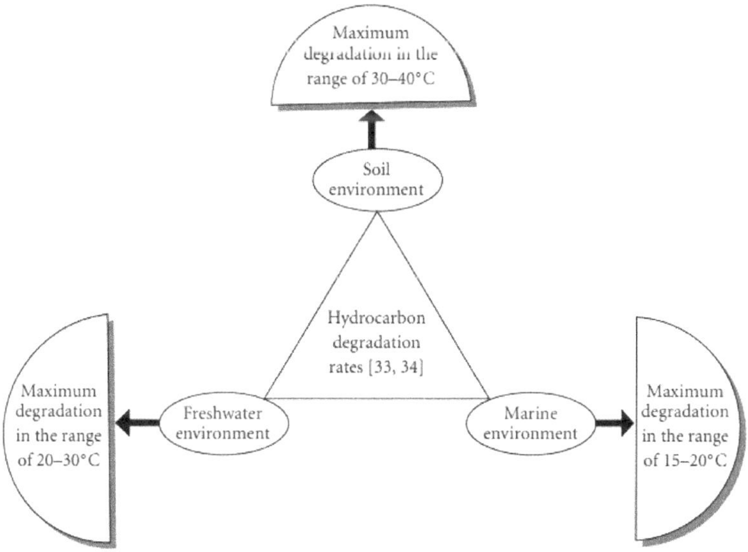


**Figure 1:** Hydrocarbon degradation rates in soil, fresh water, and marine environments.

Nutrients are very important ingredients for successful biodegradation of hydrocarbon pollutants especially nitrogen, phosphorus, and in some cases iron [34]. Some of these nutrients could become limiting factor thus affecting the biodegradation processes. Atlas [35] reported that when a major oil spill occurred in marine and freshwater environments, the supply of carbon was significantly increased and the availability of nitrogen and phosphorus generally became the limiting factor for oil degradation. In marine environments, it was found to be more pronounced due to low levels of nitrogen and phosphorous in seawater [36]. Freshwater wetlands are typically considered to be nutrient deficient due to heavy demands of nutrients by the plants [66]. Therefore, additions of nutrients were necessary to enhance the biodegradation of oil pollutant [67, 68]. On the other hand, excessive nutrient concentrations can also inhibit the biodegradation activity [69]. Several authors have reported the negative effects of high NPK levels on the biodegradation of hydrocarbons [70, 71] especially on aromatics [72]. The effectiveness of fertilizers for the crude oil bioremediation in subarctic intertidal sediments was studied by Pelletier et al. [64]. Use of poultry manure as organic fertilizer in contaminated soil was also reported [73], and biodegradation was found to be enhanced in the presence of poultry manure alone. Maki et al. [74] reported that photo-oxidation increased the biodegradability of petroleum hydrocarbon by increasing its bioavailability and thus enhancing microbial activities.

## MECHANISM OF PETROLEUM HYDROCARBON DEGRADATION

The most rapid and complete degradation of the majority of organic pollutants is brought about under aerobic conditions. Figure 2 shows the main principle of aerobic degradation of hydrocarbons [75]. The initial intracellular attack of organic pollutants is an oxidative process and the activation as well as incorporation of oxygen is the enzymatic key reaction catalyzed by oxygenases and peroxidases. Peripheral degradation pathways convert organic pollutants step by step into intermediates of the central intermediary metabolism, for example, the tricarboxylic acid cycle. Biosynthesis of cell biomass occurs from the central precursor metabolites, for example, acetyl-CoA, succinate, pyruvate. Sugars required for various biosyntheses and growth are synthesized by gluconeogenesis.

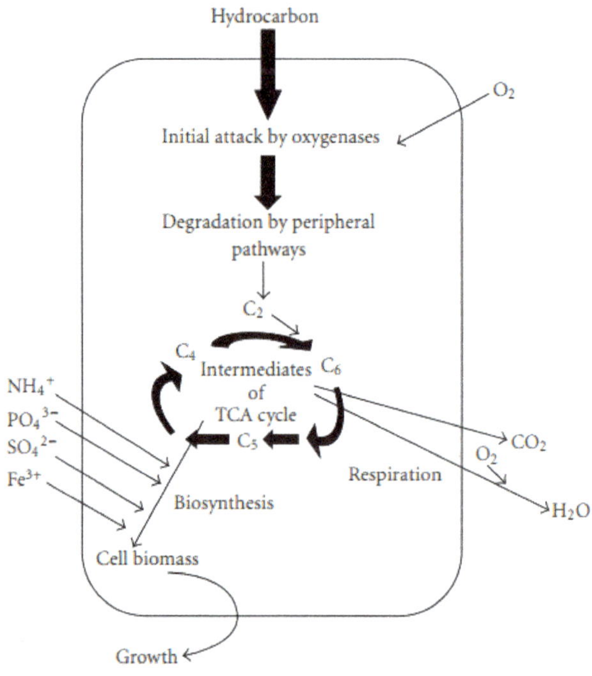


**Figure 2:** Main principle of aerobic degradation of hydrocarbons by microorganisms.

The degradation of petroleum hydrocarbons can be mediated by specific enzyme system. Figure 3 shows the initial attack on xenobiotics by oxygenases [75]. Other mechanisms involved are (1) attachment of microbial cells to the substrates and (2) production of biosurfactants [76]. The uptake mechanism linked to the attachment of cell to oil droplet is still unknown but production of biosurfactants has been well studied.

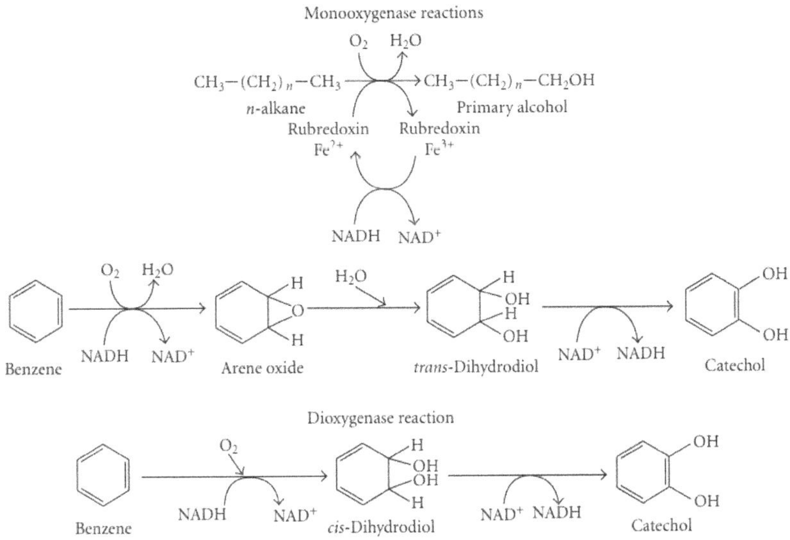


**Figure 3:** Enzymatic reactions involved in the processes of hydrocarbons degradation.

## ENZYMES PARTICIPATING IN DEGRADATION OF HYDROCARBONS

Cytochrome P450 alkane hydroxylases constitute a super family of ubiquitous Heme-thiolate Monooxygenases which play an important role in the microbial degradation of oil, chlorinated hydrocarbons, fuel additives, and many other compounds [77]. Depending on the chain length, enzyme systems are required to introduce oxygen in the substrate to initiate biodegradation (Table 1). Higher eukaryotes generally contain several different P450 families that consist of large number of individual P450 forms that may contribute as an ensemble of isoforms to the metabolic conversion of given substrate. In microorganisms such P450 multiplicity can only be found in few species [78]. Cytochrome P450 enzyme systems was found to be involved in biodegradation of petroleum hydrocarbons (Table 1). The capability of several yeast species to use n-alkanes and other aliphatic hydrocarbons as a sole source of carbon and energy is mediated by the existence of multiple microsomal Cytochrome P450 forms. These cytochrome P450 enzymes had been isolated from yeast species such asCandida maltosa, Candida tropicalis, and Candida apicola [79]. The diversity of alkaneoxygenase systems in prokaryotes and eukaryotes that are actively participating in the degradation of alkanes under aerobic conditions like Cytochrome P450 enzymes, integral membrane di-iron alkane hydroxylases (e.g., alkB), soluble di-iron methane monooxygenases, and membrane-bound copper containing methane monooxygenases have been discussed by Van Beilen and Funhoff [80].

**Table 1:** Enzymes involved in biodegradation of petroleum hydrocarbons

| Enzymes | Substrates | Microorganisms | References |
|---|---|---|---|
| Soluble Methane Monooxygenases | $C_1$–$C_8$ alkanes alkenes and cycloalkanes | *Methylococcus* *Methylosinus* *Methylocystis* *Methylomonas* *Methylocella* | McDonald et al. [46] |
| Particulate Methane Monooxygenases | $C_1$–$C_5$ (halogenated) alkanes and cycloalkanes | *Methylobacter* *Methylococcus,* *Methylocystis* | McDonald et al. [46] |
| AlkB related Alkane Hydroxylases | $C_5$–$C_{16}$ alkanes, fatty acids, alkyl benzenes, cycloalkanes and so forth | *Pseudomonas* *Burkholderia* *Rhodococcus,* *Mycobacterium* | Jan et al. [47] |
| Eukaryotic P450 | $C_{10}$–$C_{16}$ alkanes, fatty acids | *Candida maltosa* *Candida tropicalis* *Yarrowia lipolytica* | Iida et al. [48] |
| Bacterial P450 oxygenase system | $C_5$–$C_{16}$ alkanes, cycloalkanes | *Acinetobacter* *Caulobacter* *Mycobacterium* | Van Beilen et al. [49] |
| Dioxygenases | $C_{10}$–$C_{30}$ alkanes | *Acinetobacter sp.* | Maeng et al. [50] |

## UPTAKE OF HYDROCARBONS BY BIOSURFACTANTS

Biosurfactants are heterogeneous group of surface active chemical compounds produced by a wide variety of microorganisms [57, 58, 60, 81–83]. Surfactants enhance solubilization and removal of contaminants [84, 85]. Biodegradation is also enhanced by surfactants due to increased bioavailability of pollutants [86]. Bioremediation of oil sludge using biosurfactants has been reported by Cameotra and Singh [87]. Microbial consortium consisting of two isolates of Pseudomonas aeruginosa and one isolate Rhodococcus erythropolis from soil contaminated with oily sludge was used in this study. The consortium was able to degrade 90% of hydrocarbons in 6 weeks in liquid culture. The ability of the consortium to degrade sludge hydrocarbons was tested in two separate field trials. In addition, the effect of two additives (a nutrient mixture and a crude biosurfactant preparation on the efficiency of the process was also assessed. The biosurfactant used was produced by a consortium member and was identified as being a mixture of 11 rhamnolipid congeners. The consortium degraded 91% of the hydrocarbon content of soil contaminated with 1% (v/v) crude oil sludge in 5 weeks. Separate use of any one additive along with the consortium brought about a 91–95% depletion of the hydrocarbon content in 4 weeks, with the crude biosurfactant preparation being a more effective enhancer of degradation. However, more than 98% hydrocarbon depletion was obtained when both additives were added together with the consortium. The data substantiated the use of a crude biosurfactant for hydrocarbon remediation.

Pseudomonads are the best known bacteria capable of utilizing hydrocarbons as carbon and energy sources and producing biosurfactants [37, 87–89]. Among Pseudomonads, P. aeruginosa is widely studied for the production of glycolipid type biosurfactants. However, glycolipid type biosurfactants are also reported from some other species like P. putida and P. chlororaphis. Biosurfactants increase the oil surface area and that amount of oil is actually available for bacteria to utilize it [90]. Table 2 summarizes the recent reports on biosurfactant production by different microorganisms. Biosurfactants can act as emulsifying agents by decreasing the surface tension and forming micelles. The microdroplets encapsulated in the hydrophobic microbial cell surface are taken inside and degraded. Figure 4 demonstrates the involvement of biosurfactant (rhamnolipids) produced by Pseudomonas sp. and the mechanism of formation of micelles in the uptake of hydrocarbons [75].

**Table 2:** Biosurfactants produced by microorganisms

| Biosurfactants | Microorganisms |
|---|---|
| Sophorolipids | *Candida bombicola* (Daverey and Pakshirajan, [55]) |
| Rhamnolipids | *Pseudomonas aeruginosa* (Kumar et al. [56]) |
| Lipomannan | *Candida tropicalis* (Muthuswamy et al. [57]) |
| Rhamnolipids | *Pseudomonas fluorescens* (Mahmound et al. [58]) |
| Surfactin | *Bacillus subtilis* (Youssef et al. [59]) |
| Glycolipid | *Aeromonas* sp. (Ilori et al. [60]) |
| Glycolipid | *Bacillus sp.* (Tabatabaee et al. [61]) |

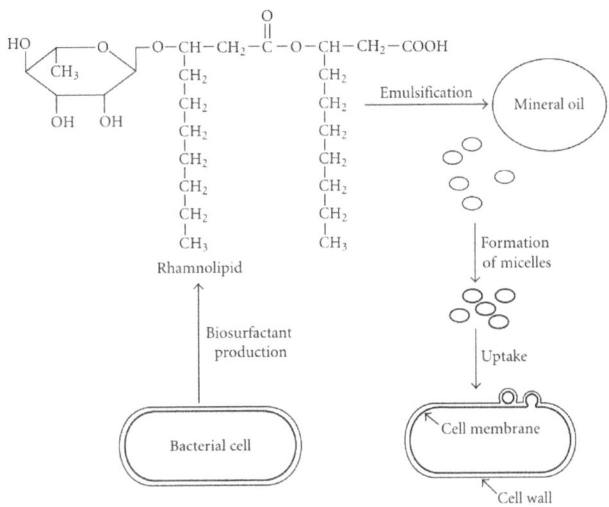


**Figure 4:** Involvement of biosurfactant (rhamnolipid) produced by Pseudomonas sp in the uptake of hydrocarbons.

## BIODEGRADATION OF PETROLEUM HYDROCARBONS BY IMMOBILIZED CELLS

Immobilized cells have been used and studied for the bioremediation of numerous toxic chemicals. Immobilization not only simplifies separation and recovery of immobilized cells but also makes the application reusable which reduces the overall cost. Wilsey and Bradely [91] used free suspension and immobilizedPseudomonas sp. to degrade petrol in an aqueous system. The study indicated that immobilization resulted in a combination of increased contact between cell and hydrocarbon droplets and enhanced level of rhamnolipids production. Rhamnolipids caused greater dispersion of water-insoluble n-alkanes in the aqueous phase due to their amphipathic properties and the molecules consist of hydrophilic and hydrophobic moieties reduced the interfacial tension of oil-water systems. This resulted in higher interaction of cells with solubilized hydrocarbon droplets much smaller than the cells and rapid uptake of hydrocarbon in to the cells. Diaz et al. [92] reported that immobilization of bacterial cells enhanced the biodegradation rate of crude oil compared to free living cells in a wide range of culture salinity. Immobilization can be done in batch mode as well as continuous mode. Packed bed reactors are commonly used in continuous mode to degrade hydrocarbons. Cunningham et al. [93] used polyvinyl alcohol (PVA) cryogelation as an entrapment matrix and microorganisms indigenous to the site. They constructed laboratory biopiles to compare immobilised bioaugmentation with liquid culture bioaugmentation and biostimulation. Immobilised systems were found to be the most successful in terms of percentage removal of diesel after 32 days.

Rahman et al. [94] conducted an experiment to study the capacity of immobilized bacteria in alginate beads to degrade hydrocarbons. The results showed that there was no decline in the biodegradation activity of the microbial consortium on the repeated use. It was concluded that immobilization of cells are a promising application in the bioremediation of hydrocarbon contaminated site.

## COMMERCIALLY AVAILABLE BIOREMEDIATION AGENTS

Microbiological cultures, enzyme additives, or nutrient additives that significantly increase the rate of biodegradation to mitigate the effects of the discharge were defied as bioremediation agents by U.S.EPA [95]. Bioremediation agents are classified as bioaugmentation agents and biostimulation agents based on the two main approaches to oil spill bioremediation. Numerous bioremediation products have been proposed and promoted by their vendors, especially during early 1990s, when bioremediation was popularized as "the ultimate solution" to oil spills [96].

The U.S. EPA compiled a list of 15 bioremediation agents [95, 97] as a part of the National Oil and Hazardous Substances Pollution Contingency Plan (NCP) Product Schedule, which was required by the Clean Water Act, the Oil Pollution Act of 1990, and the National Contingency Plan (NCP) as shown in Table 3. But the list was modified, and the number of bioremediation agents was reduced to nine.

**Table 3:** Bioremediation agents in NCP product schedule (Adapted from USEPA, 2002)

| Name or Trademark | Product Type | Manufacture |
|---|---|---|
| BET BIOPETRO | MC | BioEnviro Tech, Tomball, TX |
| BILGEPRO | NA | International Environmental Products, LLC, Conshohocken, PA. |
| INIPOL EAP 22 | NA | Societe, CECA S.A., France |
| LAND AND SEA | NA | Land and Sea Restoration LLC, San Antonio, TX |
| RESTORATION MICRO-BLAZE | MC | Verde Environmental, Inc., Houston, TX |
| OIL SPILL EATER II | NA/EA | Oil Spill Eater International, Corporation, Dallas, TX |
| OPPENHEIMER FORMULA | MC | Oppenheimer Biotechnology, Inc., Austin, TX |
| PRISTINE SEA II | MC | Marine Systems, Baton Rouge, LA |
| STEP ONE | MC | B & S Research, Inc., Embarrass, MN |
| SYSTEM E.T. 20. | MC | Quantum Environmental Technologies, Inc(QET), La Jolla, CA |
| VB591TMWATER, VB997TMSOIL, AND BINUTRIX | NA | BioNutraTech, Inc., Houston,TX |
| WMI-2000 | MC | WMI International, Inc |

Abbreviations of product type:
MC: Microbial Culture
EA: Enzyme Additive
NA: Nutrient Additive.

Studies showed that bioremediation products may be effective in the laboratory but significantly less so in the field [14, 17, 18, 98]. This is because laboratory studies cannot always simulate complicated real world conditions such as spatial heterogeneity, biological interactions, climatic effects, and nutrient mass transport limitations. Therefore, field studies and applications are the ultimate tests or the most convincing demonstration of the effectiveness of bioremediation products.

Compared to microbial products, very few nutrient additives have been developed and marketed specifically as commercial bioremediation agents for oil spill cleanup. It is probably because common fertilizers are inexpensive, readily available, and have been shown effective if used properly. However, due to the limitations of common fertilizers (e.g., being rapidly washed out due to tide and wave action), several organic nutrient products, such as oleophilic nutrient products, have recently been evaluated and marketed as bioremediation agents. Four agents, namely, Inipol EAP22, Oil Spill

Eater II (OSE II), BIOREN 1, and BIOREN 2, listed on the NCP Product Schedule have also been put into this category.

Inipol EAP22 (Societe, CECA S.A., France) is listed on the NCP Product Schedule as a nutrient additive and probably the most well-known bioremediation agent for oil spill cleanup due to its use in Prince William Sound, Alaska. This nutrient product is a microemulsion-containing urea as a nitrogen source, sodium laureth phosphate as a phosphorus source, 2-butoxy-1-ethanol as a surfactant, and oleic acid to give the material its hydrophobicity. The claimed advantages of Inipol EAP22 include (1) preventing the formation of water-in-oil emulsions by reducing the oil viscosity and interfacial tension; (2) providing controlled release of nitrogen and phosphorus for oil biodegradation; (3) exhibiting no toxicity to flora and fauna and good biodegradability [99].

Oil Spill Eater II (Oil Spill Eater International, Corp.) is another nutrient product listed on the NCP Schedule [97]. This product is listed as a nutrient/enzyme additive and consists of "nitrogen, phosphorus, readily available carbon, and vitamins for quick colonization of naturally occurring bacteria". A field demonstration was carried out at a bioventing site in a Marine Corps Air Ground Combat Center (MCAGCC) in California to investigate the efficacy of OSEII for enhancing hydrocarbon biodegradation in a fuel-contaminated vadose zone [106].

Researchers from European EUREKA BIOREN program conducted a field trial in an estuary environment to evaluate the effectiveness of two bioremediation products (BIOREN 1 and 2) [114, 115]. The two nutrient products were derived from fish meals in a granular form with urea and super phosphate as nitrogen and phosphorus sources and proteinaceous material as the carbon source. The major difference between the two formulations was that BIOREN 1 contained a biosurfactant. The results showed that the presence of biosurfactant in BIOREN 1 was the most active ingredient which contributed to the increase in oil degradation rates whereas BIOREN 2 (without biosurfactant) was not effective in that respect. The biosurfactant could have contributed to greater bioavailability of hydrocarbons to microbial attack.

## PHYTOREMEDIATION

Phytoremediation is an emerging technology that uses plants to manage a wide variety of environmental pollution problems, including the cleanup of soils and groundwater contaminated with hydrocarbons and other hazardous substances. The different mechanisms, namely, hydraulic control, phytovolatilization, rhizoremediation, and phytotransformation. could be utilized for the remediation of a wide variety of contaminants.

Phytoremediation can be cost-effective (a) for large sites with shallow residual levels of contamination by organic, nutrient, or metal pollutants, where contamination does not pose an imminent danger and only "polishing treatment" is required; (b) where vegetation is used as a final cap and closure of the site [116].

Advantages of using phytoremediation include cost-effectiveness, aesthetic advantages, and long-term applicability (Table 4). Furthermore, the use of phytoremediation as a secondary or polishing in situ treatment step minimizes land disturbance and eliminates transportation and liability costs associated with offsite treatment and disposal.

**Table 4:** Advantages and disadvantages of phytoremediation over traditional technologies

| Advantages | Disadvantages |
|---|---|
| Relatively low cost | Longer remediation times |
| Easily implemented and maintained | Climate dependent |
| Several mechanisms for removal | Effects to food web might be unknown |
| Environmentally friendly | Ultimate contaminant fates might be unknown |
| Aesthetically pleasing | Results are variable |
| Reduces landfilled wastes | |
| Harvestable plant material | |

Research and application of phytoremediation for the treatment of petroleum hydrocarbon contamination over the past fifteen years have provided much useful information that can be used to design effective remediation systems and drive further improvement and innovation. Phytoremediation could be applied for the remediation of numerous contaminated sites. However, not much is known about contaminant fate and transformation pathways, including the identity of metabolites (Table 4). Little data exists on contaminant removal rates and efficiencies directly attributable to plants under field conditions.

The potential use of phytoremediation at a site contaminated with hydrocarbons was investigated. The Alabama Department of Environmental Management granted a site, which involved about 1500 cubic yards of soil of which 70% of the baseline samples contained over 100 ppm of total petroleum hydrocarbon (TPH). After 1 year of vegetative cover, approximately 83% of the samples were found to contain less than 10-ppm TPH. Removal of total petroleum hydrocarbon (TPH) at several field sites contaminated with crude oil, diesel fuel, or petroleum refinery wastes, at initial TPH concentrations of 1,700 to 16,000 mg/kg were also investigated [117,118]. Plant growth was found to vary depending upon the species. Presence of some species led to greater TPH disappearance than with other species or in unvegetated soil. Among tropical plants tested for use in Pacific Islands, three coastal trees, kou (Cordia subcordata), milo (Thespesia populnea), and kiawe (Prosopis pallida) and the native shrub beach naupaka tolerated field conditions and facilitated cleanup of soils contaminated with diesel fuel [119]. Grasses were often planted with trees at sites with organic contaminants as the primary remediation method. Tremendous amount of fine roots in the surface soil was found to be effective at binding and transforming hydrophobic contaminants such as TPH, BTEX, and PAHs. Grasses were often planted between rows of trees to provide soil stabilization and protection against wind-blown dust that could move contaminants offsite. Legumes such as alfalfa (Medicago sativa), alsike clover (Trifolium hybridum), and peas (Pisum sp.) could be used to restore nitrogen to poor soils. Fescue (Vulpia myuros), rye (Elymus sp.), clover (Trifolium sp.), and reed canary grass (Phalaris arundinacea) were used successfully at several sites, especially contaminated with petrochemical wastes. Once harvested, the grasses could be disposed off as compost or burned.

Microbial degradation in the rhizosphere might be the most significant mechanism for removal of diesel range organics in vegetated contaminated soils [120]. This occurs because contaminants such as PAHs are highly hydrophobic, and their sorption to soil decreases their bioavailability for plant uptake and phytotransformation.

## GENETICALLY MODIFIED BACTERIA

Applications for genetically engineered microorganisms (GEMs) in bioremediation have received a great deal of attention to improve the degradation of hazardous wastes under laboratory conditions. There are reports on the degradation of environmental pollutants by different bacteria. Table 5 shows some examples of the relevant use of genetic engineering technology to improve bioremediation of hydrocarbon contaminants using bacteria. The genetically engineered bacteria showed higher degradative capacity. However, ecological and environmental concerns and regulatory constraints are major obstacles for testing GEM in the field. These problems must be solved before GEM can provide an effective clean-up process at lower cost.

**Table 5:** Genetic engineering for biodegradation of contaminants

| Microorganisms | Modification | Contaminants | Reference |
|---|---|---|---|
| *Pseudomonas. putida* | pathway | 4-ethylbenzoate | Ramos et al. [100] |
| *P. putida* KT2442 | pathway | toluene/benzoate | Panke and Sanchezromero [101] |
| *Pseudomonas* sp.FRI | pathway | chloro-, methylbenzoates | Rojo et al. [102] |
| *Comamonas. testosteroni* VP44 | substrate specificity | *o*-, *p*-monochlorobiphenyls | Hrywna et al. [103] |
| *Pseudomonas sp.* LB400 | substrate specificity | PCB | Erickson and Mondello [104] |
| *P. pseudoalcaligenes* KF707-D2 | substrate specificity | TCE, toluene, benzene | Suyama et al. [105] |

The use of genetically engineered bacteria was applied to bioremediation process monitoring, strain monitoring, stress response, end-point analysis, and toxicity assessment. Examples of these applications are listed in Table 6. The range of tested contaminants included chlorinated compounds, aromatic hydrocarbons, and nonpolar toxicants. The combination of microbiological and ecological knowledge, biochemical mechanisms, and field engineering designs are essential elements for successful in situ bioremediation using genetically modified bacteria.

**Table 6:** Application of genetically modified bacteria for assessing the biodegradation process efficiency

| Microorganisms | Application | Contaminants | Reference |
|---|---|---|---|
| *A. eutrophus* H850Lr | process monitoring | PCB | Van Dyke et al. [107] |
| *P. putida* TVA8 | process monitoring | TCE, BTEX | Applegate et al. [108] |
| *P. fluorescens* HK44 | process monitoring | naphthalene, anthracene | Sayler and Ripp [109] |
| *B. cepacia* BRI6001L | strain monitoring | 2,4-D | Masson et al. [110] |
| *P. fluorescens* 10586s/pUCD607 | stress response | BTEX | Sousa et al. [111] |
| *Pseudomonas* strain Shk1 | toxicity assessment | 2, 4-dinitrophenol hydroquinone | Kelly et al. [112] |
| *A. eutrophus* 2050 | end point analysis | non polar narcotics | Layton et al. [113] |

## CONCLUSION

Cleaning up of petroleum hydrocarbons in the subsurface environment is a real world problem. A better understanding of the mechanism of biodegradation has a high ecological significance that depends on the indigenous microorganisms to transform or mineralize the organic contaminants. Microbial degradation process aids the elimination of spilled oil from the environment after critical removal of large amounts of the oil by various physical and chemical methods. This is possible because microorganisms have enzyme systems to degrade and utilize different hydrocarbons as a source of carbon and energy.

The use of genetically modified (GM) bacteria represents a research frontier with broad implications. The potential benefits of using genetically modified bacteria are significant. But the need for GM bacteria may be questionable for many cases, considering that indigenous species often perform adequately but we do not tap the full potential of wild species due to our limited understanding of various phytoremediation mechanisms, including the regulation of enzyme systems that degrade pollutants.

Therefore, based on the present review, it may be concluded that microbial degradation can be considered as a key component in the cleanup strategy for petroleum hydrocarbon remediation.

## REFERENCES

1. K. A. Kvenvolden and C. K. Cooper, "Natural seepage of crude oil into the marine environment," Geo-Marine Letters, vol. 23, no. 3-4, pp. 140–146, 2003
2. C. Holliger, S. Gaspard, G. Glod, C. Heijman, W. Schumacher, R. P. Schwarzenbach, and F. Vazquez, "Contaminated environments in the subsurface and bioremediation: organic contaminants," FEMS Microbiology Reviews, vol. 20, no. 3-4, pp. 517–523, 1997.
3. P. J. J. Alvarez and T. M. Vogel, "Substrate interactions of benzene, toluene, and para-xylene during microbial degradation by pure cultures and mixed culture aquifer slurries," Applied and Environmental Microbiology, vol. 57, no. 10, pp. 2981–2985, 1991.
4. J. I. Medina-Bellver, P. Marín, A. Delgado, A. Rodríguez-Sánchez, E. Reyes, J. L. Ramos, and S. Marqués, "Evidence for in situ crude oil biodegradation after

the Prestige oil spill," Environmental Microbiology, vol. 7, no. 6, pp. 773–779, 2005. · · View at PubMed

5. T. M. April, J. M. Foght, and R. S. Currah, "Hydrocarbon-degrading filamentous fungi isolated from flare pit soils in northern and western Canada," Canadian Journal of Microbiology, vol. 46, no. 1, pp. 38–49, 2000.
6. W. Ulrici, "Contaminant soil areas, different countries and contaminant monitoring of contaminants," inEnvironmental Process II. Soil Decontamination Biotechnology, H. J. Rehm and G. Reed, Eds., vol. 11, pp. 5–42, 2000.
7. J. G. Leahy and R. R. Colwell, "Microbial degradation of hydrocarbons in the environment,"Microbiological Reviews, vol. 54, no. 3, pp. 305–315, 1990.
8. C. E. Zobell, "Action of microorganisms on hydrocarbons," Bacteriological Reviews, vol. 10, pp. 1–49, 1946.
9. R. M. Atlas, "Microbial degradation of petroleum hydrocarbons: an environmental perspective,"Microbiological Reviews, vol. 45, no. 1, pp. 180–209, 1981.
10. R. M. Atlas, Ed., Petroleum Microbiology, Macmillion, New York, NY, USA, 1984.
11. R. M. Atlas and R. Bartha, "Hydrocabon biodegradation and oil spill bioremediation," Advances in Microbial Ecology, vol. 12, pp. 287–338, 1992.
12. J. M. Foght and D. W. S. Westlake, "Biodegradation of hydrocarbons in freshwater," in Oil in Freshwater: Chemistry, Biology, Countermeasure Technology, J. H. Vandermeulen and S. R. Hrudey, Eds., pp. 217–230, Pergamon Press, New York, NY, USA, 1987.
13. R. M. Atlas and R. Bartha, "Fundamentals and applications," in Microbial Ecology, pp. 523–530, Benjamin/Cummings, San Francisco, Calif, USA, 4th edition, 1998.
14. A. J. Mearns, "Cleaning oiled shores: putting bioremediation to the test," Spill Science and Technology Bulletin, vol. 4, no. 4, pp. 209–217, 1997.
15. R. C. Prince, "Petroleum spill bioremediation in marine environments," Critical Reviews in Microbiology, vol. 19, no. 4, pp. 217–242, 1993.
16. R. P. J. Swannell, K. Lee, and M. Mcdonagh, "Field evaluations of marine oil spill bioremediation,"Microbiological Reviews, vol. 60, no. 2, pp. 342–365, 1996.
17. A. D. Venosa, M. T. Suidan, and M. T. Suidan, "Bioremediation of an experimental oil spill on the shoreline of Delaware Bay," Environmental Science and Technology, vol. 30, no. 5, pp. 1764–1775, 1996.·
18. A. D. Venosa, D. W. King, and G. A. Sorial, "The baffled flask test for dispersant effectiveness: a round Robin evaluation of reproducibility and repeatability," Spill Science and Technology Bulletin, vol. 7, no. 5-6, pp. 299–308, 2002.
19. R. R. Colwell, J. D. Walker, and J. J. Cooney, "Ecological aspects of microbial degradation of petroleum in the marine environment," Critical Reviews in Microbiology, vol. 5, no. 4, pp. 423–445, 1977.
20. J. J. Cooney, S. A. Silver, and E. A. Beck, "Factors influencing hydrocarbon

degradation in three freshwater lakes," Microbial Ecology, vol. 11, no. 2, pp. 127–137, 1985.

21. S. Barathi and N. Vasudevan, "Utilization of petroleum hydrocarbons by Pseudomonas fluorescensisolated from a petroleum-contaminated soil," Environment International, vol. 26, no. 5-6, pp. 413–416, 2001.
22. J. J. Perry, "Microbial metabolism of cyclic alkanes," in Petroleum Microbiology, R. M. Atlas, Ed., pp. 61–98, Macmillan, New York, NY, USA, 1984.
23. R. Atlas and J. Bragg, "Bioremediation of marine oil spills: when and when not—the Exxon Valdez experience," Microbial Biotechnology, vol. 2, no. 2, pp. 213–221, 2009. · ·
24. R. M. Atlas, "Petroleum microbiology," in Encyclopedia of Microbiology, pp. 363–369, Academic Press, Baltimore, Md, USA, 1992.
25. O. O. Amund and N. Nwokoye, "Hydrocarbon potentials of yeast isolates from a polluted Lagoon,"Journal of Scientific Research and Development, vol. 1, pp. 65–68, 1993.
26. B. Lal and S. Khanna, "Degradation of crude oil by Acinetobacter calcoaceticus and Alcaligenes odorans,"Journal of Applied Bacteriology, vol. 81, no. 4, pp. 355–362, 1996.
27. D. M. Jones, A. G. Douglas, R. J. Parkes, J. Taylor, W. Giger, and C. Schaffner, "The recognition of biodegraded petroleum-derived aromatic hydrocarbons in recent marine sediments," Marine Pollution Bulletin, vol. 14, no. 3, pp. 103–108, 1983.
28. S. A. Adebusoye, M. O. Ilori, O. O. Amund, O. D. Teniola, and S. O. Olatope, "Microbial degradation of petroleum hydrocarbons in a polluted tropical stream," World Journal of Microbiology and Biotechnology, vol. 23, no. 8, pp. 1149–1159, 2007.
29. J. Jones, M. Knight, and J. A. Byron, "Effect of gross population by kerosene hydrocarbons on the microflora of a moorland soil," Nature, vol. 227, p. 1166, 1970.
30. Y. Pinholt, S. Struwe, and A. Kjoller, "Microbial changes during oil decomposition in soil," Holarctic Ecology, vol. 2, pp. 195–200, 1979.
31. S. L. Hollaway, G. M. Faw, and R. K. Sizemore, "The bacterial community composition of an active oil field in the Northwestern Gulf of Mexico," Marine Pollution Bulletin, vol. 11, no. 6, pp. 153–156, 1980.
32. G. J. Mulkins Phillips and J. E. Stewart, "Distribution of hydrocarbon utilizing bacteria in Northwestern Atlantic waters and coastal sediments," Canadian Journal of Microbiology, vol. 20, no. 7, pp. 955–962, 1974.
33. R. Bartha and I. Bossert, "The treatment and disposal of petroleum wastes," in Petroleum Microbiology, R. M. Atlas, Ed., pp. 553–578, Macmillan, New York, NY, USA, 1984.
34. J. J. Cooney, "The fate of petroleum pollutants in fresh water ecosystems," in Petroleum Microbiology, R. M. Atlas, Ed., pp. 399–434, Macmillan, New York,

NY, USA, 1984.

35. R. M. Atlas, "Effects of hydrocarbons on micro-organisms and biodegradation in Arctic ecosystems," inPetroleum Effects in the Arctic Environment, F. R. Engelhardt, Ed., pp. 63–99, Elsevier, London, UK, 1985.View at Google Scholar
36. G. Floodgate, "The fate of petroleum in marine ecosystems," in Petroleum Microbiology, R. M. Atlas, Ed., pp. 355–398, Macmillion, New York, NY, USA, 1984.
37. K. S. M. Rahman, T. J. Rahman, Y. Kourkoutas, I. Petsas, R. Marchant, and I. M. Banat, "Enhanced bioremediation of n-alkane in petroleum sludge using bacterial consortium amended with rhamnolipid and micronutrients," Bioresource Technology, vol. 90, no. 2, pp. 159–168, 2003.
38. R. J. W. Brooijmans, M. I. Pastink, and R. J. Siezen, "Hydrocarbon-degrading bacteria: the oil-spill clean-up crew," Microbial Biotechnology, vol. 2, no. 6, pp. 587–594, 2009.
39. M. M. Yakimov, K. N. Timmis, and P. N. Golyshin, "Obligate oil-degrading marine bacteria," Current Opinion in Biotechnology, vol. 18, no. 3, pp. 257–266, 2007. · ·View at PubMed
40. K. Das and A. K. Mukherjee, "Crude petroleum-oil biodegradation efficiency of Bacillus subtilis andPseudomonas aeruginosa strains isolated from a petroleum-oil contaminated soil from North-East India,"Bioresource Technology, vol. 98, no. 7, pp. 1339–1345, 2007. · ·View at PubMed
41. M. Throne-Holst, A. Wentzel, T. E. Ellingsen, H.-K. Kotlar, and S. B. Zotchev, "Identification of novel genes involved in long-chain n-alkane degradation by Acinetobacter sp. strain DSM 17874," Applied and Environmental Microbiology, vol. 73, no. 10, pp. 3327–3332, 2007. · · View at PubMed
42. F. Chaillan, A. Le Flèche, E. Bury, Y.-H. Phantavong, P. Grimont, A. Saliot, and J. Oudot, "Identification and biodegradation potential of tropical aerobic hydrocarbon-degrading microorganisms," Research in Microbiology, vol. 155, no. 7, pp. 587–595, 2004. · · View at PubMed
43. A. J. Daugulis and C. M. McCracken, "Microbial degradation of high and low molecular weight polyaromatic hydrocarbons in a two-phase partitioning bioreactor by two strains of Sphingomonas sp,"Biotechnology Letters, vol. 25, no. 17, pp. 1441–1444, 2003.
44. H. Singh, Mycoremediation: Fungal Bioremediation, Wiley-Interscience, New York, NY, USA, 2006.
45. E. Bogusławska-Was and W. Dąbrowski, "The seasonal variability of yeasts and yeast-like organisms in water and bottom sediment of the Szczecin Lagoon," International Journal of Hygiene and Environmental Health, vol. 203, no. 5-6, pp. 451–458, 2001.
46. I. R. McDonald, C. B. Miguez, G. Rogge, D. Bourque, K. D. Wendlandt, D. Groleau, and J. Murrell, "Diversity of soluble methane monooxygenase-containing methanotrophs isolated from polluted environments," FEMS Microbiology

Letters, vol. 255, no. 2, pp. 225–232, 2006.

47. B. Jan, V. Beilen, M. Neuenschwunder, T. H. M. Suits, C. Roth, S. B. Balada, and B. Witholt, "Rubredoxins involved in alkane degradation," The Journal of Bacteriology, vol. 184, no. 6, pp. 1722–1732, 2003.
48. T. Iida, T. Sumita, A. Ohta, and M. Takagi, "The cytochrome P450ALK multigene family of an n-alkane-assimilating yeast, Yarrowia lipolytica: cloning and characterization of genes coding for new CYP52 family members," Yeast, vol. 16, no. 12, pp. 1077–1087, 2000.
49. J. B. Van Beilen, E. G. Funhoff, and E. G. Funhoff, "Cytochrome P450 alkane hydroxylases of the CYP153 family are common in alkane-degrading eubacteria lacking integral membrane alkane hydroxylases,"Applied and Environmental Microbiology, vol. 72, no. 1, pp. 59–65, 2006.
50. J. H. O. Maeng, Y. Sakai, Y. Tani, and N. Kato, "Isolation and characterization of a novel oxygenase that catalyzes the first step of n-alkane oxidation in Acinetobacter sp. strain M-1," Journal of Bacteriology, vol. 178, no. 13, pp. 3695–3700, 1996.
51. J. D. Walker, R. R. Colwell, Z. Vaituzis, and S. A. Meyer, "Petroleum degrading achlorophyllous algaPrototheca zopfii," Nature, vol. 254, no. 5499, pp. 423–424, 1975.
52. C. E. Cerniglia, D. T. Gibson, and C. Van Baalen, "Oxidation of naphthalene by cyanobacteria and microalgae," Journal of General Microbiology, vol. 116, no. 2, pp. 495–500, 1980.
53. M. L. Brusseau, "The impact of physical, chemical and biological factors on biodegradation," inProceedings of the International Conference on Biotechnology for Soil Remediation: Scientific Bases and Practical Applications, R. Serra, Ed., pp. 81–98, C.I.P.A. S.R.L., Milan, Italy, 1998.
54. R. M. Atlas, "Effects of temperature and crude oil composition on petroleum biodegradation," Journal of Applied Microbiology, vol. 30, no. 3, pp. 396–403, 1975.
55. A. Daverey and K. Pakshirajan, "Production of sophorolipids by the yeast Candida bombicola using simple and low cost fermentative media," Food Research International, vol. 42, no. 4, pp. 499–504, 2009.·
56. M. Kumar, V. León, A. De Sisto Materano, O. A. Ilzins, and L. Luis, "Biosurfactant production and hydrocarbon-degradation by halotolerant and thermotolerant Pseudomonas sp," World Journal of Microbiology and Biotechnology, vol. 24, no. 7, pp. 1047–1057, 2008.
57. K. Muthusamy, S. Gopalakrishnan, T. K. Ravi, and P. Sivachidambaram, "Biosurfactants: properties, commercial production and application," Current Science, vol. 94, no. 6, pp. 736–747, 2008.
58. A. Mahmound, Y. Aziza, A. Abdeltif, and M. Rachida, "Biosurfactant production by Bacillus strain injected in the petroleum reservoirs," Journal of Industrial Microbiology & Biotechnology, vol. 35, pp. 1303–1306, 2008.
59. N. Youssef, D. R. Simpson, K. E. Duncan, M. J. McInerney, M. Folmsbee, T.

Fincher, and R. M. Knapp, "In situ biosurfactant production by Bacillus strains injected into a limestone petroleum reservoir,"Applied and Environmental Microbiology, vol. 73, no. 4, pp. 1239–1247, 2007.

60. M. O. Ilori, C. J. Amobi, and A. C. Odocha, "Factors affecting biosurfactant production by oil degradingAeromonas spp. isolated from a tropical environment," Chemosphere, vol. 61, no. 7, pp. 985–992, 2005.·

61. A. Tabatabaee, M. M. Assadi, A. A. Noohi, and V. A. Sajadian, "Isolation of biosurfactant producing bacteria from oil reservoirs," Iranian Journal of Environmental Health Science & Engineering, vol. 2, no. 1, pp. 6–12, 2005.

62. J. M. Foght, D. W. S. Westlake, W. M. Johnson, and H. F. Ridgway, "Environmental gasoline-utilizing isolates and clinical isolates of Pseudomonas aeruginosa are taxonomically indistinguishable by chemotaxonomic and molecular techniques," Microbiology, vol. 142, no. 9, pp. 2333–2340, 1996.

63. A. D. Venosa and X. Zhu, "Biodegradation of crude oil contaminating marine shorelines and freshwater wetlands," Spill Science and Technology Bulletin, vol. 8, no. 2, pp. 163–178, 2003. ·

64. E. Pelletier, D. Delille, and B. Delille, "Crude oil bioremediation in sub-Antarctic intertidal sediments: chemistry and toxicity of oiled residues," Marine Environmental Research, vol. 57, no. 4, pp. 311–327, 2004.

65. D. Delille, F. Coulon, and E. Pelletier, "Effects of temperature warming during a bioremediation study of natural and nutrient-amended hydrocarbon-contaminated sub-Antarctic soils," Cold Regions Science and Technology, vol. 40, no. 1-2, pp. 61–70, 2004. ·

66. W. J. Mitsch and J. G. Gosselink, Wetlands, John Wiley & Sons, New York, NY, USA, 2nd edition, 1993.

67. S.-C. Choi, K. K. Kwon, J. H. Sohn, and S.-J. Kim, "Evaluation of fertilizer additions to stimulate oil biodegradation in sand seashore mesocosms," Journal of Microbiology and Biotechnology, vol. 12, no. 3, pp. 431–436, 2002.

68. S.-J. Kim, D. H. Choi, D. S. Sim, and Y.-S. Oh, "Evaluation of bioremediation effectiveness on crude oil-contaminated sand," Chemosphere, vol. 59, no. 6, pp. 845–852, 2005.

69. F. Chaillan, C. H. Chaîneau, V. Point, A. Saliot, and J. Oudot, "Factors inhibiting bioremediation of soil contaminated with weathered oils and drill cuttings," Environmental Pollution, vol. 144, no. 1, pp. 255–265, 2006.

70. J. Oudot, F. X. Merlin, and P. Pinvidic, "Weathering rates of oil components in a bioremediation experiment in estuarine sediments," Marine Environmental Research, vol. 45, no. 2, pp. 113–125, 1998.·

71. C. H. Chaîneau, G. Rougeux, C. Yéprémian, and J. Oudot, "Effects of nutrient concentration on the biodegradation of crude oil and associated microbial populations in the soil," Soil Biology and Biochemistry, vol. 37, no. 8, pp. 1490–1497, 2005.

72. L. M. Carmichael and F. K. Pfaender, "The effect of inorganic and organic

supplements on the microbial degradation of phenanthrene and pyrene in soils," Biodegradation, vol. 8, no. 1, pp. 1–13, 1997.

73. J. C. Okolo, E. N. Amadi, and C. T. I. Odu, "Effects of soil treatments containing poultry manure on crude oil degradation in a sandy loam soil," Applied Ecology and Environmental Research, vol. 3, no. 1, pp. 47–53, 2005.

74. H. Maki, T. Sasaki, and S. Haramaya, "Photooxidation of biodegradable crude oil and toxicity of the photooxidized products," Chemosphere, vol. 44, pp. 1145–1151, 2005.

75. W. Fritsche and M. Hofrichter, "Aerobic degradation by microorganisms," in Environmental Processes- Soil Decontamination, J. Klein, Ed., pp. 146–155, Wiley-VCH, Weinheim, Germany, 2000.

76. R. K. Hommel, "Formation and phylogenetic role of biosurfactants," Journal of Applied Microbiology, vol. 89, no. 1, pp. 158–119, 1990.

77. J. B. Van Beilen and E. G. Funhoff, "Alkane hydroxylases involved in microbial alkane degradation,"Applied Microbiology and Biotechnology, vol. 74, no. 1, pp. 13–21, 2007.

78. T. Zimmer, M. Ohkuma, A. Ohta, M. Takagi, and W.-H. Schunck, "The CYP52 multigene family ofCandida maltosa encodes functionally diverse n-alkane-inducible cytochromes p450," Biochemical and Biophysical Research Communications, vol. 224, no. 3, pp. 784–789, 1996.

79. U. Scheuer, T. Zimmer, D. Becher, F. Schauer, and W.-H. Schunck, "Oxygenation cascade in conversion of n-alkanes to α,ω-dioic acids catalyzed by cytochrome P450 52A3," Journal of Biological Chemistry, vol. 273, no. 49, pp. 32528–32534, 1998.

80. J. B. Van Beilen and E. G. Funhoff, "Expanding the alkane oxygenase toolbox: new enzymes and applications," Current Opinion in Biotechnology, vol. 16, no. 3, pp. 308–314, 2005.

81. M. O. Ilori, S. A. Adebusoye, and A. C. Ojo, "Isolation and characterization of hydrocarbon-degrading and biosurfactant-producing yeast strains obtained from a polluted lagoon water," World Journal of Microbiology and Biotechnology, vol. 24, no. 11, pp. 2539–2545, 2008.

82. G. S. Kiran, T. A. Hema, R. Gandhimathi, J. Selvin, T. A. Thomas, T. Rajeetha Ravji, and K. Natarajaseenivasan, "Optimization and production of a biosurfactant from the sponge-associated marine fungus Aspergillus ustus MSF3," Colloids and Surfaces B, vol. 73, no. 2, pp. 250–256, 2009.

83. O. S. Obayori, M. O. Ilori, S. A. Adebusoye, G. O. Oyetibo, A. E. Omotayo, and O. O. Amund, "Degradation of hydrocarbons and biosurfactant production by Pseudomonas sp. strain LP1," World Journal of Microbiology and Biotechnology, vol. 25, no. 9, pp. 1615–1623, 2009.

84. M. L. Brusseau, R. M. Miller, Y. Zhang, X. Wang, and G. Y. Bai, "Biosurfactant and cosolvent enhanced remediation of contaminated media," ACS Symposium Series, vol. 594, pp. 82–94, 1995.

85. G. Bai, M. L. Brusseau, and R. M. Miller, "Biosurfactant-enhanced removal of residual hydrocarbon from soil," Journal of Contaminant Hydrology, vol. 25, no. 1-2, pp. 157–170, 1997.

86. T. Barkay, S. Navon-Venezia, E. Z. Ron, and E. Rosenberg, "Enhancement of solubilization and biodegradation of polyaromatic hydrocarbons by the bioemulsifier alasan," Applied and Environmental Microbiology, vol. 65, no. 6, pp. 2697–2702, 1999.

87. S. S. Cameotra and P. Singh, "Bioremediation of oil sludge using crude biosurfactants," International Biodeterioration and Biodegradation, vol. 62, no. 3, pp. 274–280, 2008. ·

88. R. Beal and W. B. Betts, "Role of rhamnolipid biosurfactants in the uptake and mineralization of hexadecane in Pseudomonas aeruginosa," Journal of Applied Microbiology, vol. 89, no. 1, pp. 158–168, 2000.

89. O. Pornsunthorntawee, P. Wongpanit, S. Chavadej, M. Abe, and R. Rujiravanit, "Structural and physicochemical characterization of crude biosurfactant produced by Pseudomonas aeruginosa SP4 isolated from petroleum—contaminated soil," Bioresource Technology, vol. 99, no. 6, pp. 1589–1595, 2008.·

90. M. Nikolopoulou and N. Kalogerakis, "Biostimulation strategies for fresh and chronically polluted marine environments with petroleum hydrocarbons," Journal of Chemical Technology and Biotechnology, vol. 84, no. 6, pp. 802–807, 2009.

91. N. G. Wilson and G. Bradley, "The effect of immobilization on rhamnolipid production by Pseudomonas fluorescens," Journal of Applied Bacteriology, vol. 81, no. 5, pp. 525–530, 1996.

92. M. P. Díaz, K. G. Boyd, S. J. W. Grigson, and J. G. Burgess, "Biodegradation of crude oil across a wide range of salinities by an extremely halotolerant bacterial consortium MPD-M, immobilized onto polypropylene fibers," Biotechnology and Bioengineering, vol. 79, no. 2, pp. 145–153, 2002.

93. C. J. Cunningham, I. B. Ivshina, V. I. Lozinsky, M. S. Kuyukina, and J. C. Philp, "Bioremediation of diesel-contaminated soil by microorganisms immobilised in polyvinyl alcohol," International Biodeterioration and Biodegradation, vol. 54, no. 2-3, pp. 167–174, 2004. ·

94. R. N. Z. A. Rahman, F. M. Ghazali, A. B. Salleh, and M. Basri, "Biodegradation of hydrocarbon contamination by immobilized bacterial cells," Journal of Microbiology, vol. 44, no. 3, pp. 354–359, 2006.

95. W. J. Nichols, "The U.S. Environmental Protect Agency: National Oil and Hazardous Substances Pollution Contingency Plan, Subpart J Product Schedule (40 CFR 300.900)," in Proceedings of the International Oil Spill Conference, pp. 1479–1483, American Petroleum Institute, Washington, DC, USA, 2001.

96. R. Z. Hoff, "Bioremediation: an overview of its development and use for oil spill cleanup," Marine Pollution Bulletin, vol. 26, no. 9, pp. 476–481, 1993. ·

97. U.S. EPA, "Spill NCP Product Schedule," 2002, http://www.epa.gov/oilspill.

98. K. Lee, G. H. Tremblay, J. Gauthier, S. E. Cobanli, and M. Griffin,

"Bioaugmentation and biostimulation: a paradox between laboratory and field results," in Proceedings of the International Oil Spill Conference, pp. 697–705, American Petroleum Institute, Washington, DC, USA, 1997.

99. A. Ladousse and B. Tramier, "Results of 12 years of research in spilled oil bioremediation: inipol EAP 22," in Proceedings of the International Oil Spill Conference, pp. 577–581, American Petroleum Institute, Washington, DC, USA, 1991.

100. J. L. Ramos, A. Wasserfallen, K. Rose, and K. N. Timmis, "Redesigning metabolic routes: manipulation of TOL plasmid pathway for catabolism of alkylbenzoates," Science, vol. 235, no. 4788, pp. 593–596, 1987.

101. S. Panke, J. M. Sánchez-Romero, and V. De Lorenzo, "Engineering of quasi-natural Pseudomonas putidastrains toluene metabolism through an ortho-cleavage degradation pathway," Applied and Environmental Microbiology, vol. 64, no. 2, pp. 748–751, 1998.

102. F. Rojo, D. H. Pieper, K.-H. Engesser, H.-J. Knackmuss, and K. N. Timmis, "Assemblage of ortho cleavage route for simultaneous degradation of chloro- and methylaromatics," Science, vol. 238, no. 4832, pp. 1395–1398, 1987.

103. Y. Hrywna, T. V. Tsoi, O. V. Maltseva, J. F. Quensen III, and J. M. Tiedje, "Construction and characterization of two recombinant bacteria that grow on ortho- and para-substituted chlorobiphenyls,"Applied and Environmental Microbiology, vol. 65, no. 5, pp. 2163–2169, 1999.

104. B. D. Erickson and F. J. Mondello, "Enhanced biodegradation of polychlorinated biphenyls after site-directed mutagenesis of a biphenyl dioxygenase gene," Applied and Environmental Microbiology, vol. 59, no. 11, pp. 3858–3862, 1993.

105. A. Suyama, R. Iwakiri, N. Kimura, A. Nishi, K. Nakamura, and K. Furukawa, "Engineering hybridpseudomonads capable of utilizing a wide range of aromatic hydrocarbons and of efficient degradation of trichloroethylene," Journal of Bacteriology, vol. 178, no. 14, pp. 4039–4046, 1996.

106. T. C. Zwick, E. A. Foote, A. J. Pollack, et al., "Effects of nutrient addition during bioventing of fuel contaminated soils in an arid environment," in In-Situ and On-Site Bioremediation, pp. 403–409, Battelle Press, Columbus, Ohio, USA, 1997.

107. M. I. Van Dyke, H. Lee, and J. T. Trevors, "Survival of luxAB-marked Alcaligenes eutrophus H850 in PCB-contaminated soil and sediment," Journal of Chemical Technology and Biotechnology, vol. 65, no. 2, pp. 115–122, 1996. ·

108. B. M. Applegate, S. R. Kehrmeyer, and G. S. Sayler, "A chromosomally based tod-luxCDABE whole-cell reporter for benzene, toluene, ethybenzene, and xylene (BTEX) sensing," Applied and Environmental Microbiology, vol. 64, no. 7, pp. 2730–2735, 1998.

109. G. S. Sayler and S. Ripp, "Field applications of genetically engineered microorganisms for bioremediation processes," Current Opinion in Biotechnology, vol. 11, no. 3, pp. 286–289, 2000. ·

110. L. Masson, B. E. Tabashnik, A. Mazza, G. Préfontaine, L. Potvin, R. Brousseau,

and J.-L. Schwartz, "Mutagenic analysis of a conserved region of domain III in the Cry1ac toxin of Bacillus thuringiensis,"Applied and Environmental Microbiology, vol. 68, no. 1, pp. 194–200, 2002. ·

111. C. Sousa, V. De Lorenzo, and A. Cebolla, "Modulation of gene expression through chromosomal positioning in Escherichia coli," Microbiology, vol. 143, no. 6, pp. 2071–2078, 1997.

112. C. J. Kelly, C. A. Lajoie, A. C. Layton, and G. S. Sayler, "Bioluminescent reporter bacterium for toxicity monitoring in biological wastewater treatment systems," Water Environment Research, vol. 71, no. 1, pp. 31–35, 1999. ·

113. A. C. Layton, B. Gregory, T. W. Schultz, and G. S. Sayler, "Validation of genetically engineered bioluminescent surfactant resistant bacteria as toxicity assessment tools," Ecotoxicology and Environmental Safety, vol. 43, no. 2, pp. 222–228, 1999.

114. S. Le Floch, F.-X. Merlin, M. Guillerme, C. Dalmazzone, and P. Le Corre, "A field experimentation on bioremediation: bioren," Environmental Technology, vol. 20, no. 8, pp. 897–907, 1999.

115. S. Le Floch, F. X. Merlin, M. Guillerme, et al., "Bioren: recent experiment on oil polluted shoreline in temperate climate," in In-Situ and On-Site Bioremediation, pp. 411–417, Battelle Press, Columbus, Ohio, USA, 1997.

116. J. L. Schnoor, L. A. Licht, S. C. McCutcheon, N. L. Wolfe, and L. H. Carreira, "Phytoremediation of organic and nutrient contaminants," Environmental Science and Technology, vol. 29, no. 7, pp. 318A–323A, 1995.

117. D. Hecht and G. Badiane, "Phytoremediation," New Internationalist, June 1998.

118. K. V. Nedunuri, R. S. Govindaraju, M. K. Banks, A. P. Schwab, and Z. Chen, "Evaluation of phytoremediation for field-scale degradation of total petroleum hydrocarbons," Journal of Environmental Engineering, vol. 126, no. 6, pp. 483–490, 2000.

119. U.S. Army Corps of Engineers, Agriculturally Based Bioremediation of Petroleum-Contaminated Soils and Shallow Groundwater in Pacific Island Ecosystems, 2003.

120. R. K. Miya and M. K. Firestone, "Enhanced phenanthrene biodegradation in soil by slender oat root exudates and root debris," Journal of Environmental Quality, vol. 30, no. 6, pp. 1911–1918, 2001.

# 9
CHAPTER

# BIODEGRADATION OF CRUDE OIL BY FUNGI ISOLATED FROM GULF OF MEXICO

Hussein Al-Nasrawi

Fulbright visiting scholar, Florida state university, USA

## ABSTRACT

**In the present study, sand samples contaminated with oil spill were collected from Pensacola beach (Gulf of Mexico) and tested to isolates fungal diversity associated with beach sands and investigates the ability of isolated fungi for crude oil biodegradation.**

**From sixteen fungal strains, four strains were confirmed for biodegradation ability of crude oil, the isolated fungi belongs to Aspergillus niger with higher activity followed by Penicillium documbens, Cochliobolus lutanus and Fusarium solani. Aspergillus niger recorded the highest weight loss of 8.6%, Penicillium documbens (7.9 %) and Cochliobolus lutanus (4.7%) whereas the lowest weight loss was demonstrated by Fusarium solani strain 421502 (1.9%).**

## INTRODUCTION

The dominance of petroleum products in the world economy creates the conditions for distributing large amounts of complex compounds consist of hundreds of different hydrocarbon molecules, and a huge volume of oily sludge, a carcinogenic and a potent immunotoxicant [1,2]. Oil spillage is the accidental discharge or pouring of crude oil into the environment. It involves the contamination of any part of the environment with any liquid hydrocarbon. These spills endanger public health, imperil drinking water, devastate natural resources, and disrupt the economy [3]. Crude oil is a naturally occurring complex mixture of hydrocarbon and non-hydrocarbon compounds which at appropriate concentration, possesses a measurable toxicity towards living systems.

**Citation:** Al-Nasrawi H (2012) Biodegradation of Crude Oil by Fungi Isolated from Gulf of Mexico. J Bioremed Biodegrad 3:147. doi:10.4172/2155-6199.1000147.

The toxicity of crude oil or petroleum products varies widely, depending on their composition, concentration, environmental factors and on the biological state of the organisms at the time of the contamination [4].

Although oil spills from tankers and pipelines release crude oil particles to the water surface and move it to the beaches and contaminates living and nonliving organisms, microorganisms specially fungi have a higher tolerance to the toxicity of hydrocarbons due to their physiology and adaptation to such variations in the environment and have the mechanism for the elimination of spilled oil from the environment [5,6].

The effect of oil on microbial populations depends upon the chemical composition of the oil and on the species of microorganisms present. Populations of some microbes increase; typically, such microbes use the petroleum hydrocarbons as nutrients. The same crude oil can favor different genera at different temperatures [7].

In the aquatic ecosystems, fungi plays an important role during their ability in removing hazardous compounds from the water, whereas sediment particles contaminated with crude oil from oil spills is one of the desired ecological niche to fungi which inhabits such substrate and use carbon source from hydrocarbons in polluted sediment particles to biodegrade crude oil from the sediments in the beaches. Fungi have been found to be better degraders of petroleum than traditional bioremediation techniques including bacteria, and although hydrocarbon degraders may be expected to be readily isolated from a petroleum oil- associated environment, the same degree of expectation may be anticipated for microorganisms isolated from a totally unrelated environment [8,9].

Several authors have made lists containing bacteria and fungi genera that are able to degrade a wide spectrum of pollutants, proceeding from marine atmosphere as well as the soil [10-12].

Recently, many researchers studied the role of fungi in biodegradation process of petroleum products and the most common fungi which have been recorded as a biodegrades belongs to following genera: Alternaria, Aspergillus, Candida, Cephalosporium, Cladosporium, Fusarium, Geotrichum, Gliocladium, Mucor, Paecilomyces, Penicillium, Pleurotus, Polyporus, Rhizopus, Rhodotolura, Saccharomyces, Talaromyces and Torulopsis [3,4,13-20].

The aim of the present study is to isolation fungi from polluted beach sand in Gulf of Mexico and test the ability of isolated fungi in biodegradation of crude oil.

## MATERIALS AND METHODS

### Sample Collection and Culture Methods

Soil samples (400 g) from surface soil (0-15 cm depth) were collected from different localities in Pensacola beach (Gulf of Mexico) which was contaminated with crude oil. Sample were made from 3-4 random locations per plot, mixed and transferred into sterile bottles using sterile spatula for microbiological quality determination and stored in ice box to a void contamination. In the lab, stones and other unwanted soil debris were removed by using 2.5 mm sieve, one gram of each sorted soil sample was homogenously mixed with 1 drop (0.1 ml) of Tween 80 and a loopful (3 mm) of it was collected and inoculated by sprinkling method onto SDA and Czapek agar plates, respectively. Soil fungi were estimated by soil dilution plate count method. Sodium chloride 0.85 % was used as diluent for inoculum preparation. 1.0 g of homogenized, 2 mm sieved soil sample was aseptically transferred, using a flame-sterilized steel spatula, into a sterile test tube containing 9.0 mL of the diluent. This gave 10-1 dilution. Subsequently, three-fold (103) serial solutions were prepared from the 10-1 dilution. 1 ml of dilution was poured on Czapak Dox Agar (CDA) plates (30 g Powder of Czapak Dox Agar (CDA) and 49.5 gm of Malt Extract Agar (MEA) were added to 1000 ml of distilled water respectively. Streptomycin (500 mg/l) as antibiotic inhibit bacterial growth was added to the media after sterilization process.

Bushnell-Haas broth medium was used for the screening test which composed of: $MgSO_4$ (0.2 g/l), $CaCl_2$ (0.02 g/l), $KH_2PO_4$ (1 g/l), $KH_2PO_4$ (1 g/l), $FeCl_2$ (0.05 g/l) and $NH_4NO_3$ (1 g/l). Tween 80 (0.1%), redox reagent (2% 2, 6-dichlorophenol indophenols) and crude oil (1%) were incorporated into the broth.

## Identification of fungal isolates

Fungal genera were identified according to morphology characters and classified according to taxonomy keys in many literatures [21-28].

Species were identified by using DNA sequence method. A suitable mass of inoculum of fungal isolate was prepared with carefully removing the upper surface of the isolate without agar medium, the DNA extraction technique used to remove inhibitory materials, i.e. polysaccharides, proteins, mineral salts, etc., which limit the sensitivity of the different reactions in which isolated DNA is applied [29].

Genomic DNA was extracted from fungal isolates using a Mo-Bio Power Soil DNA extraction kit following manufacturer's protocol (Mo- Bio, Carlsbad CA, USA). Approximately 0.5 g of fungal hyphae were scraped off Petri dishes and transferred to bead tubes provided in the kit. Mechanical lysis was enhanced using a Talboys High Throughput Homogenizer (Troemner, Thorofare, NJ, USA) at 1600 rpm for 3 minutes. DNA extracts were assessed using a Nano drop ND-1000 Spectrophotometer (Thermo Scientific, Wilmington, DE, USA).

Fungal 18S rRNA genes were PCR amplified using NS-1 [5'-GTA GTC ATA TGC TTG TCT-3'] and FR-1 [AIC CAT TCA ATC GGT AIT]. Reactions were performed in 50 µl volumes containing 0.5 mM dNTPS, 0.5 µM of each primer, 1X DreamTaqTM Green Buffer, 1.5 U DreamTaqTM polymerase, and 10 ng template DNA. Thermo cycling conditions consisted of an initial denaturation stage of 95°C for 5 minutes followed by 35 cycles of 95°C for 45 seconds, 55°C for 90 seconds, and 72°C for 90 seconds, and a final stage of 72°C for 10 minutes.

Amplicons were cleaned using a Mo-Bio Ultra Clean® PCR Cleanup Kit and sent for sequencing on an Applied Biosystems 3130xl Genetic Analyzer (Applied Biosystems Inc., Foster City CA, USA). Raw sequence data were processed to remove low quality sequence data using the software package Sequencer (Gene Codes, Ann Arbor, MI, USA). The basic local alignment search tool-BLAST was used to classify and identify closely related fungal sequences.

## Screening Procedure

The biodegradability of isolated fungi was verified using the modified technique based on the redox indicator 2, 6-dichlorophenol indophenol (DCPIP) [30]. From 7 days fungal isolate old two plugs ( 1 cm2 for each plug ) were picked from the peripheral area of Petri dish and transferred carefully to inoculate into 50 ml Bacto Bushnell - Haas broth medium using 250 ml Conical flask. 0.1% (v/v) Tween 80 and 1% (v/v) crude oil and 0.008 mg/50 ml of redox indicator as a powder were added to the Bacto Bushnell-Haas broth medium. All flasks incubated in room temperature using a shaker with 180 rev/min for seven days. Changing in color of inoculated media in the flasks from deep blue to colorless indicates the ability of fungi to biodegradation of crude oil.

Inoculum of 0.2 ml of fungal isolate was added to essay tubes (triplicates) that contained 10 ml sterile Bushnell-Hass (BH) medium and 1% v/v of crude oil. The concentration of DCPIP was 0.16 mg/ ml. The tubes were kept under agitation 60 rpm at 28.0 ± 1.0°C. Biodegradation activity of fungi observed during the change of blue color of DCPIP to colorless.

To measure weight loss of fungal strains, weight loss method of Bartha and Bossert [31] was used in the present study. 10 ml of crude oil broth was prepared in a test tube and inoculated with 0.1 ml of filtrate from a soil sample. The test tube was then incubated at 35 ± 2°C being shaken on a mechanical shaker for 10 minutes at 200 rpm. After 48 hours, the remaining crude oil was separated using a separator funnel, and weighed. The percentage degradation of the crude oil was then calculated as described by Ijah and Ukpe [32].

$$\frac{\text{Weight of crude oil (initial)} - \text{Weight of crude oil (after treatment)}}{\text{Weight of crude oil (initial)}} \text{x}100$$

## Statistical analyses

The present study conducted an ANOVA (analysis of variance), which was performed on all the treatments and done using the SPSS (version 10.0) package to determine whether or not, a significance difference exists between the weight loss of degraded crude oil by the microorganisms over time. There is a significant effect of time (percent weight loss of crude oil increases over time) and there is also a species*time interaction (the percent weight loss of crude oil differs by fungal species) as shown in table 1.

**Table 1:** Results of Repeated Measures Analysis of Variance for weight loss of crude oil degraded by isolates

| | | | | | |
|---|---|---|---|---|---|
| N | | | | | 32 |
| DFE | | | | | 16 |
| All Between | | | | | |
| Test | Value | Exact F | NumDF | DenDF | Prob>F |
| F Test | 22590.437 | 24096.467 | 15 | 16 | <.0001 |
| Intercept | | | | | |
| Test | Value | Exact F | NumDF | DenDF | Prob>F |
| F Test | 5890.5625 | 94249 | 1 | 16 | <.0001 |
| Species | | | | | |
| Test | Value | Exact F | NumDF | DenDF | Prob>F |
| F Test | 22590.437 | 24096.467 | 15 | 16 | <.0001 |
| Within Subjects | | | | | |
| Contrast | | | | | |
| All With in Interactions | | | | | |
| Test | Value | Approx. F | NumDF | DenDF | Prob>F |
| Wilks' Lambda | 1.05e-7 | 3078.4312 | 30 | 30 | <.0001 |
| Time | | | | | |
| Test | Value | Exact F | NumDF | DenDF | Prob>F |
| F Test | 2286.1875 | 17146.406 | 2 | 15 | <.0001 |
| Time*Species | | | | | |
| Test | Value | Approx. F | NumDF | DenDF | Prob>F |
| Wilks' Lambda | 1.05e-7 | 3078.4312 | 30 | 30 | <.0001 |

## RESULTS AND DISCUSSION

Results of this study revealed that sixteen fungal strains were isolated from beach sand in Gulf of Mexico area, as shown in table 2, the mentioned fungal isolates were tested for insure their ability to biodegrade crude oil, three fungal strains demonstrated perfect biodegradation ability, Aspergillus niger, Cochliobolus lutanus and Penicillium documbcns as shown in figure 1 and 2.

**Table 2:** Fungal taxa isolated from polluted sands in Pensacola beach identified by length of DNA sequences

| Sequence ID | Sequence | Length | Top Blast | % Similarity | Class | Order | Family |
|---|---|---|---|---|---|---|---|
| >1 | TGGCTCA | 800 | Eupenicillium javanicum isolate AFTOL-ID 429 | 99 | Eurotiomycetes | Eurotiales | Trichocomaceae |
| >2 | CTCATTA | 795 | Fusariun solani strain 421502 | 99 | Sardariomycetes | Hypocreales | |
| >3 | TACCTTA | 858 | Cochliobolus lunatus | 99 | Dothideomycetes | Pleosporales | Pleosporaceae |
| >4 | CTTTACTA | 767 | Hypocrea lixii strain TXL051 | 100 | Sordariomycetes | Hypocreales | Hypocreaceae |
| >5 | AACTGCG | 877 | Aspergillus niger isolate 6 | 99 | Eurotiomycetes | Eurotiales | Trichocomaceae |
| >6 | CATTATA | 878 | Cordyceps sinensis | 99 | Sordariomycetes | Hypocreales | Ophiocordycipitaceae |
| >7 | CATTATACA | 723 | Gibberella fujikoroi | 98 | Sordariomycetes | Hypocreales | Nectriaceae |
| >8 | AATACTTTA | 764 | Hypocrea lixii | 100 | Sordariomycetes | Hypocreales | Hypocreaceae |
| >9 | TAAATCAGT | 398 | Pencillium decumbens | 94 | Eurotiomycetes | Eurotiales | Trichocomaceae |
| >10 | ATACTTTAC | 836 | Sphaerodes retispora | 99 | Sordariomycetes | Hypocreales | Ceratostomataceae |
| >11 | TTTATTTGA | 848 | Aspergillus niger | 99 | Eurotiomycetes | Eurotiales | Trichocomaceae |
| >12 | AGTACCTTA | 771 | Pencillum decumbens | 99 | Eurotiomycetes | Eurotiales | Trichocomaceae |
| >13 | ACCTTACTA | 754 | Alternaria alternata | 99 | Dothideomycetes | Pleosporales | Pleosporaceae |
| >14 | TTATTTGAT | 831 | Sphaerodes retispora | 99 | Sordariomycetes | Hypocreales | Ceratostomataceae |
| >15 | ATAATACCT | 759 | Pleospora herbarum | 98 | Dothideomycetes | Pleosporales | Pleosporaceae |
| >16 | TATCGTTTA | 811 | Eupenicillium javanicum | 100 | Eurotiomycetes | Eurotiales | Trichocomaceae |

**Figure 1:** Biodegradation ability of crude oil by fungal strains (from the left - A.niger strain 3, Cochliobolus lutanus strain 7 and control).

**Figure 2:** Biodegradation ability of crude oil by fungal strains (from the left, control and*Penicillium documbens* strain 4).

The rate of occurrence among the fungi isolates obtained from the soil samples is shown in Table 3. Aspergillus niger had the highest rate of occurrence, being 23.5 % while Cordyceps sinensis and Pleospora herbarum had the least occurrence rate of 1.9 %.

**Table 3:** Frequency of occurrence of the fungal isolates

| soil sample | Frequency of occurrence | | | | | |
|---|---|---|---|---|---|---|
| | *Alternaria alternata* | *Aspergillus niger* | *Cochliobolus lunatus* | *Cordyceps sinensis* | *Eupanicillium javanicum* | *Fusarium solani* |
| 1 | 1 | 3 | 0 | 0 | 1 | 3 |
| 2 | 2 | 1 | 0 | 1 | 0 | 1 |
| 3 | 3 | 2 | 0 | 0 | 0 | 4 |
| 4 | 1 | 2 | 1 | 0 | 1 | 2 |
| 5 | 1 | 1 | 0 | 0 | 0 | 0 |
| 6 | 2 | 3 | 1 | 0 | 0 | 1 |
| Total | 10 | 12 | 2 | 1 | 2 | 11 |
| Frequency (%) | 19.6 | 23.5 | 3.9 | 1.9 | 3.9 | 21.6 |

A)

| soil sample | Frequency of occurrence | | | | |
|---|---|---|---|---|---|
| | *Gibberella fujikuroi* | *Hypocrea lixii* | *Penicillium decumbens* | *Pleospora herbarum* | *Sphaerodes retispora* |
| 1 | 0 | 0 | 1 | 0 | 1 |
| 2 | 0 | 0 | 0 | 0 | 0 |
| 3 | 0 | 1 | 1 | 0 | 1 |
| 4 | 1 | 1 | 2 | 0 | 0 |
| 5 | 0 | 1 | 1 | 0 | 1 |
| 6 | 1 | 0 | 0 | 1 | 0 |
| Total | 2 | 3 | 5 | 1 | 3 |
| Frequency (%) | 3.9 | 5.9 | 9.8 | 1.9 | 5.9 |

B)

Table 4 shows the ability of fungal isolates in biodegradation of crude oil, by flask and test tube experiment

**Table 4:** Biodegradability experiment (DCPIP) by flasks and test tubes

| Fungal strain | decolourization in flasks | decolourization in test tubes |
|---|---|---|
| *Eupanicillium javanicum* strain1 | - ( until 15 days) | - ( until 21 days) |
| *Fusarium solani* strain 421502 | + ( after 7 days ) | + ( after 7 days ) |
| *Cochliobolus lunatus* | +++ ( after 3 days ) | +++ ( after 3 days ) |
| *Hypocrea lixii* strain TXL051 | - ( until 15 days) | - ( until 21 days) |
| *Aspergillus niger* strain 3 | +++ ( after 3 days ) | +++ ( after 3 days ) |
| *Cordyceps sinensis* | - ( until 15 days) | - ( after 15 days) |
| *Gibberella fujikuroi* | - ( until 15 days) | - ( until 21 days) |
| *Hypocrea lixii* strain 2 | - ( until 15 days) | - ( until 21 days) |
| *Penicillium decumbens* strain 1 | - ( until 15 days) | - ( until 21 days) |
| *Sphaerodes retispora* strain 1 | - ( until 15 days) | - ( until 21 days) |
| *Aspergillus niger* strain 1 | - ( until 15 days) | - ( until 21 days) |
| *Penicillium decumbens* strain 4 | +++ ( after 3 days ) | +++ ( after 3 days ) |
| *Alternaria alternata* | - ( until 15 days) | - (until 21 days) |
| *Sphaerodes retispora* strain 2 | - ( until 15 days) | - ( until 21 days) |
| *Pleospora herbarum* | - ( until 15 days) | - ( until 21 days) |
| *Eupanicillium javanicum* strain 2 | - ( until 15 days) | - ( until 21 days) |

-= no change in colour  + = mild  +++ = high

The screening method used in the present study depends on changing in the color of fungal isolates treated with redox indicator technique, so the fungal isolates which has the ability to degrade crude oil in the presence of redox indicator, confirmed the ability of the three fungi, Cochliobolus lutanus, Aspergillus niger and Penicillium documbens to biodegrade crude oil whereas low ability was observed by Fusarium solani strain 421502. Although Aspergillus and Penicillium species were recorded in former studies as crude oil biodegrades, the present study confirmed that the fungus Cochliobolus lutanus demonstrated as a new record fungus in biodegradation of crude oil.

The ability to analyses crude oil compounds to its components leads to oxidation of the carbon source in the crude oil components. There are three indicators leads to the ability of these fungi in biodegradation process, the first one is changes in color of culture media from blue to colorless, the second is disappearance of crude oil from the medium and the third is developing a mass of fungal growth in the bottom of the culture medium. Mechanism of biodegradation of crude oil occurred by incorporating an electron acceptor such as DCPIP to the culture medium, it is possible to ascertain the ability of the fungi to utilize the substrate by observing the color change of DCPIP from blue (oxidized) to colorless (reduced) [30].

In the present work Aspergillus niger and Penicillium documbens were the perfect fungal isolates demonstrated active ability to biodegrades crude oil, this result agree with results of Gesinde et al. [3] who indicated that Aspergillus niger have very active degradation capabilities of four kinds of oil compounds, Durb oil, Escravos light, Arabian light and Bonny light. Furthermore, in comparison with eight other genera, Aspergillus, Penicillium and Fusarium species were the most efficient metabolizers of hydrocarbons [4,33].

In a previous review, Bartha and Atlas [34] listed 14 genera of fungi isolated from an aquatic environment which had been demonstrated to contain members which utilize petroleum hydrocarbon. The evolution of the hydrocarbon mixture depends on the nature of the oil, microbial community, and environmental factors which impact microbial activities.

An interesting demonstration generated in this work which shows an increase in rates of fungal growth in the media containing crude oil compared with inoculated media without crude oil, this might be due to the fact that fungi use crude oil as a substrate for their survival growth using extra cellular enzymes to break down the recalcitrant hydrocarbon molecules, by dismantling the long chains of hydrogen and carbon, thereby, converting petroleum into simpler forms or products that can be absorbed for the growth and nutrition of the fungi [13].

According to table 5, there are different weight losses of fungal strains after 3 weeks of incubation. Aspergillus niger was demonstrated the highest weight loss (8.6% ) with Penicillium documbens (7.9%) and Cochliobolus lutanus (4.7%) whereas the lowest weight loss was demonstrated by Fusarium solani strain 421502 (1.9% ), this result agree with results of Gesinde et al. [3] who confirmed the ability of Aspergillus niger to biodegrade crude oil with 18% weight loss and penicillium notatum with 11.2% weight lost, this indicates that these isolates have the potential to utilize crude oil as a carbon source. Various workers have reported the similar results by micoorganisms utilize crude oil [34-38]. The significant difference in the weights of the crude oil samples before and after degradation by the fungal isolates confirms their performances, statistical analysis revealed a significant difference at 95% confidence level between weights of crude oil samples before and after exposure to fungal isolates.

**Table 5:** Percent weight loss of crude oil degraded by isolates, after inoculation (days)

| Fungal species | Percent weight loss after inoculation | | | | | |
|---|---|---|---|---|---|---|
| | 7 days | | 14 days | | 21 days | |
| | 1 | 2* | 1 | 2 | 1 | 2 |
| *Alternaria alternate* (Fr.) Keissl. | 0 | 0 | 0 | 0 | 0 | 0 |
| *Aspergillus niger* Tiegh. strain 1 | 0 | 0 | 0 | 0 | 0 | 0 |
| *Aspergillus niger* Tiegh. strain 3 | 3.1±0.3 | 3.1±0.3 | 6.1±0.8 | 6.2±0.8 | 7.8±0.6 | 7.9±0.6 |
| *Cochliobolus lunatus* R.R. Nelson & F.A. Haasis strain 7 | 1.5±1.0 | 1.5±1.0 | 2.1±0.0 | 2.2±0.0 | 4.7±0.5 | 4.0±0.5 |
| *Cordyceps sinensis* (Berk.) Sacc. | 0 | 0 | 0 | 0 | 0 | 0 |
| *Eupanicillium javanicum* (J.F.H. Beyma) Stolk & D.B. Scott strain1 | 0 | 0 | 0 | 0 | 0 | 0 |
| *Eupanicillium javanicum* (J.F.H. Beyma) Stolk & D.B. Scott | 0 | 0 | 0 | 0 | 0 | 0 |
| *Fusarium solani* (Mart.) Sacc. strain421502 | 0.7±0.3 | 0.7±0.3 | 1.9±0.6 | 1.9±0.6 | 1.9±0.5 | 1.9±0.5 |
| *Gibberella fujikuroi* (Sawada) Wollenw | 0 | 0 | 0 | 0 | 0 | 0 |
| *Hypocrea lixii* Pat. 1891 strain 1 | 0 | 0 | 0 | 0 | 0 | 0 |
| *Hypocrea lixii* Pat. strain TXL051 | 0 | 0 | 0 | 0 | 0 | 0 |
| *Penicillium decumbens* Thom strain 1 | 0 | 0 | 0 | 0 | 0 | 0 |
| *Penicillium decumbens* Thom strain 4 | 2.5±0.2 4 | 2.5±0.2 4 | 4.9±0.7 | 4.9±0.7 | 8.7±0.3 | 8.6±0.3 |
| *Pleospora herbarum* (Pers.) Rabenh. | 0 | 0 | 0 | 0 | 0 | 0 |
| *Sphaerodes retispora* (Udagawa & Cain) P.F. Cannon & D. Hawksw. strain 1 | 0 | 0 | 0 | 0 | 0 | 0 |
| *Sphaerodes retispora* (Udagawa & Cain) P.F. Cannon & D. Hawksw. strain 2 | 0 | 0 | 0 | 0 | 0 | 0 |

* = replicate

## CONCLUSIONS

Oil spills consider one of the critical problems faces global nature due to impact of discharge pollutants which cause decline of environment health. Currently the nature became more familiar with biological control solutions to remove hazardous from the environment. Although disposal methods can become prohibitively expensive when the amounts of pollutants are huge, using microbial remediation process is successful and safe way to enhance environment health in particular with low cost, technique and high public acceptance to cleaning up aquatic ecosystems from oil spills.


## ACKNOWLEDGEMENTS

I would like to thank the Kostka lab at FSU for their assistance with fungal isolation and taxonomy. This work was supported by a Fulbright scholarship to H. A-N.


## REFERENCES


1. Propst TL, Lochmiller RL, Qualls CW Jr, McBee K (1999) In situ (mesocosm) assessment of immunotoxicity risks to small mammals inhabiting petrochemical waste sites. Chemosphere, 38: 1049-1067.
2. Ojumu TV, Bello OO, Sonibare JA, Solomon BO (2004) Evaluation of microbial systems for bioremediation of petroleum refinery effluents in Nigeria. Afr J Biotechnol 4:31-35.
3. Gesinde AF, Agbo EB, Agho MO, Dike EFC (2008) Bioremediation of Some Nigerian and Arabian Crude Oils by Fungal Isolates. Int Jor P App Scs 2: 37-44.
4. Obire O, Anyanwu EC (2009) Impact of various concentrations of crude oil on fungal populations of soil. Int J Environ Sci Technol 6: 211-218.
5. Dibble JT, Bartha R (1979) The effect of environmental parameters on the biodegradation of oily sludge. Appl Environ Microbiol 37: 729-739.

6. Atlas RM (1995) Petroleum biodegradation and Oil Spill bioremediation. Mar Pollut Bull 31: 178-182.

7. Westlake DW, Jobson A, Phillippe R, Cook FD (1974) Biodegradability and crude oil composition. Can J Microbiol 20: 915-928.

8. Batelle CD (2000) Mushrooms: Higher Macrofungi to clean up the environment. Environmental Issues, Fall 2000.

9. Ojo OA (2005) Petroleum-hydrocarbon utilization by native bacterial population from a Waste water canal Southwest Nigeria. Afr J Biotechnol 5: 333-337.

10. Bouchez M, Blanchet D, Haeseler F, Vandecasteele JP (1996) Les hydrocarbures aromatiques polycycliques dans l'environnement. Revue de l'Institut franais du ptrole 51: 797-828.

11. Yateem A, Balba MT, Al-Awadhi N (1998) White rot fungi and their role in remediating oil-contaminated soil. Environ Int 24: 181-187.

12. Juhasz AL, Naidu R (2000) Bioremediation of high molecular weight polycyclic aromatic hydrocarbons: a review of the microbial degradation of benzo[a]pyrene. Int Biodeterior Biodegradation 45: 57-88.

13. Adekunle AA, Adebambo OA (2007) Petroleum Hydrocarbon Utilization by Fungi Isolated From DetariumSenegalense (J. F Gmelin) Seeds. Journal of American Science 3: 69-76.

14. Hadibarata T, Tachibana S (2009) Microbial Degradation of n-Eicosane by Filamentous Fungi. Interdisciplinary Studies on Environmental Chemistry, Environmental Research in Asia.

15. Hadibarata T, Tachibana S (2009) Microbial Degradation of Crude Oil by Fungi Pre-Grown on Wood Meal. Interdisciplinary Studies on Environmental Chemistry, Environmental Research in Asia.

16. Adekunle AA, Uaboni-Egbenni PO, Ajayi T (2004) Biodegradation of petroleum products by Saccharomyces cerevisae isolated from the Lagos, lagoon 17: 83- 94.

17. Atagana HI, Haynes RJ, Wallis FW (2006) Fungal bioremediation of creosote-contaminated soil: A laboratory scale bioremediation study using indigenous soil fungi. Water Air Soil Pollut 172: 201-219.

18. Husaini A, Roslan HA, Hii KSY, Ang CH (2008) Biodegradation of aliphatic hydrocarbon by indigenous fungi isolated from used motor oil contaminated sites. World J Microbiol Biotechnol 24: 2789-2797.

19. Romero MC, Urrutia MI, Reinoso HE, Kiernan MM (2010) Benzo[a]pyrene degradation by soil filamentous fungi. J yeast Fungal Res 1: 025-029.

20. Saraswathy A, Hallberg R (2002) Degradation of pyrene by indigenous fungi from a former gasworks site. FEMS Microbiol Lett 210: 227-232.

21. Booth C (1971) The genus Fusarium. Commonwealth Mycological Institute, Kew, Surrey, England.

22. Cai L, Zhang K, Hyde KD (2003) Freshwater Ascomycetes. Fungal diversity Research series 10: 275-324.

23. Ellis M B (1971) Dematiaceous Hyphomycetes. Commonwealth Mycological Institute, Kew, Surrey, England.

24. Ellis MB (1976) More dematiaceous Hyphomycetes. Commonwealth Mycological Institute, Kew, Surrey, England.

25. Goh TK, Clement KMT (2003) Key to common dematiaceous hyphomycetes from freshwater. Fungal diversity Research series 10: 325-343.

26. Hoog GS, Guarro J (1995) Atlas of clinical fungi. Gentra albureau Voor Schimmel cultures, Universitat Rovera in Virgili, Spain.

27. Hyde KD, Sarma VV (2000) Pictorial key to higher marine fungi. In Marine Mycology-A practical Approach, Fungal diversity Research series, Hong Kong.

28. Kohlmeyer J, Volkmann-Kohlmeyer B (1991) Illustrated key to the filamentous higher marine fungi. Botanica Marina 34: 1-61.

29. Sambrook J, Fritch EF, Maniatis T (1989) Molecular cloning: A laboratory manual. (2ndedn), Cold Spring Harbor Laboratory, Cold Spring Harbor, NY.

30. Hanson KG, Desai JD, Desai AJ (1993) A rapid and simple screening technique for potential crude oil degrading microorganisms. Biotechnol Tech 7: 745-748.

31. Bartha R, Bossert I (1984) The fate of petroleum in soil ecosystems. In: Petroleum Microbiology, Macmillan Publishing Company, New York.

32. Ijah UJJ, Ukpe LI (1992) Biodegradation of crude oil by Bacillus strains 28A and 61B isolated from oil spilled soil. Waste Manag 12: 55-60.

33. De N, Bello YM, Saleh M (2000) Biodegradation of crude oil by Fusarium sp. and Trichoderma sp. Isolated from oil contaminated soil in different auto mechanic garages.

34. Bartha R (1977) The microbiology of aquatic oil spills. Adv Appl Microbiol 22: 225-266.

35. Davies JS, Westlake DWS (1979) Crude oil utilization by fungi. Can J Microbiol 25: 146-156.

36. Amanchukwu SC, Obafemi A, Okpokwasili GC (1989) Hydrocarbon degradation and utilization by a palm- wine yeast isolate. FEMS Microbiol Lett 57: 151-154.

37. Okpokwasili GC, Nwosu AI (1990) Degradation of aldrin by bacterial isolates. Nig J Tech Res 2: 1-6.

38. Obire O, Wembedo SA (1998) Effect of oilfield wastewater on microbial population of a soil in Nigeria. Niger Delta Biologia 1: 77-85.

# 10
CHAPTER

# RHIZODEGRADATION OF HYDROCARBON FROM OILY SLUDGE

Asghari Bano[1], Asim Shahzad[1] and Samina Siddiqui[2]

[1]Department of Plant Scineces Quaid-i-azam University Islamabad, Pakistan
[2]Centre of Excellence in Geology University of Peshwar, Peshawar, Pakistan

## ABSTRACT

**Rhizoremediation involves the use of plant growth promoting rhizobacteria (PGPR) to remove organic pollutants from terrestrial environment. The main objective of this study was to evaluate the efficiency of rhizoremediation through inoculation of hydrocarbon degrading bacterial strains with and without nutrients in the rhizosphere sporadically become contaminated with oily sludge. Bacterial strains viz. Bacillus cereus (Acc KF859972), Bacillus altitudinis (Acc KF859970), comamonas (Acc KF859971) and Stenotrophomonas maltophilia (Acc KF859973) were isolated from various oily waste pits of oil fields. Autoclaved soil was treated with oily sludge at 7:3 ratio and Alfalfa plant was inoculated with individual plant growth promoting rhizobacteria (PGPR) with and without diammonium phosphate and ammonium nitrate. The different saturates and total hydrocarbon was extracted by Soxhelt extraction and was analyzed by GC-FID at 0, 5 and 10 d of incubation. The inoculation 30% sludge with B. altitudinis accelerated the rate of degradation of n-alkanes and some methyl branched than un-inoculated soils. Oily sludge inoculated with B. altitudinis indicated that more than 80% of hydrocarbons were degraded at day 5 of incubation. No further degradation was observed till the end of the incubation period. B. altitudinis is capable of degrading straight chain hydrocarbons rapidly than any other strains. Alfalfa is an N-symbiotic plant and increased nutrient content under the rhizosphere promote degradation of n-alkanes. All other bacterial strains also showed significant**

**Citation:** Bano A, Shahzad A, Siddiqui S (2015) Rhizodegradation of Hydrocarbon from Oily Sludge. J Bioremed Biodeg 6: 289. doi:10.4172/2155- 6199.1000289.

**interaction with alfalfa for the biodegradation of oily sludge but the rate of degradation remained slow than with B. altitudinis. Inoculation with B. altitudinis accelerated the rehabilitation process and within 5 days the rehabilitation occurs.**

## INTRODUCTION

Over the last several decades, large quantity of oily sludge has been removed during drilling and stored in open pits within the vicinity of oil fields. The entry of such waste to the terrestrial or aquatic environment has threat to the agro-environmental ecosystems.

Oily sludge is composed of a wide range of organic compounds such as alkanes saturate, aromatics and aspaltenes [1]. Short chain hydrocarbons such as $nC_{10}$ to $nC_{12}$ are considered to be phytotoxic even at low concentration in the oily sludge. Moreover, such hydrocarbons are degraded rapidly however the rate of degradation primarily depends on the concentration of these hydrocarbons in oily sludge and secondarily on the length of the exposure of contamination to soil with such hydrocarbons [2]. Rahman et al. [3] reported that the addition of inoculation of bacterial strains, biosurfactant and fertilizers enhanced the degradation of *n*-alkanes. The *n*-Alkanes in the range of$n$-$C_8$-$nC_{11}$ were completely degraded followed by $nC_{12}$-$nC_{21}$, $nC_{22}$-$nC_{31}$ and $nC_{33}$- $nC_{40}$ with % degradation of 100, 83-98, 80-85 and 57-73% respectively in 10 and 20% sludge at 56 days of incubation. Zand et al. [4] reported that 97% of total hydrocarbon was removed by planting Flax plant.

Oily sludge acts as a carbon substrate for diverse variety of microorganisms [5]. Siddiqui et al. [6] reported that bacterial population increased to 100 fold as compared to soil without previous history of hydrocarbon contamination. Fulekar [7] investigated two different soils and found that an increase in total microbial count was experienced with oily sludge contaminated soils, but clay soil showed no change after oil contamination.

Previous literature reported that there is a lag phase prior to utilize oily sludge as a carbon source by indigenous microorganisms [8-10]. These lag phases varies and depend on the type and concentration of hydrocarbons present in the oily sludge.

Numerous genera of bacteria such as *Aeromonas, Alcaligenes,Acinetobacter, Arthobacter, Bacillus, Brevibacterium, Geobacillus, Mycobacterium, Pseudomonas, Rhodococuss, Sphingomonas, Thermus* and *Xainthomonas* species has been isolated from oily sludge using PCR and DNA-DNA hyberdization [11,12]. Cerqueira et al. [13] isolated *Stenotrophomonasacidaminiphila, Bacillus megaterium* and*Bacillus cibi*, from *petrochemical oily* sludge and *Pseudomonas aeruginosa and B. cereus* capable of degradation of hydrocarbons from oily sludge of Brazil. Jiann-Hong et al. [14] isolated *B. altitudinis* from oily sludge and found that inoculation of such bacteria to soil become contaminated with benzene can accelerate the degradation of benzene.

Rhizoremediation includes both bioremediation and phytoremediation. Bioremediation means inoculation of oily sludge contaminated sites with hydrocarbon degrading bacterial consortium and nutrients more likely to reduce the lag phase required for indigenous bacterial population to degrade hydrocarbons into biomass, $CO_2$ and intermediate products. In addition to that phytoremediation means to use hydrocarbon tolerance plants such as alfalfa, soyabean, perennial ryegrass, fescue or kaller grasses or others to degrade hydrocarbons of oily sludge contaminated sites. Plants or rhizosphere will provide a unique environment for hydrocarbon degrading microorganisms to grow and increase in number and because of their combine effect of release of root exudates and microorganisms the rate of degradation of hydrocarbons is more rapid than in non rhizosphere environment of oily sludge contaminated sites. Through rhizoremediation natural rehabilitation process of degradation of oily sludge is more rapid than either through bioremediation or phytoremediation.

Pradhan et al. [15] reported that the degradation of poly aromatic hydrocarbons (PAHs) were greater in soil where switchgrass (*Panicumvirgatum*) and little bluestem grass (*Schizachyriumscoparium*) were grown. They found that around 57 and 47% of PAHs were degraded than uncontaminated soils. Omotayo et al. [16] reported that inoculation of 50 mg of crude oil per g of soil with consortium of hydrocarbon degrading bacteria such as *Arthrobacter sp., Bacillus pumilus, Bacillus sphaericus* and *Serratiamarcescens* and *perennial rye grass* has increased the rate of degradation of hydrocarbons and around 87.7% of total hydrocarbons were degraded from the soil at 45 days of incubation.

The study was aimed to enhance the natural rehabilitation process of oily sludge contaminated soils inoculated with plant growth promoting bacteria isolated from the host oily sludge and to understand the role of alfalfa on the degradation of hydrocarbons in oily sludge contaminated soils.

## MATERIALS AND METHODS

### Geology and Samples Description from Potwar Plateau

The Kohat-Potwar Plaeatu is situated in the south of the Himalaya and Karakorum Mountains. It is located between latitude 32° and 34°N and longitude 70° and 74°E. It is bounded by the Salt Range and Trans-Indus Range in the south, Kala chitta Range in the north. The Plateau opens up towards Jehlum River in the east. The west of the Plateau is bounded with Kurram-Parachinar Range [17]. Potwar plateau is an oil prone area. Up till now 40 structures have been drilled for oil exploration among which 10 structures have a potential of oil production [18]. The oily sludge from three oil fields within Potwar plateau were selected because of the difference in the reservoir depth, geological formation and age, API gravity, extent of degradation and crude oil composition (Table 1).

### Sample Collection

Soil samples were collected within the vicinity of Missa kasswal oil field of Potwar plateau. The root samples from Missa kasswal oil field were collected by the method of shaking and washing of root system as described by Berge et al. [19]. About 500 g of oily sludge was collected from oil waste pits (100×100 m) within oil fields and were brought to the laboratory for further analysis. Oily sludge was spread over plastic tray covered with aluminum foil. Around 50 mL of phosphate buffer saline solution was mixed thoroughly with glass spatula and the oily sludge was placed in glass bottles (250 mL). All glass bottles were incubated in the refrigerator at 4 °C for further analysis.

## DNA Extraction

Extraction of genomic DNA of bacterial strain was carried out by using the DNA GenElute Bacterial Genomic DNA Kit and DNA was visualized on agarose gel method as described by Kate [20].

## PCR (Polymerase Chain Reaction)

The genomic DNA of isolated strains was amplified by the method as described by Weisburg et al. [21]. The Polymerase chain reaction was performed by using two universal primers fd1 fD1=(AGAGTTTGATCCTGGCTCAG) and rd1=(AAGGAGGTGATCCAGCC).

## Preparation of 30% Sludge and Treatment Application

Soil was autoclaved at 121°C and sludge was mixed to soil. Sludge (30%) was prepared by mixing oily sludge with soil in 3:7 ratios. After mixing sludge with soil 1 kg samples of sludge treated soil was transferred to 1 kg plastic containers. The pots were covered with loosely fitted perforated lids and were incubated at room temperature for 60 days to allow sufficient time for most hydrocarbons to breakdown.

The nutrient broth (Nutrient Broth OXOID UK) was inoculated with pure culture of isolated PGPR strains and incubated in shaker incubator (EXCELLA E24 Germany) for 48-72 h at 150 rpm. Thereafter, the culture was centrifuged at 3000 rpm. The pellet was resuspended in autoclaved distilled water and the O.D was adjusted to 1 at 660 nm. The fertilizer solutions were made in distilled water by adding Ammonium nitrate and Diammonium phosphate (DAP). at the rate of N=70 µg/ml and P=7 µg/ml. Sludge (300 g) was filled in small pots measuring 3 cm × 3 $cm^2$ for single strain experiment in which case the single strain (20 ml) was added to the sludge and incubated at room temperature.

Nutrients such as N ($NH_4NO_3$) and $(NH_4)_2HPO_4$ were added at the rates of N and P i.e. N=70 µg/g and P=7 µg/g to each control and 30% sludge. The rates of N and P were N=70 µg/g and P=7 µg/g. Aqueous solutions (30 ml) of ($NH_4NO_3$) and $(NH_4)_2HPO_4$ solution were pipetted over the sludge and mixed well by hand several times. The PGPR inocula as single strain were added to the sludge. The Inocula was prepared by inoculating pure cultures of four PGPR of individual strain having O.D 1 at 660 nm. The samples were kept in large plastic containers and were covered with loosely fitted perforated lids and incubated at 22°C for 60 days.

## Determination of Total Bacterial Population

For viable count of aerobic heterotrophs, the pour plate method was used [22].

## Breakdown of Diesel Hydrocarbons

Triplicates (5 g) were taken after 0, 5 and 10 days of incubation from the soil/treatments and were placed into cellulose thimble of known weight. These were placed into Soxhelt extraction apparatus. Anthracene (Sigma chemical company, Germany), was used as an internal standard. This was added to bulk sample of around 50 g of soil at the rate of 1 mg/g and mixed carefully before taking 10 g subsamples. Re-distilled dichloromethane (100 ml) was poured into a clean, preweight 250 ml round bottomed flask and were soxhelt extracted as described by Song and Bartha [22]. The extract was analysed by GCFID (QC 20-Schmadzu) coupled to a flame ionization detector.

**Table 1:** The difference in the depth, formation/age, API gravity and biodegradation of crude oil within Potwar plateau

| Sr. No. | Oil wells | Reservoir depth | Geological formation/age | API gravity | Extent of contamination (years) | Biodegradation level |
|---|---|---|---|---|---|---|
| 1 | ChakNaurang (CNG 1-A) | 2687 m | Sakesar/chorgali/Khewra/Cambrian | 18.4 | 27 | Biodegraded (3) |
| 2 | Dhakni | 4,844 | Lockhart/shakesar/chorgali | 39.31 | 24 | Biodegraded (3) |
| 3 | MissaKeswal | 2187m | Chorgali/lower cambrian | 36.2 | 21 | Non biodegraded (0) |

## Crude Oil Fractionation into Saturates (*n*-alkanes and isopreniods), Aromatics and Naphthlenes (isoprenoids)

Separation of saturates (*n-alkanes and isopreniods*), aromatics and polar (risens and asphaltenes) from crude oil was carried out by silica gel 60 column chromatorgraphy as described by Asif et al. [23] For column chromatography, a glass column (40 × 0.9 cm i.d.) with cotton wool at bottom was rinsed with 20 ml of dichloromethane. Around 200 g of silica gel was placed in 250 ml beaker and was activated at 120°C for 24 hours prior to be used for column chromatography. About 10 g of activated silica gel 60 with mesh size 35-70 mm (Fluka-Germany) was removed from the oven and was poured over the column with cotton wool at the bottom. The silica gel 60 was saturated with n-hexane (50 ml). Once the column bed is prepared around 50 mg of crude oil was transferred to the column from Pasteur pipette. The saturate fraction of crude oil was eluted with n-hexane (35 mL) and was collected in 50 ml of bottles. Once saturate fraction is obtained fractionation of aromatic compounds was carried out by adding a mixture of n-hexane:dichloromethane (35 mL, 7:3) which was collected in glass bottle (50 mL) prior to be analyzed by GCFID. Polar compounds were fractionated with 35 mL of methanol: dichloromethane (1:1). The solvent in each fraction was evaporated through roto-evaporator at 60°C temperature.

## Gas Chromatography-Flame Ionization Detector (GC-FID)

GC-FID analysis was performed using a QC 2010 Shimadzu GCFID. A 30 m × 0.25 mm ID capillary column coated with a 0.25 μm 5% phenyl 95% methyl polysiloxane stationary phase (DB-5 MS, J & W scientific) was used for the analysis. 1 μL of the saturated or aromatic fractions (1 mg/mL in n-hexane) was introduced into the split/splitless injector using the QC 2010 auto-sampler. The injector was operated at 320° in pulsed splitless mode. Helium maintained at a constant flow rate of 1.1 mL/min was used as carrier gas. The GC oven was programmed from 75°C to 320°C at 10°C/min with initial and final hold times of 28.83 minutes, respectively.

The peak area of each standard on GC trace was converted to mg $kg^{-1}$ of the hydrocarbon. The response factor of each external standard was calculated by taking the mean of response factor of 5 step dilutions calibration curve using following formulas.

The calibration curve is determined by the analysis of 3 calibration levels, i.e. 0.1, 0.25 and 0.5 mg kg-1. Peak area was calculated and calibration curve was plotted for concentration vs response area or peak area. The calibration curves were best fitted to a linear curve. The correlation coefficients (R) were 0.9947. The quantification was performed from the mean of two calibration curves surrounding the samples.

### Statistical Analysis of DATA

The data of isolates were analyzed by complete randomized design using LSD and compare means by statistix 8.1.

## RESULTS AND DISCUSSION

### Alignment of 16S rRNA Sequence

For the isolate obtained from oily sludge of Missa kasswal the total length of sequence with 1497 nucleotide was obtained. The comparison of the nucleotide sequence with data nucleotide bank showed highest 99% sequence similarity for 1463/1466 nucleotide bases with that of B. cereus strain partial sequence (Acc KF859972).

For the isolate obtained from oily sludge of Missal kasswal the total length of sequence with 1535 nucleotide was obtained. The comparison of the nucleotide sequence with data nucleotide bank showed 99% sequence similarity for 1477/1485 nucleotide bases with that*Commamonas* bacterium clone EK_An354 16S ribosomal RNA gene, partial sequence (Acc KF859971).

For the isolate obtained from oily sludge of Chak Naurang the total length of sequence with 1501 nucleotide was obtained. The comparison of the nucleotide sequence with data nucleotide bank showed highest 99% sequence similarity for 1477/1482 nucleotide bases with that of *S. maltophilia* strain E56 16S ribosomal RNA gene, partial sequence (Acc KF859973)

For the isolate obtained from Dhakni oil field total length of sequence with 1506 nucleotide was obtained. The comparison of the nucleotide sequence with data nucleotide bank showed highest 99% sequence similarity for 1466/1475 nucleotide bases with that of *B. altitudinis* strain: 41KF2b 16S ribosomal RNA, partial sequence (Acc KF859970).

Several genera of bacteria were isolated from oily sludge such as *Pseudomonas, Bacillus, Proteobacteria*(with a predominance of*Alphaproteo bacteria*and*Gammaproteo bacteria*), *Actinobacteria, Cytophaga-Flavobacterium-Bacteroides, Firmicutes. Spirochaetes* [24- 26]*S. maltophilia and B. cereus* were isolated from oily sludge (*Chak Naurang*) and (*Missa Kasswal*). *Yoshida* et al. [27], *Erdoğan* et al. [28],*Saeideh Rajaei* et al. [29] reported *S. maltophilia* a gram –ive bacteria causing human bacteria because it causes human infection which is very difficult to treat. However this is also common bacteria found in oily sludge. This is the first study reporting the presence of*Comamonadaceae* sp in oily sludge of Missa kasswal.

## Bacterial Population during degradation

The bacterial count was lower in untreated sludge (30%). Inoculating sludge with PGPR increased the proliferation of the PGPR as evidenced by the log cfu/g sludge. The *B. cereus* being the most effective and showed 79% increase over that of untreated control. Addition of fertilizer decreased the cfu of *B. cereus* but had no significant effect on cfu of *S. maltophilia.* The cfu of *B. altitudines* was higher 67% in presence of fertilizer. At 10 d of inoculation the cfu was decreased even in untreated sludge but this decrease was 55% less in *B. cereus* and 32% in *S. maltophilia. B. altitudinies* showed 92% significant increase in presence of fertilizer when compared with untreated sludge (Figure 1a).

The cfu of B. cereus was 21% higher than untreated sludge in presence of alfalfa but it was 43% less than *B. cereus* used alone. The*commamonas* had 16% higher cfu in presence of alfalfa but the presence of alfalfa had decreased the cfu of *S. maltophilia* and *B. altitudinies*by 93 and 98% respectively. Addition of fertilizer had significantly increased the cfu of *B. cereus* in the presence of alfalfa. Even at 10 d after inoculation the cfu of *B. cereus* was higher 88% in presence of alfalfa.

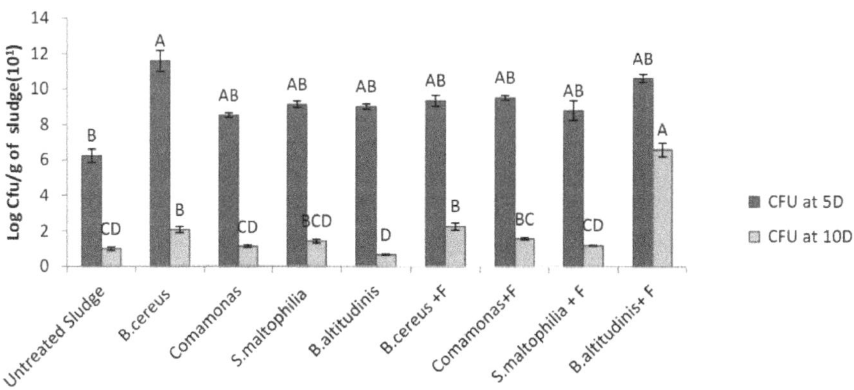


(a)

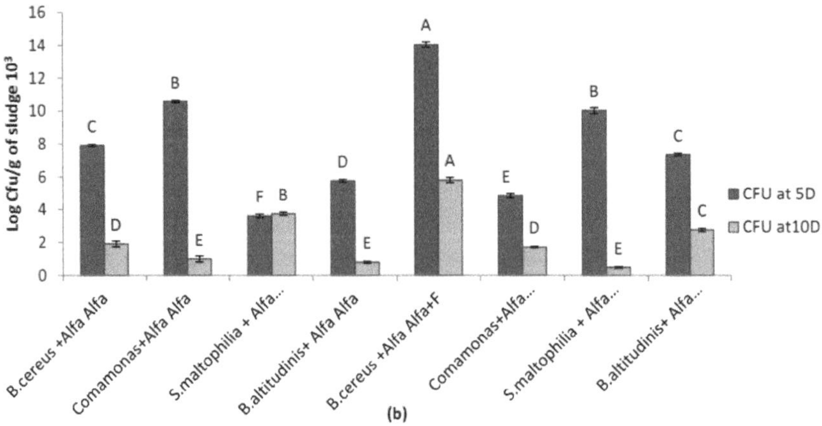


**Figure 1:** S. maltophilia. B. altitudinies showing 92% significant increase in presence of fertilizer when compared with untreated sludge.

Addition of fertilizer also increased the cfu of *S. maltophilia* and *B. altitudinies* over that of single inoculation in presence of alfalfa but the value was lower than that of *Commamonas* + alfalfa without fertilizer. However, this increase was of much lower magnitude than that of B. cereus alone. At 10 d of inoculation in presence of alfalfa variable responses observed being stimulatory for *S. maltophilia* and *B. cereus* in presence of fertilizer (Figure 1b).

Results revealed that all the PGPR bioinoculants isolated from oily sludge used to reinoculate the sludge showed increase in cfu of respective PGPR strain. Addition of fertilizer may be beneficial for certain bacterial strains perhaps acting as C/N source eg in *B. altitudines*however, the addition of fertilizer and growing alfalfa may be beneficial for other strain e.g. *B. cereus* both for short term and long term incubation. The B. cereus can survive well in the oily sludge and show significant cfu even at 5 d and *B. altitudines* needed fertilizer to have higher cfu and maintained the increased cfu even at 10 d.

Planting alfalfa can assist *Commamonas* to proliferate in oily sludge whereas *B. cereus* needed fertilizer as C/N source along with alfalfa to have significant increase in cfu both in long term (10 d) and short term (5 d) of incubation in sludge (30%). Different species of bacteria exhibit different mechanism of survival in oily sludge. In presence of alfalfa the *S. maltophilia* exhibited lower cfu at 5 d which increased at 10 d possibly the microbe has to adjust with the root exudates of the plant at early stage but may outcompete at later stages. Whereas, *B. cereus* and *Commamonas* show inhibition in cfu when fertilizer was added along with the plant alfalfa. The existence of plants inspired microbial activities and bacteria populations and the microbial populations in soil were enhanced more greatly by strain inoculated without alfalfa and Indian mustard [30], the bacterial count was increased with all the strain when used in association of alfalfa and this increased in bacterial population is might be due to extra nutrition provided by root exudates of alfalfa as compared to strain used alone. This statement supports our result findings.

## Degradation of *n*-alkanes

The fertilizer has significant effect on the degradation of ($nC_{13}$- $nC_{16}$ series) when oily sludge was inoculated with *B. cereus*. Similar was the case with commamonas and *S. maltophilia* However *B. altitudines* decreased the % degradation by 20% but *B. cerus* has no significant difference in the degradation of n-alkanes in presence of alfalfa which showed 22% degradation in presence of alfalfa in contrast to 36% without alfalfa. At 10 d of inoculation 50% or greater degradation occurred with the addition of fertilizer even within 5 d of inoculation the % degradation was earlier and almost similar to that of 10 d of inoculation without fertilizer and no significant effect of further incubation period after inoculation was detected (Figure 2a and 2b).

Growing alfalfa has further improved the % degradation in *B. cereus* and *S. maltophilia* inoculated sludge but delayed and reduced the % degradation in *B. altitudinis.* At day 10 the effect was similar to that of single inoculation. Addition of both alfalfa and fertilizer enhanced the degradation rate in *B. cereus, commamonas, S. maltophilia and B. altitudinis* (Figure 2c and 2d).

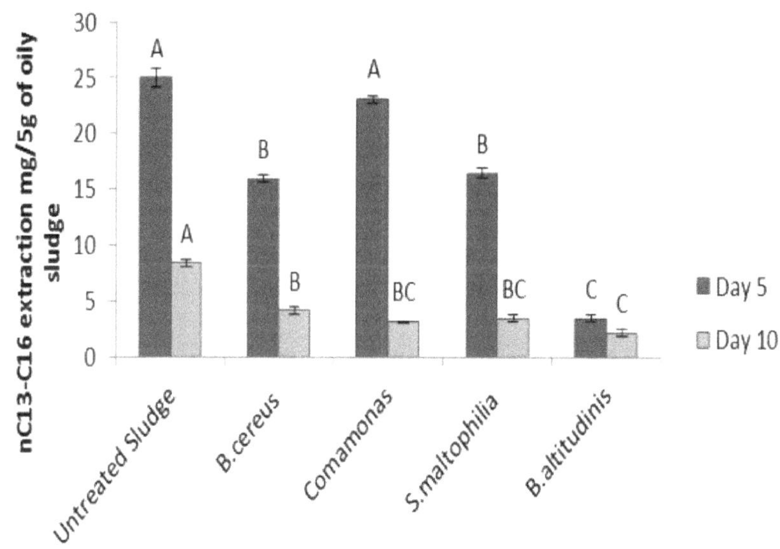


(a)

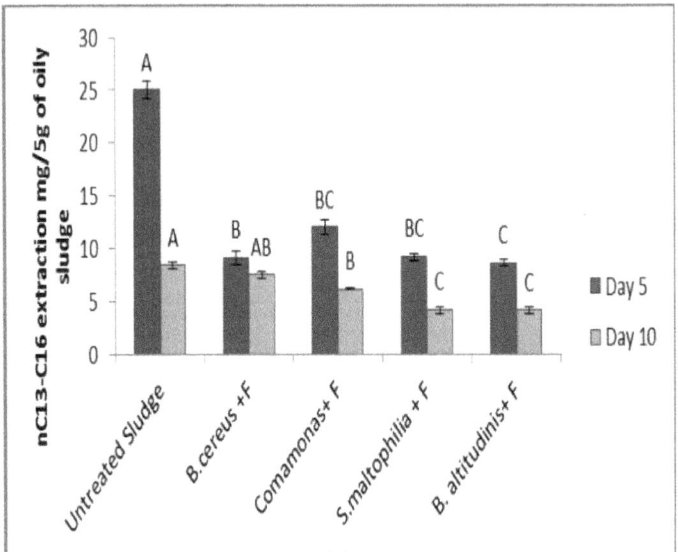
nC13-C16 extraction mg/5g of oily sludge
30
25
20
15
10
5
0
A
A
B
AB
BC
B
BC
C
C
C
Untreated Sludge
B.cereus +F
Comamonas+ F
S.maltophilia + F
B. altitudinis+ F
Day 5
Day 10

(b)

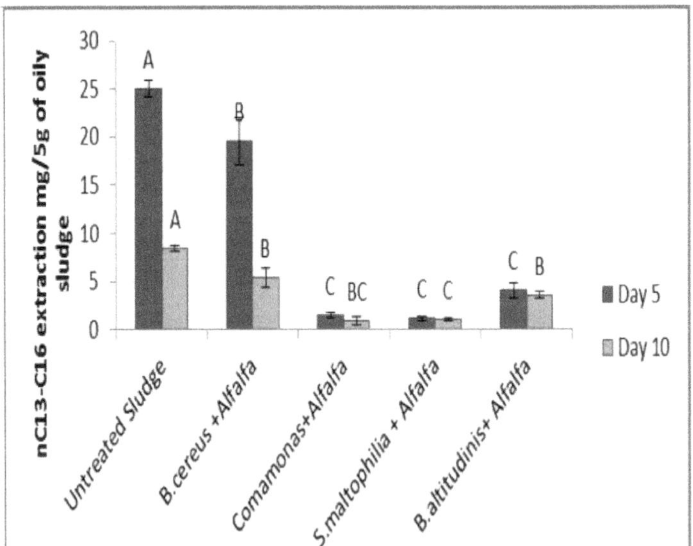
nC13-C16 extraction mg/5g of oily sludge
30
25
20
15
10
5
0
A
A
B
B
C
BC
C
C
C
B
Untreated Sludge
B.cereus +Alfalfa
Comamonas+Alfalfa
S.maltophilia + Alfalfa
B.altitudinis+ Alfalfa
Day 5
Day 10

(c)

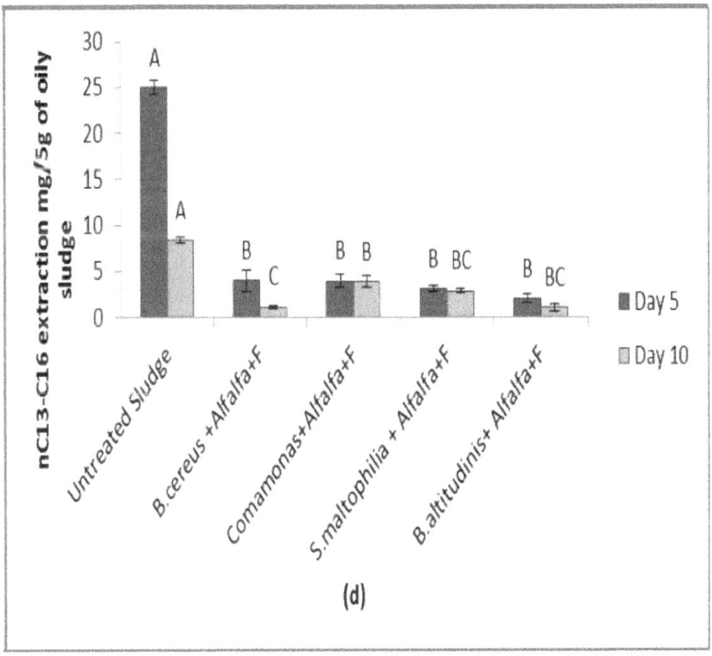


**Figure 2:** Effect of further incubation period after inoculation.

The concentration of $nC_{19}$-$nC_{29}$ was 84 % higher at 5 d but was 64 %less in 30% oily sludge at 10 d of incubation as compared to $nC_{17}$-$nC_{18}$. B. altitudinis followed by B. cereus were most effective and showed 82 to 64% degradation at 5 d over that of untreated sludge (Figures 3 and 4).

Growing alfalfa significantly increased the rate and % degradation such that negligible amount of $nC_{19}$-$nC_{29}$ were deleted in inoculated sludge. Addition of fertilizer to alfalfa + inoculation had no such marked effect except that in *S maltophilia* and *B. altitudines* which showed early degradation. In presence of both alfalfa and fertilizer the *S. maltophilia* inoculation resulted in maximum degradation over that of single inoculation (Figure 4c and 4d).

The results (Figure 5) revealed that degradation of pristane (isoprenoid) was enhanced by all the PGPR isolates., but the most effective were*B. altitudines* followed by *Commamonas* which enhanced the process of degradation, degrading about 50 to 80 % within 5 d of incubation. *B. cereus* took 10 d; alfalfa plantation was more effective for *B. cereus* and *S. maltophilia* than the fertilizer treatment. Whereas, combined treatment of fertilizer and alfalfa were not much effective to promote the effect of inoculation than that of alfalfa +inoculation. Addition of fertilizer significantly suppressed the process of biodegradation of long chain alkane's $nC_{30}$-$nC_{33}$ both at 5 and 10 d of incubation. Instead growing alfalfa increased the rate of degradation in *B. cereus, Commamonas* and *S. maltophilia* but has suppressed the efficiency of *B. altitudinis* (Figure 6).

The degradation of $nC_{13}$-$nC_{16}$ hydrocarbon (alkane series) was faster in presence of fertilizer and fertilizer +alfalfa but the rate and % degradation varied with the PGPR species. All the 4 PGPR used showed enhanced degradation in presence of fertilizer but planting alfalfa in oily sludge increased the % and rate of degradation in *Commamonas* and *S. maltophilia* because of the fact that alfalfa assist phytoremediation and degradation of hydrocarbon. The other fact might be due to indigenous microbes of oily sludge were not efficient enough to degrade multiple range of complex compounds of hydrocarbons and therefore, introduction of efficient hydrocarbon degrading bacteria can be useful to effectively degrade most of the hydrocarbons in a complex petroleum mixture [31] (Figure 7).

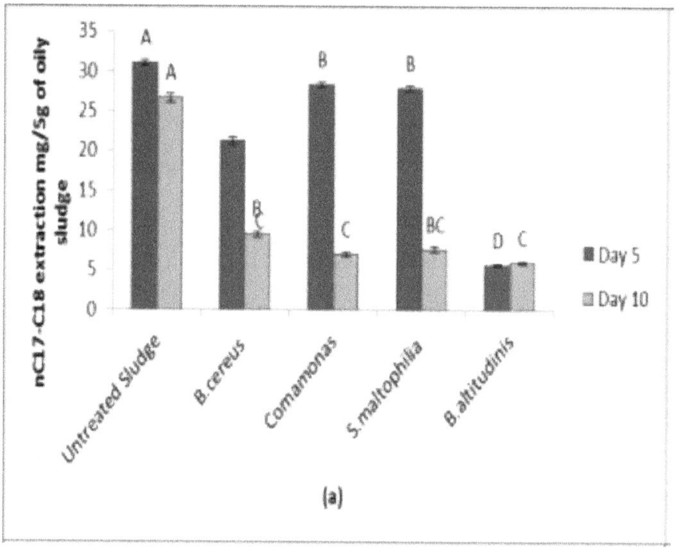


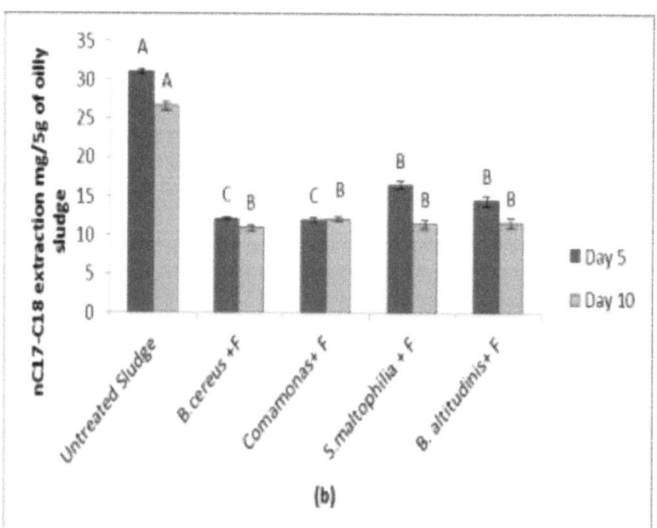

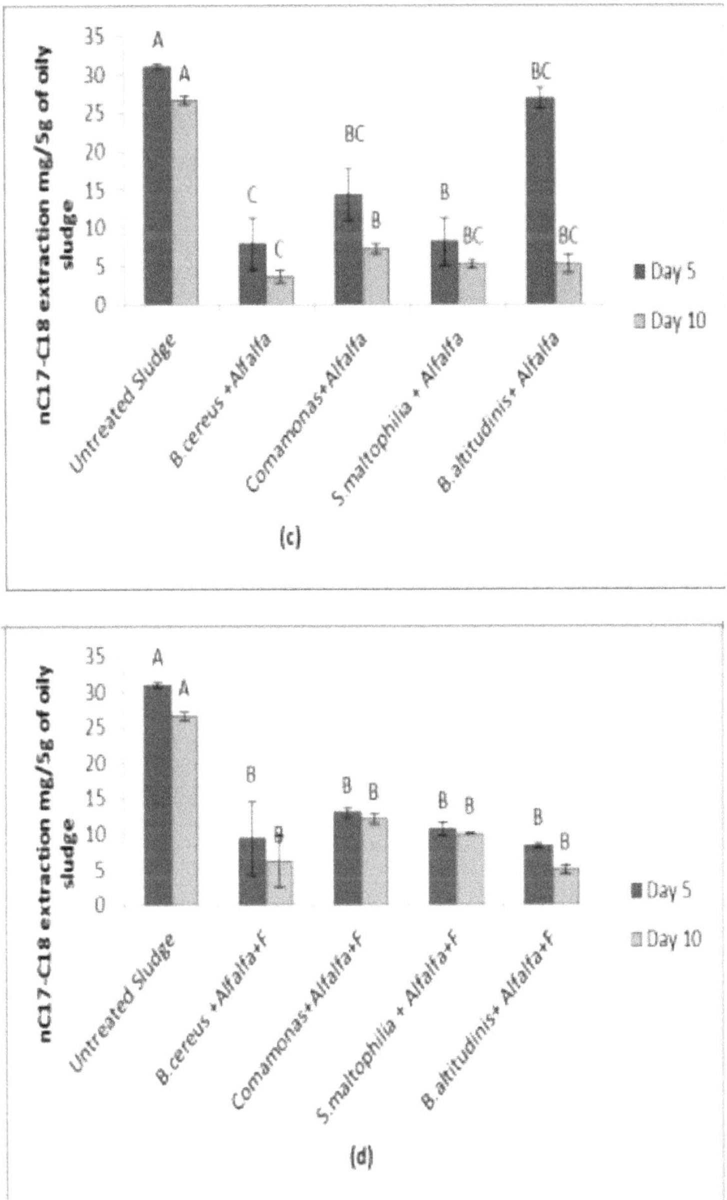


**Figure 3:** The concentration of $nC_{19}$-$nC_{29}$ was 84 % higher at 5 d but was 64 % less in 30% oily sludge at 10 d of incubation as compared to $nC_{17}$-$nC_{18}$. B. altitudinis.

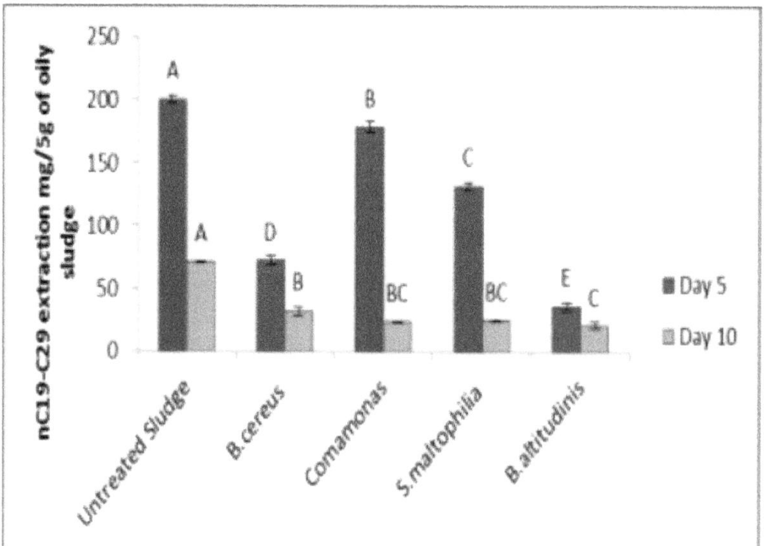
nC19-C29 extraction mg/5g of oily sludge
250
200
150
100
50
0
A
A
D
B
B
BC
C
BC
E
C
Day 5
Day 10
Untreated Sludge
B.cereus
Comamonas
S.maltophilia
B.altitudinis
(a)

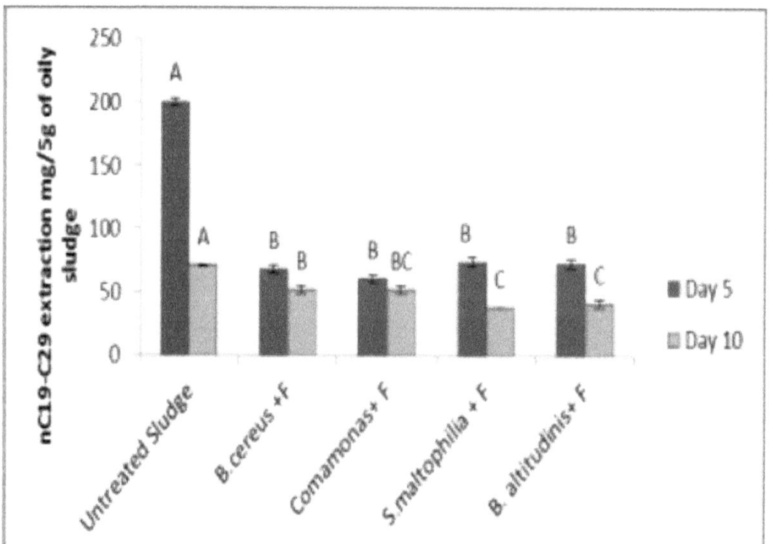
nC19-C29 extraction mg/5g of oily sludge
250
200
150
100
50
0
A
A
B
B
B
BC
B
C
B
C
Day 5
Day 10
Untreated Sludge
B.cereus +F
Comamonas+ F
S.maltophilia + F
B. altitudinis+ F
(b)

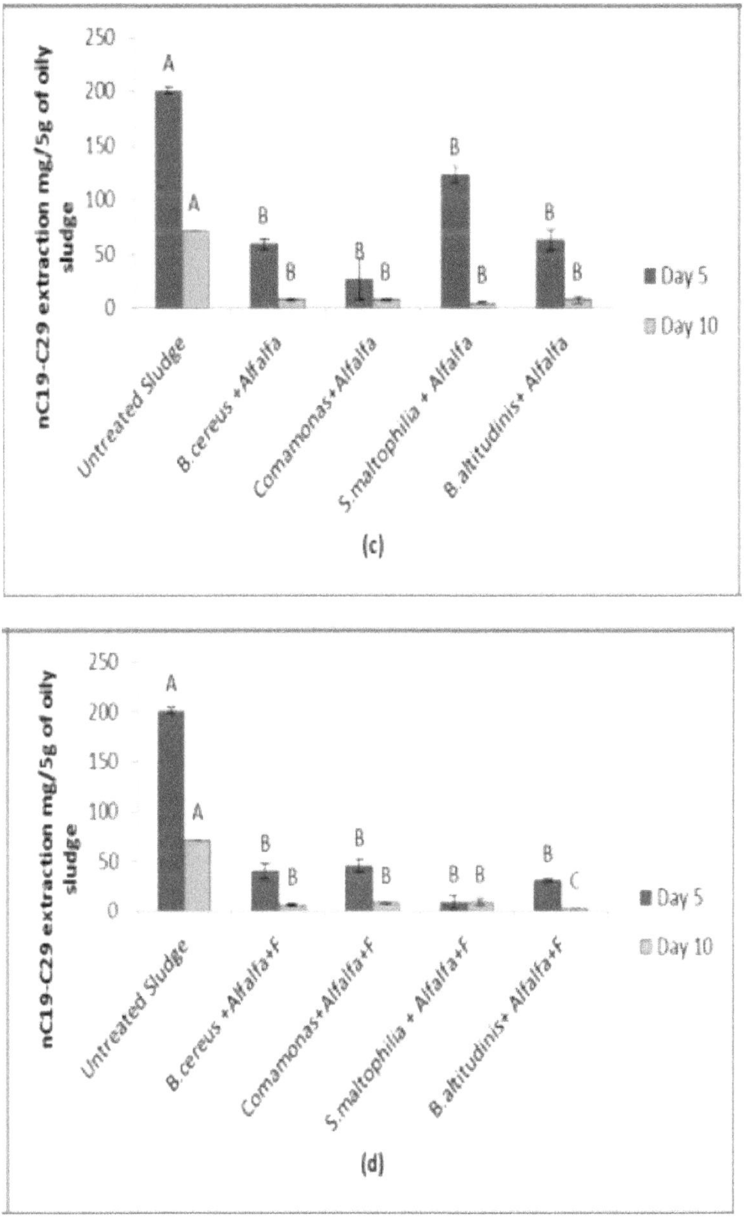


**Figure 4:** B. cereus were most effective and showed 82 to 64% degradation at 5 d over that of untreated sludge.

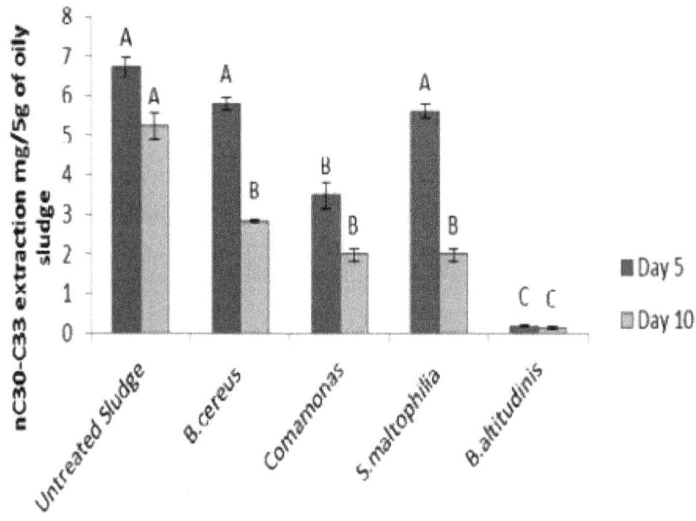
nC30-C33 extraction mg/5g of oily sludge
8
7
6
5
4
3
2
1
0
A
A
A
B
B
B
A
B
C C
Day 5
Day 10
Untreated Sludge
B.cereus
Comamonas
S.maltophilia
B.altitudinis

(a)

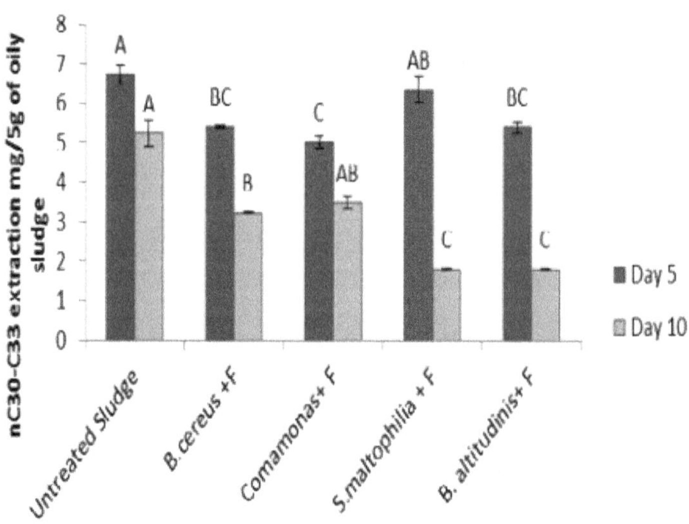
nC30-C33 extraction mg/5g of oily sludge
8
7
6
5
4
3
2
1
0
A
A
BC
B
C
AB
AB
C
BC
C
Day 5
Day 10
Untreated Sludge
B.cereus +F
Comamonas+ F
S.maltophilia + F
B. altitudinis+ F

(b)

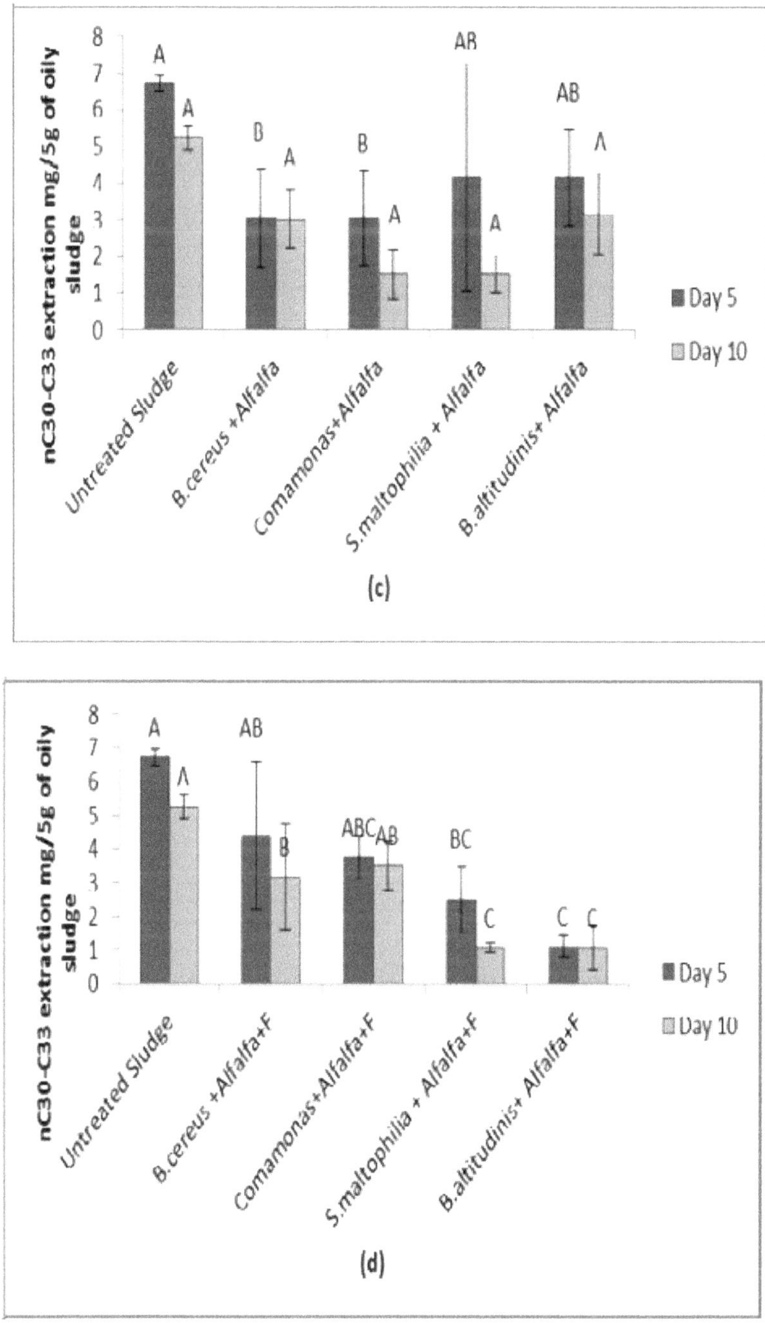


**Figure 5:** Growing alfalfa increased the rate of degradation in B. cereus, Commamonas and S. maltophilia but has suppressed the efficiency of B. altitudinis.

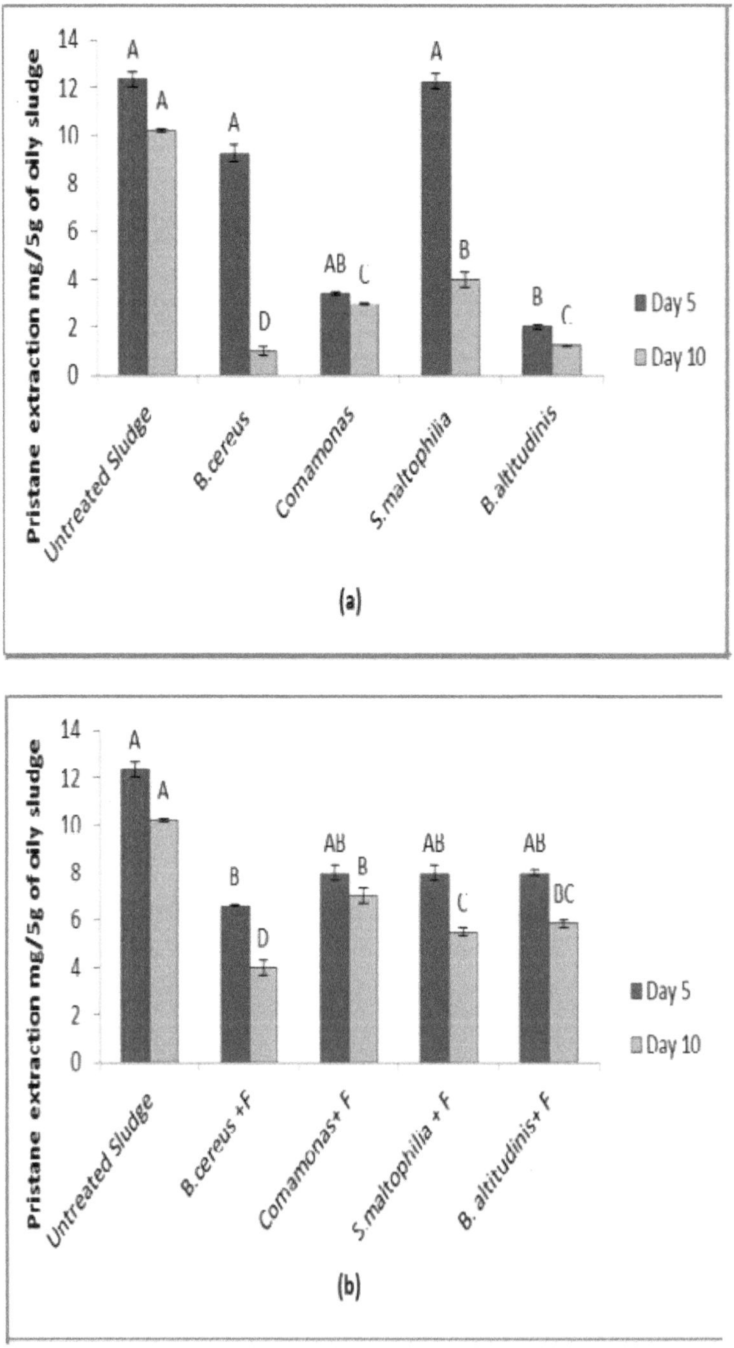

Pristane extraction mg/5g of oily sludge
14
12
10
8
6
4
2
0
A
A
A
D
AB
C
A
B
B
C
Day 5
Day 10
Untreated Sludge
B.cereus
Comamonas
S.maltophilia
B.altitudinis
(a)
Pristane extraction mg/5g of oily sludge
14
12
10
8
6
4
2
0
A
A
B
D
AB
B
AB
C
AB
BC
Day 5
Day 10
Untreated Sludge
B.cereus +F
Comamonas+ F
S.maltophilia + F
B. altitudinis+ F
(b)

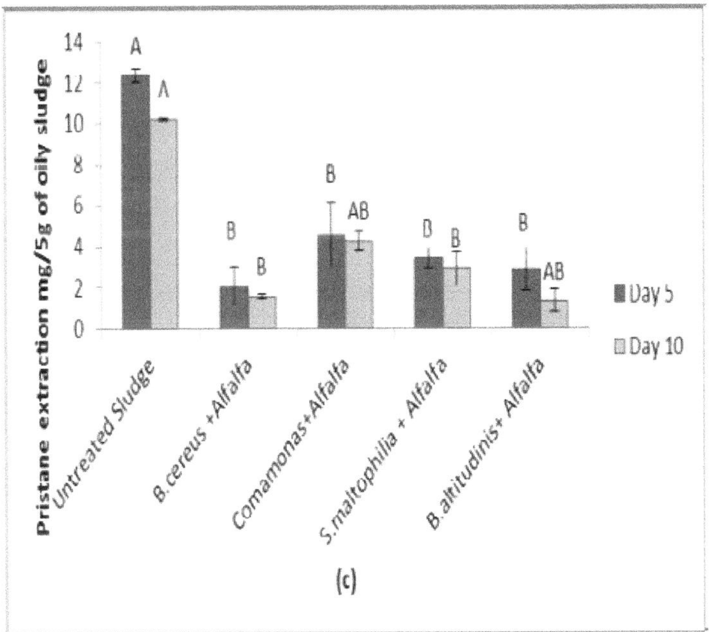


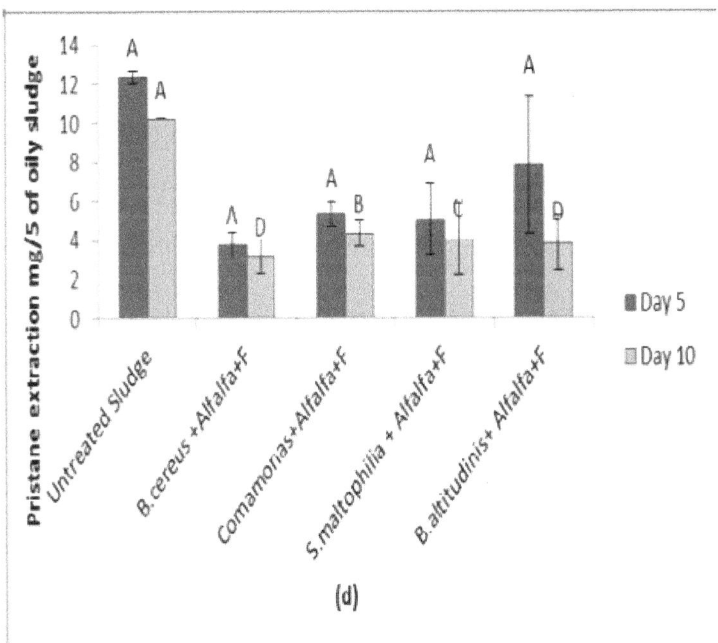


**Figure 6:** Degradation of pristane (isoprenoid) was enhanced by all the PGPR isolates.

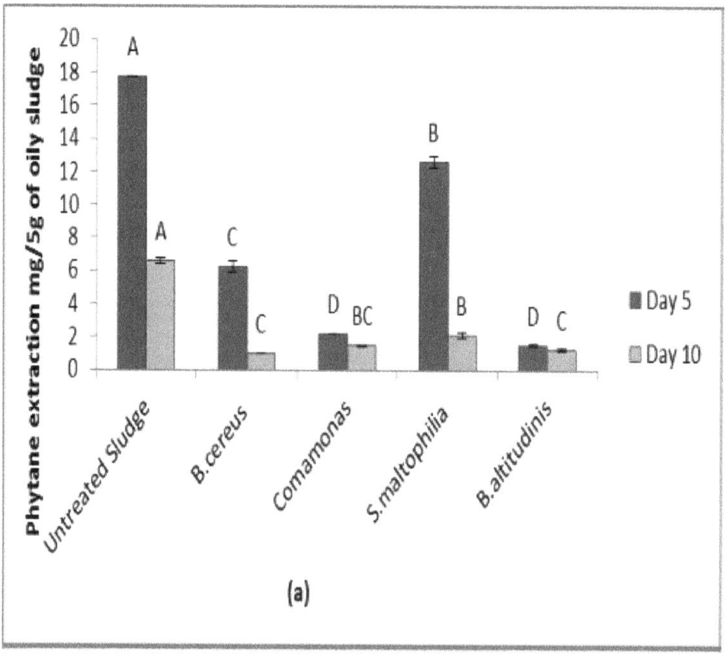
Phytane extraction mg/5g of oily sludge
20
18
16
14
12
10
8
6
4
2
0
A
A
C
C
D
BC
B
B
D
C
Untreated Sludge
B.cereus
Comamonas
S.maltophilia
B.altitudinis
Day 5
Day 10
(a)

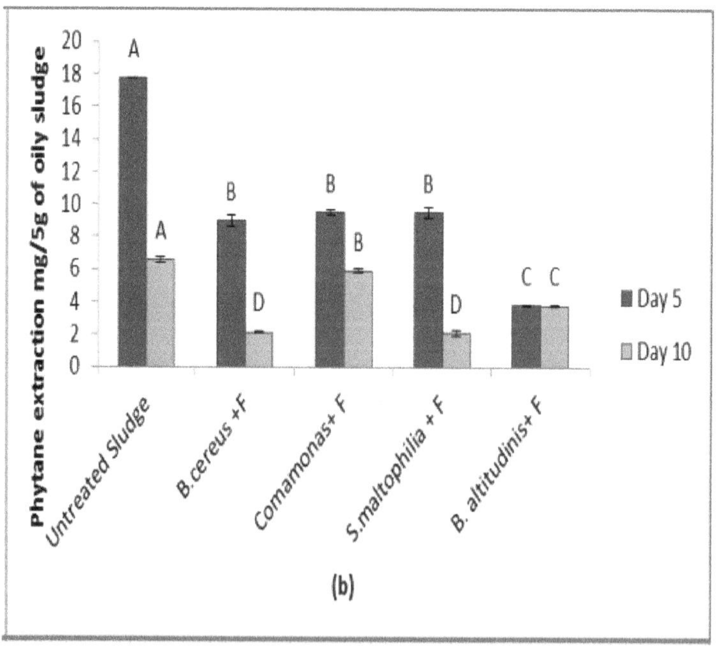
Phytane extraction mg/5g of oily sludge
20
18
16
14
12
10
8
6
4
2
0
A
A
B
D
B
B
B
D
C
C
Untreated Sludge
B.cereus +F
Comamonas+ F
S.maltophilia + F
B. altitudinis+ F
Day 5
Day 10
(b)

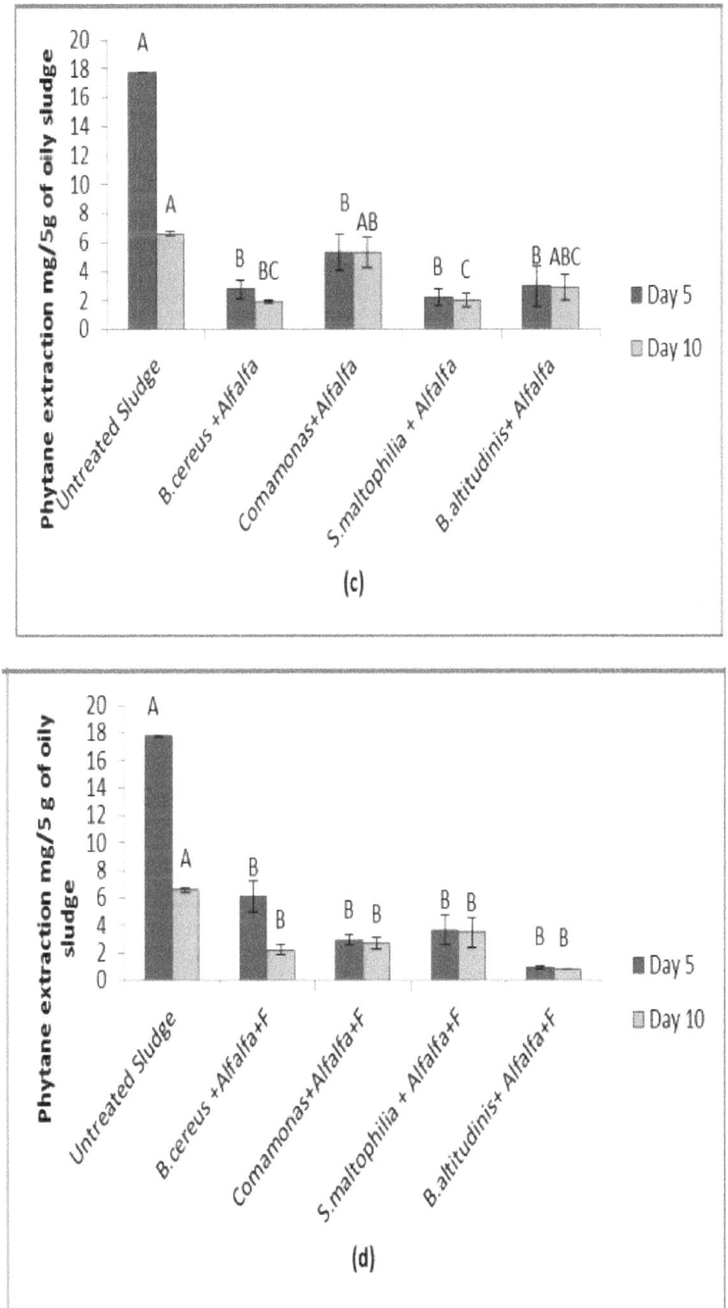


**Figure 7:** Phytane extraction mg/5 g of oily sludge.

For the degradation of $nC_{17}$-$nC_{18}$ PGPR followed the similar pattern of response in oily sludge but the difference lies in the rate of degradation. The % degradation reached a maximum at 5 d and no further increased occurred at 10 d after addition of fertilizer and inocula. But the alfalfa plantation was not as effective as that for $nC_{13}$-$nC_{16}$. Neverthless *B. altitudines* inoculation has similar % and rate i. e. 80% of control (untreated sludge).

Noteworthy, the alfalfa plantation degraded the $nC_{19}$-$nC_{29}$ nhydrocarbon series almost completely but took 10 d; further addition of fertilizer appeared effective for S. maltophilia even at 5 d and for *B. altitudines* at 10 d after inoculation. Also for degradation of long chain hydrocarbon $nC_{30}$-$nC_{33}$ *B. altitudines* appeared most effective. The biodegradation potential of alfalfa was was assisted by the B. cereus and*Commamonas* further for the degradation of pristane B. altitudines>*Commamonas* were effective; the *B. cereus* took longer (10 d) incubation period.

Alfalfa augmented the degradation in conjunction with PGPR inocula. Alfalfa increased both the rate and % degradation of phytan, *B. altitudines* was exceptional and performed well alone and particularly in presence of alfalfa. Fertilizer addition further enhanced the rate of degradation by 5 d. The *B. altitudines* being most responsive to alfalfa as well to the fertilizer for total degradation of hydrocarbon. It is inferred from the present result that *B. cereus* and *B. altitudinis* can be the putative candidate for biodegradation of hydrocarbon, the latter was superior. The effects of *B. cereus* can be further augmented by fertilizer (N and P) and alfalfa. Bacterial population increases in soils contaminated with oily sludge than that of without the previous history of oily sludge contamination [5,6,32,33]. Natural rehabilitation of hydrocarbon contaminated soils is a slow process which can be accelerated when such soils were inoculated with plant growth promoting bacteria. This study revealed the contention that natural rehabilitation process of degradation of oily sludge when the contamination is only 30 % of sludge can be accelerated with *B. altitudinis*. Only 5 days are required to rehabilitate the soil with B. altitudinis alone and n-alkanes such as straight chain alkanes ($nC_{13}$- $nC_{16}$) and long chain alkanes ($nC_{26}$-$nC_{33}$) and disappeared from sludge after 5 days of incubation. The traces of isoprenoids and middle chain alkanes ($nC_{17}$-$nC_{25}$) were still present till the end of the incubation period. This suggested that B. altitudinis was best suitable bacterial strain for bioremediation of 30 % of oily sludge when n-alkanes ($nC_{13}$ to $nC_{33}$) were dominant in contaminated soils. However bioremediation of oily sludge contaminated soils with *B. altitudinis* and nutrients were delayed and required 10 d to rehabilitate. The results of this study are in agreement with the findings of Muthuswamy et al. [34] who reported that Bacillus genera is capable of degrading short carbon chain length in hydrocarbon contaminated soils. Amund and Nwokoye [35] and Lal and Khanna [36], reported that *Arthrobacter, Burkholderia, Mycobacterium, Pseudomonas, Sphingomonas, and Rhodococcus* were capable of degrading alkyl aromatic hydrocarbons (Pseudomonas aeruginosa strain WatGonce inoculated in hydrocarbon contaminated soils cannot degrade n-hexatriacontane ($nC_{36}$) or n-tetracontane ($nC_{40}$). Acinetobacter calcoaceticus S30 and Alcaligenesodorans P20 are able to degrade paraffins [36]. Sakai et al. [37] reported an Acinetobacter strain can degrade very long chain n-alkanes (up to $nC_{44}$). Sakai et al. [37] demonstrated that *P.*

*aeruginosa* K1 and Rhodococcusequi P1 were able to degrade alkanes of chain lengths from $nC_7$ to $nC_{28}$ [38,39].

## Total Hydrocarbon degradation of oily sludge

The inoculation with B. cereus enhanced the process of degradation showing 56% total hydrocarbon by 5 d. The *B. altitudinis* degraded 80% of total hydrocarbon at 5 d and increased from 80-91% during 5 d and 10 d of incubation respectively. Inoculation of *Commamonas*and *S. maltophilia* resulted in the degradation of hydrocarbon from 22-86% and 16-85% respectively during 5 d to 10 d of incubations respectively (Figure 8a) With the addition of the fertilizer (Figure 8b) the %degradation of hydrocarbon increased by 58-63%, 22-63%, and 17-64% in *B. cereus, Comamonas* and *S. maltophilia* respectivly but was decreased in *B. altitudinis*. At 10 d of incubation all the strains showed significant difference in the %degradation of hydrocarbon. The %degradation increased from 63-81% in oily sludge treated with *B. cereus, Comamonas, S. maltophilia and B. altitudinis* respectively, when compared from 5 to 10 d of incubation. Planting alfalfa with and without inoculation of PGPR also showed degradation. The Figure 8c revealed that inoculation of alfalfa with B. cereus and B. altitudinis showed 78% and 79% greater degradation of total hydrocarbon respectively than untreated sludge at 5 d of incubation. The %degradation increased from 78-79%, 67-72% and 56-63% in oily sludge treated with *B. cereus, Comamonadas* and *S. maltophilia* respectively, when compared from 5 d to 10 of incubation. While with the addition of fertilizer and planting alfalfa inoculated with *B. cereus* showed almost similar% of degradation 78% at 5 d of incubation (Figure 8d) but was lesser in *B. altitudinis*. At 10 d of incubation all the strains showed significant difference in the %degradation of hydrocarbon. The %degradation increased from 78- 85%, 69-90%, 68-75% and 57-74% in oily sludge treated with *B. cereus, Commamonadas and S. maltophilia and B. altitudinis* respectively, when compared from 5 to 10 d of incubation.

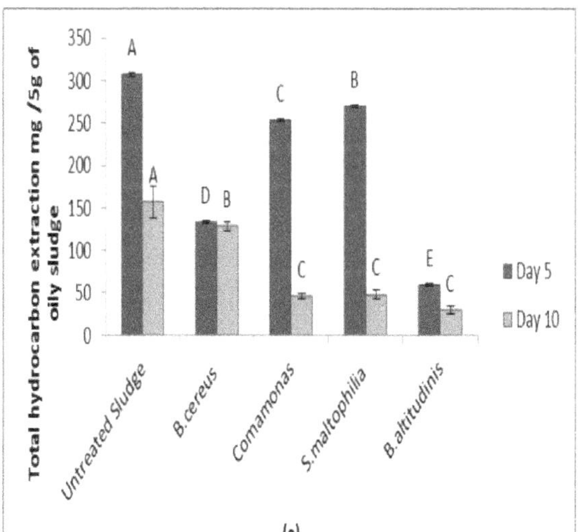


(a)

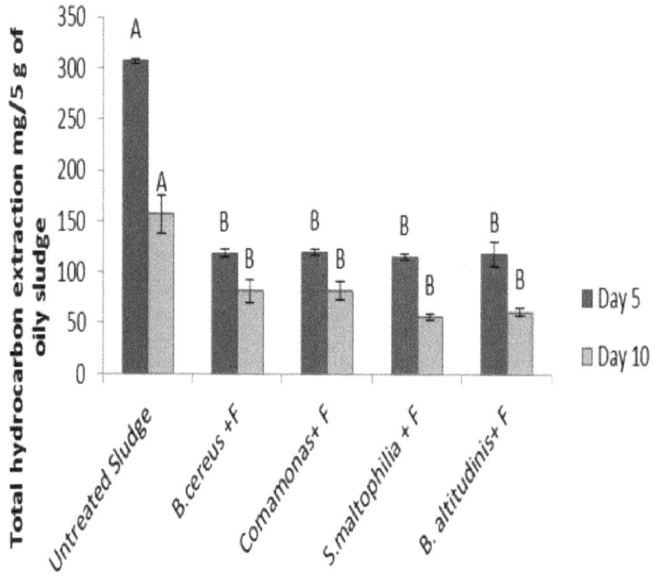
Total hydrocarbon extraction mg/5 g of oily sludge
350
300
250
200
150
100
50
0
A
A
B
B
B
B
B
B
B
B
Day 5
Day 10
Untreated Sludge
B.cereus +F
Comamonas+ F
S.maltophilia + F
B. altitudinis+ F

(b)

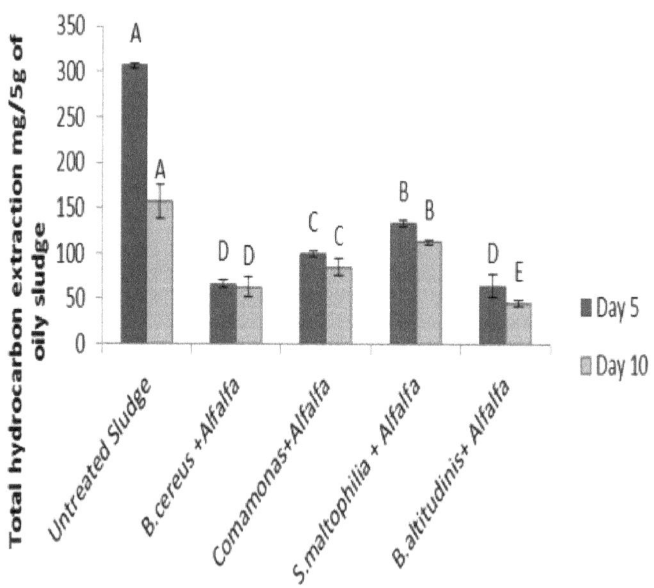
Total hydrocarbon extraction mg/5g of oily sludge
350
300
250
200
150
100
50
0
A
A
D
D
C
C
B
B
D
E
Day 5
Day 10
Untreated Sludge
B.cereus +Alfalfa
Comamonas+Alfalfa
S.maltophilia + Alfalfa
B.altitudinis+ Alfalfa

(c)

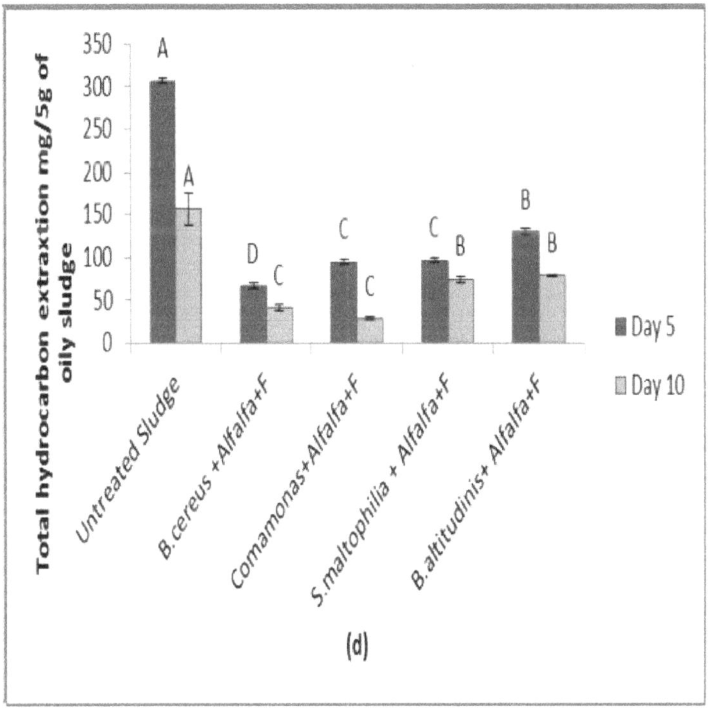


**Figure 8:** Inoculation of Commamonas and S. maltophilia resulted in the degradation of hydrocarbon from 22-86% and 16-85%.

The total hydrocarbon degradation potential of *B. altitudinis* strains at 5 d was higher from all other strains and present results findings are comparable to degradation of Nigerian crude-oil by B. subtilis and *P. aeruginosa* strains isolated from crude oil-polluted soil from Nigeria [41]. Since addition of some fertilizer as carbon source may enhance the rate of hydrocarbon degradation. Ammonium nitrate and Diammonium phosphate were added in oily sludge at 0 d of experiment to increase the rate of degradation of hydrocarbons by the seeded bacteria. The %degradation of hydrocarbons increased in all the strains except B. altitudinis indicating that fertilizer as N and P source is not required by this strain. However, these results were in close agreement with Choi et al. [42] and Kim et al. [43] who reported that the additions of nutrients were necessary to enhance the biodegradation of oil contaminants. Chaıneau et al. [44], Carmichael and Pfaender [45] have reported the negative effects of high NPK levels on the biodegradation of hydrocarbons which also support our results that *B. altitudinis*showed decreased degradation with the addition of N and P.

Oily sludge is threat to soil for cultivation and it possess numerous micro floras capable for degrading such complex hydrocarbons. Treatment of such polluted site had been a common practice in oil refineries and yields good results. The treatment is carried out by enriching specific populations from the soil and by applying them back to the contaminated soil, where the population of oil-degrading microorganisms is low [46]. In the present study, the degradation of hydrocarbons from treated pots was much higher than that in the untreated pots. This observation is in agreement with those of Barbeau et al. [47] who reported that bioaugmentation of pentachlorophenol (PCP) contaminated soil by PCP-degrading microorganisms resulted in 99% reduction of PCP concentration. It was also observed that degradation of hydrocarbons was higher during first 5 d of inoculation which might be due greater bacterial population and sufficient nutrient availability for hydrocarbon degrading microbes. The degradation slow down later might be to degradation of some fraction take longer time to degrade in total hydrocarbons This shows that the indigenous population in the control plot was not adequate to stimulate degradation of the oily sludge. Bioaugmentation thus enhanced the process of bioremediation.

Growing alfalfa with and without fertilizer enhanced the removal of total hydrocarbon at 5-10 d. The result of this study is in agreement with the findings of Liu et al. [30] who reported that the degradation rate of diesel oil was accelerated up to 69% by planting alfalfa and Indian mustard plants to hydrocarbon contaminated soils. The increase in the rate of degradation because of growing alfalfa in hydrocarbon contaminated soils is more likely because of alfalfa fibrous root systems with large root surface areas and tap root systems which can penetrate deeper down the soil hence increased the mobility and the breakdown of contaminants [48-50] Plant exudates may also provide extra nutrients for the growth of bacteria and stimulated their growth. All of these reasons possibly stimulated higher microbial populations in rhizosphere and enhanced biodegradation of diesel oil. This suggested that microbial population and activity was enhanced under the hydrocarbon contaminated alfalfa rhizosphere. Nevertheless, Inoculation of alfalfa with oil-degrading bacteria was adequately maintaining their populations in the rhizosphere even with decreasing diesel content in the soil.

Rhizoremediation of oily sludge is a low cost technique which can enhance the rehabilitation process more likely because it is a combination of bioremediation and phytoremediation. The results of this study revealed that *Commamonas, S. maltophilia* were the plant growth promoting bacteria and enhanced degradation of 30% oily sludge was observed with alfalfa.

## CONCLUSION

The present study was focused on the accelerated biodegradation of oil sludge by bacterial association with common and cost-effective plant, alfalfa. The results demonstrated that the accelerated degradation of oily sludge by fertilizer and plant-bacterial association. Different strains have different capabilities to enhance biodegradation of oily sludge. The differences in the bacterial diversity is related to the groups of hydrocarbons extracted from oily sludge. The natural rehabilitation process of hydrocarbon contaminated soil is slow however inoculation of such soil with *B. altitudinis* accelerated the rate of degradation of n-alkanes and some methyl branched than un-inoculated soils.

*B. altitudinis* is a strain which is capable of degrading straight chain hydrocarbons rapidly than any other strains. Nutrient addition to such soils had no effect on the rate of degradation of hydrocarbons demonstrate that the observation that enough nutrients were present in the soil initially on sludge. Alfalfa is an N-symbiotic plant and because of increased nutrient content under the rhizosphere promotes degradation of n-alkanes. Whereas all other bacterial strains showed significant interaction with alfalfa for the biodegradation of oily sludge but the rate of degradation remained slow than with *B. altitudinis.*

The natural rehabilitation process of degradation of toxic n-alkanes was enhanced in soil contaminated with 30% oily sludge under the alfalfa rhiosphere with sufficient nutrients are available for indigenous microbes to degrade such alkanes. Inoculation with *B. altitudinis* accelerated the rehabilitation process and within 5 days the rehabilitation occurs.

## ACKNOWLEDGEMENTS

I am grateful to the Govt of Pakistan and HEC for financial support of the project.

## REFERENCES

1. Chakraborty R, Coates JD (2004) Anaerobic degradation of monoaromatic hydrocarbons.ApplMicrobiolBiotechnol 64: 437-446.
2. Infante C, Morales F, Ehrmann EU, Hernández-Valencia I, Leon N (2010) Hydrocarbon bioremediation and phytoremediation in tropical soils: Venezuelan study case. Trends in Bioremediation and Phytoremediation. Pp 429-451.
3. Rahman KSM, Thahira-Rahman J, Kourkoutas Y, Petsas I, Marchant R, et al. (2003) Enhanced bioremediation of n-alkanes in petroleum sludge using bacterial consortium amended with rhamnolipid and micronutrients. Bioresource and Technology 90: 159-168.
4. Zand AD, Gholamreza N, Nasser M, Ravanbakhsh S, Alireza MT (2010). Total Petroleum Hydrocarbon (TPHs) Dissipation through Rhizoremediation by Plant Species. Polish Journal of Environmental Studies 19: 115-122.
5. Siddiqui S, Adams WA (2002) The fate of diesel hydrocarbons in soils and their effect on the germination of perennial ryegrass.Environ Toxicol 17: 49-62.
6. Siddiqui S, Adams WA, Schollion J (2001) Thephytotoxicity and degradation of diesel hydrocarbons in soil. Plant Nutrition and Soil Science 164 : 631-635.
7. Fulekar MH (2010) Bioremediation Technology: Recent Advances. Springer 290.
8. Geetha, SJ, Joshi SJ, Kathrotiya S (2013) Isolation and characterization of hydrocarbon degrading bacteria isolate from oil contaminated sites. APCBEE Procedia 5: 237 241.
9. Palanisamy N, Jayaprakash R, Srilakshman K, Vasanthi NS, Preethy C, et al. (2014) Diesel biodegradation capacities of indigenous bacterial species isolated from diesel contaminated soil. Journal of Environmental Health Science & Engineering 12:142.
10. Yousefi K, Khodadadi A, Ganjidoust H, Badkoubi A, Amoozegar MA (2009) Isolaton and characterization of a novel native Bacillus strain capable of degrading diesel fuel. International Journal of Environmental Science and Technology 6:

435-442.

11. Atlas RM, Bartha R (1993) Microbial interaction with Xenbiotic and inorganic pollutants. Fundamentals and Applications in Microbial Ecology 6: 544-600.
12. Sood N, Patle S, Lal B (2010) Bioremediation of acidic oily sludge-contaminated soil by the novel yeast strain Candida digboiensis TERI ASN6.Environ SciPollut Res Int 17: 603-610.
13. Cerqueira VS, Hollenbach EB, Maboni F, Vainstein MH, Camargo FA, et al. (2011) Biodegradation potential of oily sludge by pure and mixed bacterial cultures.BioresourTechnol 102: 11003-11010.
14. Jiann H, Liu J, Prakash M, Jiin-Shuh J, Chien-Yen C, et al. (2010) World Journal of Microbiology and Biotechnology.
15. Pradhan SP, Conrad JR, Paterek JR, Srivastava VJ (1998) Potential of Phytoremediation for Treatment of PAHs in Soil at MGP Sites. Soil Sediment Contamination 7: 467-480.
16. Omotayo AE, Shonubi OO, Towuru EG, Babalola SE, Ilori MO, et al. (2014) Rhizoremediation of hydrocarbon-contaminated soil by Paspalumvaginatum (Sw.) and its associated bacteria. International Research Journal of Microbiology (IRJM) 5: 1-7.
17. Asif M, Fazeelat T (2012) Petroleum geochemistry of the Potwar Basin, Pakistan: II - Oil classification based on heterocyclic and polycyclic aromatic hydrocarbons. Appllied Geochemistry 27: 1655-166.
18. Khan MA, Ahmed R, Raza HA, Kamal A (1986) Geology of petroleum in Kohat-Potwar Depression Pakistan. Am. Assoc. Petroleum Geochemical Bulletin 7: 396-414.
19. Berge KE, Ose L, Leren TP (2006) Missense mutations in the PCSK9 gene are associated with hypocholesterolemia and possibly increased response to statin therapy. Arterioscler. Thromb. Vascular Biology 26: 1094-1194.
20. Kate W (1997) Preparation of Genomic DNA from Bacteria. Current Protocols in Molecular Biology, 2.4.1-2.4.5. John Wiley & Sons, Inc.
21. Weisburg WG, Barns SM, Pelletier DA, Lane DJ (1991) 16S ribosomal DNA amplification for phylogenetic study.J Bacteriol 173: 697-703.
22. Song HG, Bartha R (1990) Effects of jet fuel spills on the microbial community of soil.Appl Environ Microbiol 56: 646-651.
23. Asif M, Grice K, Fazeelat T (2009) Assessment of petroleum biodegradation using stable hydrogen isotopes of individual saturated hydrocarbon and polycyclic aromatic hydrocarbon distribution in oils from the Upper Indus Basin, Pakistan. Org Geochem 40:301-311.
24. Wang C, Zöllner S, Rosenberg NA (2012) A quantitative comparison of the similarity between genes and geography in worldwide human populations.PLoS Genet 8: e1002886.
25. Zhang DC, Mörtelmaier C, Margesin R (2012) Characterization of the bacterial archaeal diversity in hydrocarbon-contaminated soil.Sci Total Environ 421-422:

184-96.

26. Ramadoss D, Lakkineni VK, Bose P, Ali S, Annapurna K (2013) Mitigation of salt stress in wheat seedlings by halotolerant bacteria isolated from saline habitats. Springerplus 2: 6.
27. Yoshida N, Yagi K, Sato D, Watanabe N, Kuroishi T, et al. (2005) Bacterial communities in petroleum oil in stockpiles.J BiosciBioeng 99: 143-149.
28. Erdogan EE, Fikrettin S, Ayten K (2012) Determination of petroleum-degrading bacteria isolated from crude oil-contaminated soil in Turkey. African Journal of Biotechnology 11: 4853-4859.
29. SaeidehR, Syed MS, Fayez R, Behrouz S, Jamshid R (2013) Characterization and Potentials of Indigenous Oil-Degrading Bacteria Inhabiting the Rhizosphere of Wild Oat (AvenaFatua L.) in South West of Iran. Iranian Journal of Biotechnology 11: 32-40.
30. Liu CW, Liu HS ( 2011) Rhodococcus erythropolis strain NTU-1 efficiently degrades and traps diesel and crude oil in batch and fed-batch bioreactors. Process Biochemistry 46: 202-209.
31. Atlas RM (1977) Stimulated petroleum biodegradation.CRC Crit Rev Microbiol 5: 371-386.
32. Haichar FZ, Marol C, Berge O, Rangel CJI, Prosser JI, et al. (2008) Plant host habitat and root exudates shape soil bacterial community structure. International Society of Microbial Ecology 2: 1221-1230.
33. Supaphol S, Panichsakpatana S, Trakulnaleamsai S, Tungkananuruk N, Roughjanajirapa P, et al. (2006) The selection of mixed microbial inocula in environmental biotechnology: example using petroleum contaminated tropical soils.J Microbiol Methods 65: 432-441.
34. Muthusamy K, Gopalakrishnan S, Ravi TK, Sivachidambaram P(2008) Biosurfactants: properties, commercial production and application, Current Science 94: 736-747.
35. Amund OO, Nwokoye N (1993) Hydrocarbon potentials of yeast isolates from a pollute Lagoon. Journal of Scientific Research and Development 1:65-68.
36. Lal B, Khanna S (1996) Degradation of crude oil by Acinetobactercalcoaceticus and Alcaligenesodorans.J ApplBacteriol 81: 355-362.
37. Sakai Y, Maeng JH, Tani Y, Kato N (1994) Use of long-chain n-alkanes (C13-C44) by an isolate, Acinetobacter sp. M-1. Bioscience, Biotechnology and Biochemistry 58: 2128-2130.
38. Abalos A, Viñas M, Sabaté J, Manresa MA, Solanas AM (2004) Enhanced biodegradation of Casablanca crude oil by a microbial consortium in presence of a rhamnolipid produced by Pseudomonas aeruginosa AT10.Biodegradation 15: 249-260.
39. Norman RS, Frontera-Suau R, Morris PJ (2002) Variability in Pseudomonas aeruginosa lipopolysaccharide expression during crude oil degradation.Appl Environ Microbiol 68: 5096-5103.

40. Mulligan CN, Young RN, Gibbs BF (2001) Surfactant enhanced remediation of contaminated soil: a review . EngGeol 60:371-380.

41. Ilori MO, Amund DI (2000) Degradation of anthracene by bacteria isolated from oil polluted tropical soils.Z Naturforsch C 55: 890-897.

42. Choi SC, Kwon KK, Sohn JH, Kim SJ (2002) Evaluation of fertilizer additions to stimulate oil biodegradation in sand seashore mescocosms. Journal of Microbiology and Biotechnology 12:431-436.

43. Kim SJ, Choi DH, Sim DS, Oh YS (2005) Evaluation of bioremediation effectiveness on crude oil-contaminated sand.Chemosphere 59: 845-852.

44. Chaineau CH, Rougeux G, Yepremian C, Oudot J (2005) Effects ofnutrient concentration on the biodegradation of crude oil and associated microbial populations in the soil. Soil. Biology and Biochemistry 37:1490-1497.

45. Carmichael LM, Pfaender FK (1997) The effect of inorganic and organic supplements on the microbial degradation of phenanthrene and pyrene in soils. Biodegradation 8: 1-13.

46. Forsyth JV, Tsao YM, Bleam RD (1995) Bioremediation: when is bioaugmentation needed? inBioaugmentation for site remediation. edsHinchee R. E., Fredrickson J., Alleman B. C. (Battelle Press, Columbus, Ohio) 1-58.

47. Barbeau C, Deschênes L, Karamanev D, Comeau Y, Samson R (1997) Bioremediation of pentachlorophenol-contaminated soil by bioaugmentation using activated soil.ApplMicrobiolBiotechnol 48: 745-752.

48. Ingham ER, Klein DA (1984) Soil fungi: relationships between hyphal activity and staining with fluorescein diacetate. Soil Biology and Biochemistyr 16: 273-278.

49. Shimp JF, Tracy JC, Davis LC, Huang W, Erickson LE, et al. (1993) Beneficial effects of plants in the remediation of soil and groundwater contaminated with organic materials. Critical Review on Environmental Science and Technology 23: 41-77.

50. Chaineau CH, Morel JL, Oudot J (2000) Biodegradation of fuel oil hydrocarbons in the rhizosphere of maize. Journal of Environmental Quality 29: 549-578.

# 11

CHAPTER

# EFFECTIVE BIOREMEDIATION STRATEGY FOR RAPID IN SITU CLEANUP OF ANOXIC MARINE SEDIMENTS IN MESOCOSM OIL SPILL SIMULATION

Maria Genovese[1], Francesca Crisafi[1], Renata Denaro[1], Simone Cappello[1], Daniela Russo[1,2], Rosario Calogero[1], Santina Santisi[1,2], Maurizio Catalfamo[1], Alfonso Modica[3], Francesco Smedile[1], Lucrezia Genovese[1], Peter N. Golyshin[4], Laura Giuliano[1,5] and Michail M. Yakimov[1]

[1]Institute for Coastal Marine Environment, CNR, Messina, Italy

[2]Department of Biological and Environmental Sciences, University of Messina, Messina, Italy

[3]Environmental Laboratory, Syndial SpA, Priolo Gargallo, Italy

[4]Environmental Genomics, School of Biological Sciences, Bangor University, Bangor, UK

[5]Mediterranean Science Commission, Monaco, Monaco

## ABSTRACT

**The purpose of present study was the simulation of an oil spill accompanied by burial of significant amount of petroleum hydrocarbons (PHs) in coastal sediments. Approximately 1000 kg of sediments collected in Messina harbor were spiked with Bunker C furnace fuel oil (6500 ppm). The rapid consumption of oxygen by aerobic heterotrophs created highly reduced conditions in the sediments with subsequent recession of biodegradation rates. As follows, after 3 months of ageing, the anaerobic sediments did not exhibit any significant levels of biodegradation and more than 80% of added Bunker C fuel oil remained buried. Anaerobic microbial community exhibited a strong enrichment in sulfate-reducing PHs-degrading and PHs-associated*Deltaproteobacteria*. As an effective bioremediation strategy to clean up these contaminated sediments, we applied a Modular Slurry System (MSS) allowing the containment of sediments and their physical–chemical**

**Citation:** Genovese M, Crisafi F, Denaro R, Cappello S, Russo D, Calogero R, Santisi S, Catalfamo M, Modica A, Smedile F, Genovese L, Golyshin PN, Giuliano L and Yakimov MM (2014) Effective bioremediation strategy for rapid in situ cleanup of anoxic marine sediments in mesocosm oil spill simulation. Front. Microbiol. 5:162. doi: 10.3389/fmicb.2014.00162.

**treatment, e.g., aeration. Aeration for 3 months has increased the removal of main PHs contaminants up to 98%. As revealed by CARD-FISH, qPCR, and 16S rRNA gene clone library analyses, addition of Bunker C fuel oil initially affected the activity of autochthonous aerobic obligate marine hydrocarbonoclastic bacteria (OMHCB), and after 1 month more than the third of microbial population was represented by *Alcanivorax-*, *Cycloclasticus-*, and*Marinobacter*-related organisms. In the end of the experiment, the microbial community composition has returned to a status typically observed in pristine marine ecosystems with no detectable OMHCB present. Eco-toxicological bioassay revealed that the toxicity of sediments after treatment was substantially decreased. Thus, our studies demonstrated that petroleum-contaminated anaerobic marine sediments could efficiently be cleaned through an *in situ* oxygenation which stimulates their self-cleaning potential due to reawakening of allochtonous aerobic OMHCB.**

## INTRODUCTION

The worldwide production of crude oil and natural gas is at the peak, with an estimated worldwide production of 89 million barrels per day in 2011 (International Energy Agency,http://omrpublic.iea.org/). Approximately, a half of this amount is transported by the sea (Gertler et al., 2010). As follows, worldwide marine coastal areas are exposed to the oil spills occurring as a result of accidents or illegal practices (Psarros et al., 2010). The release of thousands of tons of petroleum hydrocarbons (PHs) affects the marine environment and causes severe ecological and economical damage. For example, only the pollution resulting from the tanker washing or ballast water has been estimated to contribute about 2 million tons per year worldwide (Ferraro et al., 2007; Gertler et al., 2010). The recent spillage of 780,000 $m^3$ of oil into the Gulf of Mexico proved again that human activities might cause a contamination without precedents. This accident presented a huge challenge to existing oil spill treatment methods, and current technologies were not able to cope with the size and nature of the Deepwater Horizon oil spill. Therefore, there is an urgent demand for development and optimization of bioremediation techniques that can play a central role in marine oil spill response contingency plans.

One of the most important issues in bioremediation is the application (or stimulation) of autochthonous hydrocarbon-degrading microbial populations. Some marine gammaproteobacteria have a high affinity toward PHs. Species such as *Alcanivorax borkumensis* (Yakimov et al., 1998),*Cycloclasticus pugetii* (Dyksterhouse et al., 1995), *Oleispira antarctica* (Yakimov et al., 2003),*Oleiphilus messinensis* (Golyshin et al., 2002), and *Thalassolituus oleivorans* (Yakimov et al., 2004) constitute a distinct group of obligate marine hydrocarbon-degrading bacteria (OMHCB). Following a sudden oil spill event, these organisms outcompete most of the naturally occurring oligotrophic marine microorganisms (Hara et al., 2003; Yakimov et al., 2007). Growing on PHs, aerobic OMHCB use oxygen not only as the terminal electron acceptor for respiratory energy conservation, but also as an indispensable reactant in the PHs activation mechanism. Thus, the stimulation of the OMHCBs degradation activity in the contaminated site can significantly improve the self-cleaning potential. Unfortunately, due to metabolic requirements of the OMHCBs, this type of bioremediation is restricted to either seawater column or superficial sediments. Due to a high biological oxygen demand and its slow diffusion into marine sediments, these compartments below the surface are typically highly reduced (Engelen and Cypionka, 2009). The realization that activated oxygen is used to overcome the chemical sluggishness of hydrocarbons has for some decades favored the view that hydrocarbons are not biodegradable under anoxic conditions (Widdel and Rabus, 2001). Although recently, a number of strictly anaerobic microorganisms have been shown to utilize PHs as growth substrates, this process is extremely slow compared to aerobic degradation and can not be considered as rapid bioremediation scenario.

The aim of this study was to monitor both immediate and long-term responses of indigenous microbial consortia to a simulated oil spill and during bioremediation treatment. Keeping in mind that the decontamination of PHs-polluted anoxic sediments is a very sluggish process, the stimulation of indigenous aerobic OMHCBs was performed by the *in situ* aeration of sediments within a modular slurry system (MSS) to avoid the contamination of the surrounding aquifer. The succession of microbial community and efficacy of petroleum biodegradation in both untreated (anoxic) external and aerated internal sediments was monitored during 3 months after contamination. Additionally, the toxicity of sediments was controlled by application of Microtox®and amphipods *Corophium orientale* eco-toxicological bioassays.

## MATERIALS AND METHODS

### Experimental Mesocosm

The experiment was carried out in rectangular tank of 3.75 $m^3$ capacity (166 cm long, 150 cm deep, 150 cm wide). This reservoir was filled with ca. 2000 l of seawater taken directly from the harbor of Messina (38°11′42.58″N 15°34′25.19″E). Prior to use, the seawater was filtered through a 200 μm nylon mesh to remove large metazoans and detritus. Approximately 1000 kg of sandy sediments were collected at the same place and artificially contaminated with Bunker C furnace fuel oil (6500 ppm) to simulate the oil spill accident. Temperature inside the mesocosm was maintained about 20 ± 1°C for all experimental period. Mesocosm has continuously received seawater at the flow rate of 1 l $min^{-1}$. The MSS used in mesocosm experiment is shown on Figure 1. The MSS was developed especially for *in situ* aeration (20 l $min^{-1}$) of polluted sediments without their removal from contaminated side to avoid the re-contamination of adjacent aquifer. The reactor was inserted into the sediment. Sediment directly beneath the MSS were treated by the reactor, and those sediments outside of it were undisturbed and served as a control. All experimentations have been conducted for 3 months. To monitor the succession of microbial population and the efficiency of petroleum degradation, 1.5–2.0 kg of sediments (up to 10 kg in total) were sampled on fixed days ($T_0$, $T_1$, $T_{29}$, and $T_{90}$) at six different points inside and outside the MSS. Additionally, measurements of the biochemical oxygen demand ($BOD_5$), reduction potential ($E_h$), and eco-toxicological assays (Microtox® and *Corophium orientale* mortality test) were monitored. The $E_h$ of sediment was measured by a Waterproof CyberScan PCD 650 multiparameter (Eutech

Instruments) according to the manufacturer›s instructions at aforementioned time intervals. $BOD_5$ was measured using a BOD sensor (VELP Scientifica) after 5 days of incubation in accordance with the manufacturer›s instructins.

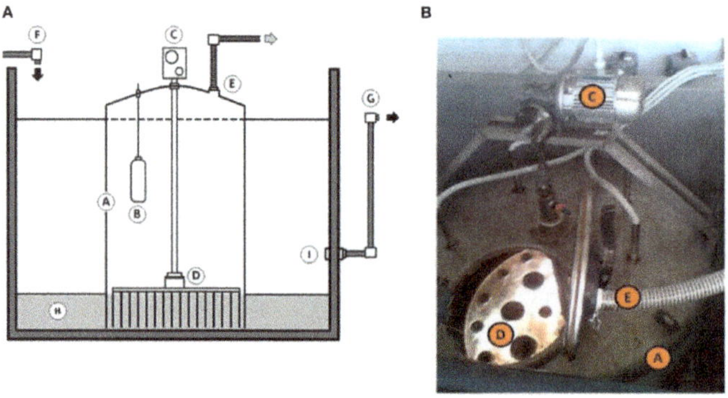


**Figure 1:** (A,B) Schematic representation (A) and detail (B) of "Modular Slurry System" used throughout this study. Abbreviation used: A, Modular Slurry System; B, temperature controller; C, external air pump; D, steel plate with needles to supply an oxygen into deep sediments; E, exhaust tube; F, seawater inlet; G, seawater outlet; H, contaminated sediments; I, overflow regulation system.

## Analysis of Petroleum Hydrocarbons

Efficiency of petroleum degradation was estimated by the analysis of total extracted and resolved hydrocarbons (TERHC). At the fixed time points, TERCH were extracted from sediments following the 3550C EPA (Environmental Protection Agency) procedure. Briefly, 500 ml mixture of $CH_2Cl_2$: $CH_3COCH_3$ (1: 1, vol/vol) was added to 1000 g of pooled and dried either internal or external sediments, collected from six different sites (see above). Mixture was sonicated for 2 min in ultrasound bath (Branson 1200 Ultrasonic Cleaner, Branson USA). Samples were further shaken at 150 rpm for 30 min, centrifuged for 10 min at 5000× g and supernatant was passed through a ceramic column filled with anhydrous $Na_2SO_4$ (Sigma-Aldrich, Milan). Same treatment of pooled and dried sediments was repeated with 500 ml of $CH_2Cl_2$ and the obtained extracts were combined and volatilized to the dryness. Residues were re-suspended in $CH_2Cl_2$ prior the gas chromatography (GC) analysis (Rocchetti et al., 2011). All measures were performed using a Master GC DANI Instruments (Development ANalytical Instruments), equipped with SSL injector and FID detector. Sample (1 μl) was injected in splitless mode at 330°C. The analytical column was a Restek Rxi-5 Sil MS with Integra-Guard, 30 m × 0.25 mm (ID × 0.25 μm film thickness). Helium carrier gas was maintained at a constant flow of 1.5 ml $min^{-1}$. TERCH were calculated using the mean response factors of *n*-alkanes, i.e., individual *n*-alkane concentrations from *n*-$C_{15}$ to *n*-$C_{40}$, pristane and phytane were calculated for each sample. To estimate the biodegradation of aliphatic fraction, the evaluation indices *n*-$C_{17}$/pristane and *n*-$C_{18}$/phytane were selected for this study. The amount of analyzed TERCH was expressed as ppm (part per million) or mg $kg^{-1}$.

## Ecotoxicological Assays

The Microtox® luminescence assay was performed on sediment pore water. Sub-samples of sediment were centrifuged (5000× g for 45 min, 4°C) and filtered (0.45 μm nitrate cellulose membrane) to remove the fine suspended particles and maintained at 4°C until used in assays. Microtox® toxicity tests were conducted according to the standard procedures EN12457 with the following modifications. As far as the Solid Phase Test deals with fine particles that affected the bioluminescence of bacteria (Bulich et al., 1992; Benton et al., 1995; Ringwood et al., 1997), each experimental sediment sample was compared with a reference sediment sample with the same granulometry, collected from a pristine site. Toxicities were reported as effective concentration of toxicant resulting in a 50% decrease in bioluminescence ($EC_{50}$). $EC_{50}$ with 95% of confidence intervals were calculated following the procedures outlined in the Microtox® System Operating Manual (Microtox, 1982). Amphipods *Corophium orientale* were delivered from CIBM (Livorno, Italy). The animals were used following the procedure reported by Onorati et al. (1999). Briefly, the juveniles and young adults, which passed through 1000 μm and retained by 710 μm mesh sieve, were selected for ecotoxicology experiments. The test was carried out inside 2.5-l glass flasks containing approximately 2 cm layer of sediments and filled with 1000 ml of filtered seawater. The seawater was aerated and kept at a constant temperature (16 ± 2°C). Flasks were illuminated during 12 h daily by a lamp system consisted of six tubes (36W, 120 cm). One hundred amphipods were randomly selected and introduced into each flask. No food was added to the test and control chambers. At the end of exposition (10 days), amount of survived organisms were counted. Missing amphipods were assumed as dead animals. The sensitivity of the populations was estimated as a fraction of dead organisms to the initial amount of added amphipods. All the experiments were replicated twice.

## Total RNA Extraction, Reverse Transcription-PCR (RT-PCR) and 16S rRNA Clone Libraries

Total RNA was extracted from sediment sample (5 g) using the FastRNA® Pro soil direct Kit (MP Biomedicals™) as previously described (Roussel et al., 2009). Extracted RNA from three different samples was pooled and further converted to cDNA using First-Strand cDNA Synthesis SuperScript™ II Reverse Transcriptase (Life Technologies). RT reaction mixtures (20 μl) contained 1 μl the random hexamer primer mix (Biolinc), 30 ng of RNA, 1 μl 10 mM dNTPs (Gibco, Invitrogen Co., Carlsbad, CA) and sterile distilled water (14 μl). After heating to 65°C for 5 min, and incubate on ice for 1 min, the mixture was supplemented with 4 μl of 5 × First-Strand Buffer, 1 μl of DTT and 1 μl of SuperScript™. The mixture was shaken and incubated at 50°C for 30 min. Finally, SuperScript™ enzyme was inactivated by heating at 70°C for 15 min. 16S rRNA genes were amplified from total cDNA using the universal primes (530F [5′-GTGCCAGCMGCCGCGG-3′] and 1492R [5′-TACGGYTACCTTGTTACGACT-3′]) (Lane, 1991). The PCR was performed in 50 μ l mixture (total volume) containing 1× solution Q (Qiagen), 1× Qiagen reaction buffer, 1 μM of each forward and reverse primer, 10 μM dNTPs (Gibco, Invitrogen

Co.), 2.0 ml (50–100 ng) of template and 2.0 U of Qiagen Taq Polymerase (Qiagen). The reaction (3 min hot-start at 95°C; 1 min at 94°C, 1 min at 50°C, 2 min at 72°C, 30 cycles; final extension 10 min at 72°C) was performed with GeneAmp 5700 (PE Applied Biosystems). The quality of amplification products was checked by agarose electrophoresis and purified using Qiaquick Gel Extraction kit (Qiagen). Purified 16S crDNA amplicons were further cloned into the pGEM T-easy Vector II (Promega), transformed into*E. coli* DH10β cells and subsequently amplified with primers, specific for the pGEM T-easy vector (M13F (5′-TGTAAAACGACGGCCAGT-3′) and M13R (5′-TCACACAGGAAACAGCTATGAC-3′). Positive products were purified and sequenced by Macrogen (Amsterdam, The Netherlands). Sequences were checked for possible chimeric origin using Pintail 1.1 software (Ashelford et al., 2005). For the 16S rRNA gene sequences, initial alignment of amplified sequences and close relatives identified with BLAST (Altschul et al., 1997) were performed using the SILVA alignment tool (Pruesse et al., 2007) and manually aligned with ARB (Ludwig et al., 2004). After alignment, the neighbor-joining algorithm of ARB package was used to generate the phylogenetic trees based on distance analysis for 16S rRNA genes. The robustness of inferred topologies was verified by bootstrap re-sampling analysis, using the same distance model (1000 replicates).

## Enumeration of Cells by Card-Fish

The catalyzed reporter deposition fluorescence *in situ* hybridization (CARD-FISH) was performed using the protocol of Pernthaler et al. (2002) with some modifications. Briefly, 5 g of pooled sediments samples were fixed with formaldehyde (2% v/v final concentrations) and left for 12 h at 4°C. Fixed sediment samples were further incubated for at least 15 min with Tween 80 (final concentration, 1 mg $l^{-1}$) and than sonicated during 20 min in an Ultrasonic cleaner (Branson 1200, Milan) for the bacterial dispersion (Kuwae and Hosokawa, 1999). Supernatant samples were then filtered onto polycarbonate membrane filters (type GTTP; pore size, 0.2 μm; diameter, 2.5 cm; Sartorius, Göttingen, Germany). Filters for CARD-FISH counts were embedded in low-melting point agarose (0.1% wt/vol, Sigma-Aldrich, Milan), dried at 37°C for 20 min, and dehydrated with 95% ethanol. Embedded cells were permeabilized by 1 h of exposition with solution A (10 mg $ml^{-1}$of lysozyme (Sigma-Aldrich, Milan); 0.5 M EDTA, 0.1 M Tris-HCl [pH 8.0]) followed by the 30 min-long incubation with achromopeptidase (60 U $ml^{-1}$; 0.01 M NaCl, 0.01 M Tris-HCl [pH 8.0]) at 37°C. Filters were cut in sections and hybridized with 5′-horseradish peroxydase (HRP)-labeled oligonucleotide probes as described by Pernthaler et al. (2002). Probes used in this work are listed in Table 1. After the hybridization and amplification steps, slides were examined using a epifluorescence with Axioplan 2 Imaging (Zeiss; Carl Zeiss Inc., Thornwood, NY) microscope. All results were expressed as number of cells per gram of sediment.

**Table 1:** CARD-FISH specific probes used in the preset study

| Probe | Sequence (5′ to 3′) of probe | FA (%) | Source |
|---|---|---|---|
| Eub338 | GCTGCCTCCCGTAGGAGT | 35 | Amann et al., 1990 |
| Marinobacter sp. | ATGCTTAGGAATCTGCCCAGTAGTG | 20 | Karner and Fuhrman, 1997 |
| Cycloclasticus sp. | GGAAACCCGCCCAACAGT | 20 | Karner and Fuhrman, 1997 |
| Alcanivorax sp. | CGACGCGAGCTCATCCATCA | 20 | Karner and Fuhrman, 1997 |

FA: percentage of formamide in hybridization buffer.

## Determination of Cell Numbers by Quantitative PCR (qPCR)

qPCR method was employed to determine the relative cell densities of *Alcanivorax*, *Marinobacter*, and *Cycloclasticus* in the sediment samples. Primers used through this study are listed in the Table2. Primers were based on the sequences of *Alcanivorax* alkane hydroxylase gene (*alkB2*),*Marinobacter* alkane hydroxylase gene (*alkB*), and *Cycloclasticus* aromatic ring-hydroxylating dioxygenase (*phnA*) gene. Primers for *alKB2* and *phnA* genes were previously designed and validated for qPCR elsewhere (McKew et al., 2007; Gray et al., 2011). The primers specific for the*alkB* gene of *Marinobacter* were designed through this study using Primer Express software Primer Express software, version 2.0 (Applied Biosystems, Foster City, Calif.) with reference to the*Marinobacter hydrocarbonoclasticus* (FO203363).

**Table 2:** qPCR primers used in the preset study.

| Gene | Forward Reverse | Organism | Amplification efficiency (E) | Source |
|---|---|---|---|---|
| alkB2 | CGCCGTGTGAATGACAAGGG<br>CGACGCTTGGCGTAAGCATG | Alcanivorax | 99.6% | McKew et al., 2007 |
| alkB | TCCTTTGGTATGGCGCAGTT<br>ACGATCCTGTTCAAGCCGAG | Marinobacter | 97.3% | This study |
| phnA | CGTTGTGCGCATAAAGGTGCGG<br>CTTGCCCTTTCATACCCCGCC | Cycloclasticus | 96.2% | McKew et al., 2007 |

Total DNA was extracted from 2.0 g of sediment samples collected at selected time scales from three different points inside the MSS by using a Bio101 FastDNA SPIN kit as described by the manufacturer (Bio101, Inc., Vista, Calif.). Extracted DNA was dissolved in 50 μ l of TE buffer (10 mM Tris-HCl, 1 mM EDTA [pH 7.5]) and quantified using a NanoDrop ND-1000 spectrophotometer (Celbio). The quality of the extracted DNA was analyzed by electrophoresis on a 1.0% agarose gel. The qPCR was performed with absolute quantification method in an ABI Prism 7300 Real-time PCR System (Applied Biosystems) in a total volume of 25 μ l, consisting 12.5 μ l of SYBR green master mix, 200 nM of each primers, and 50 ng of template DNA. The volume of each reaction was adjusted to 25 μ l by adding DNase-free water. The cycling parameters for the qPCR amplification were as follows: an initial denaturation step at 95°C for 10 min, followed by 45 cycles of denaturation at 95°C for 15 s and annealing/elongation at 60°C for 60 s. A dissociation step was added to check for primer-dimer formation. A tenfold serial dilution series of genomic DNA ranging from 10 to $10^8$ copies per reaction was used in triplicate to create the standard curve for quantification. Serial dilutions were prepared once for each target and used for real-time quantification. From the slope of each curve, PCR amplification efficiency was calculated as it described elsewhere (Rasmussen, 2001). Obtained slope values, (−3.17 for *alkB2*, −3.29 for *alkB*, and −3.37 for *phnA*) fell within the optimal range corresponding to an efficiency of 99.6, 97.3, and 96.2%, respectively. Amplicon numbers were quantified against the standard curve using automatic analysis settings for the Ct values and baseline settings. Detected target genes were converted to cell density in sediments (cells $gram^{-1}$) assuming that all three genes present as a single copy per genome.

## Statistical Analysis

For statistical analyses, clones from each 16S crDNA library were separately considered to define phylotypes, or operational taxonomic units (OTUs) at cutoff of either 95 or 97% of sequence identity (Kemp and Aller, 2004). Clustering of sequences was performed using Dotur program (Schloss and Handelsman, 2005). Various parameters for each clone library, such as diversity index, rarefaction analysis, taxa, total clones, singletons, Shannon dominance, equitability, Simpson and chao2 were calculated by PAST version 2.17c (http://folk.uio.no/ohammer/past;Hammer et al., 2001). Coverage values given as $C = 1 - (n_1/N)$, where $n_1$ is the number of clones which occurred only once in the library of N clones (Good, 1953), were calculated to determine how efficient clone libraries described the complexity of original bacterial community. The Primer 6 ecological software package developed by the Plymouth Marine Laboratory (Clarke and Gorley, 2006) was employed to perform the Hierarchical Cluster Analysis (HCA) (Clarke, 1993). We applied HCA on microbial biodiversity and Bray-Curtis similarity on relative abundance matrix of the OTUs detected at different sampling time. Significant difference of the microbial assemblages derived from both treated and control samples at the different sampling times was detected via the*P*-test significance and principal coordinates analysis (PCoA) using UniFrac program (http://bmf.colorado.edu/unifrac/index.psp, last access: 24 July 2008) (Lozupone et al., 2007). Differences in cell number per gram of sediment detected by DAPI, CARD, and qPCR between different samples was examined by analysis of variance (ANOVA) on ranks (Holm_Sindak method). Relative importance of each treatment group was investigated by Multiple Comparisons vs. Control Group comparison test. Calculations were carried out using SigmaStat software for Windows, ver. 3.1 (Copyright 1992–1995; Jandel Corporation). Differences were considered significant at $P < 0.05$.

## Nucleotide Sequence Accession Numbers

To avoid the submission of identical sequences obtained among 411 analyzed clones, only 98 distinguishing 16S rRNA gene sequences have been deposited in the DDBJ/EMBL/GenBank databases under accession numbers KF896304-KF896401.

# RESULTS

## Geochemical Properties of the Sediments

Oxygen consumption in the external superficial sediments (0–5 cm) was monitored during all 3 months of experimentation. These values were compared with BOD in the internal MSS sediments taken at the beginning ($T_0$), after 1 day ($T_1$), 1 month ($T_{29}$), and after 3 months ($T_{90}$) of experimentation (Figure 2). Sediments outside of the MSS exhibited constant BOD rates of approximately 2.5 mg $O_2$ day$^{-1}$ kg$^{-1}$ during all period of experimentation, whereas the aerated sediments inside of the MSS demonstrated a progressive increment of oxygen consumption. Maximum of oxygen demand (16.0 mg $O_2$ day$^{-1}$ kg$^{-1}$) was obtained at $T_{29}$ and afterwards the BOD values trended to diminish, reaching 10.0 mg $O_2$ day$^{-1}$ kg$^{-1}$ at the end of experiment. Outside the MSS, the amendment of the Bunker C furnace fuel oil turned initially oxygenated ($T_0$, $E_h$ = 77 ± 4 mV) sediments into highly reduced ones (Figure 3). The external sediments below 5 cm became oxygen-depleted already within 1 day after spiking, obviously due to the active respiration of aerobic heterotrophic microorganisms. The reduction potential of external sediments decreased continuously during the experiment and after 3 months reached the $E_h$ values of −345 mV in the deepest layers. In contrary, inside the MSS the sediments remained aerobic during all period of bioremediaion effort.

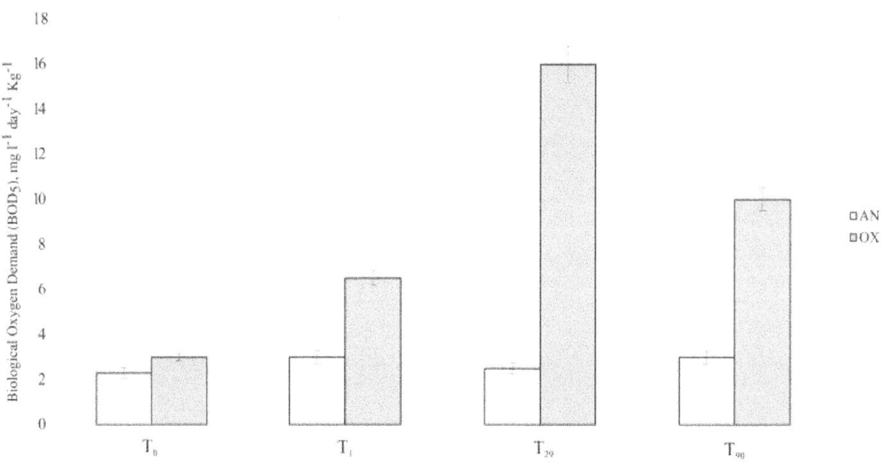


**Figure 2:** Dynamic of oxygen consumption (BOD values) measured in MSS external (untreated) superficial sediments (white bars) and internal sediments (gray bars). Error bar indicates the standard deviation of triplicate easurements.

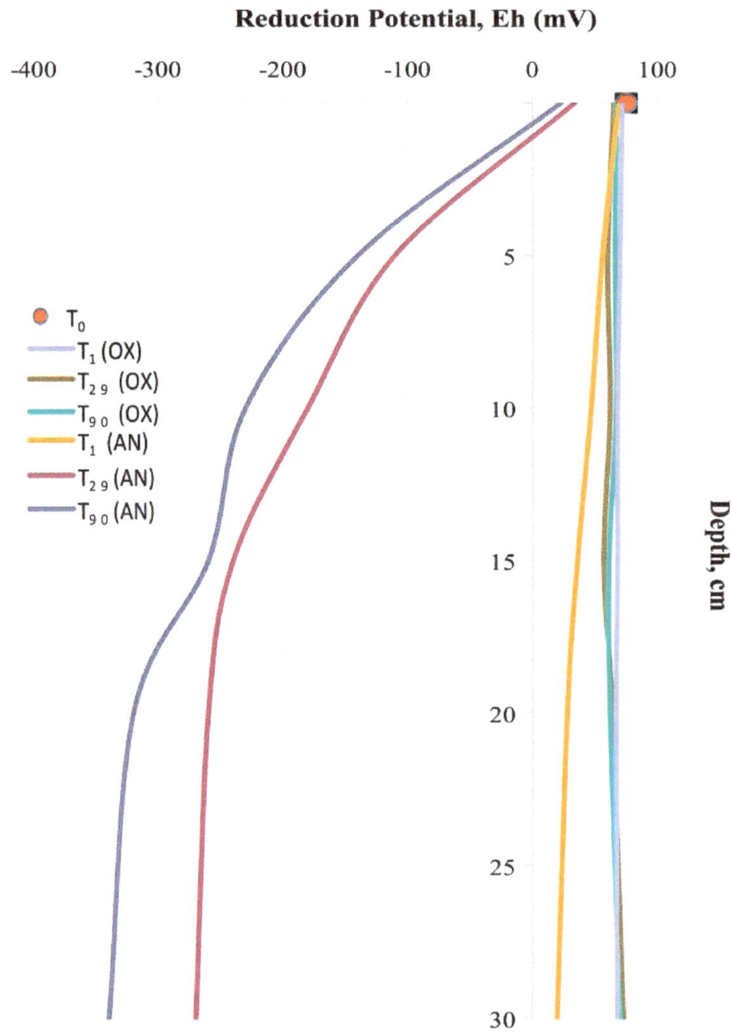


**Figure 3:** Reduction potential ($E_h$) measured in external anaerobic sediments (AN) and in internal aerated sediments (OX) during 3 months of experimentation.

## Hydrocarbon Analysis

Before the addition of 6500 ppm of Bunker C fuel oil into the mesocosm, the total hydrocarbon concentration in the original Messina harbor sediments was estimated at the level of 120 ppm. The PHs fingerprint analysis showed a clear dominance of alkyl-aromatic derivatives (97%) over aliphatic hydrocarbons (3%) (data not shown). Once Bunker C fuel oil was added to the sediments, the TERHC fraction mass balance was shifted toward the dominance of aliphatic and naphthenic hydrocarbons (70%) over aromatics (30%). The degree of the Bunker C fuel oil degradation in both, aerated

MSS and in the untreated anoxic sediments was examined at the end of experiment ($T_{90}$). The concentration of PHs, especially aliphatic hydrocarbons was normalized using the pristane/phytane ratio, and the values obtained in triplicate subsamples were averaged. In the aerated internal MSS sediments, the total degradation of Bunker C fuel oil TERHC fraction was 97.7 ± 0.9%. In contrast, external anoxic sediments contained more than 81.8 ± 1.2% of initially added Bunker C fuel oil. Additionally, we performed the gravimetrical analysis of total extracted hydrophobic fraction (TEHF) in the sediments. The TEHF values inside the MSS accounted for 780 ± 80 mg $kg^{-1}$ dry sediment weight, whereas in concordance with extremely slow biodegradation rates under anaerobic conditions, the external sediments contained more than 5400 ± 120 mg $kg^{-1}$of hydrophobic material. As shown in subsequent section, TEH fraction seems be primarily responsible for the observed toxicity of Bunker C fuel oil.

## Ecotoxicological Analysis

According to standard guidelines of Italian Institute for Environmental Protection and Research (ISPRA, 2013), eco-toxicological analysis of hydrocarbon-contaminated sediments were carried out using the Microtox® luminescence and amphipod *Corophium orientale* bioassays. Eventual decrease in Microtox® bioluminescence was measured on sediment pore water, whereas the rate of amphipods mortality was tested by direct exposition of *C. orientale* with the petroleum-contaminated sediments during 10 days. Following the EN12457 protocol, we combined the sediment pore water with sterile seawater in both 1:2 and 1:10 ratios (vol/vol) and no significant level of bioluminescence decay has been observed.

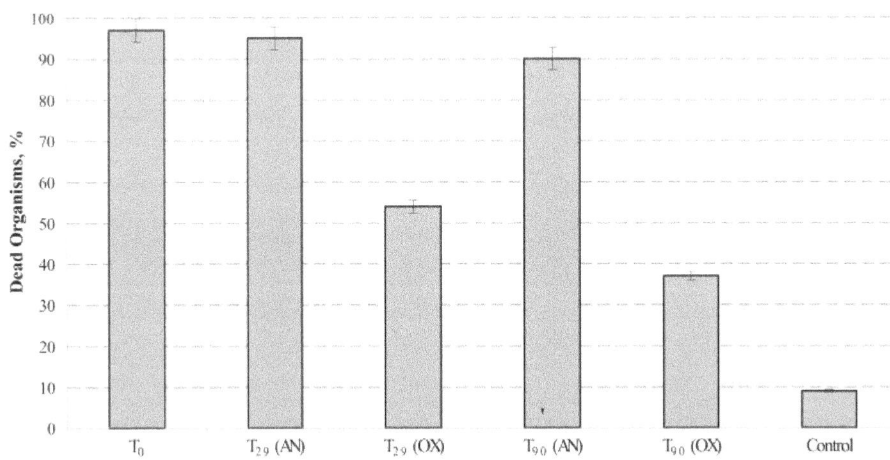


**Figure 4:** Mortality of *Corophium orientale* **organisms in polluted (AN), treated (OX), and control (native) sediments.** Error bar indicates the standard deviation of duplicate measurements.

As it reported in EN12457 protocol, Microtox®bioluminescent assay tested on sediment pore water typically exhibits an underestimated sensitivity against highly hydrophobic contaminants, such as PHs. This is mainly due to both extremely low solubility of these compounds in water and an almost irreversible adsorption to a sedimentary matrix. Corresponding to the standardized protocol described by Onorati et al. (1999), the toxicological analysis of the sediments was performed with amphipods *C. orientale*. Addition of Bunker C fuel oil to the sediments ($T_0$) caused the mortality in almost all organisms (98 ± 2%). The petroleum-contaminated external sediments remained highly toxic during all 3 months of experimentation with mortality indices exceeding 90% (Figure 4). At the same time, the toxicity of polluted internal MSS sediments dropped almost twice after 1 month of aeration ($T_{29}$) and continued to decrease till the end of bioremediation treatment, approaching the vitality of 62% of amphipods exposed to $T_{90}$ sediments. This indicated that the internal MSS sediment at $T_{90}$ was significantly less toxic than their exteral counterpart.

## Diversity and Succession of Bac Communities Duringduring Bioremediation and Natural Sediment Ageing

To monitor the succession of metabolically active microbial communities during the bioremediation treatment, five 16S rRNA transcript libraries were established. 450 clones from these libraries were randomly selected and analyzed. Among those, 411 sequences were included into phygenetic analysis ($T_0$, 96 clones; $T_1$, 83 clones; $T_{29}$, 80 clones; $T_{90}$, 90 clones and $T_{90OUT}$, 62 clones). The majority of native $T_0$ clones were affiliated with the *Gammaproteobacteria* (51%), followed by the *Alphaproteobacteria* (22%). Other proteobacteria, belonging to microaerophilic and anaerobic *Epsilon-* and *Deltaproteobacteria* were also present, although in significantly lower numbers (6 and 2% of all clones analyzed, respectively). Remaining fraction of $T_0$ microbial community consisted of the members of *Cyanobacteria* (7%), *Chloroflexi* (5%), *Verrucomicrobia*(3%), and *Bacteroidetes* (2%) (Figure 5). At the level of the Class, the derived from members of*Gammaproteobacteria* were predominant in all analyzed MSS internal sediments, with percentage ranging from 73 to 88%. There was a 50%-reduction of *Alphaproteobacteria* observed during first month of sediment treatment (decrease from 22 to 11%). Further on, their numbers returned to the initial ($T_0$) values by the end of experiment. No *Deltaproteobacteria*–related organisms were detected in the internal MSS sediments throughout the experiment. A completely different scenario was observed regarding the succession of microbial community thriving in the external anoxic sediments. As we mentioned above, after loading the Bunker C fuel oil, the sediments became highly reduced within a short period of time and were inhabited mainly by the members of*Deltaproteobacteria* (96.8%). The analysis of 16S rRNA transcripts from $T_{90OUT}$ clone library revealed a notable prevalence of hydrocarbon-degrading or hydrocarbon contamination-associated*Deltaproteobacteria* (90.3%). In particular, almost two-thirds of all clones (40 out of 62 clones analyzed) were closely related to the uncultured bacterium RII-AN044 found in anoxic polluted sediments after the Prestige oil spill (Acosta González et al., 2013). More than 16% of clones revealed >98% of similarity to the deltaproteobacterium NaphS6, capable of naphthalene and 2-methylnaphthalene degradation (Wilkes et al., 2008). Almost 10% of $T_{90OUT}$ clones were closely related to the *n*-alkanes- and *n*-alkenes-degrading strain *Desulfatibacillum aliphaticivorans*(Cravo-Laureau et al., 004) (Figure 6).

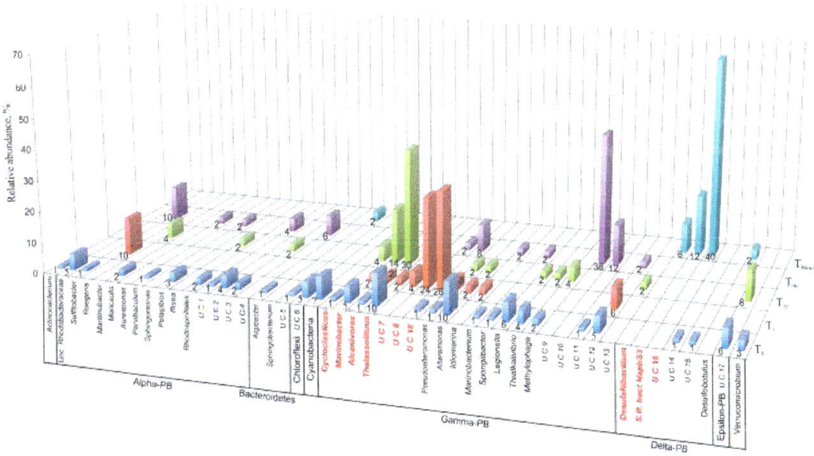


**Figure 5:** Dendrogram representation of taxonomic analysis of 16S crDNA clones retrieved from five libraries. The numbers at the base of columns represent the percentage of clones in corresponding libraries. Abbreviations used: PB, *Proteobacteria*divisions; UC, Unaffiliated cluster. Bacterial groups involved in, or associated to, hydrocarbon degradation are marke in bold and red.

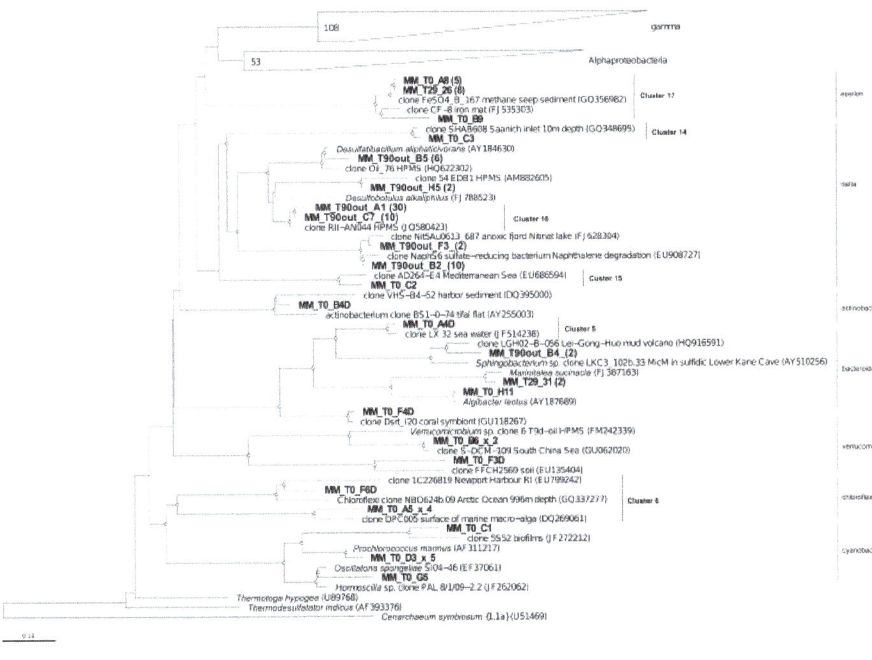


**Figure 6:** Phylogenetic affiliation of the *Eubacteria* **clones retrieved in Messina harbor native sediments and during** mesocosm experiment. Neighbor-joining analysis using 1000 bootstrap replicates was used to infer tree topology. The scale bar represents 10%

of sequence divergence. Bootstrap values ≥50 are shown as open circles.*Cenarcheum symbiosum* was used as the outgroup. The scale bar represents the expected number of changes pet nucleotide position. Sequences from this study are indicated in bold.

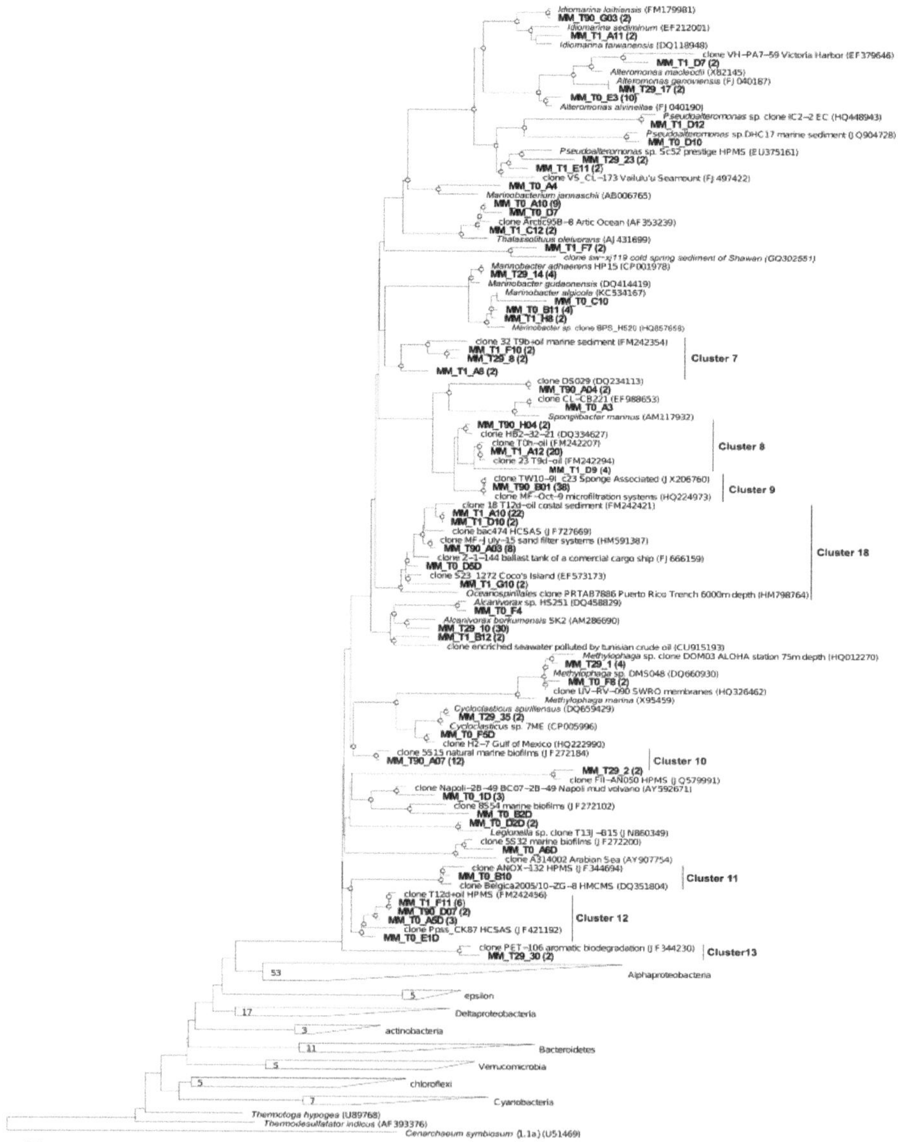


**Figure 7:.** Phylogenetic affiliation of the *Gammaproteobacteria* **clones retrieved in Messina harbor sediments and in** mesocosm experimentation. Neighbor-joining analysis using 1000 bootstrap replicates was used to infer tree topology. The scale bar represents 10% of sequence divergence Bootstrap values ≥50 are shown as open circles.*Cenarcheum symbiosum* was used as the out-group. Sequences obtained in this study ard in bold.

In accordance with the type of the contaminant load, the majority of analyzed clones had highest Blastn homologies with sequences related to PH-degrading or petroleum contamination-associated organisms belonging to the *Gamma-*, *Alpha-* and *Deltaproteobacteria* (Figures 6–8). At the genus and species level, more than 10% of initial microbial population was attributed to obligate marine hydrocarbonoclastic bacteria *Thalassolituus oleivorans* (Yakimov et al., 2004). This organism seemed to be sensitive to Bunker C fuel oil, since its abundance decreased to 2% after 1 day of oil exposition and disappeared afterwards. Three other OMHCB belonging to genera *Alcanivorax*,*Cycloclasticus*, and *Marinobacter* demonstrated similar dynamics, i.e., being in relative minority in the beginning of the experiment, they became predominant in $T_{29}$ microbial community and disappeared in the $T_{90}$ library (Figure 5). Noteworthy, the addition of Bunker C fuel oil to sediments drastically changed the structure of $T_1$ microbial community and the proportion of aforementioned *Gammaproteobacteria*-related OMHCB decreased threefold compared with the initial population. In contrast, previously undetected organisms related to dinoflagellate-associated*Rugeria* sp. and to three deep-branching clusters of *Gammaproteobacteria* accounted for 77% of all analyzed $T_1$ clones. Having only 16S rRNA gene sequences at our disposal, we cannot state that these uncultured organisms were involved in biodegradation activity, but they definitely possessed a remarkable resilience to the toxicity of Bunker C fuel oil (Païssé e8, 2010).

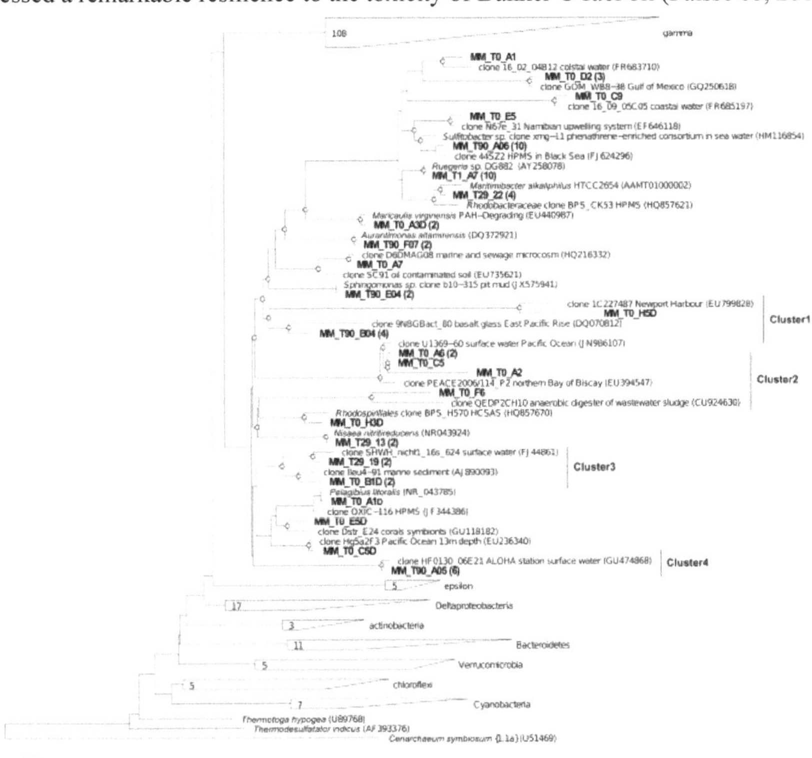


**Figure 8:.** Phylogenetic affiliation of the *Alphaproteobacteria* **clones retrieved in Messina harbor sediments and in** mesocosm experimentation. Neighbor-joining analysis

using 1000 bootstrap replicates was used to infer tree topology. The scale bar represents 10% of sequence divergence. Bootstrap values ≥50 are shown as open circles. *Cenarcheum symbiosum* was used as the outgroup. Sequences obtained in this study are indicated in bold.

Additionally to 16S rRNA-based analysis, the diversity and succession of both, total bacterial population and hydrocarbon-degrading bacteria, was assessed through the combined application of CARD-FISH and qPCR. Using the *Eubacteria*-specific probe Eub338, the concentration of CARD-positive cells at the beginning of experiment was estimated at $2.98 \pm 0.17 \times 10^6$ cells gram$^{-1}$ (Table3). Their numbers decreased by 20% within 1 day after the oil spill simulation ($P < 0.001$, $n = 10$) and reached initial values at the end of experiment ($2.79 \pm 0.12 \times 10^6$ cells gram$^{-1}$). Noteworthy, a tenfold increase in the number of CARD-positive cells ($2.95 \pm 0.11 \times 10^7$ cells gram$^{-1}$) was detected after 29 days of oil spill, which fully corresponded to the observed dynamics of the BOD and clone libraries' values (Figure 2). Before the oil spill simulation ($T_0$), the fraction of *Alcanivorax*-related cells, detected with the CARD-FISH genus-specific probe, accounted for 7.7% of all Eub338-positive cells. After the addition of Bunker C fuel oil, their abundance increased within 1 month up to 27.5%. According to the analysis of 16S crDNA clone libraries, *Alcanivorax* became extinct in microbial community thriving in the MSS internal sediments at the end of experiment. Dynamics of the *Marinobacter*-related bacteria was comparable with that of *Alcanivorax* population, with the only exception that the *Marinobacter* cells decreased their abundance by 44% at $T_1$, likely due to the higher sensitivity to the load of fuel oil. Compared to an initial density, their population increased twofold at $T_{29}$, from $4.5 \pm 0.2$ to $9.0 \pm 0.1 \times 10^5$ cells g$^{-1}$. Although due to the overwhelming growth of *Alcanivorax*, the relative abundance of *Marinobacter* during bioremediation experiment has never exceeded its initial values (15.2% of all Eub338-stained cells at $T_0$) and at $T_{29}$ (corresponding to the maximum of cell density in MSS) accounted for only 3% of total microbial community. Bacteria stained with *Cycloclasticus*-specific CARD-FISH probe, initially present in mesocosm sediments at concentration of $1.14 \pm 0.15 \times 10^5$ cells gram$^{-1}$, were not detected at $T_1$, whereas their concentration increased four times after 1 month of the oil spill simulation. Similarly to the dynamics of *Alcanivorax*, neither *Marinobacter*, nor *CycloclastEicus*were present in the MSS microbial community at the end of experiment.

**Table 3:** CARD-FISH and qPCR cell number quantification in the MSS internal sediment during the biionoremediation treatment.

| Method | Cell numbers, $10^5 \times$ g sediments$^{-1} \pm SD$ | | | | |
|---|---|---|---|---|---|
| | **Probe/primers** | $T_0$ | $T_1$ | $T_{29}$ | $T_{90}$ |
| CARD-FISH | *Eubacteria* | 29.8 ± 1.7 | 23.6 ± 1.5 | 295.0 ± 11.0 | 27.9 , 1.2 |
| | *Alcanivorax* sp. | 2.3 ± 0.1 | 3.1 ± 0.2 | 81.0 ± 1.3 | ND |
| | *Marinobacter* sp. | 4.5 ± 0.2 | 2.6 ± 0.1 | 9.0 ± 0.1 | ND |
| | *Cycloclasticus* sp. | 1.2 ± 0.2 | ND | 4.9 ± 0.1 | ND |
| qPCR* | *alkB2* | 5.17 ± 0.17 | 4.73 ± 0.18 | 94.60 ± 3.80 | ND |
| | *alkB* | 7.90 ± 0.23 | 3.50 ± 0.20 | 15.30 ± 2.00 | ND |
| | *phnA* | 2.10 ± 0.16 | $3.78 \pm 0.12 \times 10^{-3}$ | 7.13 ± 0.28 | ND |

*These values mean the average number of cells detected in triplicate from three individual subsamples of sediments collected in different parts of MSS.

Succession of hydrocarbon-degrading bacteria during the sediment bioremediation was additionally quantified by qPCR. Based on the a priori higher sensitivity of qPCR approach, obtained numbers were slightly higher than those from taxon-specific CARD-FISH data. Nevertheless, obtained results remarkably corroborated with CARD-FISH counts and the general trend in *Alcanivorax*, *Marinobacter*, and *Cycloclasticus* dynamics was identical (Table 3). None of these hydrocarbon-degrading bacteria were detected by qPCR at the end of experiment.

## Statistical Analysis of Bacterial Diversity

As it shown in Table 4, the diversity index and coverage values have been calculated for each 16S rRNA transcript library. Based on sequence similarity threshold for OTU definition ≥97% (Rosselló-Mora and Amann, 2001), a total of 93 different OTUs were discerned. The rarefaction analysis did not demonstrate the saturation in any library (data not shown), however, the coverage values varying between 0.64 ($T_0$ library) and 0.92 ($T_{90OUT}$ library) indicated that a satisfactory overview on bacterial community was obtained. As reflected by the high values of Simpson, Shannon and the number of missing species provided by Chao-2, the highest bacterial diversity was observed in original sediments ($T_0$). Bacterial biodiversity was drastically affected by Bunker C fuel oil addition, decreasing the Simpson, Shannon and Equitability values, due to strong selection for both petroleum contamination-resilient and hydrocarbon-degrading bacteria (Harayama et al., 1999; Kasai et al., 2001, 2002; Syutsubo et al., 2001). Dominance index reached the highest value at $T_{29}$ and $T_{90}$. After 1 month of aeration, the microbial population of the internal MSS sediments was dominated by OMHCBs, whereas at the end of experiment, the dominant Cluster 9 of the*Gammaproteobacteria* accounted for 42% of all $T_{90}$ clones. This cluster consists of organisms recovered from pristine seawater and marine sediments. As mentioned above, the microbial community of anaerobic external sediments $T_{90OUT}$ was totally different compared with aerated internal sediments. Both fuel oil contamination and rapid development of anaerobiosis have selected a very specialized and poorly diverse microbial population (5 OTUs; Dominance 0.44; Coverage 91%). These data were confirmed by the HCA and UniFrac PCoA analyses. At the end of the experiment, the microbial communities thriving in the MSS external ($T_{90OUT}$) and the internal ($T_{90}$) sediments were significantly different from each other (P1 = 50.85% and P2 = 31.39%). Noteworthy, both applied statistical analyses confirmed that comparing with OMHCB-enriched $T_{29}$population, the $T_{90}$ microbial community was more similar to those recovered from early stages of the experiment ($T_0$ and $T_1$ libraries) (Figure 9). This finding indirectly hints at the process of self-recovery of petroleum-contaminated sediments, which was confirmed by GC analysis of the contaminated sediments in and outside the MSS.

**Table 4:** Diversity indices calculated for the five clone libraries created at different time of aer-atedidiment agein.

| Diversity index | T0 | T1 | T29 | T90 | T90OUT |
|---|---|---|---|---|---|
| Taxa/(OTUs)_S | 46 | 16 | 14 | 12 | 5 |
| Number of clones | 96 | 83 | 80 | 90 | 62 |
| Dominance_D | 0.04 | 0.15 | 0.22 | 0.24 | 0.44 |
| Simpson_1-D | 0.96 | 0.85 | 0.78 | 0.76 | 0.55 |
| Shannon_H | 3.5 | 2.31 | 2.08 | 1.89 | 1.09 |
| Equitability_J | 0.91 | 0.84 | 0.78 | 0.76 | 0.68 |
| Chao-2 | 104.1 | 16 | 14 | 12 | 5 |
| Singletons | 31 | 1 | 0 | 0 | 0 |
| Doubletones | 7 | 10 | 8 | 6 | 2 |
| Coverage | 0.64 | 0.74 | 0.77 | 0.87 | 0.92 |

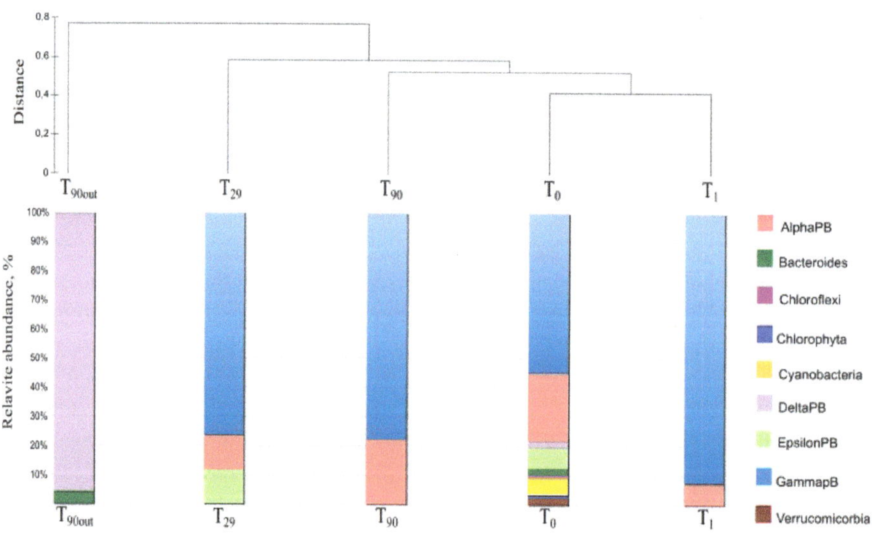


**Figure 9:.** Dendrogram of microbial biodiveristy and similarity analysis of the 16S rRNA transcripts detected at different sampling time and treatment. The UPGMA cluster analysis was obtained by using group average clustering from Euclidean distance on relative abundance matrix of OTUs detected in the analyzed libraries.

## DISCUSSION

The influence of the massive Bunker C fuel oil load upon the microbial population of coastal sediments collected in Messina harbor and their recovery was investigated during 3 months in the mesocosms experiment. These sediments chronically polluted with alkylnaphthalenes were chosen since the autochthonous microbial community is likely adapted to the steady presence of PHs and consequently may exhibit a higher biodegradation ability than those from pristine environments (Païssé et al., 2008, 2010). Accordingly, the fraction of the OMHCB genera *Alcanivorax Cycloclasticus Marinobacter* and *Thalassolituus*, usually imperceptible in pristine marine environments (Yakimov et al., 2007), accounted for a fifth part of the native microbial community of Messina harbor sediments. To simulate accidental oil spill we spiked 1000 kg sandy sediments with Bunker C fuel oil (6500 mg $kg^{-1}$). This type of heavy fuel oil is frequently used by the cargo and tanker ships and is generally more complex in composition and impurities than distilled fuels. Bunker C fuel oil includes about 25% polyaromatic, 15% aliphatic, 45% naphthenic and 15% non-hydrocarbon compounds (Clark et al., 1990) As it generally observed, after the oil spill reaches a shoreline, the oil burial is the main mechanism of the pollution dispersal. This results in intense oiling of subsurface sediments and consequent reduction in oxygen concentration due to initial activation of aerobic hydrocarbon-degrading microorganisms (Albaigés et al., 2006; Acosta González et al., 2013). Once $O_2$ concentration drops to zero, the degradation rates of both aliphatic and aromatic PHs decrease significantly. According with this observation, the oil cleanup was very marginal, and >80% of initially added Bunker C fuel oil was extracted from the anaerobic external sediments after 3 months of oil spill simulation. At the end of experiment, analysis of microbial community of the MSS external sediments revealed the overwhelming dominance of the*Deltaproteobacteria*.

The stimulation of autochthonous bacteria to tackle the pollution in contaminated environment is widely used in the remediation of aerobic sites (Harayama et al., 2004). However, this technology is hardly applicable to oxygen-depleted marine sediments. Additionally, due to high probability of contaminant spreading while removing, the application of *ex-situ* remediation technologies is severely limited. To initiate the self-cleaning process in sediments driven by indigenous aerobic OMHCB, we used *in situ* aeration of polluted anoxic sediments in specially designed MSS. It is generally assumed, that petroleum contamination induced drastic changes in the bacterial community structure associated with a decrease of diversity (Grötzschel et al., 2002; Röling et al., 2002; Yakimov et al., 2005, 2007; Head et al., 2006; Bordenave et al., 2007). These changes were referred to both toxic effect of PHs and a strong selection toward highly specialized hydrocarbon-degrading microorganisms (Harayama et al., 1999; Kasai et al., 2001; Grötzschel et al., 2002;Yakimov et al., 2005, 2007). Accordingly, the most drastic shifts in the MSS bacterial community dynamic was observed at the beginning and after 1 month of the oil spill. As revealed by 16S crDNA clone library analysis, more than a half of MSS microbial population at $T_{29}$, belonged to*Alcanivorax* (43%), *Cycloclasticus* (7%) and *Marinobacter* (5%), the genera of OMHCB, known to play a pivotal role in petroleum degradation in marine environments (Harayama et al., 1999;Röling et al., 2002, 2004; Yakimov et al., 2007). Noteworthy, the dominance of these OMHCB genera was confirmed by both CARD-FISH and qPCR analyses. At the end of the treatment, the level of TERHC degradation in the MSS internal sediments was almost 98%, and the resulting microbial community was characterized by an almost complete extinction of OMHCBs. Both HCA and UniFrac PCoA statistic analyses of the OTUs abundance matrix indicated that the structure of

$T_{90}$ microbial community was more similar to initial microbial community structures. As a consequence of successful bioremediation, the *Corophium orientale* eco-toxicological bioassay revealed that toxicity of the MSS internal sediments was substantially lower compared with the untreated external sediments. Thus, to the best of our knowledge, our studies for the first time demonstrated that petroleum-contaminated anaerobic marine sediments could be efficiently recovered by their capping and *in situ* aeration, thus stimulating the self-cleaning potential due to reawakening ofntial conflict of interest.

## ACKNOWLEDGMENTS

This work was funded by the European Community Projects ULIXES (FP7-KBBE-2010-4 Project 266473 "*Unravelling and exploiting Mediterranean Sea microbial diversity and ecology for xenobiotics' and pollutants' clean-up*") and KILL-SPILL (FP7-KBBE-2012.3.5-01-4 Project 312139 "*Integrated Biotechnological Solutions for Combating Marine Oil Spills*"). We would like to thank all partners of these projects for their useg marine bacteria. *Curr. Opin. Biotech.* 18, 257–266. doi: 10.1016/j.copbio.2007.04.006

## REFERENCES

1. Acosta González, A., Rosselló Móra, R., and Marqués, S. (2013). Characterization of the anaerobic microbial community in oil polluted subtidal sediments: aromatic biodegradation potential after the Prestige oil spill. *Environ. Microbiol.* 15, 77–92. doi: 10.1111/j.1462-2920.2012.02782.x
2. Albaigés, J., Morales-nin, B., and Villas, F. (2006). The Prestige oil spill: a scientific response. *Mar. Pollut. Bull.* 53, 205–207. doi: 10.1016/j.marpolbul.2006.03.012
3. Altschul, S. F., Madden, T. L., Schaffer, A. A., Zhang, J., Zhang, Z., Miller, W., et al. (1997). Gapped BLAST and PSI-BLAST: a new generation of protein database search programs. *Nucleic Acids Res.* 25, 3389–3402. doi: 10.1093/nar/25.17.3389
4. Amann, R. I., Krumholz, L., and Stahl, D. A. (1990). Fluorescent-oligonucleotide probing of whole cells for determinative, phylogenetic, and environmental studies in microbiology. *J. Bacteriol.* 172, 762–770.
5. Ashelford, K. E., Chuzhanova, N. A., Fry, J. C., Jones, A.J., and Weightman, A. J. (2005). At least 1 in 20 16S rRNA sequence records cur- rently held in public repositories is estimated to contain substantial anomalies. *Appl. Environ. Microbiol.* 71, 7724–773. doi: 10.1128/AEM.71.12.7724-7736.2005
6. Benton, M. J., Malott, M. L., Knight, S. S., Cooper, C. M., and Benson, W. H. (1995). Influence of sediment composition on apparent toxicity in a solid phase test using bioluminescent bacteria. *Environ. Toxicol. Chem.* 14, 411–414. doi: 10.1002/etc.5620140309
7. Bordenave. S., Goñi-Urriza, M. S., Caumette, P., and Duran, R. (2007). Effects of heavy fuel oil on the bacterial community structure of a pristine microbial mat. *Appl. Environ. Microbiol.* 73, 6089–6097. doi: 10.1128/AEM.01352-07
8. Bulich, A. A., Greene, M. W., and Underwood, S. R. (1992). "Measurement of

soil and sediment toxicity to bioluminescent bacteria when in direct contact for a fixed time period," in *Memorias de Water Environment Federation, 65th Annual Conference and Exposition* (Nueva Orleans, Louisiana), 53–64.

9. Clark, R. N., King, T. V. V., Klejwa, M., Swayze, G., and Vergo, N. (1990). High spectral resolution reflectance spectroscopy of minerals. *J. Geophys. Res.* 95, 12653–12680. doi: 10.1029/JB095iB08p12653
10. Clarke, K. R. (1993). Non parametric multivariate analyses of changes in community structure. *Aust. J. Ecol.* 18, 117–143. doi: 10.1111/j.1442-9993.1993.tb00438.x
11. Clarke, K. R., and Gorley, R. N. (2006). *PRIMER v6. User Manual and Tutorial.* Plymouth: PRIMER-E.
12. Cravo-Laureau, C., Matheron, R., Cayol, J. L., Joulian, C., and Hirschler-Réa, A. (2004). *Desulfatibacillum aliphaticivorans*gen. nov., sp. nov., an n-alkane-and n-alkene-degrading, sulfate-reducing bacterium. *Int. J. Syst. Evol. Microbiol.* 54, 77–83. doi: 10.1099/ijs.0.027170-0
13. Dyksterhouse, S. E., Gray, J. P., Herwig, R. P., Lara, J. C., and Staley, J. T. (1995). *Cycloclasticus pugetii* gen. nov., sp. nov., an aromatic hydrocarbon-degrading bacterium from marine sediments. *Int. J. Syst. Bacteriol.* 45, 116–123. doi: 10.1099/00207713-45-1-116
14. Engelen, B., and Cypionka, H. (2009). The subsurface of tidal-flat sediments as a model for the deep biosphere. *Ocean Dynam.* 59, 385–391. doi:10.1007/s10236-008-0166
15. Ferraro, G., Bernardini, A., David, M., Meyer-Roux, S., Muellenhoff, O., Perkovic, M., et al. (2007). Towards an operational use of space imagery for oil pollution monitoring in the Mediterranean basin: a demonstration in the Adriatic Sea. *Mar. Pollut. Bull.* 54, 403–422. doi: 10.1016/j.marpolbul.2006.11.022
16. Gertler, C., Yakimov, M. M., Malpass, M. C., and Golyshin, P. N. (2010). "Shipping-related accidental and deliberate release into the environment," in *Handbook of Hydrocarbon and Lipid Microbiology*, ed K. N. Timmis (Berlin, Heidelberg: Springer), 243–256. doi: 10.1007/978-3-540-77587-4_16
17. Golyshin, P. N., Chernikova, T. N., Abraham, W. R., Lünsdorf, H., Timmis, K. N., and Yakimov, M. M. (2002). *Oleiphilaceae*fam. nov., to include *Oleiphilus messinensis* gen. nov., sp nov., a novel marine bacterium that obligately utilizes hydrocarbons.*Int. J. Syst. Evol. Microbiol.* 52, 901–911. doi: 10.1099/ijs.0.01890-0
18. Good, I. J. (1953). The population frequencies of species and the estimation of population parameters. *Biometrika.* 40, 237–264. doi: 10.2307/2333344
19. Gray, S. B., Classen, A. T., Kardol, P., Yermakov, Z., and Miller, R. M. (2011). Multiple climate change factors interact to alter soil microbial community structure in an old-field ecosystem. *Soil Sci. Soc. Am. J.* 75, 2217–2226. doi: 10.2136/sssaj2011.0135
20. Grötzschel, S., Köster, J., Abed, R. M. M., and De Beer, D. (2002). Degradation

of petroleum model compounds immobilized on clay by a hypersaline microbial mat. *Biodegradation* 13, 273–283. doi: 10.1023/A:1021263009377

21. Hammer, Ø., Harper, D. A. T., and Ryan, P. D. (2001). PAST: palaeontological statistics software package for education and data analysis. *Palaeontol. Electron.* 4, 9.
22. Hara, A., Syutsubo, K., and Harayama, S. (2003). *Alcanivorax* which prevails in oil-contaminated seawater exhibits broad substrate specificity for alkane degradation. *Environ. Microbiol.* 5, 746–753. doi: 10.1046/j.1468-2920.2003.00468.x
23. Harayama, S., Kasai, Y., and Hara, A. (2004). Microbial communities in oil-contaminated seawater. *Curr. Opin. Biotechnol.* 15, 205–214. doi: 10.1016/jcopbio.2004.04.002
24. Harayama, S., Kishira, H., Kasai, Y., and Shutsubo, K. (1999). Petroleum biodegradation in marine environments. *J. Mol. Microbiol. Biotechnol.* 1, 63–70.
25. Head, I. M, Jones, D. M, and Röling, W. F. (2006). Marine microorganisms make a meal of oil. *Nat. Rev. Microbiol.* 4, 173–182. doi:10.1038/nrmicro1348
26. ISPRA. (2013). "Batterie di saggi ecotossicologici per sedimenti e acque interne," in *Manuali di Ecotossicologia*, eds ISPRA-settore editoria. ISBN: 978-88-448-0607-1.
27. Karner, M. B., and Fuhrman, J. (1997). Determination of active marine bacterioplankton: a comparison of universal 16S rRNA probes, autoradiography, and nucleoid staining. *Appl. Environ. Microbiol.* 63, 1208–1213.
28. Kasai, Y., Kishira, H., Sasaki, T., Syutsubo, K., Watanabe, K., and Harayama, S. (2002). Predominant growth of *Alcanivorax* strains in oil contaminated and nutrient supplemented seawater. *Environ. Microbiol.* 4, 141–147. doi: 10.1046/j.1462-2920.2002.00275.x
29. Kasai, Y., Kishira, H., Syutsubo, K., and Harayama, S. (2001). Molecular detection of marine bacterial populations on beaches contaminated by the Nakhodka tanker oil spill accident. *Environ. Microbiol.* 3, 246–255. doi: 10.1046/j.1462-2920.2001.00185.x
30. Kemp, P. F., and Aller, J. Y. (2004). Bacterial diversity in aquatic and other environments: what 16S rDNA libraries can tell us. *FEMS Microbiol. Ecol.* 47, 161–177. doi: 10.1016/S0168-6496(03)00257-5
31. Kuwae, T., and Hosokawa, Y. (1999). Determination of the abundance and biovolume of bacteria in sediments by dual staining with 4',6-diamidino-2-phenylindole and acridine orange: relationship to dispersion treatment and sediment characteristics. *Appl. Environ. Microbiol.* 65, 3407–3412.
32. Lane, D. J. (1991). "16/23S rRNA sequencing," in *Nucleic Acid Techniques in Bacterial Systematics*, eds E. Stackerbrandt and M. Goodfellow (New York, NY: Wiley), 115–175.
33. Lozupone, C. A., Hamady, M., Kelley, S. T., and Knight, R. (2007). Quantitative and qualitative β diversity measures lead to different insights into factors that

structure microbial communities. *Appl. Environ. Microbiol.* 73, 1576–1585. doi: 10.1128/AEM.01996-06

34. Ludwig, W., Strunk, O., Westram, R., Richter, L., Meier, H., Yadhukumar, et al. (2004). ARB: a software environment for sequence data. *Nucleic Acids Res.* 32, 1363–1371. doi: 10.1093/nar/gkh293
35. McKew, B.A., Coulon, F., Yakimov, M. M., Denaro, R., Genovese, M., Smith, C. J., et al. (2007). Efficacy of intervention strategies for bioremediation of crude oil in marine systems and effects on indigenous hydrocarbonoclastic bacteria. *Environ. Microbiol.* 9, 1562–1571. doi: 10.1111/j.1462-2920.2007.01277.x
36. Microtox System Operating Manual. (1982). *Microbics Operations*. Carlsbad, CA: Beckman Instruments, Inc.
37. Onorati, F., Bigongiari, N., Pellegrini, D., and Giuliani, S. (1999). The suitability of *Corophium orientale* (Crustacea, Amphipoda) in harbour sediment toxicity bioassessment. *Aquat. Ecosys. Health Manage.* 2, 465–473. doi: 10.1016/S1463-4988(99)00030-5
38. Païssé, S., Coulon, F., Goñi-Urriza, M., Peperzak, L., McGenity, T. J., and Duran, R. (2008). Structure of bacterial communities along a hydrocarbon contamination gradient in a coastal sediment. *FEMS Microbiol. Ecol.* 66, 295–305. doi: 10.1111/j.1574-6941.2008.00589
39. Païssé, S., Goñi-Urriza, M., Coulon, F., and Duran, R. (2010). How a bacterial community originating from a contaminated coastal sediment responds to an oil input. *Microb. Ecol.* 60, 394–405. doi: 10.1007/s00248-010-9721-7
40. Pernthaler, A., Pernthaler, J., and Amann, R. (2002). Fluorescence *in situ* hybridization and catalyzed reporter deposition for the identification of marine bacteria. *Appl. Environ. Microbiol.* 68, 3094–3101. doi: 10.1128/AEM.68.6.3094-3101.2002
41. Pruesse, E., Quast, C., Knittel, K., Fuchs, B. M., Ludwig, W. G., Peplies, J., et al. (2007). SILVA: a comprehensive online resource for quality checked and aligned ribosomal RNA sequence data compatible with ARB. *Nucleic Acids Res.* 35, 7188–7196. doi: 10.1093/nar/gkm864
42. Psarros, G., Skjong, R., and Eide, M. S. (2010). Under-reporting of maritime accidents. *Accid. Anal. Prev.* 42, 619–625. doi: 10.1016/j.aap.2009.10.008
43. Rasmussen, R. (2001). "Quantification on the LightCycler," in *Rapid Cycle Real-Time PCR, Methods and Applications*, eds S. Meuer, C. Wittwer, and K. Nakagawara (Heidelberg: Springer Press), 21–34. doi: 10.1007/978-3-642-59524-0_3
44. Ringwood, A. H., DeLorenzo, M. E., Ross, P. E., and Holland, A. F. (1997). Interpretation of Microtox® solid phase toxicity tests: the effects of sediment composition. *Environ. Toxicol. Chem.* 16, 1135–1140. doi: 10.1002/etc.5620160607
45. Rocchetti, L., Beolchini, F., Ciani, M., and Dell'Anno, A. (2011). Improvement of bioremediation performance for the degradation of petroleum hydrocarbons

in contaminated sediments. *Appl. Environ. Soil Sci.* 2011:319657. doi: 10.1155/2011/319657

46. Röling, W. F., Milner, M. G., Jones, D. M., Fratepietro, F., Swannell, R. P., Daniel, F., et al. (2004). Bacterial community dynamics and hydrocarbon degradation during a field-scale evaluation of bioremediation on a mudflat beach contaminated with buried oil. *Appl. Environ. Microbiol.* 70, 2603–2613. doi: 10.1128/AEM.70.5.2603-2613.2004
47. Röling, W. F., Milner, M. G., Jones, D. M., Lee, K., Daniel, F., Swannell, R. P., et al. (2002). Robust hydrocarbon degradation and dynamics of bacterial communities during nutrient-enhanced oil spill bioremediation. *Appl. Environ. Microbiol.* 68, 5537–5548. doi: 10.1128/AEM.68.11.5537-5548.2002
48. Rosselló-Mora, R., and Amann, R. (2001). The species concept for prokaryotes. *FEMS Microbiol. Rev.* 25, 39–67. doi: 10.1016/S0168-6445(00)00040-1
49. Roussel, E. G., Sauvadet, A. L, Allard, J., Chaduteau, C., Richard, P., Cambon Bonavita, M. A., et al. (2009). Active archaeal methane cycling communities associated with gassy subsurface sediments of marennes-oléron bay (France) *Science* 320, 1046. doi: 10.1126/science.1154545
50. Schloss, P. D., and Handelsman, J. (2005). Introducing DOTUR, a computer program for defining operational taxonomic units and estimating species richness. *Appl. Environ. Microbiol.* 71, 1501–1506. doi: 10.1128/AEM.71.3.1501-1506.2005
51. Syutsubo, K., Sinthurat, N., Ohashi, A., and Harada, H. (2001). Population dynamics of anaerobic microbial consortia in thermophilic granular sludge in response to feed composition change. *Water Sci. Technol.* 43, 59–66.
52. Widdel, F., and Rabus, R. (2001). Anaerobic biodegradation of saturated and aromatic hydrocarbons. *Curr. Opin. Biotech.* 12, 259–276. doi: 10.1016/S0958-1669(00)00209
53. Wilkes, H., Vieth, A., and Elias, R. (2008). Constraints on the quantitative assessment of in-reservoir biodegradation using compound-specific stable carbon isotopes. *Org. Geochem.* 39, 1215–1221. doi: 10.1016/j.orggeochem.2008.02.013
54. Yakimov, M. M., Denaro, R., Genovese, M., Cappello, S., D'Auria, G., Chernikova, T. N., et al. (2005). Natural microbial diversity in superficial sediments of Milazzo Harbor (Sicily) and community successions during microcosm enrichment with various hydrocarbons. *Environ. Microbiol.* 7, 1426–1441. doi: 10.1111/j.1462-5822.2005.00829.x
55. Yakimov, M. M., Giuliano, L., Denaro, R., Crisafi, E., Chernikova, T. N., Abraham, W. R., et al. (2004). *Thalassolituus oleivorans* gen. nov., sp. nov., a novel marine bacterium that obligately utilizes hydrocarbons. *Int. J. Syst. Evol. Microbiol.* 54, 141–148. doi: 10.1099/ijs.0.02424-0
56. Yakimov, M. M., Giuliano, L., Gentile, G., Crisafi, E., Chernikova, T. N., Abraham, W. R., et al. (2003). *Oleispira antarctica* gen. nov., sp. nov., a novel hydrocarbonoclastic marine bacterium isolated from Antarctic cosatal sea

water. *Int. J. Syst. Evol. Microbiol*. 53, 779–785. doi: 10.1099/ijs.0.02366-0

57. Yakimov, M. M., Golyshin, P. N., Lang, S., Moore, E. R. B., Abraham, W. R., Lünsdorf, H., et al. (1998). *Alcanivorax borkumensis* gen. nov., sp. nov., a new, hydrocarbon-degrading and surfactant-producing marine bacterium. *Int. J. Syst. Bacteriol*. 48, 339–348. doi: 10.1099/00207713-48-2-339

58. Yakimov, M. M., Timmis, K. N., and Golyshin, P. N. (2007). Obligate oil-degrading marine bacteria. *Curr. Opin. Biotech*. 18, 257–266. doi: 10.1016/j.copbio.2007.04.006

# 12
CHAPTER

# FACILITATION OF CO-METABOLIC TRANSFORMATION AND DEGRADATION OF MONOCHLOROPHENOLS BYPSEUDOMONAS SP. CF600 AND CHANGES IN ITS FATTY ACID COMPOSITION

Agnieszka Nowak and Agnieszka Mrozik

Department of Biochemistry, Faculty of Biology and Environmental Protection, University of Silesia, Jagiellońska 28, 40-032 Katowice, Poland

## ABSTRACT

**In this study, co-metabolic degradation of monochlorophenols (2-CP, 3-CP, and 4-CP) by the*Pseudomonas* sp. CF600 strain in the presence of phenol, sodium benzoate, and 4-hydroxybenzoic acid as an additional carbon source as well as the survival of bacteria were investigated. Moreover, the changes in cellular fatty acid profiles of bacteria depending on co-metabolic conditions were analyzed. It was found that bacteria were capable of degrading 4-CP completely in the presence of phenol, and in the presence of all substrates, they degraded 2-CP and 3-CP partially. The highest 2-CP and 3-CP removal was observed in the presence of sodium benzoate. Bacteria exhibited three various dioxygenases depending on the type of growth substrate. It was also demonstrated that bacteria exposed to aromatic growth substrates earlier degraded monochlorophenols more effectively than unexposed cells. The analysis of fatty acid profiles of bacteria indicated the essential changes in their composition, involving alterations in fatty acid saturation, hydroxylation, and cyclopropane ring formation. The most significant change in bacteria exposed to sodium benzoate and degrading monochlophenols was the appearance of branched fatty acids. The knowledge from this study indicates that *Pseudomonas*sp. CF600 could be a suitable candidate for the bioaugmentation of environments contaminated with phenolic compounds.**

**Citation:** Agnieszka Nowak and Agnieszka Mrozik, Facilitation of Co-Metabolic Transformation and Degradation of Monochlorophenols by Pseudomonas sp. CF600 and Changes in Its Fatty Acid Composition, DOI 10.1007/s11270-016-2775-5.

## INTRODUCTION

Developments in industry and agriculture would not be possible without chemical compounds, like solvents, wood preservatives, pesticides, dyes, or disinfectants. For the production of many of these chemicals, different chlorophenols are commonly used. The global production of chlorophenols is 90,000 t/year, including 60,000 t of lower chlorinated phenols (Veenagayathri and Vasudevan 2013). Furthermore, large amounts of these compounds, especially monochlorophenols, are released into the environment as by-products of the pulp-bleaching process in the paper industry and the chlorination of wastewater and drinking water (Roy et al.2004; Ge et al. 2006). Due to their acute toxicity, chlorophenols are considered to be priority pollutants by the World Health Organization (WHO), Unites States Environmental Protection Agency (USEPA), and the European Union (EU).

Among all chlorophenols, monochlorophenols are the most mobile in the environment because of their relatively high solubility in water (Czaplicka 2004). They can easily migrate within soil and aqueous environments and therefore contaminate soil, water, groundwater, and sediments. This is the reason why effective methods for their detoxification and/or degradation have been developed. One of them is microbiological degradation with the following three major steps: hydroxylation of monochlorophenols to chlorocatechols, aromatic ring cleavage, and further oxidation of aliphatic compounds to intermediates of central metabolic pathways. The key reaction in this process is aromatic ring cleavage catalyzed by intradiol or extradiol dioxygenases (EC 1.13.11) (Haddock 2010). Intradiol cleavage via the *ortho*-pathway is catalyzed by catechol 1,2-dioxygenase (EC 1.13.11.1). The cleavage reaction product is 3-chloromuconate transformed to *cis*-dienolactone and further to maleylacetate. *Meta*-cleavage of chlorocatechols catalyzed by catechol 2,3-dioxygenase (EC 1.13.11.2) generally results in incomplete degradation to dead end metabolite 5-chloro-2-hydroxymuconic semialdehyde (Arora and Bae 2014). To enhance the degradation of chlorophenols by bacteria, additional carbon sources are commonly applied. Co-metabolism refers to the degradation of two or more compounds, where the biotransformation of the barely degradable compound (co-metabolite) depends on the presence of an additional co-substrate serving as a carbon and energy source. This phenomenon reflects the processes which take place in a natural environment, where a single contaminant is rather rare. Traditionally, the growth substrates have been divided into the following two major groups: conventional carbon sources such as glucose (Bhatkal et al. 2012), sodium glutamate (Wang and Loh 2001), or yeast extract (Fakhruddin and Hossain 2007) and compounds structurally similar to the non-growth substrate (Baggi et al. 2002; Lee and Lee 2007). The conventional carbon sources support cell growth and increase biomass concentration, while structurally analogous compounds induce the enzymes of the co-metabolite metabolic pathways. The development of bioremediation methods based on co-metabolism could allow the counteraction of the accumulation of barely degradable compounds in the environment.

In response to exposure to harmful substances, bacteria can change their membrane lipid composition and integrity. Adaptive mechanisms such as changes in the length of fatty acids,*cis/trans* isomerization, changes in the degree of saturation of fatty acids, the increase in branched and cyclopropane fatty acid content, or the modification of the lipid to protein ratio have been well documented in many species of bacteria (Heipieper et al. 2003; Mrozik et al.2005; Segura et al. 2010).

The aim of this study was to investigate the co-metabolic degradation of monochlorophenols by the *Pseudomonas* sp. CF600 strain in the presence of phenol, sodium benzoate, and 4-hydroxybenzoic acid as an additional carbon source in batch liquid cultures. Moreover, changes in the cellular fatty acid methyl ester (FAME) profiles of bacteria depending on co-metabolic conditions were analyzed.

# MATERIALS AND METHODS

## Bacterial Strain and Culture Conditions

The bacterial strain used in this study was *Pseudomonas* sp. CF600, kindly provided by Prof. V. Shingler from the Department of Cell and Molecular Biology, Umeå University, Sweden. It is deposited in the Culture Collection of the University of Goteborg (no. 32333).

Bacteria were grown in a mineral salt medium (Mrozik et al. 2007) containing phenol (P) at a concentration of 282 mg $l^{-1}$, sodium benzoate (SB) 432 mg $l^{-1}$, or 4-hydroksybenzoic acid (4-HB) 414 mg $l^{-1}$ in 500-ml flasks on a rotary shaker (130 rpm) at 30 °C. The initial number of bacterial cells inoculated to the medium was $5 \cdot 10^8$ $ml^{-1}$. In order to examine the rate of degradation of P, SB, or 4-HB and the induction of enzymes involved in the mineralization of these compounds, cells were adapted to each substrate by transferring them three successive times on the same substrate, using each time an initial cell number adjusted to $5 \cdot 10^8$ cells $ml^{-1}$. For the co-metabolic degradation study, bacteria were grown in a mineral salt medium in the presence of SB (432 mg $l^{-1}$), 4-HB (414 mg $l^{-1}$), or P (282 mg $l^{-1}$) as a growth substrate and 2-chlorophenol (2-CP), 3-chlorophenol (3-CP), or 4-chlorophenol (4-CP) at a concentration of 130 mg $l^{-1}$ as a co-metabolite. It was the highest concentration of chlorophenols which bacteria were able to degrade. To determine which growth substrate accelerates co-metabolic degradation of monochlorophenols, bacteria were incubated in the mineral salt medium with appropriate substrate and monochlorophenol for 24 h. To estimate the influence of long-term exposure of *Pseudomonas* sp. CF600 on its ability to co-metabolically degrade monochlorophenols, cells were cultured in the medium containing a mixture of each monochlorophenol and selected growth substrate for 7 days. Results of our previous experiments on co-metabolic degradation of chlorophenols in dependence on the different molar ratio of the growth to the non-growth substrate (1:1, 2:1, 3:1, 4:1, and 5:1) indicated that the optimal ratio was 3:1.

The microbial counts were determined by the dilution plate count technique using nutrient agar. The inoculated plates were incubated at 30 °C for 48 h. The number of bacteria was expressed as log CFU $ml^{-1}$. Data are representative of three individual experiments.

## Determination of Aromatic Compound Concentration

The aromatic compounds were determined by a Merck Hitachi HPLC equipped with an Ascentis® Express C18 HPLC Column (100 × 4.6 mm), an Opti-Solw® EXP precolumn, and a DAD detector (Merck Hitachi). The mobile phase was the mixture of acetonitrile, methanol, and 1 % acetic acid (20:20:60, *v*/*v*). The flow rate was 1 ml $min^{-1}$. Chemical compounds in the supernatant were identified and quantified by comparing HPLC retention times and UV-visible spectra with external standards. The detection wavelength was set at 272 nm for the detection of P, SB, 2-CP, 3-CP, and 4-CP and at 260 nm for 4-HB.

## Enzymes Activity Assays

For the preparation of crude extract for dioxygenase activity, assay cells were harvested in the late exponential phase of growth by centrifugation at 4612*g* for 20 min at 4 °C. The cells were washed with a 50-mM phosphate buffer, pH 7.2, and re-suspended in the same buffer. Cell-free extracts were prepared by sonication (20 kHz) six times for 15 s with 30-s intervals and centrifugation (9000*g*, 30 min, 4 °C). The clear supernatant was used as crude extract for enzyme assay (Wojcieszyńska et al. 2011).

Catechol 1,2- and 2,3-dioksygenase activities were estimated by the spectrophotometric method of Hegeman (1966). Catechol 1,2-dioxygenase activity was measured by the formation of *cis,cis*-muconic acid ($\varepsilon = 16{,}800\ dm^3 \cdot mol^{-1} \cdot cm^{-1}$), while catechol 2,3-dioxygenase activity was detected by the formation of 2-hydroksymuconic acid ($\varepsilon = 36{,}000\ dm^3 \cdot mol^{-1} \cdot cm^{-1}$). The activity of protocatechuate 3,4-dioxygenase was measured by the method described by Hou et al. (1976) and expressed as the amount of oxygen consumption during the oxidation of protocatechuic acid. The specific activities of enzymes were expressed as the number of enzyme units per milligram of protein. The protein concentration in the crude extract was determined by the Bradford method (Bradford 1976) using lysozyme as a standard.

## MIDI-FAME Analysis

The whole-cell-derived fatty acids were extracted in the late exponential phase of growth from bacteria cultured with a single carbon source and on days 1, 4, and 7 from cultures with two carbon sources. Bacteria were harvested by centrifugation (4612*g*, 20 min, 4 °C). The cell pellets were washed with 0.9 % NaCl to remove any residue of the culture medium. Fatty acids were extracted according to the procedure by Sasser (1990) and identified using the Microbial Identification System (MIS; Microbial ID Inc., Newark). FAMEs were separated with a gas chromatograph (Hewlett-Packard 6890) equipped with an HP-Ultra 2 capillary column (25 and 0.22 mm ID) and hydrogen as a carrier gas. FAMEs were detected by a flame ionization detector (FID) and identified using the MIDI Microbial Identification System software (Sherlock TSBA 6.1 method and TSBA6 library; MIDI Inc., Newark, DE, USA).

## Data Analysis

The degradation rate constant ($k$) was determined using the algorithm $C_t/C_0 = e^{-kt}$, where $C_0$ was the substrate concentration in the culture at time 0 and $C_t$ was the substrate concentration in the culture at time $t$. The average degradation rates ($V$) of aromatic substrates were calculated by dividing the net amount of the degraded compound by the period of time between $t$ and 0.

Results were also evaluated by analysis of variance, and statistical analyses were performed on three replicates of data obtained from each treatment. The statistical significance ($p < 0.05$) of differences was treated statistically by two-way ANOVA, considering the effect of substrate and incubation time, and assessed by post hoc comparison of means using the lowest significant difference (LSD) test. The FAME profiles were also subjected to principal component analysis (PCA). This was performed based on the average values of three replicates. All analyses were performed using the Statistica 10.0 PL software package.

# RESULTS

### Degradation of Aromatic Compounds by *Pseudomonas* sp. CF600

The preliminary study showed that *Pseudomonas* sp. CF600 was not able to degrade monochlorophenols at a concentration of 130 mg $l^{-1}$ as a single carbon source in the batch cultures. The depletion of 2-CP, 3-CP, and 4-CP did not exceed 3 % of their initial concentration after 7 days of incubation (data not shown). In connection to this, in the next step, SB, 4-HB, and P were tested as potential growth substrates in the co-metabolic degradation of 2-CP, 3-CP, and 4-CP. As inductors of dioxygenases in bacterial cells, SB, 4-HB, and P were applied. At first, bacteria degraded three successive, identical doses of each aromatic compound as a single carbon source. They metabolized each dose of SB, 4-HB, and P for no more than 9 h. The most rapid degradation was observed in the case of 4-HB (Table 1). As the data obtained from this study indicate, there were significant differences ($p < 0.05$) in the transformation potential of successive dosages of the tested compounds. The highest value of the average rate of disappearance was recorded for the second dosage of 4-HB (130.09 mg $h^{-1}$) and the lowest for the third dosage of P. However, there were no significant differences between $V$ values for three dosages of phenol degradation (Table 1). The values of rate constant, following the first-order rate kinetics, were between 0.26 $h^{-1}$ (the third dosage of P) to 0.78 $h^{-1}$ (the first dosage of 4-HB; Table 1).

**Table 1**: Degradation rate constant (*k*) and rate of disappearance (*V*) for SB, 4-HB, and P in batch cultures inoculated with *Pseudomonas* sp. CF600

| Substrate | Dosage number | $k$, $h^{-1}$ | $V$, mg $h^{-1}$ |
|---|---|---|---|
| SB | 1 | 0.36 ± 0.03 | 68.03 ± 0.57a |
| | 2 | 0.42 ± 0.10 | 74.14 ± 5.04b |
| | 3 | 0.37 ± 0.01 | 81.69 ± 0.09c |
| 4-HB | 1 | 0.78 ± 0.04a | 73.29 ± 2.29a |
| | 2 | 0.47 ± 0.01b | 130.09 ± 14.13b |
| | 3 | 0.42 ± 0.04b | 89.25 ± 4.47a |
| P | 1 | 0.48 ± 0.00a | 50.46 ± 5.61 |
| | 2 | 0.32 ± 0.03b | 53.00 ± 6.33 |
| | 3 | 0.26 ± 0.02c | 46.79 ± 1.84 |

The data presented are means of three replicates. The plus/minus values represent standard deviation. The means within each column with different letters are significantly different ($p < 0.05$, LSD test), considering the effect of number of substrate dosages

To assess if the presence of an additional carbon source accelerates the degradation of monochlorophenols by *Pseudomonas* sp. CF600 in subsequent studies, bacteria were grown in a medium containing two carbon sources—monochlorophenol (2-CP, 3-CP, or 4-CP) and one of the growth substrates (SB, 4-HB, or P). The highest 2-CP and 3-CP removal was observed in the cultures containing SB as a growth substrate (Fig. 1a, b). During 24 h of incubation, bacteria utilized 45 and 40 % of the initial dose of 2-CP and 3-CP, respectively. At the same time, the depletion of 4-CP in the presence of SB was greater than 2-CP and 3-CP (60 % of 4-CP added), although in the presence of P, bacteria completely degraded 4-CP (Fig. 1c). In turn, in the medium with P, the losses of 2-CP and 3-CP were 18 and 17 %, respectively (Fig. 1a, b). On the basis of the obtained results, for subsequent steps of the study, SB was selected as the additional carbon source for the co-metabolic degradation of all monochlorophenols and P was chosen for the degradation of 4-CP.

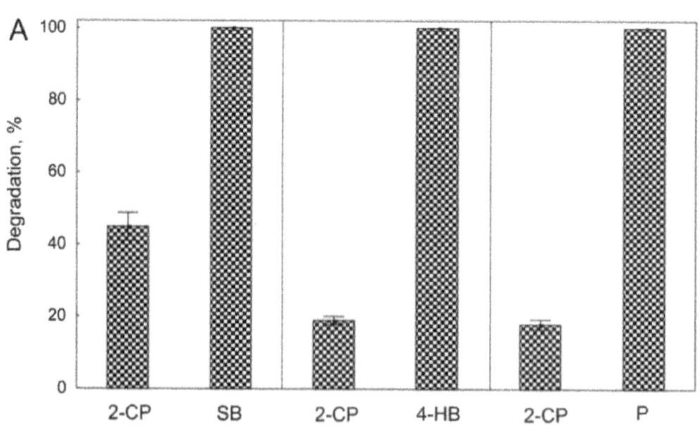

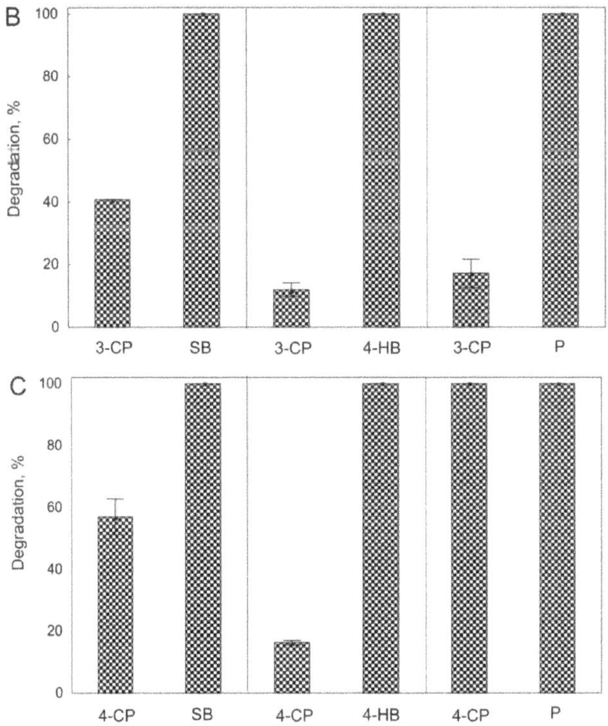


**Figure 1**: Depletion of 2-CP (a), 3-CP (b), and 4-CP (c) in bacterial culture in the presence of SB, 4-HB, and P during 24 h of incubation. The data presented are means of three replicates.

To estimate the influence of long-term exposure of *Pseudomonas* sp. CF600 to SB and P on its ability to co-metabolically degrade monochlorophenols, cells were cultured in a medium containing a mixture of each monochlorophenol and SB, as well as 4-CP and P for 7 days. In the case of each substrate depletion, its identical dosage was added to the medium. Cells that degraded three successive doses of SB or P were treated as exposed cells. It was found that the highest removal of aromatic compounds was detected in cultures with exposed cells during the first day of incubation (Fig. 2a, c, e, g). Among all monochlorophenols, only 4-CP in the presence of P was completely metabolized in that time. The second dosage of 4-CP and P bacteria were completely removed during the next 3 days (Fig. 2g). Within the next 3 days, 50 % of the third dosage of 4-CP and 77 % of P added to the medium were removed. In comparison, cells exposed to SB were not able to completely degrade monochlorophenols during 1 week of incubation. At that time, bacteria transformed 62, 40, and 75 % of 2-CP, 3-CP, and 4-CP, respectively (Fig. 2a, c, e). Data obtained from a parallel study on the co-metabolic degradation of monochlorophenols by *Pseudomonas* sp. CF600 unexposed to SB nor P indicated the lower degradative potential of such cells toward 2-CP and 4-CP in comparison with exposed cells. Unexposed bacteria in the presence of SB transformed 50 % of the initial dosage of 2-CP (Fig. 2b) and 23 % of 4-CP (Fig. 2f) within 7 days of incubation. In turn, in the presence of P, they degraded two doses of 4-CP within 4 days and during the following 3 days, utilized 30 % of the third dose (Fig. 2h). Both cells unexposed and exposed to SB removed 40 % of the initial dose of 3-CP during 7 days (Fig. 2c, d). Interestingly, the numbers of bacteria in cultures unexposed to SB or P were higher in comparison with their numbers in cultures with exposed bacteria. The highest bacterial counts were estimated in the medium inoculated with unexposed cells in the presence of 2-CP and SB. The number of cells increased in this culture by 40 % compared to the initial counts during 1 week of incubation (Fig. 2b). At the same time, the number of bacteria in the medium inoculated with cells exposed to SB and amended with 2-CP increased by 20 % in comparison with initial counts (Fig. 2a).

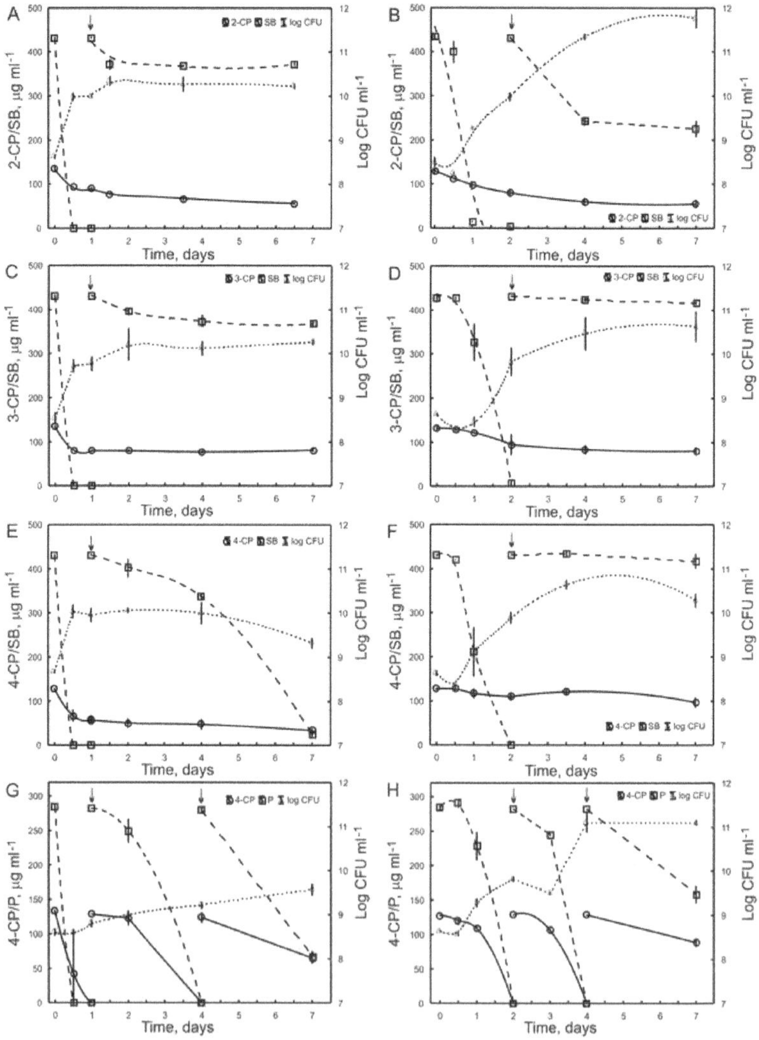


**Figure 2**: Co-metabolic transformation of 2-CP (a, b), 3-CP (c, d), and 4-CP (e–h) in the presence of SB (a–f) or P (g, h) by *Pseudomonas* sp. CF600 and survival of

bacteria. a, c, eTransformation by bacteria exposed to SB and g transformation by bacteria exposed to P. Each *symbol* represents the mean of three replicates.

## Activity of Dioxygenases

Parallel to the degradation study, the activities of catechol 1,2-dioxygenase, catechol 2,3-dioxygenase, and protocatechuate 3,4-dioxygenase in *Pseudomonas* sp. CF600 were determined. It was evidenced that the specific activities of all enzymes tested increased during the degradation of successive doses of SB, 4-HB, and P. The highest activity of protocatechuate 3,4-dioxygenase (1488.52 mU $mg^{-1}$ protein) bacteria was exhibited during degradation of the third dosage of 4-HB. Catechol 1,2-dioxygenase was induced in cells by SB, and it showed the highest activity (120.49 mU $mg^{-1}$) during the degradation of the third dose of this substrate. In turn, in the presence of P and SB, *Pseudomonas* sp. CF600 exhibited catechol 2,3-dioxygenase. Its activity in bacteria degrading the third dosage of P was 100-fold higher in comparison with the activity of this enzyme in cells growing in the presence of SB (Table 2).

**Table 2**: Activities of dioxygenases of *Pseudomonas* sp. CF600 during degradation of aromatic substrates

| | Specific activity, mU $mg^{-1}$ protein | | | | | | | | |
|---|---|---|---|---|---|---|---|---|---|
| | Catechol 1,2-dioxygenase | | | Catechol 2,3-dioxygenase | | | Protocatechuate 3,4-dioxy-genase | | |
| Inductor | SB | 4-HB | P | SB | 4-HB | P | SB | 4-HB | P |
| Dosage 1 | 20.80 ± 2.44a | 0.00 ± 0.00 | 0.00 ± 0.00 | 20.95 ± 1.84a | 0.00 ± 0.00 | 215.55 ± 2.38a | 0.00 ± 0.00 | 495.53 ± 32.11a | 0.00 ± 0.00 |
| Dosage 2 | 67.50 ± 2.87b | 0.00 ± 0.00 | 0.00 ± 0.00 | 30.35 ± 3.37b | 0.00 ± 0.00 | 280.13 ± 16.57b | 0.00 ± 0.00 | 505.78 ± 59.49a | 0.00 ± 0.00 |
| Dosage 3 | 120.49 ± 6.36c | 0.00 ± 0.00 | 0.00 ± 0.00 | 4.48 ± 0.63c | 0.00 ± 0.00 | 474.66 ± 16.07c | 0.00 ± 0.00 | 1488.52 ± 66.96b | 0.00 ± 0.00 |

The data presented are means of three replicates. The plus/minus values represent standard deviation. The different letters indicate significant differences ($p < 0.05$, LSD test), considering the effects of aromatic compound dosage

The specific activities of the tested dioxygenases were also determined under co-metabolic conditions. As shown in Table 3, monochlorophenols did not influence catechol 1,2-dioxygenase activity. It turn, the activity of protocatechuate 3,4-dioxygenase decreased by 40 % in cells growing in the presence of monochlorophenols and 4-HB in comparison with its activity when bacteria were cultured with 4-HB only. However, monochlorophenols affected the catechol 2,3-dioxygenase activity. In the presence of 2-CP, 3-CP, and 4-CP, the decline of this enzyme activity was about 80, 65, and 50 %, respectively, as compared to activity detected in cells in the presence of P (Table 3).

**Table 3**: Relative activity of dioxygenases of *Pseudomonas* sp. CF600 during transformation of 2-CP, 3-CP, and 4-CP in the presence of SB, 4-HB, and P

| **Substrate(s)** | **Relative activity, %** |
|---|---|
| Catechol 1,2-dioxygenase | |
| SB | 100.00 ± 8.44ab |
| SB + 2-CP | 111.31 ± 6.89a |
| SB + 3-CP | 87.56 ± 5.87b |
| SB + 4-CP | 96.06 ± 18.19ab |
| Protocatechuate 3,4-dioxygenase | |
| 4-HB | 100.00 ± 10.94a |
| 4-HB + 2-CP | 61.06 ± 1.78b |
| 4-HB + 3-CP | 65.11 ± 4.95b |
| 4-HB + 4-CP | 62.56 ± 0.92b |
| Catechol 2,3-dioxygenase | |
| P | 100.00 ± 4.91a |
| P + 2-CP | 20.41 ± 1.61b |
| P + 3-CP | 34.19 ± 2.49bc |
| P + 4-CP | 47.89 ± 4.63c |

The data presented are means of three replicates. The plus/minus values represent standard deviation. The different letters indicate significant differences ($p < 0.05$, LSD test), considering the effects of monochlorophenols on enzymes activity.

## FAME Analysis

To illustrate FAME variability in *Pseudomonas* sp. CF600 cells during the degradation of successive doses of SB, 4-HB, or P, whole-cell-derived fatty acids were directly extracted from bacteria. The PCA of the FAME patterns involved the most common fatty acids (Fig. 3a). FAMEs isolated from bacteria during the degradation of the first dosage of SB, 4-HB, and P were clearly distinguished along the first axis from FAMEs isolated during the degradation of the second and the third dosages of these substrates (Fig. 3b). Simultaneously, FAME profiles obtained during the degradation of the second and the third dosages of 4-HB and the third dosage of SB were distinguished along the second axis from profiles obtained during the degradation of the second and the third dosages of P and the second dosage of SB (Fig. 3b). It should be noted that during the degradation of successive doses of the tested substrates, the percentage of 16:1 ω7*c* in FAME profiles decreased, although the content of cyclopropane (17:0 *cy* and 19:0 *cy* ω8*c*) and hydroxylated (12:0 2OH and 12:0 3OH) fatty acids increased. Furthermore, during the degradation of successive doses of 4-HB, the percentage of 16:0 fatty acid increased, while during the degradation of P, the content of 14:0 increased (Fig. 3a, b).

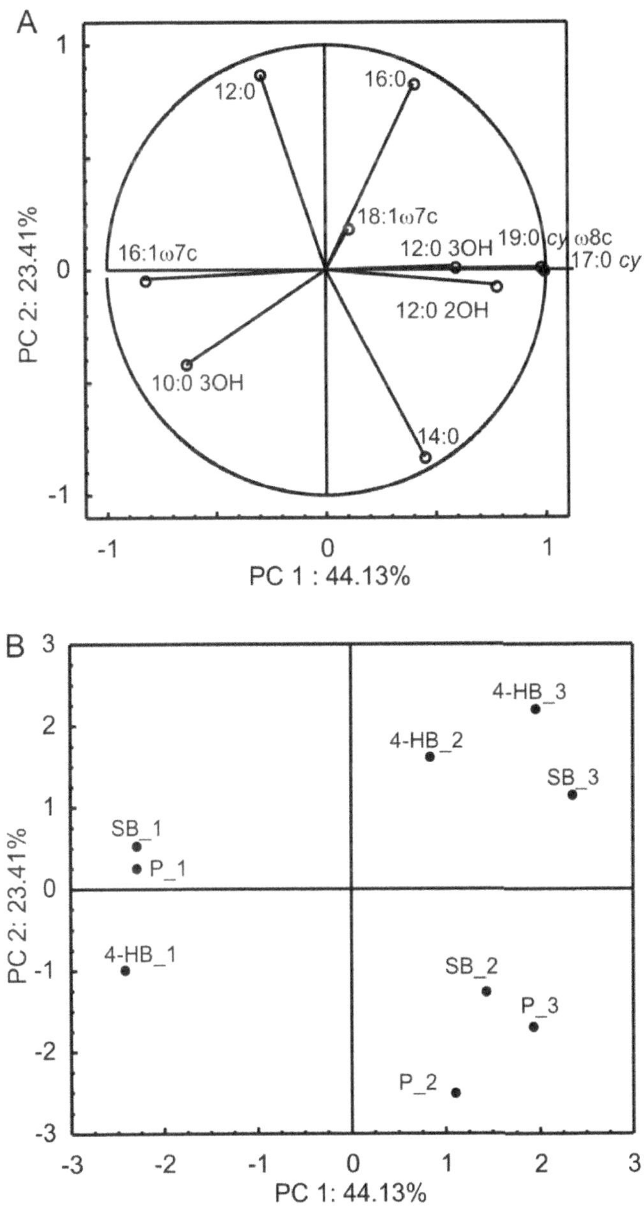


**Figure 3**: PCA of analyzed FAMEs during degradation of three successive doses of SB (SB_1, SB_2, and SB_3), 4-HB (4-HB_1, 4-HB_2, and 4-HB_3), and P (P_1, P_2, and P_3) by*Pseudomonas* sp. CF600. a Correlation of the fatty acids with PC1 and PC2 and b. projection of FAMEs on the plane defined by PC1 and PC2.

FAMEs were also isolated from *Pseudomonas* sp. CF600 cells unexposed and exposed to SB and P during the co-metabolic degradation of monochlorophenols. The PCA of FAMEs obtained from cells exposed to SB or P involved 15 fatty acids (Fig. 4) and from unexposed cells involved 8 fatty acids (Fig. 4c). The PCA analysis distinguished FAMEs isolated from cells exposed to P during the degradation of the mixture of 4-CP and P on all sampling days along the first axis from FAMEs obtained from bacteria cultured with monochlorophenols and SB on days 4 and 7 (Fig. 4b). Simultaneously, all FAMEs extracted from bacteria on day 1 were distinct in comparison with the other fatty acids along the second axis (Fig. 4b). On the basis of the obtained results, it should be evidenced that *Pseudomonas* sp. CF600 under exposure to aromatic compounds showed an increase in the degree of saturation of fatty acids. The percentage of saturated FAMEs increased during the co-metabolic degradation of monochlorophenols, although the content of unsaturated fatty acids (16:1 ω7*c* and 18:1 ω7*c*) decreased. Furthermore, it was also demonstrated that in the presence of SB as a growth substrate, the percentage of branched (15:0 *iso*, 15:0 *anteiso*, 16:0 *iso*, and 17:0 *anteiso*) and cyclopropane (17:0 *cy*) fatty acids increased, while in the presence of P, the content of straight-chain fatty acids mainly increased (12:0, 16:0, and 18:0; Fig. 4a, b). In FAME profiles of cells unexposed to SB or P but degrading monochlorophenols co-metabolically, the most distinct were FAME profiles of cells degrading 4-CP with SB or P (Fig. 4d). These profiles obtained from bacteria on day 1 were characterized by a higher percentage of unsaturated fatty acids in comparison with profiles on days 4 and 7 of the experiment, when the increase of 17:0 *cy* and 16:0 fatty acid content was recorded (Fig. 4c, d). The other FAME profiles were gathered together (Fig. 4d).

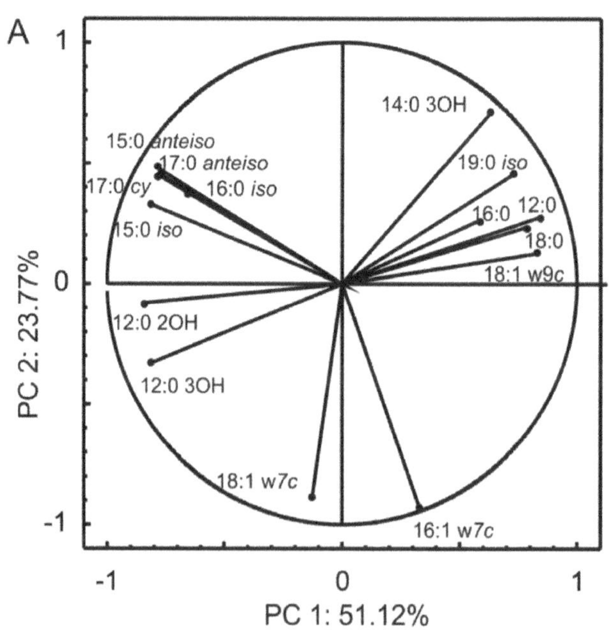

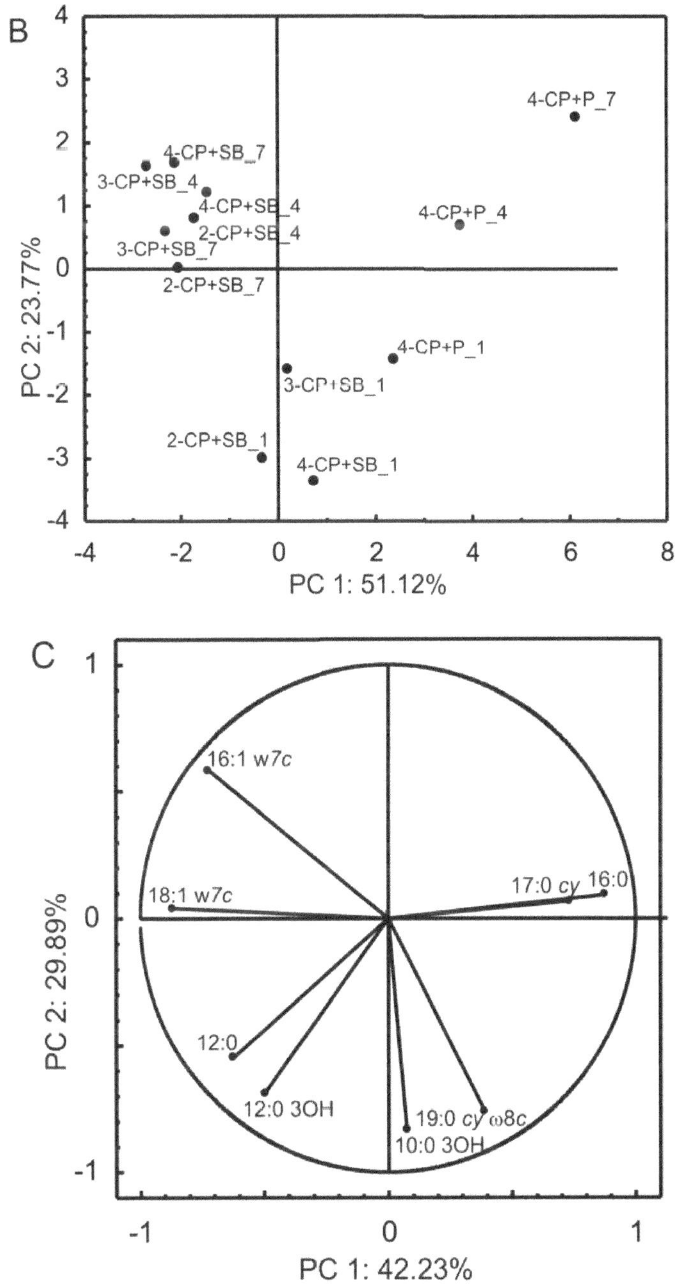
B
4-CP+P_7
4-CP+SB_7
3-CP+SB_4
4-CP+SB_4
2-CP+SB_4
3-CP+SB_7
2-CP+SB_7
4-CP+P_4
4-CP+P_1
3-CP+SB_1
2-CP+SB_1
4-CP+SB_1
PC 2: 23.77%
PC 1: 51.12%
C
16:1 w7c
18:1 w7c
17:0 cy
16:0
12:0
12:0 3OH
19:0 cy ω8c
10:0 3OH
PC 2: 29.89%
PC 1: 42.23%

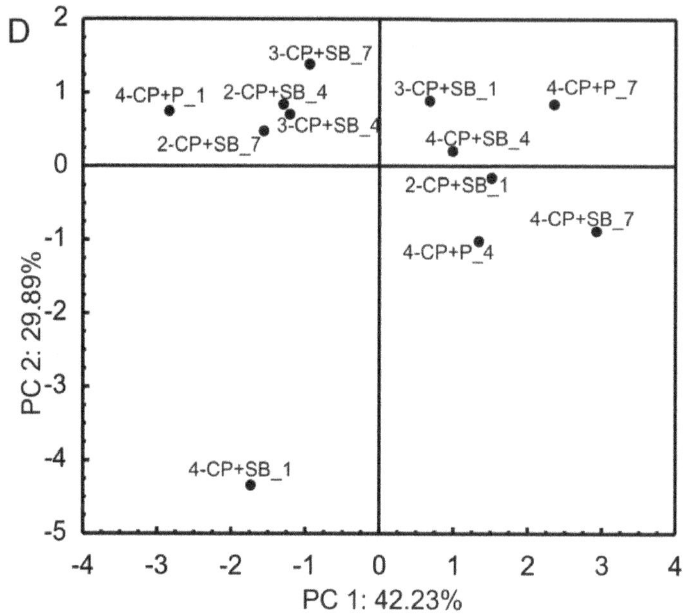


**Figure 4**: PCA of analyzed FAMEs during co-metabolic transformation of monochlorophenols in the presence of SB or P by *Pseudomonas* sp. CF600 unexposed (c, d) and exposed (a, b) to SB or P. a, c Correlation of fatty acids with PC1 and PC2. b, d Projection of FAMEs on the plane defined by PC1 and PC2. *2-CP + SB_X* FAMEs isolated from bacteria degrading 2-CP and SB, *3-CP + SB_X* FAMEs isolated from bacteria degrading 3-CP and SB, *4-CP + SB_X* FAMEs isolated from bacteria degrading 4-CP and SB, *4-CP + P_X* FAMEs isolated from bacteria degrading 4-CP and P, and *X* sampling day.

## DISCUSSION

Monochlorophenols belong to aromatic compounds with a chlorine substituent. The presence of this electron-withdrawing group on the aromatic ring makes such compounds more resistant to microbial degradation and more toxic for living organisms than un-substituted analogues (Escuder-Gilabert et al. 2001; Chrzanowski et al. 2011). One of the possibilities of removing these compounds from contaminated environments is co-metabolic degradation by microorganisms (Kim and Hao 1999; Lee and Lee 2007). In this study, their degradation by*Pseudomonas* sp. CF600 in the presence of selected compounds was investigated. As co-substrates, phenol, sodium benzoate, and 4-hydroxybenzoic acid were used. P was chosen as a structural analogue of monochlorophenols and frequent contaminant of soil and industrial wastes. Sodium benzoate (BS) is less toxic than phenol and has low potential to accumulate in living organisms (WHO 2000). In turn, 4-HB is mainly known as an allelochemical secondary plants metabolite and, similarly to sodium benzoate, is not regarded as a contaminant (Wu et al.2009). The co-metabolic degradation of monochlorophenols by bacteria was studied in the presence of glucose, sodium glutamate, and phenol as growth substrates (Wang and Loh 2000; Tarighian et al. 2003). However, there is a lack of reports about the use of sodium benzoate and 4-hydroxybenzoic acid as co-substrates. Greń et al. (2010) indicated that benzoic acid, 4-hydroxybenzoic acid, and 3,4-dihydroxybenzoic acid stimulated the biotransformation of mononitrophenols by *Stenotrophomonas maltophilia* KB2. These plant origin aromatic compounds induced catechol and protocatechuate dioxygenases involved in a broad range of aromatic compounds' degradation pathways. The un-specificity of these enzymes permit to co-metabolize compounds, which are barely biodegradable as sole carbon sources (Bugg and Ramaswamy 2008; Guzik et al. 2009, 2013). The concentrations of monochlorophenols added to the culture medium had mean values in comparison with concentrations used by other authors (Kim and Hao 1999; Wang and Loh 2000; Lee and Lee 2007).

The degradation study of the following doses of SB, 4-HB, and P by *Pseudomonas* sp. CF600 indicated that bacteria required the longest time to metabolize the first doses of these compounds, which could be connected with the induction of enzymes of adequate catabolic pathways. Among all the tested substrates, the rate of disappearance of 4-HB was the highest and the loss of P was the lowest. These differences may be connected with their toxicity to bacterial cells. Moreover, the higher value of $V$ for 4-HB in comparison with SB could be related with the presence of an additional hydroxyl group in the 4-HB structure which facilitates the aromatic ring cleavage (Greń et al. 2010). Similarly, the numbers of bacteria were the lowest during the degradation of the first dosage of SB, 4-HB, and P, as compared to their numbers during the degradation of subsequent doses.

The activities of three various dioxygenases detected during the degradation of SB, 4-HB, and P indicated that *Pseudomonas* sp. CF600 has three different metabolic pathways of aromatic compound degradation. In the presence of SB, bacteria exhibited high activity of catechol 1,2-dioxygenase and the slight activity of catechol 2,3-dioxygenase. The lack of activity of protocatechuate 3,4-dioxygenase during SB degradation may suggest that *Pseudomonas* sp. CF600 degraded SB through the catechol pathway with the simultaneous induction of additional decarboxylase (Peng et al. 2003; Wojcieszyńska et al. 2011). Furthermore, it was found that this strain metabolized P via the *meta*-cleavage pathway, which is consistent with data obtained by Powlowski and Shingler (1994). It should also be noted that this is the first report about the activity of intradiol protocatechuate 3,4-dioxygenase in *Pseudomonas* sp. CF600.

The obtained results indicated that *Pseudomonas* sp. CF600 was not able to remove monochlorophenols completely in the presence of SB or 4-HB. Despite the fact that bacteria degraded 4-HB as a sole carbon source in the shortest time and synthesized protocatechuate 3,4-dioxygenase, 4-HB was not an appropriate growth substrate for the co-metabolic degradation of monochlorophenols. In turn, the removal of 2-CP and 3-CP by *Pseudomonas* sp. CF600 was most effective in the presence of SB, although the degradation of these co-metabolites was incomplete. Among the tested isomers, bacteria were capable of degrading only 4-CP completely in the presence of P. This could be connected with the similar structure of these compounds, as well as the position of the chloride atom on the aromatic ring. Menke and Rehm (1992) reported that the substitution of the second hydroxyl group to the aromatic ring in 4-CP is easier than the substitution of 2-CP and 3-CP.

As described in results, bacteria exposed to SB or P removed 2-CP and 4-CP more effectively than unexposed ones. This may suggest that the exposition of bacteria used in the bioaugmentation of contaminated environments can accelerate the removal of xenobiotics and reduce the time of their biodegradation. The lack of differences in the degradation of 3-CP by unexposed and exposed bacteria probably resulted from the unfavorable configuration of hydroxyl and chloride substituents on the aromatic ring. This configuration affected the introduction of a second hydroxyl group onto the aromatic ring and diminished the enzymatic transformation of 3-CP (Menke and Rehm 1992).

Analyses of enzymes' activities could suggest that 4-CP was metabolized by *Pseudomonas* sp. CF600 via the *meta*-cleavage pathway, even though catechol 2,3-dioxygenase was the most sensitive to monochlorophenols in comparison with catechol 1,2-dioxygenase and protocatechuate 3,4-dioxygenase. It was also proved that the activity of catechol 1,2-dioxygenase was not inhibited by 2-CP, 3-CP, and 4-CP. The complete inhibition of catechol 1,2-dioxygenase, catechol 2,3-dioxygenase, and protocatechuate 3,4-dioxygenase in the presence of mononitrophenols was reported by Wojcieszyńska et al. (2011). The higher sensitivity of dioxygenases to mononitrophenols in comparison with monochlorophenols was likely to be connected with the various toxicities of these phenol derivatives (Michałowicz and Duda 2007). On the other hand, the presence of monochlorophenols did not inhibit the degradation of SB, 4-HB, and P by *Pseudomonas* sp. CF600 during incubation in the co-metabolic culture. The increase in the number of cells, especially unexposed to SB, 4-HB, or P, did not indicate the significant influence of monochlorophenols on their survival.

Our results demonstrated that *Pseudomonas* sp. CF600 underwent structural changes of cell-derived fatty acids when grown on SB, 4-HB, or P. The highest variability in FAME composition was observed during the degradation of the first and the second dosages of each compound. These changes involved alterations in the degree of saturation of fatty acids, hydroxylation, and cyclopropane ring formation. The increasing degree of membrane saturation is a major adaptive mechanism that enables bacterial cells to survive under aromatic hydrocarbon stress (Segura et al. 2010). It was indicated that changes in the ratio of straight-chain and hydroxylated fatty acids in cellular lipids permitted the regulation of the uptake of amphipathic compounds, reducing their toxic effect. The increase of straight-chain fatty acid content caused the reduction of polar substance uptake. The presence of hydroxy fatty acids caused the opposite effect (Donato et al. 1997; Kaczorek et al. 2013). Another mechanism of bacteria adaptation to SB, 4-HB, and P was the increase of the content of cyclopropane fatty acids. The cyclopropane ring is less reactive than the double bond of their precursors—*cis*-unsaturated fatty acids, and decreased the fluidity of the membrane. An increase of cyclopropane fatty acid content in Gram-negative bacteria was observed in various stress conditions (Chang and Cronan 1999; Mrozik et al. 2004; Muñoz-Rojas et al. 2006).

Interestingly, the earlier exposition of the bacteria tested to SB and P had an influence on the diversity of fatty acids during the co-metabolic degradation of monochlorophenols. The main difference was the appearance of branched fatty acids in bacteria exposed to SB on days 4 and 7 of the experiment. The absence of these fatty acids in FAMEs isolated from bacteria during the degradation of SB as the sole carbon source suggested that branched fatty acids were involved in the adaptation of bacteria to monochlorophenols. Despite the fact that branched fatty acids are regarded as signature lipid biomarkers of Gram-positive bacteria (Piotrowska-Seget and Mrozik 2003), their content in Gram-negative bacteria under stress conditions could also be significant in *Pseudomonas putida* and *Pseudomonas stutzeri* exposed to naphthalene (Mrozik et al. 2005). The main alteration in fatty acid composition observed in *Pseudomonas* sp. CF600 degraded 4-CP, and P was the increase of the content of straight-chain fatty acids. It should be pointed out that in bacteria unexposed to SB and P, distinct adaptive mechanisms were observed only during the degradation of 4-CP and SB, as well as 4-CP and P. In FAMEs isolated from bacteria on days 4 and 7, significant increases of straight-chain (16:0) and cyclopropane (17:0 *cy*) fatty acid content were recorded. During the co-metabolic degradation of 2-CP and 3-CP by unexposed bacteria, well-defined changes in their fatty acid profiles were difficult to find.

## CONCLUSIONS

The obtained results indicate that phenol was the most suitable growth substrate in the degradation of monochlorophenols by *Pseudomonas* sp. CF600 in comparison with sodium benzoate and 4-hydroxybenzoic acid. However, among tested enzymes, the activity of catechol 2,3-dioxygenase induced by phenol was the most inhibited by monochlorophenols. It was proved that chlorophenols in the co-metabolic cultures caused additional alterations in whole-cell fatty acid composition of bacteria as compared to changes induced in the cells by the growth substrate only.

It should be emphasized that the co-metabolic degradation of monochlorophenols by the*Pseudomonas* sp. CF600 strain has not yet been studied. This study confirmed that it was able to degrade 4-CP and P completely, which makes it a good candidate for the bioaugmentation of contaminated areas, where both compounds are frequently present as a mixture. Additionally, its capability to degrade 2-CP and 3-CP partially in the presence of additional growth substrates makes it possible to accelerate the biodegradation processes in specific cases. Moreover, the analysis of fatty acid composition expanded our knowledge on the adaptive mechanisms of cells exposed to highly toxic and persistent chlorophenols.

## ACKNOWLEDGMENTS

This work was supported by the Polish Ministry for Science and Higher Education (MNiSW).

## REFERENCES


1. Arora, P. K., & Bae, H. (2014). Bacterial degradation of chlorophenols and their derivatives.*Microbial Cell Factories, 13*, 31.
2. Baggi, G., Andreoni, V., Bernasconi, S., Cavalca, L., & Zangrossi, M. (2002). Cometabolic degradation of mixtures of monochlorophenols by phenol-degrading microorganisms. *Annals of Microbiology, 52*, 133–143.
3. Bhatkal, A., Punage, S., & Deshannavar, U. B. (2012). Biodegradation of 4-chlorophenol by*Pseudomonas putida* NCIM sp. 2650 under aerobic conditions. *Research Journal of Environmental Sciences, 6*, 238–244.
4. Bradford, M. M. (1976). A rapid and sensitive method for the quantitation of microgram quantities of protein utilizing the principle of protein-dye binding. *Analytical Biochemistry, 72*, 248–254.
5. Bugg, T. D. H., & Ramaswamy, S. (2008). Non-heme iron-dependent dioxygenases: unraveling catalytic mechanisms for complex enzymatic oxidation. *Current Opinion in Chemical Biology, 12*, 134–140.
6. Chang, Y. Y., & Cronan, J. E. (1999). Membrane cyclopropane fatty acid content is a major factor in acid resistance of *Escherichia coli. Molecular Microbiology, 33*, 249–259.
7. Chrzanowski, Ł., Owsianiak, M., Szulc, A., Marecik, R., Piotrowska-Cyplik, A., Olejnik-Schmidt, A. K., Staniewski, J., Lisiecki, P., Ciesielczyk, F., Jesionowski, T., & Heipieper, H. J. (2011). Interactions between rhamnolipid biosurfactants and toxic chlorinated phenols enhance biodegradation of a model hydrocarbon-rich effluent. *International Biodeterioration & Biodegradation, 65*, 605–611.
8. Czaplicka, M. (2004). Sources and transformations of chlorophenols in the natural environment. *Science of the Total Environment, 322*, 21–39.
9. Donato, M. M., Jurado, A. S., Antunes-Madeira, M. C., & Madeira, V. M. C. (1997). Effects of a lipophilic environmental pollutant (DDT) on the phospholipid and fatty acid contents of*Bacillus stearothermophilus. Archives of Environmental Contamination and Toxicology, 33*, 341–349.
10. Escuder-Gilabert, L., Martin-Biosca, Y., Sagrado, S., Villanueva-Camañas, R. M., & Medina-Hernández, M. J. (2001). Biopartitioning micellar chromatography to predict ecotoxicity.*Analytica Chimica Acta, 448*, 173–185.
11. Fakhruddin, A. N. M., & Hossain, M. (2007). Degradation of monochlorophenols by*Psudomonas putida* CP1 in the presence of growth supplements. *Bangladesh Journal of Microbiology, 24*, 115–118.
12. Ge, F., Zhu, L., & Chen, H. (2006). Effect of pH on the chlorination process of phenols in drinking water. *Journal of Hazardous Materials B, 133*, 99–105.
13. Greń, I., Wojcieszyńska, D., Guzik, U., Perkosz, M., & Hupert-Kocurek, K. (2010). Enhanced biotransformation of mononitrophenols by *Stenotrophomonas maltophilia* KB2 in the presence of aromatic compounds of plant origin. *World Journal of Microbiology and Biotechnology, 26*, 289–295.

14. Guzik, U., Greń, I., Wojcieszyńska, D., & Łabużek, S. (2009). Isolation and characterization of a novel strain of *Stenotrophomonas maltophilia* possessing various dioxygenases for monocyclic hydrocarbon degradation. *Brazilian Journal of Microbiology, 40*, 285–291.

15. Guzik, U., Hupert-Kocurek, K., & Wojcieszyńska, D. (2013). Intradiol dioxygenases—the key enzymes in xenobiotics degradation. In R. Chamy & F. Rosenkrand (Eds.), *Biodegradation of hazardous and special products*. Rijeka, Croatia: InTech.

16. Haddock, J. D. (2010). Aerobic degradation of aromatic hydrocarbons: enzyme structures and catalytic mechanisms. In K. N. Timmis (Ed.), *Handbook of hydrocarbon and lipid microbiology*. Berlin, Heidelberg: Springer-Verlag.

17. Hegeman, G. D. (1966). Synthesis of the enzymes of the mandelate pathway by *Pseudomonas putida. Journal of Bacteriology, 91*, 1140–1154.

18. Heipieper, H. J., Meinhardt, F., & Segura, A. (2003). The *cis-trans* isomerase of unsaturated fatty acids in *Pseudomonas* and *Vibrio*: biochemistry, molecular biology and physiological function of a unique stress adaptive mechanism. *FEMS Microbiology Letters, 229*, 1–7.

19. Hou, C. T., Lillard, M. O., & Schwartz, R. D. (1976). Protocatechuate 3,4-dioxygenase from*Acinetobacter calcoaceticus. Biochemistry, 15*, 582–588.

20. Kaczorek, E., Sałek, K., Guzik, U., Jesionowski, T., & Cybulski, Z. (2013). Biodegradation of alkyl derivatives of aromatic hydrocarbons and cell surface properties of a strain of*Pseudomonas stutzeri. Chemosphere, 90*, 471–478.

21. Kim, M. H., & Hao, O. J. (1999). Cometabolic degradation of chlorophenols by *Acinetobacter*species. *Water Research, 33*, 562–574.

22. Lee, C. Y., & Lee, Y. P. (2007). Degradation of 4-chlorophenol by enriched mixed cultures utilizing phenol and glucose as added growth substrate. *World Journal of Microbiology and Biotechnology, 23*, 383–391.

23. Menke, B., & Rehm, H. J. (1992). Degradation of mixtures of monochlorophenols and phenol as substrates for free and immobilized cells of *Alcaligenes* sp. A7-2. *Applied and Environmental Microbiology, 37*, 655–661.

24. Michałowicz, J., & Duda, W. (2007). Phenols transformations in the environment and living organisms. *Current Topics in Biophysics, 30*, 24–36.

25. Mrozik, A., Piotrowska-Seget, Z., & Łabużek, S. (2004). Changes in whole cell-derived fatty acids induced by naphthalene in bacteria from genus *Pseudomonas. Microbiological Research, 159*, 87–95.

26. Mrozik, A., Łabużek, S., & Piotrowska-Seget, Z. (2005). Changes in fatty acid composition in*Pseudomonas putida* and *Pseudomonas stutzeri* during naphthalene degradation.*Microbiological Research, 160*, 149–157.

27. Mrozik, A., Piotrowska-Seget, Z., & Łabużek, S. (2007). FAME profiles in *Pseudomonas vesicularis* during catechol and phenol degradation in the presence of glucose as an additional carbon source. *Polish Journal of Microbiology, 56*, 157–164.

28. Muñoz-Rojas, J., Bernal, P., Duque, E., Godoy, P., Segura, A., & Ramos, J. L. (2006). Involvement of cyclopropane fatty acids in the response of *Pseudomonas putida* KT2440 to freeze-drying. *Applied and Environmental Microbiology, 72*, 472–477.

29. Peng, X., Misawa, N., & Harayama, S. (2003). Isolation and characterization of thermophilic*Bacilli* degrading cinnamic, 4-coumaric and ferulic acids. *Applied and Environmental Microbiology, 69*, 1417–1427.

30. Piotrowska-Seget, Z., & Mrozik, A. (2003). Signature lipid biomarker (SLB) analysis in determining changes in community structure of soil microorganisms. *Polish Journal of Environmental Studies, 12*, 669–675.

31. Powlowski, J., & Shingler, V. (1994). Genetics and biochemistry of phenol degradation by*Pseudomonas* sp. CF600. *Biodegradation, 5*, 219–236.

32. Roy, M., Chakrabarti, S. K., Bharadwaj, N. K., Chandra, S., Kumar, S., Singh, S., Bajpai, P. K., & Jauhari, M. B. (2004). Characterization of chlorinated organic material in Eucalyptus pulp bleaching effluents. *Journal of Scientific & Industrial Research, 63*, 527–535.

33. Sasser, M. (1990). *Identification of bacteria by gas chromatography of cellular fatty acids. MIDI Technical Note 101*. Newark: Microbial ID. Inc.

34. Segura, A., Bernal, P., Pini, C., Krell, T., Daniels, C., & Ramos, J. L. (2010). Membrane composition and modifications in response to aromatic hydrocarbons in gram-negative bacteria. In K. N. Timmis (Ed.), *Handbook of hydrocarbon and lipid microbiology*. Berlin, Heidelberg: Springer-Verlag.

35. Tarighian, A., Hill, G., Headley, J., & Pedras, S. (2003). Enhancement of 4-chlorophenol biodegradation using glucose. *Clean Technologies and Environmental Policy, 5*, 61–65.

36. Veenagayathri, K., & Vasudevan, N. (2013). Degradation of 4-chlorophenol by a moderately halophilic bacterial consortium under saline conditions. *British Microbiology Research Journal, 3*, 513–524.

37. Wang, S. J., & Loh, K. C. (2000). New cell growth pattern on mixed substrates and substrate utilization in cometabolic transformation of 4-chlorophenol. *Water Research, 34*, 3786–3794.

38. Wang, S. J., & Loh, K. C. (2001). Biotransformation kinetics of *Pseudomonas putida* for cometabolism of phenol and 4-chlorophenol in the presence of sodium glutamate.*Biodegradation, 12*, 189–199.

39. WHO. (2000). Concise international chemical assessment document 26. Benzoic acid and sodium benzoate.

40. Wojcieszyńska, D., Guzik, U., Greń, I., Perkosz, M., & Hupert-Kocurek, K. (2011). Induction of aromatic ring: cleavage dioxygenases in *Stenotrophomonas maltophilia* strain KB2 in cometabolic systems. *World Journal of Microbiology and Biotechnology, 27*, 805–811.

41. Wu, H. S., Shen, S. H., Han, J. M., Liu, Y. D., & Liu, S. D. (2009). The effect in vitro of exogenously applied *p*-hydroxybenzoic acid on *Fusarium oxysporum* f. sp. *niveum. Phytopathologia Mediterranea, 48*, 439–446.

# 13

CHAPTER

# BIODEGRADATION OF 2,6-DICHLOROPHENOL WASTEWATER IN SOIL COLUMN REACTOR IN THE PRESENCE OF PINEAPPLE PEELS-DERIVED ACTIVATED CARBON, PALM KERNEL OIL AND INORGANIC FERTILIZER

Samuel E. Agarry, Mujidat O. Aremu, Oluwafunmilayo A. Aworanti

Biochemical Engineering and Biotechnology Laboratory, Department of Chemical Engineering, Ladoke Akintola University of Technology, Ogbomoso, Nigeria.

## ABSTRACT

**In this study, the potential effects of palm kernel oil (PKO), pineapple peels derived-activated carbon (PPAC) and NPK fertilizer (20:10:10) as amendment agents on the natural bioattenuation of 2,6-dichlorophenol (2,6-DCP) in tropical agricultural soil were investigated. The effect of PPAC dosage on 2,6-DCP biodegradation was also studied. Column reactors containing soil were spiked with 2,6-dichlorophenol (2,6-DCP) wastewater (300 mg/l) and amended with PKO, NPK fertilizer and PPAC alone or in combinations. The rates of 2,6-DCP biodegradation were studied for a remediation period of 42 days under laboratory conditions. The results showed that there was a positive relationship between the rate of 2,6-DCP biodegradation, bacterial growth rate and presence of NPK fertilizer and PPAC (alone or in combination) in soil column microcosms contaminated with 2,6-DCP. The 2,6-DCP biodegradation data fitted well to the first-order kinetic model. The model revealed that 2,6-DCP contaminated-soil microcosms amended with NPK fertilizer and PPAC (alone or in combination) had higher biodegradation rate constants (k) as well as lower half-life times ($t_{1/2}$) than soil column microcosms amended with PKO and unamended soil (natural attenuation) remediation system. Thus, the use of combined NPK fertilizer and activated carbon (NPK + PPAC) to enhance 2,6-DCP degradation in the soil could be one of the severally sought bioremediation strategies of remediating natural ecosystem (environment) contaminated with organic chemicals.**

**Citation:** S. Agarry, M. Aremu and O. Aworanti, "Biodegradation of 2,6-Dichlorophenol Wastewater in Soil Column Reactor in the Presence of Pineapple Peels-Derived Activated Carbon, Palm Kernel Oil and Inorganic Fertilizer," Journal of Environmental Protection, Vol. 4 No. 6, 2013, pp. 537-547. doi: 10.4236/jep.2013.46063.

## INTRODUCTION

The contamination of land and aquatic environment by hydrocarbons has become a serious problem in major parts of the world and Nigeria in particular. This represents a global concern for the potential consequences on ecosystem and human health [1]. Phenolic compounds including chlorophenols are frequently encountered components of industrial wastewaters and a major concern for environmental engineers. They comprise 10 out of 114 EPA-designated organic priority pollutants [2,3]. Chlorophenols constitute an important class of pollutants because of their wide use in the production of wood preservers, pesticides and biocides [4-6]. These compounds are present in the wastewater generated from industrial activities such as petrochemical, pharmaceutical, wood preserving, plastic, rubber proofing, pesticide, iron steel, textile, and paper and cellulose bleaching industries [7,8]. Chlorophenols as environmental pollutants are highly toxic and carcinogenic with strong odour emission, not readily biodegradable and persistent in the environment and thus poses a serious ecological problem and public health risk causing problems with the human respiratory and nervous system [9-12]. For this reason an increasing attention has been directed toward the research of new strategies and environmental-friendly technologies to be applied for the remediation of ecosystems contaminated by hydrocarbons. Among these, biotechnological strategies based on the autochthonous microbial communities to biodegrade organic pollutants are of particular relevance.

Biological treatment has proved to be the most promising and economical method for the removal of phenolic compounds from the environment. It is believed to lead to complete mineralization of organic compound [13] and can handle a wide range of concentrations [14]. Microbial degradation of hydrocarbons is a predominant mechanism for natural attenuation at hydrocarbon-contaminated sites, but contaminant loss may also be attributed to abiotic processes such as sorption, volatilization, dilution, and dispersion [15]. Bioremediation techniques often include the addition of supplements or amendments to enhance bioavailability and biodegradation of petroleum hydrocarbons in natural soils/sediments by indigenous or inoculated microorganisms. The supplements can be divided into three general categories according to their impact on the microbial processes: surfactants that enhance the availability of the pollutants [16], specific cosubstrates that can play the role of alternate carbon source that promote the growth of a particular group of degraders [17-20] or act as mass transfer agent between the liquid phase of the soil and the solid matrix [21,22] and organic and inorganic nutrients that increase the size and activity of the soil microbial community [23].

Vegetable oils are renewable and low cost co-substrates, used to enhance process efficiency without increasing significantly bioremediation costs. Vegetable oils such as peanut, sunflower and rapeseed oils have been found to be efficient in enhancing polycyclic aromatic hydrocarbons (PAHs) removal from contaminated soils [24-26]. The use of vegetable oils as co-substrates in the remediation of soil contaminated with monoaromatic hydrocarbons such as phenol and its substituted compounds have not been investigated. In many natural ecosystems a combination of biological and adsorption processes is a common phenomenon. Thus, combined methods of adsorption and biodegradation are nowadays gaining importance. The recent literature on methods of hydrocarbons removal from contaminated soil focuses on adsorption and microbial degradation process [27,28]. Recently, a growing interest in the sorption of organic contaminants to carbonaceous materials has drawn attention to treat contaminated soil/ sediment, with the purpose of finding cheaper and functionally acceptable materials for the removal of contaminants. Previous studies have demonstrated that carbonaceous materials, such as peat [29], soybean stalk [30], coke [31], and activated carbon [32- 34] can effectively reduce organic contaminants in soil/ sediment.

Activated carbon is most widely used as adsorbent in the removal of heavy metals and hazardous organic chemicals [35] because of its hydrophobicity, high specific surface (8,001,200 $m^2/g$), and microporous structure. In recent years, extensive researches are focused on the development of alternative and economic activated carbon from agricultural by-products [35-39] that are ubiquitous green waste in the environment and which may cause some serious environmental pollution when filling at a fixed site. However, the use of activated carbon derived from agricultural waste materials as adsorbents for the enhancement of hydrocarbon biodegradation rate in soil is very limited [28].

Therefore, the objective of this study were to use laboratory soil column reactor to investigate the catabolic potential of natural soil microbial communities to biodegrade target hydrocarbon pollutants (2,6-dichlorophenol) and also to evaluate the efficacy of adsorption and biostimulation during hydrocarbon degradation by natural microbial communities augmented with pineapple peels derived-activated carbon, palm kernel oil and NPK fertilizer additions.

## MATERIALS AND METHODS

### Collection of Samples

The soil sample used for the study was collected from the top surface soil (0 - 15 cm) of Ladoke Akintola university of Technology (LAUTECH) agricultural farm land, Ogbomoso, Nigeria. The soil samples were air dried, homogenized, passed through a 2-mm (pore size) sieve and stored in a polythene bag and kept in the laboratory prior to use. The chemicals 2,6-dichlorophenol (2,6-DCP) and n-hexane (analytical grade) being products of SigmaAldrich Chemical Co., were bought from a chemical store at Ibadan while NPK fertilizer used as nutrient supplement was purchased from an agrochemical store in Ogbomoso, Nigeria. Pineapple peels, a waste product of pineapple pulp used for the production of activated carbon were obtained from pineapple fruits bought from a local market at Ogbomoso, Nigeria.

### Characterization of Soil Sample

The soil sample was characterized for total carbon (TOC), total nitrogen (N), total phosphorus, moisture content, and pH according to standard methods. The pH was determined according to the modified method of McLean [40]; total organic carbon was determined by the modified wet combustion method [41] and total nitrogen was determined by the semi-micro-Kjeldhal method [42]. Available phosphorus was determined by Brays No.1 method [43] and moisture content was determined by the dry weight method. The total hydrocarbon degrading bacteria (THDB) populations was determined by the vapor phase transfer method [44]. The physicochemical characterized parameters are presented in Table 1.

## Preparation of Activated Carbon

The pineapple peels as the precursor material for activated carbon preparation was well washed with water several times, sun-dried for two weeks and oven dried at 110°C for 3 h. The dried sample was milled into small particle sizes. A known amount (1 kg) of the dried pineapple peels was pyrolyzed in a furnace (Vecstra, Model 184 A, Italy) at a temperature of 300°C for 2 h. The resulting biochar was collected and cooled at room temperature. The biochar material was then subjected to chemical activation (using phosphoric acid ($H_3PO_4$)) by weighing samples of the carbonized material into beakers and then soaked in excess phosphoric acid ($H_3PO_4$) agitated for 3 h. After the agitation, the mixtures were then charged inside an oven at a temperature of 200°C for 24 h to ensure proper adsorbate drying. The biochar samples were then removed from the oven and allowed to cool for 2 h after which they were washed with distilled water until leachable impurities due to free acid and adherent powder were removed. The samples were further soaked in 2% (w/v) sodium bicarbonate ($NaHCO_3$) to remove any residual acid left. The resulting mixture were further washed with distilled water to bring the pH to 7.0 and finally drained and dried overnight in an oven at 110°C to get the final activated carbon product.

## Preparation of Simulated Dichlorophenol Wastewater

A known amount (2 g) of 2,6-DCP was weighed and added to 1 L of de-ionized water in a glass bottle which was properly shaken for thorough dissolution to produce a stock simulated wastewater of 2 g/l 2,6-DCP. This was kept in a refrigerator prior to use.

## Slurry-Phase Biodegradation Experimental Design

A constructed column reactor using a 20 mm thick plastic glass (transparent material) with the dimensions 20.32 cm height by 10.16 cm width by 8.5 cm length and a volume capacity of 10 L was used to simulate reaction conditions occurring in the depth from which soil samples were collected. The aerobic biodegradation studies were conducted by spiking 1 kg of LAUTECH agricultural farm soil with 300 mg/l of the simulated dichlorophenol wastewater. Different amount of the pineapple peels-derived activated carbon (PPAC), PKO and NPK fertilizer was respectively added to the contaminated soil and thoroughly mixed together (Table 2). A control experiment which did not contain any of the amendments was set up alongside the experimental samples following the same procedure.

**Table 1**: Physicochemical and microbiological properties of soil

| Parameter | Value |
|---|---|
| Organic carbon (%) | 1.05 ± 0.01 |
| Total nitrogen (%) | 0.77 ± 0.03 |
| Phosphorus (%) | 0.06 ± 0.02 |
| Potassium (%) | 0.25 ± 0.01 |
| pH | 7 ± 0.1 |
| Moisture (%) | 11.4 ± 0.02 |
| Total hydrocarbon degrading bacteria (THDB) CFU/g | $0.02 \pm 2.45 \times 10^6$ |

Note: Each value is a mean of three replicates and ± indicates standard deviation among them.

**Table 2**: Experimental design

| Soil Column Microcosm | Amount of Amendments |
|---|---|
| A | Soil + 2,6-DCP + 1% PKO |
| B | Soil + 2,6-DCP + 5 g NPK |
| C | Soil + 2,6-DCP + 10 g PPAC |
| D | Soil + 2,6-DCP + 5 g NPK + 1% PKO |
| E | Soil + 2,6-DCP + 10 g PPAC + 1% PKO |
| F | Soil + 2,6-DCP + 5 g NPK + 10 g PPAC |
| G | Soil + 2,6-DCP + 5 g NPK + 1% PKO + 10 g PPAC |
| H | Soil + 2,6-DCP + 15 g PPAC |
| I | Soil + 2,6-DCP + 20 g PPAC |
| J (Control) | Soil + 2,6-DCP (No amendment) |

The contaminated soil was then packed into the column reactor and incubated at ambient conditions for six weeks (42 days). The study was carried out during the rainy season between the month of October and November when the climate is cool and the temperature is usually not very high. The water (moisture) content of soil in each column reactor was adjusted every week by addition of sterile distilled water to a moisture holding capacity of 50%. In order to avoid anaerobic conditions, contents of the column reactors were aerated by mixing every 3 days. Samples were taken every week and analyzed for residual dichlorophenol and total hydrocarbon-degrading bacteria (THDB), respectively. The experiments were carried out in triplicates.

## Determination of Residual Dichlorophenol

For the determination of residual chlorophenol compound, soil-slurry samples were extracted with n-hexane, centrifuged for 20 min at 3000 rpm, and filtered through No. 2 Whatman filter paper. The clear supernatant organic phase was analyzed by means of a gas chromatograph, HP 6890 adapted to a mass spectrometer detector, model 5972 (GC-MS) and injector 7973. The GC-MS was equipped with a HP5MS and HP1 column of 30 m length, 0.25 mm internal diameter and 0.25 lm of film thickness. Helium was used as the carrier gas at a flow rate of 2.9 $ml \cdot min^{-1}$. The conditions for analysis were: split 1:20; range of mass scan 50 - 500; with temperatures as follows: injector 260°C, detector 280°C; gradient program: 65°C to 96°C (4°C·min$^{-1}$), 96°C to 160°C (8°C·min$^{-1}$) and up to 230°C (12°C·min$^{-1}$).

## Statistical Analysis

One-way ANOVA (Tukey's test) was used to compare 2,6-DCP biodegradation in soil among different amendment agents and in unamended soil. Statistical analyses were performed using the Statistical Package Social Science (SPSS) software (Version 15.0; SPSS, Inc.).

## Biodegradation Kinetics and Estimation of Half-Life Time

Kinetic analysis is a key factor for understanding biodegradation process, bioremediation speed measurement and development of efficient clean up for an organic chemical contaminated environment. Biodegradability of hydrocarbon can be explained by first order kinetics [45,46] and this is given as in Equation (2):

$$C_t = C_o e^{-kt} \qquad (2)$$

where $C_o$ is the initial 2, 6-DCP content in soil (mg/kg), $C_t$ is the residual 2,6-DCP content in soil at time t, (mg/kg), k is the biodegradation rate constant (day$^{-1}$) and t is time (day).

The biodegradation half-life is the time taken for a substance to lose half of its amount. Biodegradation half life times ($t_{1/2}$) are calculated using Equation (3) [46]:

$$t_{1/2} = \frac{\ln 2}{k} \qquad (3)$$

where k is the biodegradation rate constant (day$^{-1}$). The half life model is based on the assumption that the biodegradation rate of hydrocarbons positively correlated with the hydrocarbon pool size in soil [47].

# RESULTS AND DISCUSSION

## Effect of Amendment Agents on Natural Attenuation

A general increase in the percentage degradation of 2,6-DCP was observed with time as shown in Figures 1 and 2, respectively. These results indicate that 2,6-DCP naturally degraded in the unamended slurry soil due to the presence of microorganisms present in the soil.

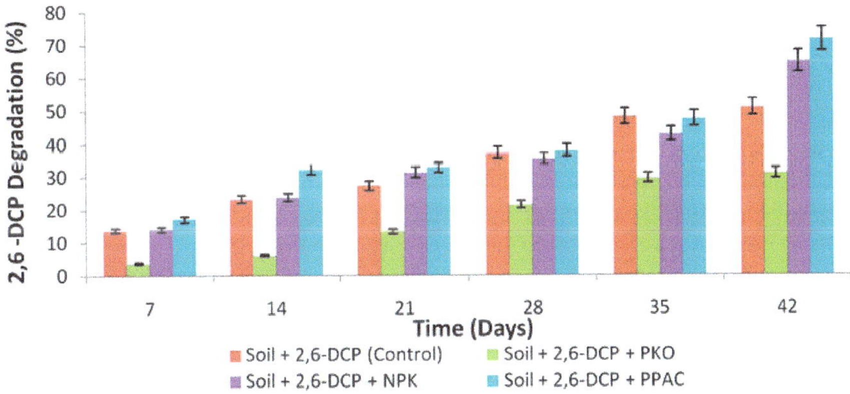


**Figure 1**: Percentage degradation of 2,6-DCP as a function of time in slurry soil amended with PKO, NPK, PPAC and unamended soil.

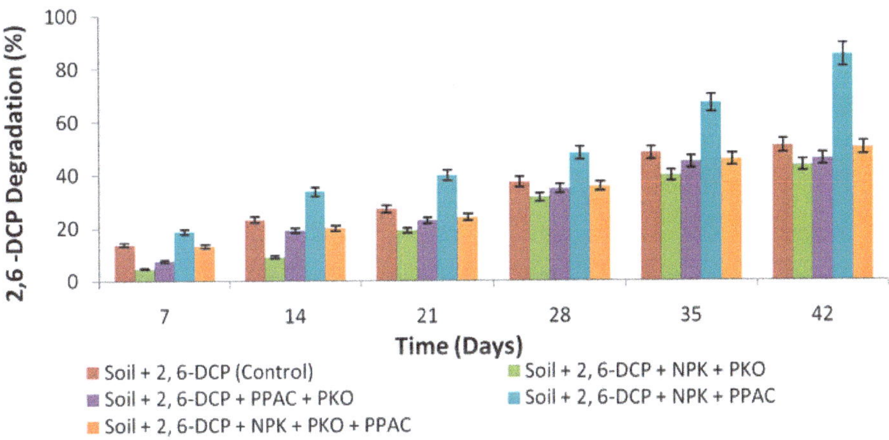


**Figure 2**: Percentage degradation of 2,6-DCP as a function of time in slurry soil amended with NPK + PKO, PPAC + PKO, NPK + PPAC, NPK + PKO + PPAC and unamended soil.

The results in Figure 1 showed that the biodegradation rate of 2,6-DCP was greater in slurry soils amended respectively with PPAC and NPK fertilizer (with higher percentage degradation) than the unamended soil (natural attenuation). This observation could be due to the fact that PPAC acted as an adsorbent in the soil as well as cometabolic carbon substrate while the NPK fertilizer provided inorganic nutrients to the autochthonous microorganisms in the soil. Enhanced bioremediation of soil highly contaminated with herbicide propanil (3',4'-dichloropropionanilide) in the Krasnodar region of Russia, 3,4-dichloroaniline, polychlorinated biphenyls (PCB) and 2,4,6-trinitrotoluene (TNT) in the presence of activated carbon have been reported [48-52]. Accelerated biodegradation of anthracene in the presence of groundnut shellderived activated carbon has been reported by Owabor and Aluyor [28] while Lou et al. [53] have also observed accelerated biodegradation of pentachlorophenol in soil in the presence of bamboo biochar.

However, the percentage degradation of 2,6-DCP in PPAC-amended soil is relatively greater than the NPK fertilizer-amended soil. While the percentage degradation in soil amended with PKO was lower than that of unamended soil (natural attenuation) and this indicates that rate of biodegradation was faster in the latter than the former. Vegetable oils have been proposed as natural, cost-effective and non-toxic solvents. Nevertheless, the lower percentage degradation observed in PKO-amended soil suggests that the PKO was inhibitory to the soil microorganisms' metabolic functions. This inhibitory effect may be due to the high amount of oil used in this work which restrained the biological processes and inhibited the degradation of 2,6-DCP [54]. Pizzul et al. [54] have reported the inhibitory effect of rapeseed oil on indigenous soil microorganism which resulted in lower percentage degradation of phenanthrene and pyrene when compared to the control soil without rapeseed oil. They attributed this observation as due to high amount of rapeseed oil (1% w/w) that was used. However, these same workers have also reported the stimulatory effect of rapeseed oil which led to higher degradation values for anthracene and benzo(a)pyrene as compared to the control soil without rapeseed oil. Pannu et al. [55] reported the effects of small amounts of vegetable oil on the biodegradation of polycyclic aromatic hydrocarbon (PAH). The presence of 0.2% (v/v) of peanut oil had a stimulatory effect on the degradation of polycyclic aromatic hydrocarbon (PAH) in a soil slurry bioreactor system. The beneficial effect was attributed to the solubilization of the PAH in the oil and subsequent transfer to the bacterial cells associated with the oil drops and to the promotion of biosurfactant production.

The influence of two or three level combination of the amendment agents on 2,6-DCP biodegradation in soil is shown in Figure 2. It could be seen that among the two or three level combination of amendment agents, the slurry soil amended with the combination of NPK fertilizer and pineapple peel-derived activated carbon (NPK + PPAC) displayed the highest percentage degradation of 2,6-DCP as compared to percentage degradation in soil amended with NPK + PKO, PPAC + PKO, and the unamended soil (natural attenuation). The percentage degradations of 2,6-DCP in soils amended with NPK + PKO and PPAC + PKO are respectively lower than that in unamended soil. On the other hand, the percentage degradation of 2,6-DCP in soil amended with NPK + PKO and PPAC + PKO was respectively higher than the percentage degradation in soil amended with PKO only. This improvement may be due to the supply of nutrients by the NPK fertilizer as well as due to the adsorptive capacity of PPAC and ability to act as co-metabolic carbon substrate to the autochthonous microorganisms in the soil. However, there is no statistical significant difference ($p < 0.05$) in the percentage degradation of 2,6-DCP in soil amended with NPK + PKO and PPAC + PKO.

## Effect of Amendment on the Microbial Growth

Figure 3 shows the growth profile of the hydrocarbon degrading bacterial population in the slurry soil. An initial hydrocarbon-degrading bacterial population of about $1.1 \times 10^6$, $1.8 \times 10^6$, $2.0 \times 10^6$, and $1.6 \times 10^6$ CFU/g was observed respectively before the start of remediation for slurry soil amended with PKO, NPK fertilizer, PPACand unamended soil as shown in Figure 3. As seen fromFigure 3, at the end of incubation (day 42), the THDB count had increased to a maximum bacterial population of $2.8 \times 10^6$, $7.9 \times 10^6$, $12.5 \times 10^6$, and $5.4 \times 10^6$ CFU/g which corresponded to 60.7%, 77.2%, 84% and 70.4% for slurry soil amended with PKO, NPK, PPAC, and unamended soil, respectively. However, the growth rate of THDB in soil amended with PKO was lower (resulting in lower percentage degradation of 2,6-DCP) as compared to the growth rate of THDB in soil amended with NPK fertilizer, PPAC and unamended soil, respectively. This observation may be due to the inhibitory effect of PKO on the autochthonous microorganisms in the soil.

Similarly, as shown in Figure 4, the THDB count in- creased from $1.4 \times 10^6$, $1.3 \times 10^6$, $1.7 \times 10^6$, and $1.5 \times 10^6$ CFU/g (day 0) to $3.4 \times 10^6$, $3.7 \times 10^6$, $14.8 \times 10^6$, and $4.6 \times 10^6$ CFU/g (day 42) corresponding to a growth increase of 58.8%, 64.9%, 88.5% and 70.2% for slurry soil amended with NPK + PKO, PPAC + PKO, NPK + PPAC, and NPK + PKO + PPAC, respectively.

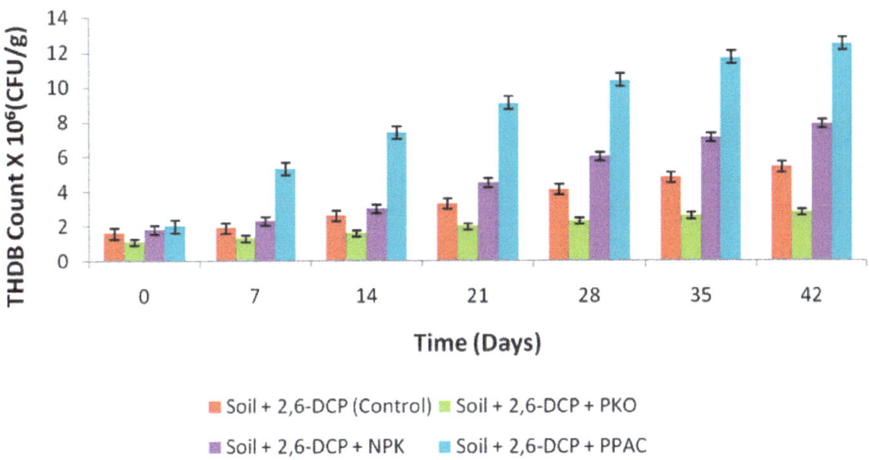


**Figure 3**: Bacterial growth on 2,6-DCP as a function of time in slurry soil amended with PKO, NPK, PPAC and unamended soil.

This observation indicated that the growth rate of the hydrocarbon degrading bacteria was higher in slurry soil amended with NPK + PPAC (resulting in higher percentage degradation of 2,6-DCP) than that amended with NPK + PKO, PPAC + PKO, NPK + PKO + PPAC, and the unamended soil (natural attenuation), respectively. Furthermore, the rate of bacteria growth in soil amended with NPK + PKO and PPAC + PKO are respectively lower than that in the unamended soil but higher than the growth rate in soil singly amended with PKO. The higher growth rate observed in soil amended with NPK + PKO and PPAC + PKO than that in soil amended with only PKO is due to the presence of inorganic nutrient provided by NPK fertilizer and the adsorption mechanism provided by PPAC, respectively. Nevertheless, there is no statistical significant difference ($p < 0.05$) in the growth rate of bacteria (THDB) in slurry soil amended with NPK + PKO and PPAC + PKO as well as in slurry soil amended with NPK + PKO + PPAC and the unamended soil.

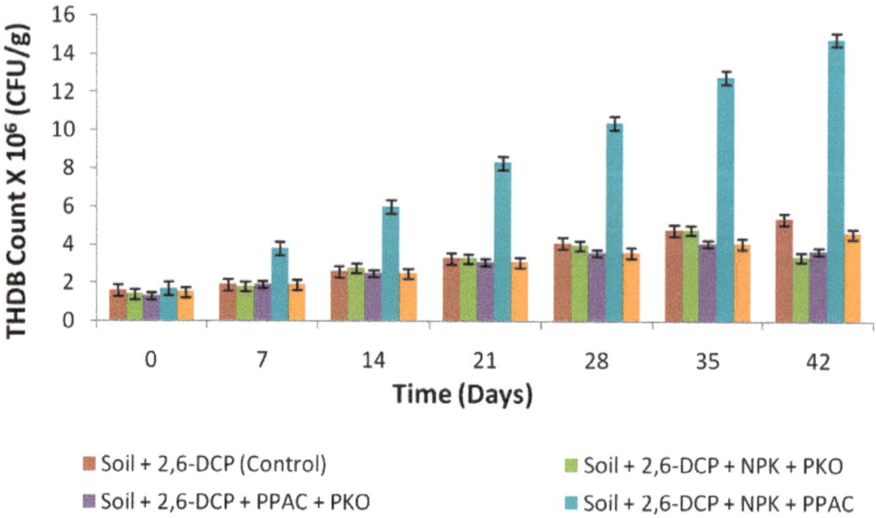


**Figure 4**: Bacterial growth on 2,6-DCP as a function of time in soil amended with NPK + PKO, PPAC + PKO, NPK + PPAC, NPK + PKO + PPAC and unamended soil.

## Biodegradation Kinetics and Half-Life

Fitting the biodegradation data obtained for each soil treatment to first-order kinetic model using the non-linear regression routine of MATLAB 7.0 software package made possible further evaluation and comparison of the applicability of the various amendment strategies. The parameter "k" of the model, being the specific degradation rate constants for biodegradation of 2,6-DCP in the soil, was determined for soil treatment amended singly with PKO, NPK fertilizer, and PPAC as well as for those amended with two or three level combinations (NPK + PKO, PPAC + PKO, NPK + PPAC, and NPK + PKO + PPAC), respectively. The results are presented in Table 3. The results indicate that the "k" value (0.0211 $day^{-1}$) for the biodegradation of 2,6-DCP in soil amended with PPAC was comparatively higher than the value for soil amended with NPK (0.0189 $day^{-1}$) and PKO (0.0095 $day^{-1}$), respectively. The soil treatment with PKO only had the smallest "k" value (0.0095 $day^{-1}$) while the $k$ value for the biodegradation of 2,6-DCP in unamended soil (natural attenuation) is 0.0170 $day^{-1}$. More also, the k value (0.0309 $day^{-1}$) for the biodegradation of 2,6-DCP in soil amended with NPK + PPAC was relatively higher than that amended with NPK + PKO (0.0142 $day^{-1}$), PPAC + PKO (0.0156 $day^{-1}$) and NPK + PKO + PPAC (0.0161 $day^{-1}$), respectively. However, the k values for 2,6-DCP biodegradation in soil amended with NPK + PKO and PPAC + PKO are comparatively lower than that in unamended soil (natural attenuation).

The biodegradation half-life times was calculated using Equation (2) and the values are presented in Table 3. Lower the half-life value and faster the rate of degradation. The soil treatment with the lower half-life has the best performance. As seen from Table 3, half-life of 40.72 days was observed for the biodegradation of 2,6- DCP in unamended contaminated soil (control soil or natural attenuation). This was reduced to lesser number of days in 2,6-DCP-contaminated soil amended with NPK fertilizer and PPAC (either singly or in combination), respectively. Nevertheless, 2,6-DCP contaminated soil amended with PKO had the highest half-life (73.9 days). Therefore, values of the kinetic parameter and half-life showed that the degree of effectiveness of these amendment strategies in the cleanup of soil contaminated with 2,6-DCP is in the following order: PKO < NPK + PKO < PPAC + PKO < NPK + PKO + PPAC < natural attenuation < NPK < PPAC < NPK + PPAC.

## Effect of Activated Carbon Dose

Figure 5 shows the effect of PPAC dosage on the percentage degradation of 2,6-DCP in soil. It was observed that the percentage degradation of 2,6-DCP in soil increased with the increase in the PPAC dosage. A similar observation has been reported for the use of bamboo biochar on pentachlorophenol leachability and bioavailability in agricultural soil [53].

**Table 3**: Specific degradation rate constants (k), correlation coefficient ($R^2$) and half-life ($t_{1/2}$) for 2, 6-DCP biodegradation in soil

| Soil Treatment | $k$ (day$^{-1}$) | $R^2$ | $t_{1/2}$ (days) |
|---|---|---|---|
| Soil + 2,6-DCP + PKO | 0.0095 | 0.9581 | 73.14 |
| Soil + 2,6-DCP + NPK | 0.0189 | 0.9343 | 36.55 |
| Soil + 2,6-DCP + PPAC | 0.0211 | 0.8981 | 32.78 |
| Soil + 2,6-DCP + NPK + PKO | 0.0142 | 0.9589 | 48.77 |
| Soil + 2,6-DCP + PPAC + PKO | 0.0156 | 0.9808 | 44.48 |
| Soil + 2,6-DCP + NPK + PPAC | 0.0309 | 0.9380 | 22.46 |
| Soil + 2,6-DCP + NPK + PKO + PPAC | 0.0161 | 0.9812 | 43.02 |
| Soil + 2,6-DCP (Natural attenuation) | 0.0170 | 0.9867 | 40.72 |

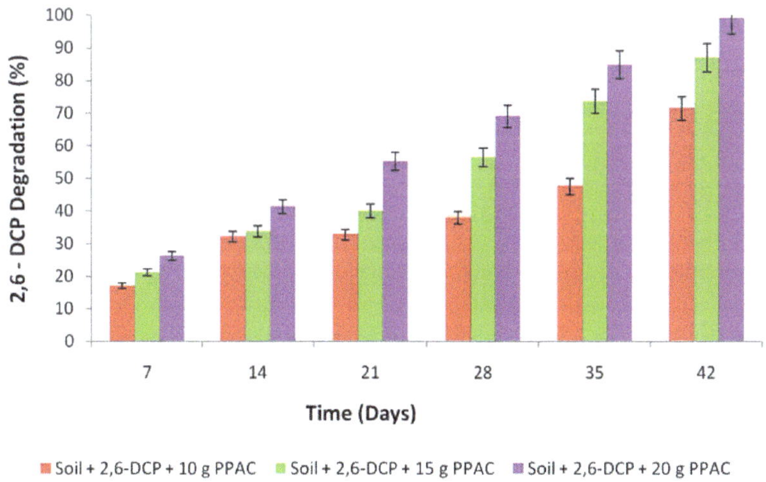


**Figure 5**: Effect of activated carbon dosage on percentage degradation of 2,6-DCP.

## CONCLUSION

From this present study, it can be concluded that the addition of activated carbon, prepared from pineapple peels waste (PPAC) as well as the addition of NPK fertilizer alone or in combination, significantly ($P < 0.05$) enhanced the biodegradation of 2,6-DCP and the rate of growth of autochthonous microorganisms in soil columns and thus decrease the time for complete bioremediation. While the addition of co-metabolic carbon substrate in the form of vegetable oil (PKO) slows down the biodegradation rate of 2,6-DCP and the growth rate of autochthonous microorganisms in soil. The soil treatment under NPK + PPAC amendment (combined biostimulation and adsorptive bioremediation) exhibited the highest degree of degradation and the soil treatment under PKO amendment the least degradation. Furthermore, because of the high efficiency, beneficial ecological effects and abundant sources of pineapple peels, it is a potentially attractive material for activated carbon development which can be used for in situ remediation to control organic pollution in soil and to protect surface and ground water from organic pollutants. Thus, the use of combined NPK fertilizer and activated carbon (biostimulation and adsorption-biodegradation) to enhance 2,6-DCP degradation in the soil could be one of the severally sought bioremediation strategies of remediating natural ecosystem (environment) contaminated with organic chemicals.

## REFERENCES

1. N. E. Onwurah, V. N. Ogugua, N. B. Onyike, A. E. Ochonogor and O. F. Otitoju, "Crude Oil Spills in the Environment, Effects and Some Innovative Clean-Up Biotechnologies," International Journal of Environmental Research, Vol. 1, No. 4, 2007, pp. 307-320.
2. W. W. Eckenfelder Jr., "Industrial Water Pollution Controll," 2nd Edition, McGraw-Hill Publishers, New York, 1989.
3. B.-H. Bae, R. L. Autenrieth and J. S. Bonner, "Kinetics of Multiple Phenolic Compounds Degradation with a Mixed Culture in a Continuous Flow Reactor," Water Environmental Research, Vol. 67, No. 2, 1995, pp. 215-223. doi:10.2175/106143095X131376
4. H. Bae, T. Yamagishi and Y. Suwa, "Evidence for Degradation of 2-Chlorophenol by Enrichment Cultures under Denitrifying Conditions," Microbiology, Vol. 148, 2002, pp. 221-227.
5. A. P. Annachhatre and S. H. Gheewala, "Biodegradation of Chlorinated Phenolic Compounds," Biotechnology Advances, Vol. 14, No. 1, 1996, pp. 35-56. doi:10.1016/0734-9750(96)00002-X
6. K. A. Short, J. D. Doyle, R. J. King, R. J. Seidler, G. Stotzky and R. H. Olsen, "Effects of 2, 4-Dichlorophenol, a Metabolite of a Genetically Engineered Bacterium, and 2, 4-Dichlorophenoxyacetate on Some MicroorganismsMediated Ecological Processes in Soil," Applied Environmental Microbiology, Vol. 57, No. 2, 1991, pp. 412- 418.
7. N. Fattahi, Y. Assadi and M. R. Hosaseini, "Determination of Chlorophenols in Water Sample Using Simultaneous Dispersive Liquid-Liquid Microextraction and

Derivatization Followed by Gas Chromatography-ElectronCapture Detection," Journal of Chromatography A, Vol. 1157, No. 1-2, 2007, pp. 23-29.

8. B. K. Hamad, A.-Md. Noor and A.A. Rahim, "Removal of 4-Chloro-2-Methoxyphenol from Aqueous Solution by Adsorption to Oil Palm Shell Activated Carbon Activated with $K_2CO_3$," Journal of Physical Sciences, Vol. 22, No. 1, 2011, pp 39-55.

9. Agency for Toxic Substance and Disease Registry, "Toxicological Profile for Chlorophenols," US Department of Health and Human Services, Public Health Service, Atlanta, 1989.

10. D. C. Yee and T. K. Wood, "2,4-Dichlorophenol Degradation Using Streptomyces Viridosporus T7A Lignin Peroxidase," Biotechnology Progress, Vol. 13, No. 1, 1997, pp. 53-59. doi:10.1021/bp960091x

11. P. M. Armenante, D. Kafkewitz, G. A. Lewandowski and C. J. Jou, "Anaerobic-Aerobic Treatment of Halogenated Phenolic Compounds," Water Research, Vol. 33, No. 3, 1999, pp. 681-692. doi:10.1016/S0043-1354(98)00255-3

12. I. A. W. Tan, A. L. Ahmad and B. H. Hameed, "Adsorption Isotherm, Kinetics, Thermodynamics and Desorption Studies of 2,4,6-Trichlorophenol on Oil Palm Empty Fruit Bunch-Based Activated Carbon," Journal of Hazardous Material, Vol. 164, No. 2-3, 2009, pp. 473-482. doi:10.1016/j.jhazmat.2008.08.025

13. A. Nuhoglu and B. Yalcin, "Modeling of Phenol Removal in a Batch Reactor," Process Biochemistry, Vol. 40, No. 3-4, 2005, pp. 1233-1239. doi:10.1016/j.procbio.2004.04.003

14. M. H. El-Naas, S. A. Al-Muhtaseb and S. Makhlouf, "Biodegradation of Phenol by Pseudomonas Putida Immobilized in Polyvinyl Alcohol (PVA) Gel," Journal of Hazardous Material, Vol. 164, No. 2-3, 2009, pp. 720-725. doi:10.1016/j.jhazmat.2008.08.059

15. P. B. Bedient H. S. Rifai and C. J. Newell, "Groundwater Contamination, Transport and Remediation," Prentice Hall, Englewood Cliffs, 1994.

16. P. P. E. Carriere and F. A. Mesania, "Enhanced Biodegradation of Creosote-Contaminated Soil," Waste Management, Vol. 15, No. 8, 1995, pp. 579-583.

17. J. W. C. Wong, K. M. Lai, K. K. Ma and M. Fang, "Isolation and Optimization of PAH-Degradative Bacteria from Contaminated Soil for PAHs Bioremediation," Water Air Soil Pollution, Vol. 139, No. 1-4, 2002, pp. 1-13. doi:10.1023/A:1015883924901

18. P. H. Lee, K. J. Doick and K. T. Semple, "The Development of Phenanthrene Catabolism in Soil Amended with Transformer Oil," FEMS Microbiology Letters, Vol. 228, No. 2, 2003, pp. 217-223. doi:10.1016/S0378-1097(03)00751-1

19. D. Sarkar, M. Ferguson, R. Datta and S. Birnbaum, "Bioremediation of Petroleum Hydrocarbons in Contaminated Soils: Comparison of Biosolids Addition, Carbon Supplementation, and Monitored Natural Attenuation," Environmental Pollution, Vol. 136, No. 1, pp. 187-195. doi:10.1016/j.envpol.2004.09.025

20. S. Llado, N. Jime´nez, M. Vin˜as and A. M. Solanas, "Microbial Populations

Related to PAH Biodegradation in an Aged Biostimulated Creosote-Contaminated Soil," Biodegradation, Vol. 20, No. 5, 2009, pp. 593-601. doi:10.1007/s10532-009-9247-1

21. X. D. Huang, Y. El-Alawi, D. M. Penrose, B. R. Glick and B. M. Grenberg, "Responses of Three Grass Species to Cresote during Phytoremediation," Environmental Pollution, Vol. 130, No. 3, 2004, pp. 453-464. doi:10.1016/j.envpol.2003.12.018
22. M. Mellendorf, G. Soja, H. Martin, M. H. Gerzabek and A. Watzinger, "Soil Microbial Community Dynamics and Phenanthrene Degradation as Affected by Rape Oil Application," Applied Soil Ecology, Vol. 46, No. 3, 2010, pp. 329-334. doi:10.1016/j.apsoil.2010.10.008
23. L. M. Carmichael and F. K. Pfaender, "The Effect of Inorganic and Organic Supplements on the Microbial Degradation of Phenanthrene and Pyrene in Soils," Biodegradation, Vol. 8, No. 1, 1997, pp. 1-13. doi:10.1023/A:1008258720649
24. J. H. Berg Schuur and B. Mattiasson, "Separation of Coal-Tar Constituents from Soil Particles in a Two-Liquid Phase Slurry System," Environmental Technology, Vol. 24, No. 6, 2003, pp. 755-765. doi:10.1080/09593330309385612
25. J. K. Pannu, A. Singh and O. P. Ward, "Vegetable Oil as a Contaminated Soil Remediation Amendment: Application of Peanut Oil for Extraction of Polycyclic Aromatic Hydrocarbons from Soil," Process Biochemistry, Vol. 39, No. 10, 2004, pp. 1211-1216.doi:10.1016/S0032-9592(03)00254-1
26. Z. Gong, K. Alef, B.-M. Wilke and P. Li, "Dissolution and Removal of PAHs from a Contaminated Soil Using Sunflower Oil," Chemosphere, Vol. 58, No. 3, 2005, pp. 291-298. doi:10.1016/j.chemosphere.2004.07.035
27. S. B. Kim, D. J. Kim, S. Lee and W. A. Jury, "Effect of Sorption on Benzene Biodegradation in Sandy Soil," Environmental Toxicology and Chemistry, Vol. 22, No. 10, 2003, pp. 2306-2311. doi:10.1897/02-511
28. C. N. Owabor and E. O. Aluyor, "Application of Adsorbent as a Novel Technique during Biodegradation of a Polycyclic Aromatic Hydrocarbon (Anthracene)," African Journal of Biotechnology, Vol. 7, No. 18, 2008, pp. 3321- 3325.
29. B. Wen, R. J. Li, S. Z. Zhang, X. Q. Shan, J. Fang, K. Xiao and S. U. Khan, "Immobilization of Pentachlorophenol in Soil Using Carbonaceous Material Amendments," Environmental Pollution, Vol. 157, No. 3, 2009, pp. 968-974. doi:10.1016/j.envpol.2008.10.015
30. J. Tang and W. J. Weber, "Development of Engineered Natural Organic Sorbents for Environmental Applications. 2. Sorption Characteristics and Capacities with Respect to Phenanthrene," Environmental Science and Technology, Vol. 40, No. 5, 2006, pp. 1657-1663. doi:10.1021/es051665+
31. J. R. Zimmerman, U. Ghosh, R. N. Millward, T. S. Bridges and R. G. Luthy, "Addition of Carbon Sorbents to Reduce PCB and PAH Bioavailability in Marine Sediments: Physicochemical Tests," Environmental Science and Technology, Vol. 38, No. 20, 2004, pp. 5458-5464. doi:10.1021/es034992v

32. S. P. Pradhan and V. J. Srivastava, "A Pilot-Scale Demonstration of an Innovative Soil Remediation Process: Air Emissions Quality," Journal of Air Waste Management Association, Vol. 47, No. 6, 1997, pp. 710-715. doi:10.1080/10473289.1997.10463928

33. A. P. Khodadoust, G. A. Sorial, G. J. Wilson, M. T. Suidan, R. A. Griffiths and R. C. Brenner, "Integrated System for Remediation of Contaminated Soils," Journal of Environmental Engineering—ASCE, Vol. 125, No. 11, 1999, pp. 1033-1041. doi:10.1061/(ASCE)0733-9372(1999)125:11(1033)

34. R. N. Millward, T. S. Bridges, U. Ghosh, J. R. Zimmerman and R. G. Luthy, "Addition of Activated Carbon to Sediments to Reduce PCB Bioaccumulation by a Polychaete (Neanthes arenaceodentata) and an Amphipod (Leptocheirus plumulosus)," Environmental Science and Technology, Vol. 39, No. 8, 2005, pp. 2880-2887. doi:10.1021/es048768x

35. B. H. Hameed, "Spent Tea Leaves: A New Non-Conventional and Low-Cost Adsorbent for Removal of Basic Dye from Aqueous Solution," Journal of Hazardous Material, Vol. 161, No. 2-3, 2009, pp. 253-259. doi:10.1016/j.jhazmat.2008.04.019

36. Y. Li, B. Chen and L. Zhu, "Enhanced Sorption of Polycyclic Aromatic Hydrocarbons from Aqueous Solution by Modified Pine Bark," Bioresource Technology, Vol. 101, No. 19, 2010, pp. 7307-7313. doi:10.1016/j.biortech.2010.04.088

37. B. Chen, M. Yuana and H. Liu, "Removal of Polycyclic Aromatic Hydrocarbons from Aqueous Solution Using Plant Residue Materials as a Biosorbent," Journal of Hazardous Material, Vol. 188, No. 1-3, 2011, pp. 436-442. doi:10.1016/j.jhazmat.2011.01.114

38. A. O. Alade, O. S. Amuda, T. J. Afolabi and A. A. Okoya, "Adsorption of Naphthalene onto Activated Carbons Derived from Milk Bush Kernel Shell and Flamboyant Pod," Journal of Environmental Chemistry and Ecotoxicology, Vol. 4, No. 7, 2012, pp. 124-132.

39. S. E. Agarry and C. N. Owabor, "Evaluation of the Adsorption Potential of Rubber (Hevea brasiliensis) Seed Pericarp-Activated Carbon in Abattoir Wastewater Treatment and in the Removal of Iron (iii) Ions from Aqueous Solution," Nigerian Journal of Technology, Vol. 31, No. 3, pp. 1-14.

40. E. O. McLean, "Soil pH and Lime Requirement," In: C. A. Black, Ed., Methods in Soil Analysis: Chemical and Microbiological Properties, WI American Society of Agronomy, Madison, 1982.

41. D. W. Nelson and L. E. Sommers, "Determination of Organic Carbon," In: A. L. Page, R. H. Miller and D. R. Keeney, Eds., Method of Soil Analysis, American Society of Agronomy, Madison, 1982, p. 539.

42. J. M. Bremner and C. S. Mulvaney, "Total Nitrogen Determination," In: A. L. Page, R. H. Miller and D. R. Keeney, Eds., Method of Soil Analysis, Vol. 2, American Society of Agronomy, Madison, 1982, p. 595.

43. S. R. Olsen and L. E. Sommers, "Determination of Available Phosphorus," In: A.

L. Page, R. H. Miller and D. R. Keeney Eds., Method of Soil Analysis, American Society of Agronomy, Madison, 1982, p. 403.

44. S. C. Amanchukwu, A. Obafemi and G. C. Okpokwasili, "Hydrocarbon Degradation and Utilization by a Palm Wine Yeast Isolates," FEMS Microbiology Letters, Vol. 57, No. 2, 1989, pp. 151-154. doi:10.1111/j.1574-6968.1989.tb03290.x

45. S. E. Agarry, C. N. Owabor and R. O. Yusuf, "Bioremediation of Soil Artificially Contaminated with Petroleum Hydrocarbon Mixtures: Evaluation of the use of Animal Manure and Chemical Fertilizer," Bioremediation Journal, Vol. 14, No. 4, 2010, pp. 189-195. doi:10.1080/10889868.2010.514965

46. M. A. Zahed, H. Abdul Aziz, M. H. Isa, L. Mohajeri, S. Mohajeri and S. R. M. Kutty, "Kinetic Modeling and Half Life Study on Bioremediation of Crude Oil Dispersed by Corexit 9500," Journal of Hazardous Material, Vol. 185, No. 2-3, 2011, pp. 1027-1031.doi:10.1016/j.jhazmat.2010.10.009

47. P. Y. Yeung, R. L. Johnson and J. G. Xu, "Biodegradation of Petroleum Hydrocarbons in Soil as Affected by Heating and Forced Aeration," Journal of Environmental Quality, Vol. 26, 1997, pp. 1511-1576. doi:10.2134/jeq1997.00472425002600060009x

48. G. K. Vasilyeva, L. P. Bakhaeva and E. G. Surovtseva, "The Use of in Situ Soil Adsorptive Bioremediation Following an Accidental Spill of Propanil in the Krasnodar Region of Russia," Land Contamination and Reclamation, Vol. 4, No. 4, 1996, pp. 263-268.

49. G. K. Vasilyeva, V. D. Kreslavski, P. J. Shea and B.-T. Oh, "Potential of Activated Carbon to Decrease 2,4,6-Trinitritoluene Toxicity and Accelerate Soil Decontamination," Environmental Toxicology and Chemistry, Vol. 20, No. 5, 2001, pp. 965-971.doi:10.1002/etc.5620200505

50. G. K. Vasilyeva, V. D. Kreslavski, P. J. Shea and J.-M. Bollag, "Accelerated Transformation and Binding of 2,4,6- Trinitrotoluene in Soil Amended with Activated Carbon," In: A. Violante, P. M. Huang, J. M. Bollag and L. Gianfreda, Eds., Soil Mineral-Organic Matter-Microorganism Interactions and Ecosysstems Health, Development in Soil Science 28 B, Elsevier, Amsterdam, 2002, pp. 157-168. doi:10.1016/S0166-2481(02)80014-9

51. G. K. Vasilyeva, L. P. Bakhaeva, E. R. Strijakova and P. J. Shea, "Biodegradation of 3,4-Dichloroaniline and 2,4,6- Trinitrotoluene in Soil in the Presence of Natural Adsorbents," Environmental Chemistry Letters, Vol. 1, No. 3, 2003, pp. 176-183. doi:10.1007/s10311-003-0033-8

52. G. K. Vasilyeva, E. R. Strijakova and P. J. Shea, "Use of Activated Carbon for Soil Bioremediation In: T. Wardowska, et al., Eds., Soil and Water Pollution Monitoring, Protection and Remediation, Springer Publishers, New York, 2006, pp. 3-23.doi:10.1007/978-1-4020-4728-2_20

53. T. Xu, L. Lou, L. Luo, R. Cao, D. Duan and Y. Chen, "Effect of Bamboo Biochar on Pentachlorophenol Leachability and Bioavailability in Agricultural

Soil," Science Total Environment, Vol. 414, 2012, pp. 727-731. doi:10.1016/j.scitotenv.2011.11.005

54. L. Pizzul, Å. Sjögren, Md. P. Castillo and M. J. Stenstro, "Degradation of Polycyclic Aromatic Hydrocarbons in Soil by a Two-Step Sequential Treatment," Biodegradation, Vol. 18, No. 5, 2007, pp. 607-616. doi:10.1007/s10532-006-9093-3

55. J. Pannu, A. Singh and O. Ward, "Influence of Peanut Oil on Microbial Degradation of Polycyclic Aromatic Hydrocarbons," Canadian Journal of Microbiology, Vol. 49, No. 8, 2003, pp. 508-513. doi:10.1139/w03-068

# 14

CHAPTER

# PLASTIC DEGRADING MICROORGANISMS AS A TOOL FOR BIOREMEDIATION OF PLASTIC CONTAMINATION IN AQUATIC ENVIRONMENTS

Gabriella Caruso

Institute for Coastal Marine Environment (IAMC), Italian National Research Council (CNR), Spianata S. Raineri, 86, 98122 Messina, Italy

## INTRODUCTION

Plastic contamination of aquatic environments from waste discharges, industrial raw materials, manufactured pellets or fragments of fishing nets, is becoming a future research priority, since it has been recognized to be an emerging global threat for its multiple (social and environmental) implications. Each year, approximately 140 million tonnes of synthetic polymers are estimated to be produced and several studies aim at investigating their global impact and interactions with organisms at several trophic levels [1-5]. Additional sources of plastic pollution in the ocean are given by secondary microplastics originating from the breakdown of larger plastic items. In fact, although plastics are extremely persistent in the marine environment, exposure to physical, chemical and biological processes results in their fragmentation down into smaller pieces. Plastics present in surface waters are more prone to degradation compared to those on the seafloor, for which decomposition is made longer because of the cold water temperature and reduced sunlight (UV) penetration [5]. The products of this fragmentation process are microplastics, i.e. plastic particles having a size lower than 5 mm in diameter, which are a global concern since these small particles can be ingested by many organisms including zooplankton, mussels, fish, seabirds and cetaceans. To solve this important environmental problem, many operational clean-up programs have started; for example, UNEP has engaged more than 40 million people from 120 Countries, setting educational measures to promote reduction of plastic use, waste recycling or disposal facilities. In November 2014, the European Union agreed binding

**Citation:** Gabriella Caruso, Plastic Degrading Microorganisms as a Tool for Bioremediation of Plastic Contamination in Aquatic Environments, doi:10.4172/2375-4397.1000e112.

targets for Member States to reduce the use of thin plastic bags by 2019, but other kind of plastic products may continue to be a threat for the sea. So, which actions could be performed? Apart from innovations in solid waste management and sewage treatment to trap smaller particles before discharging of effluents into the sea, an interesting approach could be provided by the use of microbes able to degrade plastics. Biodegradation is the capacity of one or more strains of microorganisms to utilize a synthetic polymer as the sole source of carbon and energy; some types of plastics such as polyhydroxyalkanoates (polyhydroxybutyrate PHB), and polylactic acid (PLA) are highly biodegradable, while synthetic polymers such as polyethylene (PE) polycaprolactone (PCL) and polystyrene (PS) have low biodegradability [6]. The degree of polymer biodegradation in natural ecosystems is affected by several factors, such as the nature of the substrate to be degraded and by environmental and microbiological factors [7].

The microbial degradation of plastics has been reviewed by some studies [8-10], which have addressed the abiotic and biotic (microbial) degradation of a wide array of synthetic polymers. The actinomycete Rhodococcus ruber [11] and the fungus Penicillium simplicissimum [12] were shown to produce extracellular enzymes able to degrade PE, but also thethermophylic bacterium Brevibacillus borstelensis [13] and Streptomyces sp. [14]. Polyhydroxyalkanoates (PHA), among which polyhydroxybutyrate (PHB), are metabolized by several microorganisms; PHA depolymerases have been detected in Pseudomonas stutzeri, Alcaligenes faecalis and Streptomyces sp. [8,10,15,16]. PHA-degrading fungi have been isolated from soil and marine environments and belong mostly to Basidiomycetes, Deuteromycetes (Penicillium and Aspergillus) and Ascomycetes [7]. Polycaprolactone (PCL) is a synthetic polyester easily degraded by microorganisms, among which the bacterium Alcaligenes faecalis [17] and Clostridium botulinum [10] and the fungus Fusarium [7,18]. Polylactic acid (PLA) is a polymer frequently used in biodegradable plastics; its degradation by a thermophilic bacterium (Bacillus brevis) was reported [19], as well as by only two fungal strains of Fusarium moniliforme and by Penicillium roqueforti [7,10]. Compared to other polymers, PLA degradation is slow and less susceptible to microbial attack [10].

Polyurethane is degraded by several fungal species, such as Fusarium solani, Aureobasidium pullulans sp., although its biodegradation is frequently incomplete [8,20]; a polyester PUR degrading enzyme produced by the bacterium Pseudomonas chlororaphis was isolated [21]. Polyvinyl chloride (PVC) is degraded by the bacterium Pseudomonas putida [22], polystyrene by the actinomycete Rhodococcus ruber [23].

In order to improve the process of plastic biodegradation, the mechanisms involved in the plastic degradation have also been explored [24]. There are two different processes: a direct action, in which the deterioration of plastics provides a trophic resource for microbial growth, or an indirect action, in which metabolic products of microorganisms affect the plastic structure. The main groups of microorganisms and the degradative pathways involved in polymer degradation often depend on the environmental conditions. At least two categories of enzymes are involved in biological degradation, the extracellular and the intracellular dehydrogenases. Enzymes secreted by microbes for biodegradation of plastics are mostly lipase, proteinase K, dehydrogenases [10].

In conclusion, microbial degradation of plastic is a promising ecofriendly strategy which represents a great opportunity to manage waste plastic materials with no adverse impacts. Thanks to the progress in blue technologies, new plastic degrading microorganisms have recently been discovered, that have great biotechnological potential and could help the natural bioremediation processes, favoring the natural cleaning of natural ecosystems. Further advances in biochemistry and biotechnological fields could offer new perspectives the bioremediation of plastic contamination and should be encouraged to select the most active microbial consortia in the plastics degradation process.

## REFERENCES

1. Andrady AL (2011) Microplastics in the marine environment. Mar Poll Bull 62: 1596-1605.
2. Cole M, Lindeque P, Halsband C, Galloway TS (2011) Microplastics as contaminants in the marine environment: a review. Mar Poll Bull 62: 2588-2597.
3. Eerkes-Medrano D, Thompson RC, Aldridge DC (2015) Microplastics in freshwater systems: a review of the emerging threats, identification of knowledge gaps and prioritisation of research needs. Water Res 75: 63-82.
4. Caruso G (2015) Microplastics in marine environments: possible interactions with the microbial assemblage. J Pollut Eff Cont 3.
5. UNEP (2014) Plastic debris in the ocean. UNEP Year Book 2014 emerging issues update. Available at http://www.unep.org/yearbook/2014/PDF/UNEP_YearBook_2014.pdf
6. Sivan A (2011) New perspectives in plastic biodegradation. CurrOpinBiotechnol 22: 422-426.
7. Kim DY, Rhee YH (2003) Biodegradation of microbial and synthetic polyesters by fungi. Appl Microbiol Biotechnol 61: 300-308.
8. Shimao M (2001). Biodegradation of plastics. Curr. Opinion Biotechnol. 12: 242-247.
9. Shah AA, Hasan F, Hameed A, Ahmed S (2008) Biological degradation of plastics: a comprehensive review. Biotechnol Adv 26: 246-265.
10. Ghosh SK, Pal S, Ray S (2013) Study of microbes having potentiality for biodegradation of plastics. Environ Sci Pollut Res Int 20: 4339-4355.
11. Gilan O, Hadar Y, Sivan A (2004). Colonization, biofilm formation and biodegradation of polyethylene by a strain of Rhodococcus ruber. Appl Microbiol Biotechnol 65: 97-104.
12. Yamada-Onodera K, Mukumoto H, Katsuyaya Y, Saiganji A, Tani Y (2001) Degradation of polyethylene by a fungus Penicillus simplicissimus YK. Polym Degrad Stab 72: 323-327.
13. Hadad D, Geresh S, Sivan A (2005) Biodegradation of polyethylene by the thermophilic bacterium Brevibacillusborstelensis. J Appl Microbiol 98: 1093-1100.

14. Pometto AL, Lee B, Johnson KE (1992) Production of an extracellular polyethylene-degrading enzyme by Streptomyces sp. Appl Environ Microbiol 58: 731-733.
15. Mabrouk MM, Sabry SA (2001) Degradation of poly (3hydroxybutyrate) and its copolymer poly (3-hydroxybutyrate-co3-hydroxyvalerate) by a marine Streptomyces sp. SNG9. Microbiol Res 156: 323-335.
16. Kato N (1997) Cloning of poly(3-hydroxybutyrate) depolymerase from a marine bacterium, Alcaligenes faecalis AE122, and characterization of its gene product. Biochim Biophys Acta 1352:113-122.
17. Oda Y, Oida N, Urakami T, Tonomura K (1997) Polycaprolactonedepolymerase produced by the bacterium Alcaligenes faecalis. FEMS Microbiol Lett 152: 339-343.
18. Murphy CA, Cameron JA, Huang SJ, Vinopal RT (1996) Fusarium polycaprolactone depolymerase is cutinase. Appl Environ Microbiol 62: 456-460.
19. Tomita K, Kuroki Y, Nagai K (1999) Isolation of thermophiles degrading poly (L-lactic acid). J Biosci Bioeng 87: 752-755.
20. Nakajima-Kambe T, Shigeno-Akutsu Y, Nomura N, Onuma F, Nakahara T (1999) Microbial degradation of polyurethane, polyester polyurethanes and polyether polyurethanes. Appl Microbiol Biotechnol 51: 134-140.
21. Howard GT, Ruiz C, Hillard NP (1999) Growth of Pseudomonas chlororaphis on a polyester-polyurethane and the purification and characterization of a polyurethane-esterase enzyme. Int Biodeter Biodegrad 43: 7-12.
22. Anthony SD, Meizhong L, Christopher EB, Robin LB, David LF (2004 ) Involvement of linear plasmids in aerobic biodegradation of vinyl chloride. Appl Environ Microbiol 70: 6092-6097.
23. Mor R, Sivan A (2008) Biofilm formation and partial biodegradation of polystyrene by the actinomycete Rhodococcus ruber. Biodegradation 19: 851-858.
24. Singh B, Sharma N (2008) Mechanistic implications in plastic degradation. Polym Degrad Stab 93: 564-581.

# 15
CHAPTER

# REMOVAL OF PHENOL FROM SYNTHETIC AND INDUSTRIAL WASTEWATER BY POTATO PULP PEROXIDASES

Katarzyna Kurnik[1], Krzysztof Treder[2], Monika Skorupa-Kłaput[1], Andrzej Tretyn[1, 3], and Jarosław Tyburski[1, 3]

[1]Chair of Plant Physiology and Biotechnology, Nicolaus Copernicus University, Lwowska 1, 87-100 Toruń, Poland

[2]Laboratory of Molecular Diagnostic and Biochemistry, Department of Potato Protection and Seed Science in Bonin, Plant Breeding and Acclimatization Institute-National Research Institute, 76-009 Bonin, Poland

[3]Centre for Modern Interdisciplinary Technologies, Nicolaus Copernicus University, Wileńska 4, 87-100 Toruń, Poland

## ABSTRACT

**Plant peroxidases have strong potential utility for decontamination of phenol-polluted wastewater. However, large-scale use of these enzymes for phenol depollution requires a source of cheap, abundant, and easily accessible peroxidase-containing material. In this study, we show that potato pulp, a waste product of the starch industry, contains large amounts of active peroxidases. We demonstrate that potato pulp may serve as a tool for peroxidase-based remediation of phenol pollution. The phenol removal efficiency of potato pulp was over 95 % for optimized phenol concentrations. The potato pulp enzymes maintained their activity at pH 4 to 8 and were stable over a wide temperature range. Phenol solutions treated with potato pulp showed a significant reduction in toxicity compared with untreated phenol solutions. Finally we determined that this method may be employed to remove phenol from industrial effluent with over 90 % removal efficiency under optimal conditions.**

**Citation:** Katarzyna Kurnik, Krzysztof Treder, Monika Skorupa-Kłaput, Andrzej Tretyn, and Jarosław Tyburski, Removal of Phenol from Synthetic and Industrial Wastewater by Potato Pulp Peroxidases, DOI 10.1007/s11270-015-2517-0.

## INTRODUCTION

Phenol and its derivatives are major organic pollutants, produced by various industrial activities, such as coal conversion, petroleum refining, resin, plastic and pharmaceutical production, wood preservation, metal coating, and textile dyeing (Michałowicz and Duda 2007; Hamid and Rehman 2009; Varsha et al. 2011). Moreover, pesticide production and degradation release huge quantities of phenols into the environment (Michałowicz and Duda 2007).

Exposure to most phenolic compounds poses a serious risk to human health, due to their toxic, mutagenic, carcinogenic, and teratogenic effects (Michałowicz and Duda 2007). Therefore, phenolic compounds need to be eliminated from wastewater before the water is discharged into the environment (Hamid and Rehman 2009). Conventional, physico-chemical methods for phenol removal from industrial wastewaters include membrane filtration; coagulation/flocculation; ion exchange; electrolysis, adsorption on activated carbon; advanced oxidation processes (chlorination, bleaching, ozonation, Fenton oxidation, photocatalytic oxidation); and chemical reduction. All these methods suffer from serious shortcomings due to limited effectiveness and high operational costs (Radushev et al. 2008; Qayyum et al. 2009). Thus, considerable work has aimed to develop alternative methods for the removal of phenolic pollutants from industrial waste streams and the environment. Biodegradation has emerged as one of the most promising approaches for processing water contaminated with phenols. Several reports show partial or complete biodegradation of such chemicals by cultures of bacteria, fungi, or algae (Aitken et al. 1989; Longoria et al. 2008). However, using microbes for treating pollutants is still an expensive process, due to the high costs of producing the microbial culture. Moreover, inhibitors present in the effluent may suppress growth of the culture and the expression of enzymes involved in the degradation of phenolic compounds may vary over time, or depend on the growth phase of the organism (Qayyum et al. 2009).

Enzymatic treatment with peroxidase has been proposed as another strategy for the remediation of phenol-contaminated wastewater. Peroxidases catalyze the oxidative transformation of a variety of organic and inorganic compounds at the expense of peroxide, usually $H_2O_2$. Peroxidases oxidize phenolic compounds to highly reactive radical species, which spontaneously produce insoluble oligo- or polymeric derivatives. These insoluble, polymerized products may be removed from solutions by simple filtration or sedimentation techniques (Regalado et al. 2004). Peroxidases that have been used for treatment of phenolic contaminants include fungal class II peroxidases, such as lignin peroxidase (LiP) or manganese peroxidase (MnP), or plant class III peroxidases, including horseradish peroxidase (HRP), soybean peroxidase, turnip peroxidase (Regalado et al. 2004), or bitter gourd peroxidase (Ashraf and Husain 2010). Among class III peroxidases, HRP has been most commonly used for studies of phenol removal. However, the high cost of purified HRP and its high susceptibility to inactivation by various side reactions during treatment diminish the usefulness of HRP for large-scale bioremediation (González et al. 2008). Soybean peroxidase, considered as an alternative to HRP, is cheaper than HRP, and displays some favorable characteristics, such as higher thermo-tolerance and higher conformational stability over a wide pH range. However, like HRP, soybean peroxidase is still quite susceptible to inactivation during phenol oxidation and polymerization (Feng et al. 2013).

Besides the preparations of purified enzyme (immobilized or free), several peroxidase-producing systems have been considered for use in phenol remediation. Hairy root cultures have been tested for their ability to remove phenol or dichlorophenol (González et al. 2006,2008, 2012; Singh et al. 2006; Paisio et al. 2010; Sosa Alderete et al. 2012; Jha et al. 2013). Hairy roots offer an attractive system for this purpose due to their ability to produce large quantities of exudates that are rich in peroxidase and chelating agents. Moreover, hairy roots have fast rates of biomass production, providing an extensive area of contact with contaminated solution (Suza et al. 2008). Most studies on phenol remediation by hairy roots have been conducted on a laboratory scale. However, recent work has successfully scaled up the removal of 2,4-dichlorophenol using *Brassica napus* hairy roots in stirred tank reactor, thus opening future prospects for employing hairy roots on an industrial-scale for biodepollution of wastewater (Angelini et al. 2011).

Besides hairy root cultures, some other sources of peroxidase have been considered for utilization in wastewater decontamination. Crude enzyme extracts from *Raphanus sativus* roots efficiently remove phenolic compounds from wastewater from the leather industry (Diao et al.2011). Juice from cut roots of *Raphanus* was also able to decontaminate phenol-polluted synthetic wastewater (Naghibi et al. 2003). The peroxidase-containing homogenate from onion solid waste, obtained after peeling the bulbs, was examined for its ability to oxidize caffeic acid, the main contaminant in wastewater from olive mills (El Agha et al. 2008). Minced shepherd's purse (*Capsella bursa-pastoris*) roots were used for the remediation of soil contaminated with 2,4-dichlorophenol (Park et al. 2006).

One prerequisite for the large-scale application of plant peroxidases for bioremediation is the availability of large quantities of cheap material containing highly active, robust, and inactivation-resistant enzyme. In this work, we assess the potential use of potato pulp for phenol removal. Potato pulp is a waste product obtained during starch production. It constitutes the residue that remains after washing out nearly all the starch from potato mash (Lesiecki et al. 2012) and is composed of juice water (90 %), cellulose, hemicelluloses, pectin, starch, proteins, free amino acids, and salts (Mayer and Hillebrandt 1997; Mayer 1998). The starch manufacturing industry produces huge amounts of potato pulp, running into the hundreds of thousands of tons per year (Mayer 1998). The discharge of potato pulp has become an environmental pollution problem and a cost factor, partly due to the huge quantities of this waste product (Wang et al. 2010).

Here, we demonstrate that potato pulp, obtained from a potato starch factory, contains active peroxidases. Then, we evaluated this industrial waste for its efficiency in removing phenol and established the optimum conditions to obtain the maximum decontamination efficiency. Then we used a phytotest to evaluate the toxicity of phenol solutions before and after treatment with potato pulp. Finally, we assessed the potential use of potato pulp to reduce phenol content in the effluent from the fine mechanics industry.

# MATERIALS AND METHODS

## Plant Material

Potato pulp was obtained from the starch company PPZ Trzemeszno, Poland. The raw material was portioned, frozen, and stored at −20 °C. Directly before use, the pulp was thawed in room temperature.

## Potato Pulp Peroxidase Isolation and Activity

Peroxidase (POX, EC 1.11.1.7) activity of potato pulp was assessed by measuring the rate of oxidation of pyrogallol to purpurogallin, by following the change in absorbance at 420 nm. The assay medium was composed of 100 mM phosphate buffer (pH 6.0), 39 mM pyrogallol, 23.5 mM $H_2O_2$, and 5–60 mg of potato pulp in a total volume of 1 ml. The components of the reaction mixture were mixed in a 2-ml centrifuge tube. The reaction was initiated by adding $H_2O_2$ and conducted for 1 min. After that time, it was stopped by adding $H_2SO_4$ to each tube, to a final concentration of 2 %. The $A_{420}$ was measured immediately after adding $H_2SO_4$ to the medium, in 1-cm quartz cuvettes. Enzyme-free controls were run along with the test samples to correct the assay for non-enzymatic pyrogallol oxidation. POX activity was expressed as the micromoles purpurogallin per minute

To extract the soluble peroxidases and the peroxidases that were ionically bound to the cell wall, the samples consisting of 2.5–10 g of potato pulp were incubated for 3 h in 25 ml of McIlvaine buffer (127 mM $Na_2HPO_4$, 36.9 mM citric acid), pH 6.0, supplemented with 1 M NaCl, under constant agitation. After that, the samples were filtered through nylon mesh and centrifuged for 20 min at 10,000×g at 4 °C to remove the solid phase. The extract was dialyzed overnight against three changes of 0.25× McIlvaine buffer. Then, 30 μl of dialyzed extract was subjected to native electrophoresis on a polyacrylamide gel. The samples were subjected to discontinuous polyacrylamide gel electrophoresis under non-denaturing, non-reducing conditions, as described by Laemmli (1970) using a stacking gel containing 4 % (*w*/*v*) acrylamide and a separating gel containing 15 % (*w*/*v*) acrylamide. After the completion of electrophoresis, the gels were stained for POX activity. To visualize POX activity, the gels were placed in Pierce 1-Step TMB-Blotting Substrate Solution (Pierce) until the bands appeared. Peroxidase isoforms stained dark blue. Along with the samples, molecular mass markers (11–245 kDa, 3-Colour Prestained Protein Marker, DNA Gdańsk) were subjected to electrophoretic separation.

## Phenol Removal Conditions from Synthetic Wastewater

The reactions for phenol removal were carried out in 15-ml polypropylene tubes containing the reaction mixture composed of aqueous phenol solution, potato pulp, and $H_2O_2$, in a total volume of 2.5 ml. The reactions were incubated for 2 or 3 h at 20±2 °C in an orbital shaker at 200 rpm. After incubation, the residual phenol was measured spectrophotometrically. Following the measurement, the percentage of phenol removal (removal efficiency) was calculated. If not indicated otherwise, tap water was used as the reaction medium. The stock solutions of phenol and $H_2O_2$ were prepared using distilled water. To optimize the conditions of phenol removal, the effect of several variables, such as the weight of potato pulp inoculum, $H_2O_2$, phenol concentration, the effect of PEG 3350, pH of the reaction mixture, the incubation temperature, and shaking rate were investigated. To optimize the mass of the potato pulp inoculum, we determined the efficiency of phenol removal in reactions performed with 200, 300, 400, or 500 mg of potato pulp. The samples were added to 2.5 ml of 1 mM phenol solution and the reactions were initiated by supplementing the mixture with $H_2O_2$ to a final concentration of 2.59 mM. To determine the optimal $H_2O_2$ concentration, the mixture of 1 mM phenol solution with 400 mg potato pulp was supplemented with $H_2O_2$ to final concentrations of 1.29 to 3.24 mM. The efficiency of the removal reaction was tested within the range of initial phenol concentrations from 1 to 6 mM. The phenol solutions were mixed with 400 mg of potato pulp and then the reaction was initiated by adding $H_2O_2$ to a concentration of 2.59 mM. Simultaneously, we tested the effect of polyethylene glycol (PEG) 3350 on the efficiency of the phenol removal reaction, by supplementing the reaction mixtures with 100 mg/l PEG.

## Control Variants

To assess the involvement of the native proteins in phenol removal, we performed control reactions replacing the raw potato pulp with autoclaved potato pulp. The reactions with autoclaved material were carried out in the presence of 1–3 mM phenol and 2.59 mM $H_2O_2$. Furthermore, controls without potato pulp were performed by measuring residual phenol concentration after 2–3 h incubation of 1 to 3 mM phenol solutions in the presence or absence of $H_2O_2$. The reaction mixtures were incubated at room temperature with 200 rpm shaking.

## pH Range

To assess the effect of pH on the efficiency of phenol removal, the reactions were performed using buffered solutions of different pH values (pH 2 to 10) as the reaction medium. The reaction buffer consisted of 0.2 M acetic acid, 0.2 M phosphoric acid, and 0.2 M boric acid. Concurrently, tap water (pH 7.3) was tested. The assay was conducted as described above.

## Effect of Temperature

To assess the effect of the incubation temperature on the efficiency of phenol removal, the reactions were carried out, in reaction mixtures composed of 1–3 mM phenol, 400 mg potato pulp, and 2.59 mM $H_2O_2$, at temperatures from 10 to 60 °C, for 2 or 3 h. After the reaction was finished, residual phenol was measured spectrophotometrically.

## Effect of Shaking

Reactions were performed in optimized conditions, with shaking rate as the variable. We tested shaking rates from 0 to 250 rpm. The removal reaction was carried out in the presence of 1–3 mM phenol, 400 mg potato pulp, 2.59 mM $H_2O_2$, and incubated for 2 or 3 h. After that, the residual phenol was measured spectrophotometrically.

## Phenol Removal from Industrial Effluent

The conditions of phenol removal were essentially the same as for assay with the synthetic wastewater, except the phenol solution was replaced by industrial wastewater from fine mechanics industry of Sohbi Craft Poland. The potato pulp samples of 50 to 200 mg and 0.65 to 3.88 mM $H_2O_2$ were added to the reaction mixtures. Control assays without $H_2O_2$ were performed to estimate the peroxidase-independent phenol removal. The effect of the incubation temperature and pH on the efficiency of phenol removal was determined as described above. The efficiency of phenol removal was calculated by comparing phenol concentration in post-reaction mixture to the phenol concentration in the untreated sample.

## Phenol Determination

The phenol concentration in synthetic or industrial wastewater was measured spectrophotometrically using the 4-aminoantipyrine method, as described by Kinsley and Nicell (2000). In brief, 1 ml of post-reaction solution was mixed with 5 μl of $NH_4OH$, 5 μl of 2 % 4-aminoantipyrine, and 10 μl of 8 % potassium hexacyaniferrate. After 5 min, the absorbance of the mixture was determined at 510 nm and compared with the calibration curve in the phenol concentration range of 5–40 mg/l.

## $H_2O_2$ Determination

The unreacted $H_2O_2$ remaining after the reaction was measured using 1 M potassium iodide. In brief, 500 μl of tap water and 1000 μl of 1 M KI were sequentially added to 500 μl of post-reaction solution. Then, the absorbance of the mixture was measured at 390 nm and compared with the calibration curve in the range of 0–1.75 mM $H_2O_2$.

## Garden Cress (*Lepidium sativum*) Toxicity Test

*Lepidium sativum* seeds were preincubated for 1 h in tap water and then distributed on Petri dishes lined with wet lignin. Ten seeds were placed into each plate. Thereafter, the seeds were treated with the post-reaction phenol solutions and incubated for 24 or 48 h at room temperature. As negative controls, untreated phenol solutions or phenol solutions incubated with potato pulp without hydrogen peroxide were used. For the positive control, phenol solution was replaced with water. To test the toxicity of hydrogen peroxide alone, the phenol solution was replaced with water and 0.55 mM $H_2O_2$ (the $H_2O_2$ concentration, determined to be present in the mixture after reaction when 1 mM phenol was used) was added to the mixture. After 24 or 48 h, the length of the roots of the seedlings was measured.

## Statistics

All of the experiments were conducted as three independent replicates. Mean and standard deviation were calculated. To determine whether the differences between the results were statistically significant, we performed one-way ANOVA with a significance threshold of 0.05. All the statistical analyses were performed using SigmaPlot 11.0 (Systat Software). The figures show representative results for each experiment.

# RESULTS AND DISCUSSION

## Peroxidase Activity and Isoenzyme Analysis

The peroxidase activity of potato pulp was assayed by testing its ability to oxidize pyrogallol to purpurogallin in the presence of $H_2O_2$. Increasing the mass of the potato pulp in the reaction mixture resulted in a progressive increase in the production of purpurogallin (Fig. 1a). This finding demonstrates that the peroxidase activity is associated with potato pulp. When the protein preparation, obtained from potato pulp by extraction with NaCl-supplemented buffer, was subjected to native electrophoresis and monitored for peroxidase activity, seven peroxidase isoforms were detected. Among the peroxidase isoforms, the band corresponding to a 72 kDa protein was distinguished by its high staining intensity (Fig. 1b). However, further studies are required to determine whether the band represents a single protein or multiple polypeptides.

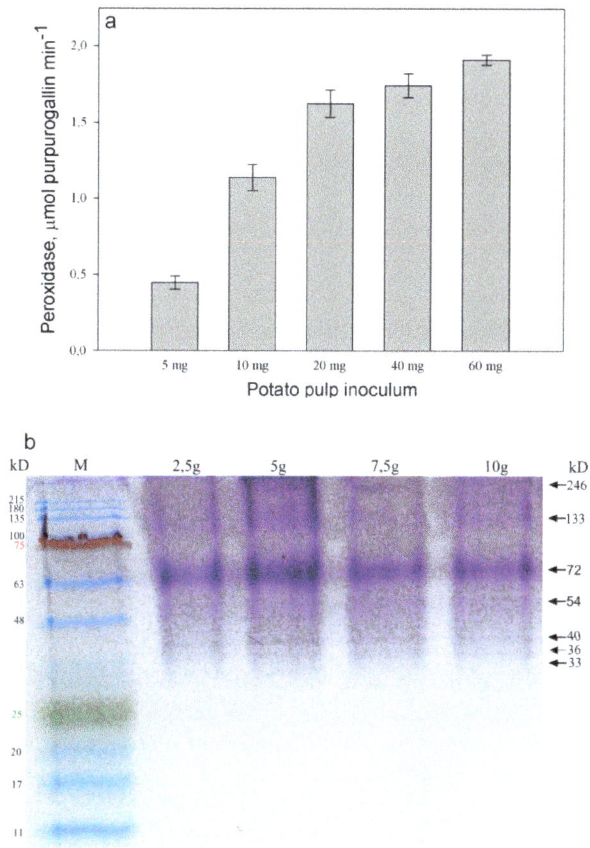


**Figure 1**: Peroxidative activity of potato pulp. A series of sample weights was tested for the ability to oxidize pyrogallol to purpurogallin in the presence of $H_2O_2$ (a). The

isoenzyme pattern of potato pulp peroxidases. Protein preparations obtained from 2.5, 5, 7.5, or 10 g of potato pulp were separated electrophoretically and stained for peroxidase activity. Since the proteins were subjected to non-denaturing PAGE, the protein molecular mass could not be precisely determined. Therefore, the molecular masses (kDa) of peroxidase isoforms indicated represent an approximation (b).

## Phenol Removal from Synthetic Wastewater

To test the capacity of the potato pulp to remove phenol from a water solution, 400 mg of potato pulp was suspended in 2.5 ml of phenol solution. The range of phenol concentrations tested was from 1 to 6 mM. The mixtures were supplemented with $H_2O_2$ to a final concentration of 2.59 mM. Simultaneously, a set of samples representing each phenol concentration was supplemented with PEG 3350. Reaction mixtures with and without PEG were accompanied by control samples without $H_2O_2$. The reactions were incubated at room temperature under constant agitation. The residual phenol was measured after 2 or 3 h of incubation. After both 2 and 3 h of reaction, we observed very high efficiencies of phenol removal, reaching 99 %, in reaction mixtures with initial phenol concentrations from 1 to 3 mM. These results are similar to those obtained by González et al. (2012), where phenol at concentrations of 0.1 to 2.65 mM was removed from the solution by hairy roots of *Brassica napus* with 100–80 % efficiency, respectively. However, our results show higher phenol removal efficiency than that obtained using *Raphanus sativus* roots (Naghibi et al. 2003).

After 2 h of reaction, we observed no significant differences in the efficiency of phenol removal between the variants with initial phenol concentrations from 1 to 3 mM. After 3 h of incubation, we observed that the reaction mixtures with 3 mM initial phenol concentration displayed significantly lower phenol removal efficiencies than the 1 or 2 mM variants. Phenol solutions with higher concentration were degraded with significantly lower efficacy. Approximately 80 or 70 % phenol was removed from solutions with initial concentrations of 4 and 6 mM, respectively (Fig. 2a, b). However, those results still demonstrate higher removal efficiencies than those obtained by Sosa Alderete et al. (2012), who employed tobacco hairy roots to eliminate phenol from solution. The authors showed that treating 4.25 or 6.37 mM phenol, in the presence of 2.5 mM $H_2O_2$, resulted in the removal of 51.7 or 42.2 % of the pollutant, respectively (Sosa Alderete et al. 2012). The high efficiency of phenol removal by potato pulp was strongly dependent on the presence of $H_2O_2$ in the assay medium. In contrast to the reaction mixtures supplemented with $H_2O_2$, the phenol removal from $H_2O_2$-free samples did not exceed 55 %, irrespective of the initial phenol concentration. This finding strongly supports the idea that the highly efficient phenol removal is possibly due to peroxidases from the potato pulp (Fig. 2).

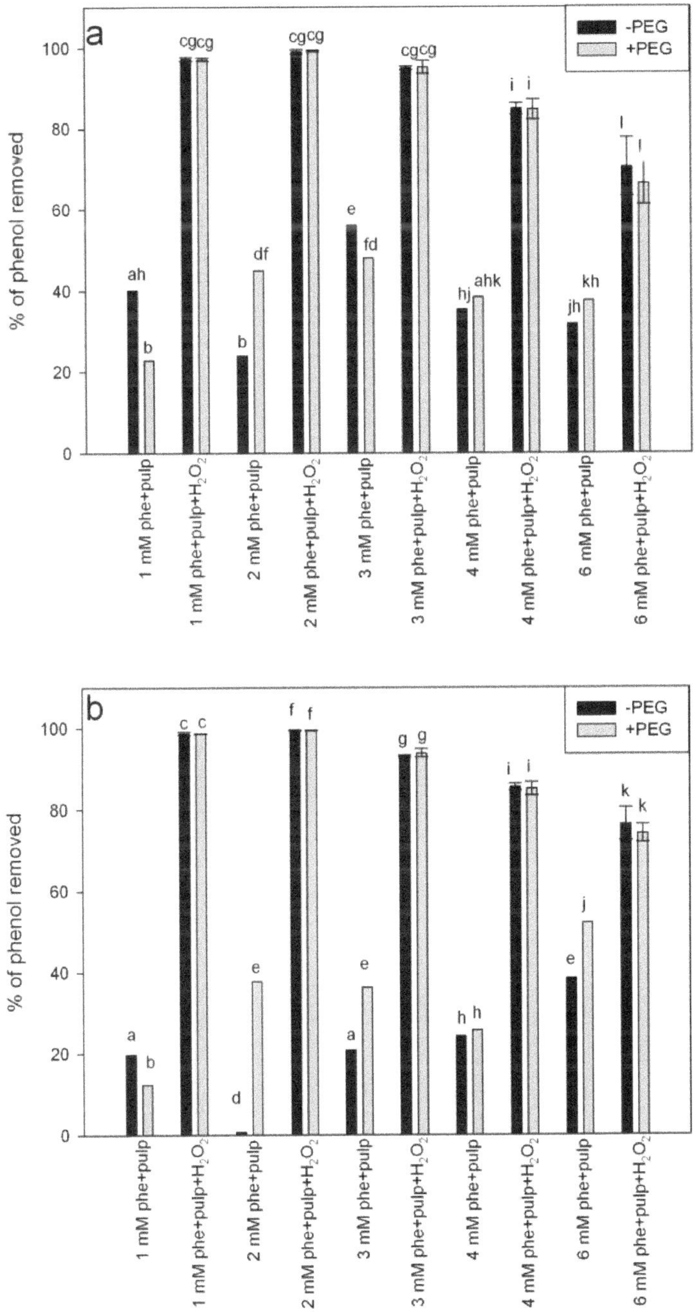


**Figure 2**: Removal efficiencies at different phenol (phe) concentrations. The efficiency of phenol removal was assessed in assay mixtures composed of phenol solution, potato

pulp inoculum, and $H_2O_2$ and in corresponding control mixtures containing phenol and potato pulp but not $H_2O_2$. The efficiency of phenol removal was determined after 2 (a) or 3 (b) hours of incubation in a reaction containing 1–6 mM phenol, 400 mg potato pulp, and 2.59 mM $H_2O_2$ in tap water as a reaction medium. Reaction mixtures were either supplemented (*gray bars*) or not supplemented (*black bars*) with 100 mg/l PEG. Different *letters* denote significant differences at $p \leq 0.05$.

To provide further evidence for the enzymatic nature of the phenol degradation process, potato pulp was autoclaved to inactivate enzymes prior to composing the reaction mixture. Phenol concentrations of 1, 2, or 3 mM were selected for this control assay, since they were the most efficiently removed if the native potato pulp was used. The reactions with the autoclaved potato pulp showed much lower removal efficiencies compared to the experiments with native pulp. In contrast to native potato pulp, the average phenol removed by the autoclaved material did not exceed 40 % (Fig. 3a, b). There were no significant differences in phenol removal efficiency between the tested initial phenol concentration variants (Fig. 3a). Partial phenol removal by autoclaved potato pulp might be caused by the adsorption of phenol on the surface of potato pulp. We also observed that the phenol removal efficiencies were negligible if the potato pulp was not added to the reaction, either in the absence or in the presence of $H_2O_2$ (Fig. 3b, c). These findings suggest that the decrease in the phenol concentration observed in our experiments was not due to spontaneous degradation and/or oxidant-dependent polymerization of a non-enzymatic nature.

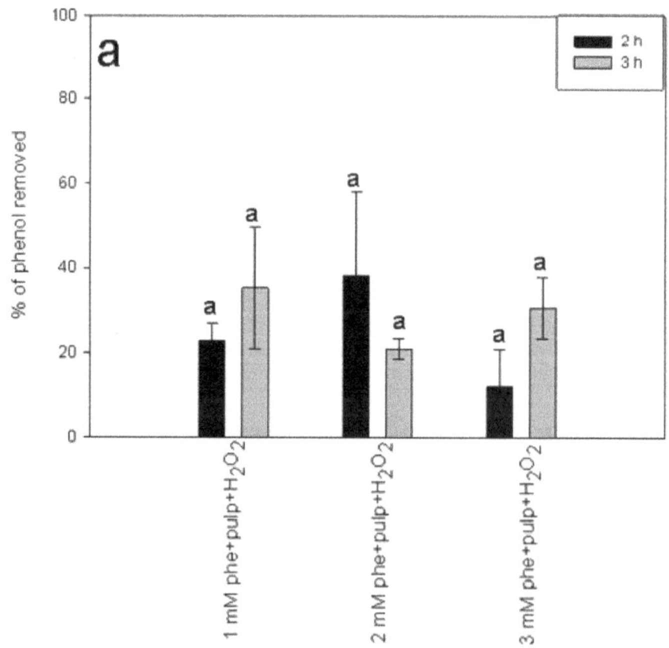

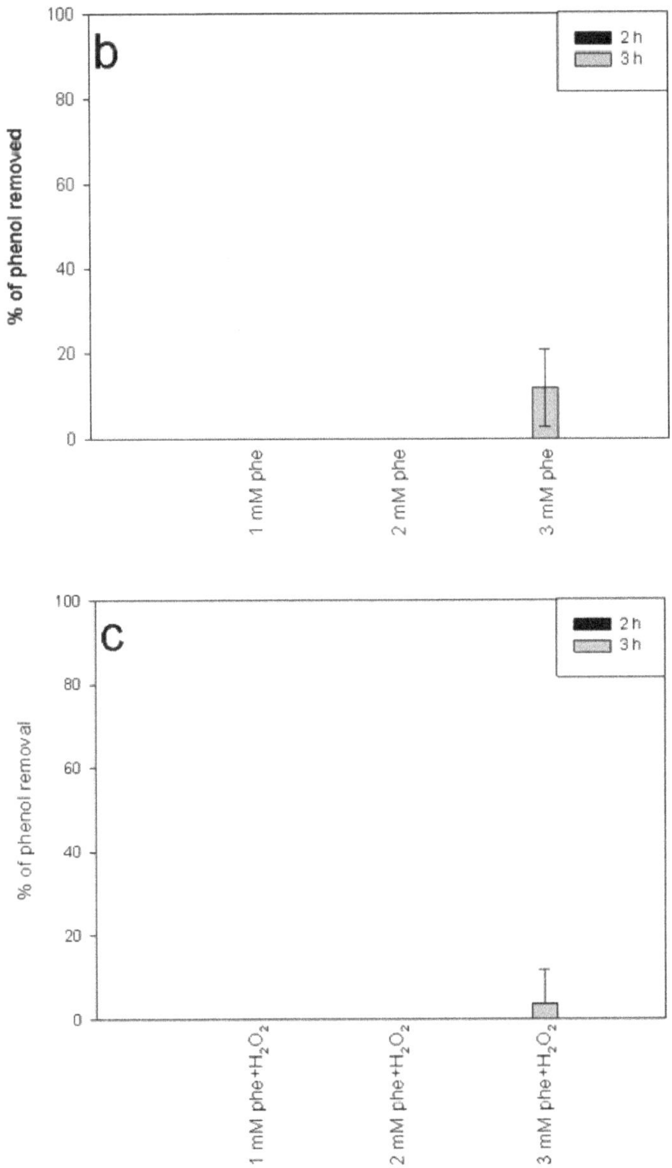


**Figure 3**: Control variants. The percentage of phenol (phe) removed after the incubation of 1–3 mM phenol solutions with autoclaved potato pulp and 2.59 mM $H_2O_2$ for 2 (*black bars*) or 3 (*gray bars*) hours (a). The percentage of phenol removed after the incubation of water solutions of phenol for 2 or 3 h (b). The percentage of phenol removed after the incubation of 1–3 mM phenol solutions supplemented with 2.59 mM $H_2O_2$ for 2 or 3 h (c). Different *letters* denote significant differences at $p \leq 0.05$.

Enzyme inactivation is considered to be the main disadvantage of peroxidase-based phenol removal, significantly elevating the cost of treatment (Feng et al. 2013). Three mechanisms of enzyme inactivation have been proposed as follows: (1) the reaction of an active enzyme intermediate with excess peroxide; (2) destruction of heme by strong reagents, such as free radicals generated during the enzymatic reaction; or (3) adsorption of the enzyme on the precipitated polymerization products, which occludes the enzyme's active site (Mao et al. 2013).

The rate of enzyme inactivation may be decreased by supplementing the reaction medium with a substance that decreases the adsorption of polymers onto the enzyme's active site, using substances such as polyethylene glycol (PEG). Several authors reported a stimulatory effect of PEG on the efficiency of phenol removal by HRP or soybean peroxidase (Kinsley and Nicell2000). PEG substitutes for the enzyme in adsorption onto the solid polymer products, thus preventing the enzyme from co-precipitation with the polymeric products (Sakurai et al. 2003). PEG also efficiently suppressed HRP inactivation via heme destruction (Mao et al. 2013). Consequently, the addition of PEG helps to maintain peroxidase activity in the reaction mixture and can reduce the amount of the enzyme required (Qayyum et al. 2009).

When potato pulp was used in the removal process, we observed very high efficiencies of decontamination in the absence of any protective factors and the addition of PEG 3350 did not change phenol depletion at the tested concentrations (Fig. 2). Similarly, phenol removal by tomato hairy roots displayed comparable efficiency in PEG-supplemented and PEG-unsupplemented reaction mixtures. By contrast, adding PEG to the assay medium significantly increased phenol removal by rapeseed hairy roots (Paisio et al. 2010). Similarly, González et al. (2008) reported that PEG supplementation improves the efficiency of the reaction with tomato hairy roots. Wu et al. (1997), using HRP, observed the same improvement in phenol removal after supplementing the reaction medium with PEG. Also, Cooper and Nicell (1996) reported that adding polyethylene glycol resulted in a 22-fold decrease in the amount of enzyme required to effectively perform the removal reaction. The lack of difference between PEG-supplemented and PEG-unsupplemented samples observed in our experiments is consistent with the results of Caza et al. (1999), who found that adding PEG did not drastically improve the efficiency of phenol removal by soybean peroxidase.

Another approach to improve the performance of peroxidase in the detoxification of pollutants is the use of enzymes immobilized on natural or synthetic supports. Many materials and different methods, such as glass beads, magnetic beads (Bayramoğlu and Arica 2008), polymers, ion exchange resins, magnetite, aluminum-pillared clay, or sodium alginate encapsulation, have been used for peroxidase immobilization (Alemzadeh and Nejati 2009; Husain and Ulber 2011). Immobilized enzymes usually have a longer operational lifetime, being more stable to physical, chemical, and biological denaturing agents (Rao et al. 2010). Immobilization on glutaraldehyde-activated aminopropyl glass beads improved the stability of HRP and the efficiency of phenol removal (Gómez et al. 2006). HRP immobilized by bioaffinity-layering could be successfully used five times, with 100 % conversion of *p*-chlorophenol (Dalal and Gupta 2007). HRP entrapped in alginate capsules could be reused up to four times without any changes in retention activity. Media containing immobilized enzymes are more suitable when large amounts of wastewater need to be processed (Alemzadeh and Nejati 2009). A possible advantage of potato pulp as a tool for the remediation of phenol-contaminated water may be that some of the enzyme is ionically bound to cell wall polymers. Given that enzymes display improved resistance to inactivation when immobilized on a solid support, this fixation of the peroxidase to a solid fraction of the potato pulp may, at least partly, account for the robustness and stability of these enzymes in the assay medium.

## Residual $H_2O_2$

Since $H_2O_2$ is toxic for many organisms, it is necessary to determine the residual concentration of $H_2O_2$ before releasing the post-reaction mixtures into the environment (Paisio et al. 2010). We assessed the $H_2O_2$ concentration in post-reaction mixtures initially supplemented with 1, 2, or 3 mM phenol and 2.59 mM $H_2O_2$ and found, that in the post-reaction solution initially supplemented with 1 mM initial phenol, the residual $H_2O_2$ was $0.546 \pm 0.09$ mM. For the mixtures supplemented with 2 or 3 mM initial phenol, the residual $H_2O_2$ was present at much lower levels, $0.066 \pm 0.002$ or $0.020 \pm 0.001$ mM, respectively. Studies on $H_2O_2$ toxicity in an oligodendrocyte cell line indicated that $H_2O_2$ concentrations greater than 0.1 mM ($10^{-4}$ M) induce cell toxicity (Bhat and Zhang 1999; Twiner et al. 2001). However, the threshold of $H_2O_2$toxicity to rainbow trout (*Oncorhynchus mykiss*) fry and fingerlings is on the order of 10 mM ($10^{-2}$ M) or 1 mM ($10^{-3}$ M) $H_2O_2$, for 30 or 60 min of exposure, respectively (Arndt and Wagner1997). Thus we conclude that the residual $H_2O_2$ in the post-reaction solution should not produce toxic effects when released into the environment if the initial phenol levels were high enough to consume the bulk of the $H_2O_2$ in the reaction. However, the residual $H_2O_2$ may rise to toxic levels for reactions with low initial phenol concentrations. In that case, the $H_2O_2$concentration in the reaction mixture should be adjusted to assure its efficient scavenging during the incubation.

## Effect of the Weight of the Potato Pulp Inoculum and Hydrogen Peroxide Concentration on Phenol Removal

To optimize the mass of the potato pulp inoculum, we tested weights from 200 to 500 mg per 2.5 ml of reaction mixture. We observed the highest efficiency of phenol removal (from $93.2 \pm 1.1$ to $96.3 \pm 0.2$ %), for both 2 and 3 h of reaction time, when we used 300 to 500 mg of potato pulp per reaction. There were no statistically significant differences between samples from 300 to 500 mg. Supplementing the reaction mixture with more than 500 mg of pulp hampered manipulating the samples because the pulp swelled in the solution. Decreasing the weight to 200 mg still resulted in high ($89 \pm 0.2$ % after 3 h of incubation), but statistically lower phenol removal efficiency compared to the other sample weights (Fig. 4a). The potato pulp tested displayed a higher volume of enzyme source per volume of reaction medium than was described for other systems. Naghibi et al. (2003) tested 0.6 g of *Raphanus sativus* root per 20 ml of reaction medium, but obtained only 58 % phenol removal efficiency. However, González et al. (2006, 2012) and Sosa Alderete et al. (2012) demonstrated efficiencies ranging up to 100 % phenol removal, using 0.4 g of hairy roots per 10 ml of reaction medium. We also tested a series of $H_2O_2$ concentrations to assess the effect of the oxidant concentration on phenol removal efficiency. We found over 90 % phenol removal in the solutions supplemented with all the $H_2O_2$ concentrations tested. However, with 1.29 mM $H_2O_2$, there was no phenol removal after 2 h of incubation (Fig. 4b). Removal efficiencies with 1.94, 2.59, or 3.24 mM $H_2O_2$ were comparable when measured after 2 and 3 h of incubation. However, the smallest variation in the removal efficiency, as inferred from the standard deviation values, was observed for 2.59 mM $H_2O_2$; hence, this $H_2O_2$ concentration was chosen for further experiments (Fig. 4b).

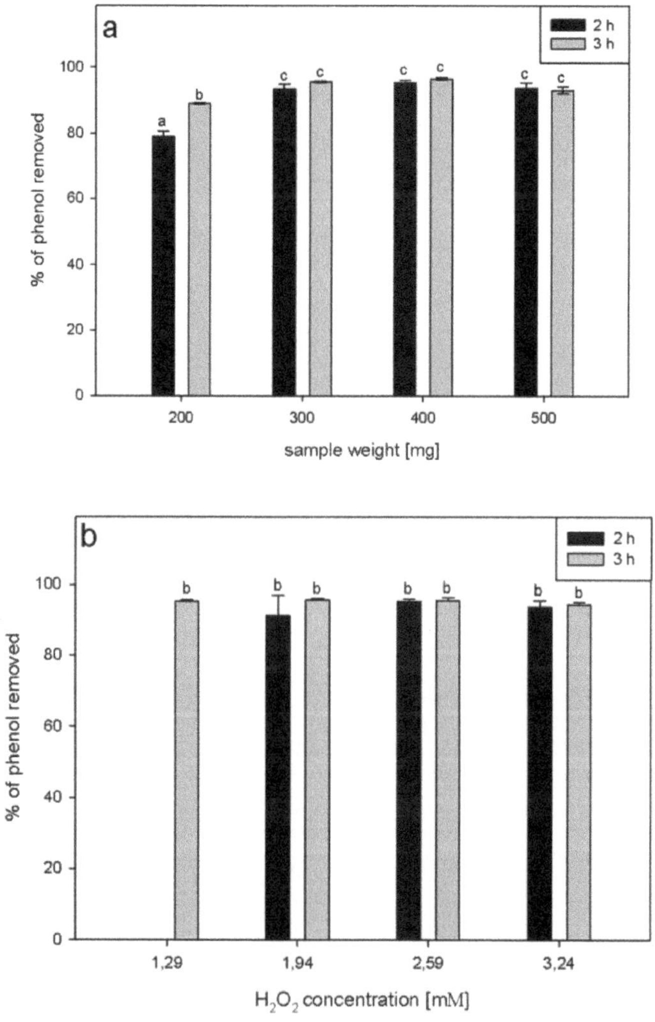


**Figure 4**: The effect of sample weight (a) and $H_2O_2$ concentration on phenol (phe) removal. The reactions were performed for 2 (*black bars*) or 3 (*gray bars*) hours with 1 mM phenol. Different *letters* denote significant differences at $p \leq 0.05$.

## Effect of the Reaction Mixture pH on Phenol Removal

The pH range tested was from pH 2.0 to 10.0 for initial phenol concentrations from 1 to 4 mM (Fig. 5). The efficiency of phenol removal was measured after 2 or 3 h of incubation. Each time, $H_2O_2$-free controls were run along with the reactions to monitor peroxidase-independent changes in phenol concentration. At pH 2.0, the phenol was either not removed from the solution or the removal efficiency was low (Fig. 5a).

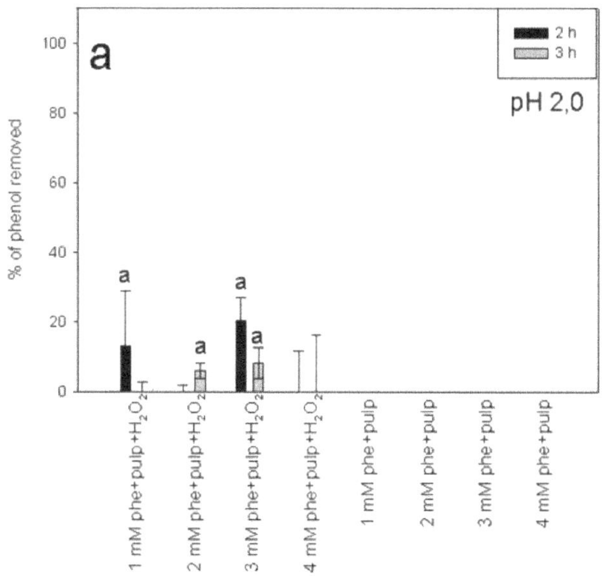


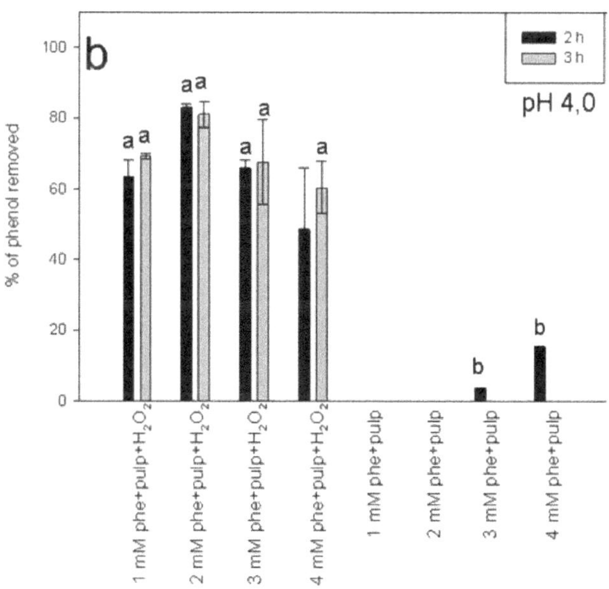

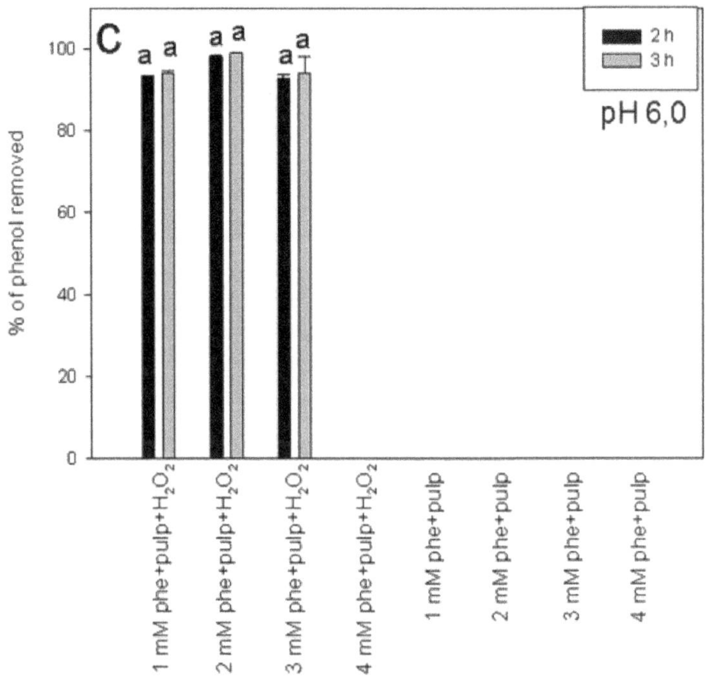
c
2 h
3 h
pH 6,0
% of phenol removed
1 mM phe+pulp+H2O2
2 mM phe+pulp+H2O2
3 mM phe+pulp+H2O2
4 mM phe+pulp+H2O2
1 mM phe+pulp
2 mM phe+pulp
3 mM phe+pulp
4 mM phe+pulp

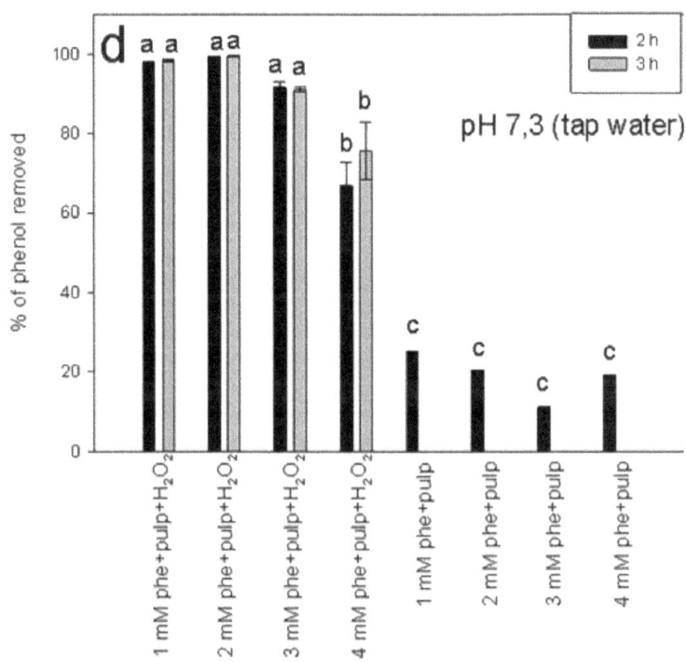
d
2 h
3 h
pH 7,3 (tap water)
% of phenol removed
1 mM phe+pulp+H2O2
2 mM phe+pulp+H2O2
3 mM phe+pulp+H2O2
4 mM phe+pulp+H2O2
1 mM phe+pulp
2 mM phe+pulp
3 mM phe+pulp
4 mM phe+pulp

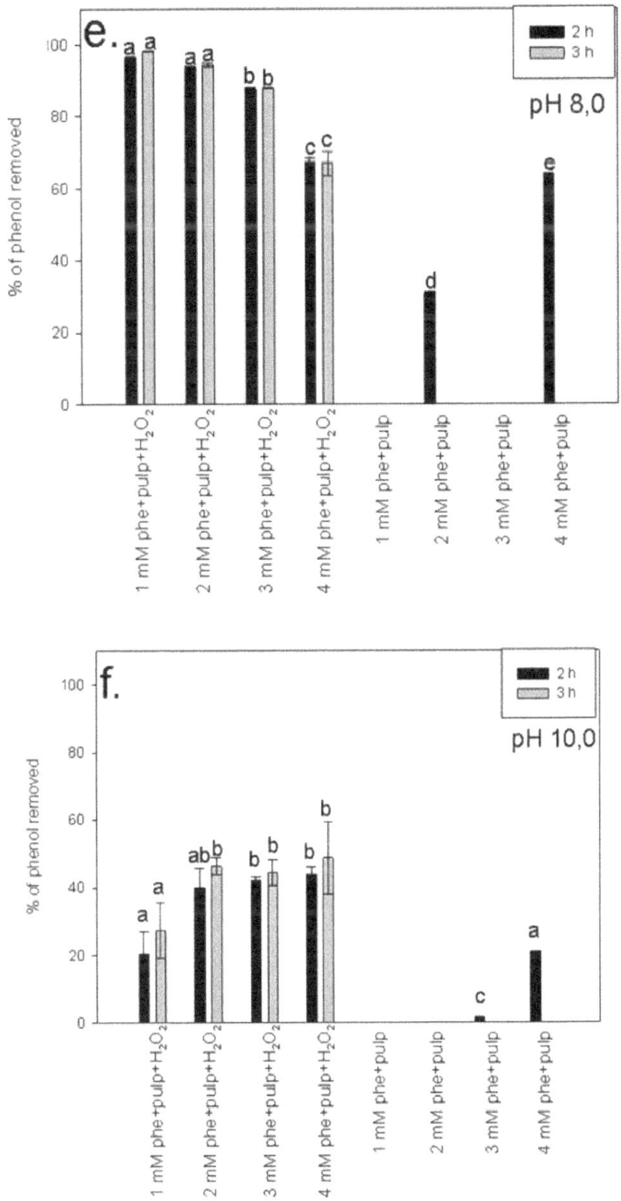


**Figure 5**: The effect of pH on the efficiency of phenol (phe) removal. Efficiency of phenol removal was assessed in assay mixtures composed of phenol solution, potato pulp inoculum, and $H_2O_2$ and in corresponding control mixtures containing phenol and potato pulp but not $H_2O_2$. The assays were carried on at pH 2.0 (a), pH 4.0 (b), pH 6.0 (c), pH 7.3 (tap water) (d), pH 8.0 (e), or pH 10.0 (f). The reactions were incubated for 2 (*black bars*) or 3 (*gray bars*) hours. Different *letters* denote significant differences at $p \leq 0.05$.

At pH 4.0, we observed better phenol removal efficiencies, with the highest reaching over 80 % for the 2 mM initial phenol concentration, whereas at the same pH, González et al. (2006) obtained phenol removal efficiency of about 45 %. Our results are similar to those of Caza et al. (1999), who found that at this pH, about 70 % of the phenol was removed from the reaction medium by soybean peroxidase. For all the phenol concentrations tested, the removal efficiencies at higher pH were higher than at pH 2.0. However, we did not observe statistically significant differences among the efficiencies of phenol removal reactions differing in their initial phenol concentrations (Fig. 5b).

At pH 6.0, a typical optimum pH for peroxidase activity, reactions resulted in very high phenol removal efficiencies, up to almost 99 % for 1, 2, or 3 mM phenol, but at a phenol concentration of 4 mM, we did not observe any phenol removal. There were no statistically significant differences in removal efficiencies between the concentration variants where phenol removal occurred (Fig. 5c). For the same pH, 1 mM phenol was removed with an efficiency of about 70 and 90 % by tomato hairy roots (González et al. 2006) and soybean peroxidase (Caza et al.1999), respectively.

The phenol removal efficiency in the tap water (pH 7.3) as a reaction medium was 99 % for 1, 2, or 3 mM initial phenol concentrations, whereas the removal efficiency at 4 mM initial phenol concentration was 75 % after 3-h reaction (Fig. 5d). Those findings are in agreement with results obtained by Naghibi et al. (2003), who achieved over 90 % phenol removal efficiency in unbuffered tap water.

At pH 8.0, we also observed the removal of over 90 % of phenol from the treated solution (Fig. 5e). However, at higher phenol concentrations, the removal efficiencies were lower than in tap water. The highest phenol removal was observed at the 1 mM phenol concentration and reached over 96 %, whereas at this pH, hairy roots and soybean peroxidase removed about 70 and 85 % of the phenol, respectively (González et al. 2006; Caza et al. 1999). The removal efficiency gradually decreased with increasing phenol concentration to 66 % at 4 mM phenol (Fig. 5e). At pH 10, which was the highest pH tested, we observed a significant decrease in phenol removal, irrespective of the initial phenol concentration in the reaction mixture (Fig. 5f).

The results of the assays on the effect of pH show that phenol can be efficiently removed from the medium at pH 6 to 8, and the buffer may be replaced with tap water without compromising the removal efficiency. Irrespective of the pH, the efficiency of phenol removal was either significantly reduced, or did not occur, unless $H_2O_2$ was added to the reaction mixture (Fig. 5a–f). Our results are consistent with those of González et al. (2006) who reported that, using tomato hairy roots, high phenol removal efficiencies were observed in the pH range from 4 to 9, with an optimal pH of 7.5. Similarly, Caza et al. (1999) and Wu et al. (1997) observed peroxidase activity and resulting phenol removal at a wide range of pH values, from 4 to 9, and from 4 to 10, respectively, with the neutral pH as optimal. By contrast, Kalaiarasan and Palvannan (2014) observed that stabilized horseradish peroxidase displayed the highest phenol removal efficiency at pH 4.2.

## Effect of Temperature on Phenol Removal

Peroxidases usually have a wide temperature range within which their activity is maintained (González et al. 2006). Here, we studied the efficiency of phenol removal at temperatures from 10 to 60 °C. We observed that phenol removal displayed high efficiency within the tested temperature range. In the presence of 1 mM phenol, the removal efficiency did not drop below 95 %, even at the highest temperature tested (Fig. 6a, b). Similarly, the removal efficiency for 2 mM phenol was 97 % at 10 °C after 2 h of incubation and did not significantly change over the entire temperature range (Fig. 6c, d). The most distinct differences in phenol removal efficiencies at different temperatures were observed if the reaction was carried out in the presence of 3 mM phenol. In that case, at both 10 and 60 °C, we observed a slight inhibitory effect on the reaction; however, the removal efficiencies were still over 88 % and we detected no significant differences within the tested temperature range (Fig. 5e, f). For comparison, Kalaiarasan and Palvannan (2014) reported that native HRP lost its activity at 40 °C. However, the enzyme's capacity for phenol removal was maintained at higher temperatures (65 °C) if dextran and sodium alginate were added to the reaction mixture. By contrast, González et al. (2006) observed the highest phenol removal efficiency at 40 and 60 °C, using tomato hairy roots.

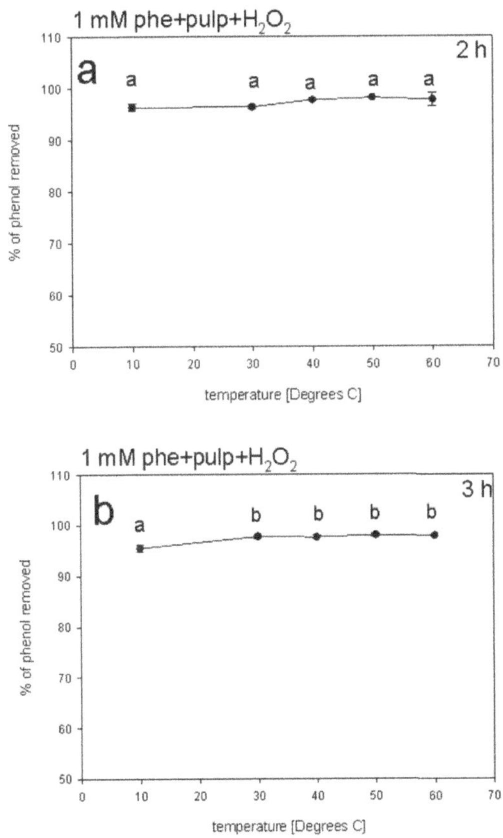

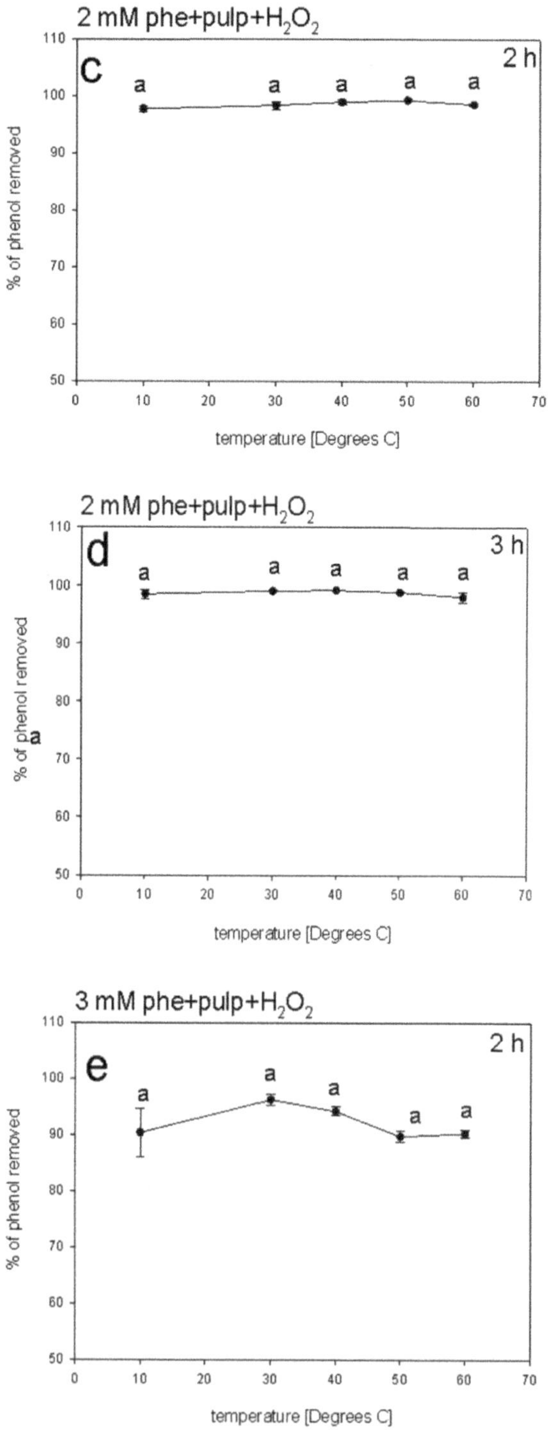
2 mM phe+pulp+H2O2
c
2 h
% of phenol removed
temperature [Degrees C]
2 mM phe+pulp+H2O2
d
3 h
% of phenol removed
temperature [Degrees C]
3 mM phe+pulp+H2O2
e
2 h
% of phenol removed
temperature [Degrees C]

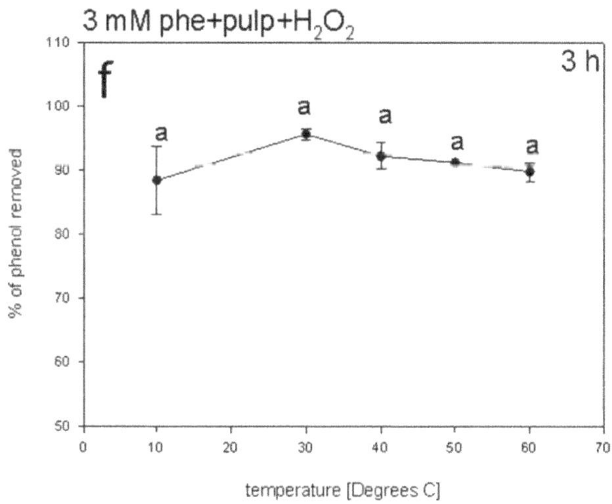


**Figure 6**: The effect of temperature on the efficiency of phenol (phe) removal. Efficiency of phenol removal was assessed in assay mixtures composed of phenol solution, potato pulp inoculum, and $H_2O_2$. The assays were incubated at different temperatures. The percentage of phenol removal was determined after 2 (a, c, e) or 3 (b, d, f) hours of incubation. The assay solutions were initially supplemented with 1 (a, b), 2 (c, d), or 3 (e,f) mM phenol. Different *letters* denote significant differences at $p \leq 0.05$.

## Effect of Shaking Rate on Phenol Removal

Most studies testing the ability of peroxidases to detoxify phenol solutions used shaking rates from 50 to 200 rpm to achieve optimal results (González et al. 2008, 2013; Paisio et al. 2010; Gholizadeh et al. 2013; Asadgol et al. 2014; Kalaiarasan and Palvannan 2014; Malani et al.2013; Abdallah 2013; Angelini et al. 2014). However, to the best of our knowledge, assessing the effect of various shaking rates on the efficiency of phenol removal by peroxidase has not yet been tested as a distinct experiment. To assess the effect of shaking rate on phenol removal by potato pulp, the reaction mixtures subjected to shaking rates from 50 to 200 rpm were compared to the control, which was mixed after the reaction mixture was composed, and then left unshaken. The differences in the removal efficiency between the shaken samples compared to the control were the most obvious for the 1 mM phenol solutions. The variants that were not shaken displayed phenol removal efficiencies of 70 %, whereas shaking the samples enabled almost 100 % phenol removal within the entire range of shaking intensities (Fig. 7a, b). In the case of 2 mM phenol, the unshaken controls displayed removal efficiencies of 85 %, significantly lower than in the agitated variants. Agitating samples at 50 rpm increased phenol removal up to 98 %. The other variants also displayed removal efficiencies close to 100 % (Fig. 7c, d). After 2 and 3 h of reaction, there were no statistically significant differences in phenol removal efficiencies at shaking rates from 50 to 200 rpm. Slightly lower removal efficiencies were detected for reaction mixtures with 3 mM phenol. The unshaken controls displayed removal efficiencies of 87 %, and the efficiencies did not significantly change over the agitation rates applied, except the samples subjected to 150 rpm, after 2-h incubation, rising to over 90 % phenol removal (Fig. 7e, f).

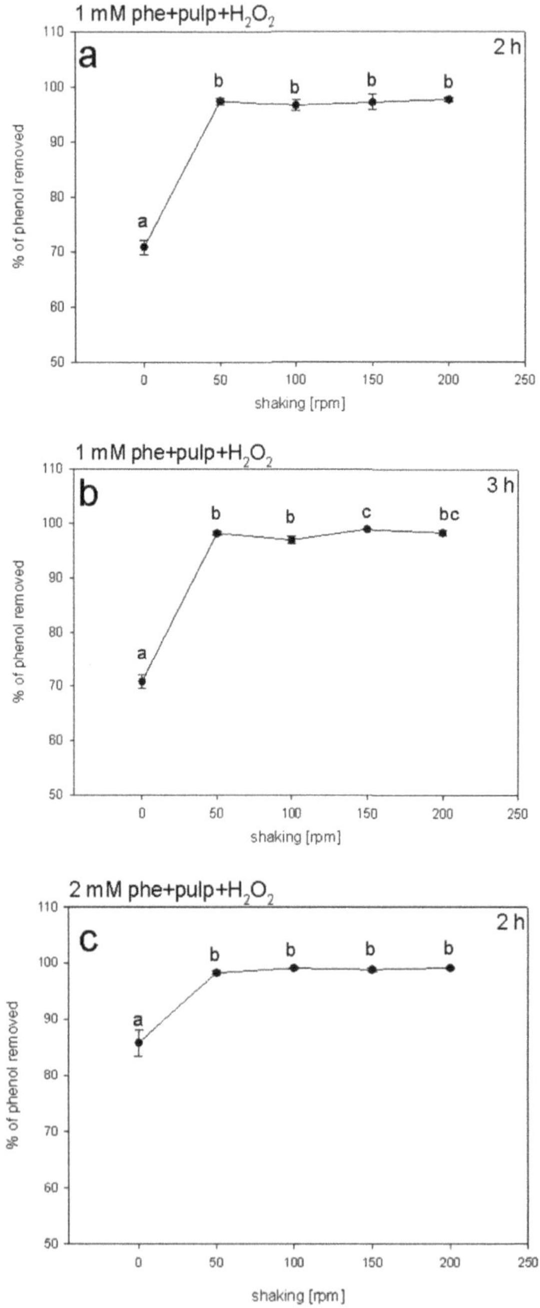

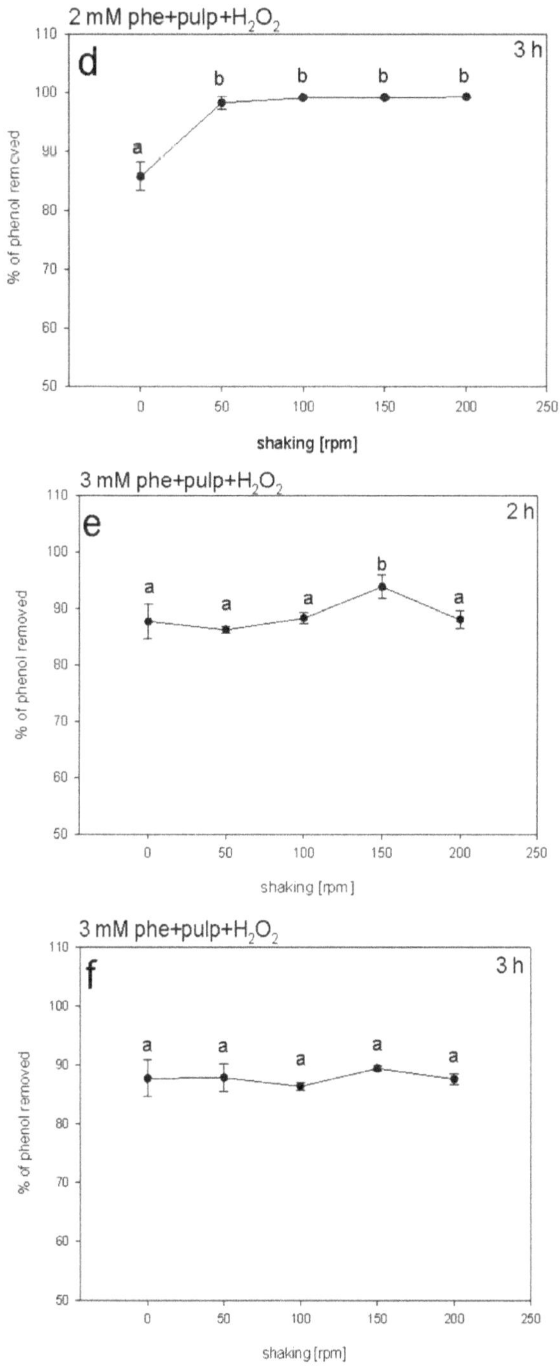


**Figure 7**: The effect of shaking rate on the efficiency of phenol (phe) removal. Efficiency of phenol removal was assessed in assay mixtures composed of phenol solution,

potato pulp inoculum, and $H_2O_2$. The assays were conducted at different shaking rates. The percentage of phenol removed was determined after 2 (a, c, e) or 3 (b, d, f) hours of incubation. The assay solutions were initially supplemented with 1 (a, b), 2 (c, d), or 3 (e,f) mM phenol. Different *letters* denote significant differences at $p \leq 0.05$.

We show that the efficiency of the removal reaction is dependent on agitating the samples, for reaction mixtures with 1 or 2 mM phenol, whereas shaking does not affect phenol removal at higher concentrations of the pollutant. Moreover, the reaction's efficiency remained unchanged throughout the range of shaking intensities tested (Fig. 7). These finding may provide clues for the optimization of energy usage by future technology employing potato pulp for phenol removal from wastewater.

## Lepidium sativum Toxicity Test

To find out whether the removal of phenol catalyzed by the potato pulp peroxidase reduced the toxicity of the phenol solution, we compared the root growth response of *Lepidium sativum*seedlings treated with the post-reaction mixtures to seedlings treated with unreacted phenol solution, for initial phenol concentrations of 1–4 mM. Concurrently, separate sets of seedlings were treated with $H_2O_2$-unsupplemented mixtures of phenol and potato pulp, to discriminate between the effects of peroxidase-dependent and peroxidase-independent detoxification. Additionally, to assess the toxicity of the residual $H_2O_2$, a set of seedlings was incubated with $H_2O_2$ solution at a concentration equivalent to that of the post-reaction mixture. The length of seedling roots was determined after 24 or 48 h of incubation (Fig. 8).

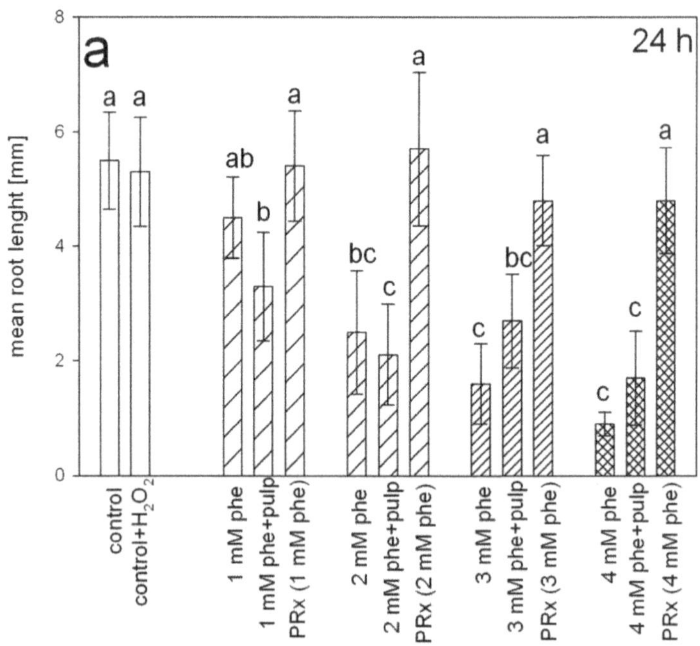

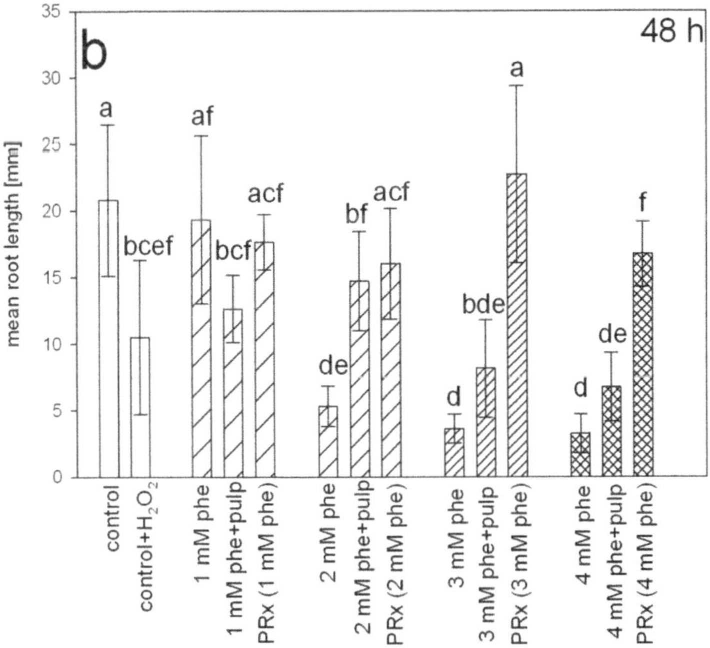


**Figure 8**: *Lepidium sativum* toxicity test. The length of the seedling root is shown. *L. sativum* seeds were treated for 24 (a) or 48 (b) hours with distilled water (control), 0.55 mM $H_2O_2$(control + $H_2O_2$), 1–4 mM phenol (phe) solutions, phenol solution incubated with potato pulp but not provided with $H_2O_2$ (phe + pulp) or the corresponding post-reaction solution (PRx). Different *letters* denote significant differences at $p \leq 0.05$.

When the *Lepidium sativum* seedlings were incubated for 24 h in the presence of phenol, we observed progressive inhibition of seedling root growth. Treatment with 1 mM phenol did not significantly reduce seedling root length, whereas, higher phenol concentrations displayed a gradually increasing inhibitory effect on root growth. Roots of seedlings treated with $H_2O_2$-free, phenol-pulp mixture remained approximately at the same length as those incubated in the presence of unreacted phenol. By contrast, roots of seedlings treated with post-reaction solution were significantly longer than roots of seedlings treated with unreacted phenol or phenol incubated with potato pulp, being comparable to the roots of control plants, irrespective of the initial phenol concentration (Fig. 8).

Similar to the seedlings subjected to the 24-h incubation, those treated for 48 h with the phenol solutions displayed strong reductions in the root length if 2, 3, or 4 mM phenol was used. Seedlings treated with 1 mM phenol did not display statistically significant differences in root growth rate, compared to those subjected to treatment with the relevant post-reaction solution. However, there were significant reductions in root length between seedlings incubated with 1 mM phenol or the post-reaction solution, when compared to those treated with 1 mM phenol mixed with potato pulp. The reduction in the root growth, due to presence of 2–4 mM phenol, was fully abolished if phenol solution was replaced by the relevant post-reaction mixture (Fig. 8).

Seedlings grown for 48 h in the presence of a mixture of pulp and 3 or 4 mM phenol (but devoid of $H_2O_2$) produced slightly longer roots than those treated with unreacted phenol. However, the differences did not pass the significance test. In the case of seedlings incubated with 2 mM phenol-potato pulp mixture, root lengths were similar to those incubated with 2 mM post-reaction solution. Having analyzed the results of the 24 h incubation, we observed that seedling growth was not arrested due to the presence of residual $H_2O_2$. Based on that observation, we concluded that the concentration of hydrogen peroxide used had no toxic effect on *Lepidium sativum* seedlings at the initial phase of root growth (Fig. 8a). However, the 48-h exposure to this reagent resulted in a significant decrease in root length. This suggests that prolonged presence of oxidant concentrations similar to those remaining in the post-reaction mixtures may have a toxic effect. However, any growth inhibition was not observed if the post-reaction mixture was used instead of the $H_2O_2$ solution (Fig. 8). This finding could be explained by the scavenging of the residual $H_2O_2$ at the expense of the residual phenol by secretory peroxidase of*Lepidium sativum* during seedling incubation.

Our results are consistent with those obtained by Angelini et al. (2014), who demonstrated, using *Lactuca sativa* seedlings, that the toxicity caused by 2,4-dichlorophenol was completely abolished after removal of the pollutant with *Brassica napus* hairy roots. Paisio et al. (2010) reported that *Rhinella arenarum* embryos treated with post-reaction solution from rapeseed hairy roots displayed significantly reduced mortality for initial phenol concentrations from 1 to 2.65 mM, compared to the corresponding concentrations of unreacted phenol. PEG 3350 in the assay medium significantly enhanced the reduction of phenol toxicity during the phenol removal reaction (Paisio et al. 2010). The results of toxicity tests with onion (*Allium cepa*) bulbs performed by González et al. (2012) show that phenol removal by tomato hairy root peroxidase significantly reduced the toxicity of the solution originally supplemented with 2.65 mM phenol. Onion bulbs treated with phenol post-reaction solutions displayed an increase of 34 % in root length, compared with the phenol solution of the same concentration (González et al. 2012).

## Phenol Removal from Industrial Effluent

The phenol concentration in the samples of the industrial effluent used in this study varied between 0.02 and 0.1 mM. Since the phenol concentrations in the industrial effluent were much lower compared to the synthetic wastewater prepared in the laboratory, we anticipated that smaller inocula of potato pulp will be required for phenol removal. Therefore, 50, 100, or 200 mg of potato pulp samples were assessed for their ability to decrease phenol content in the industrial wastewater. It was determined that using 100 mg of potato pulp resulted in the highest phenol removal from the industrial effluent. In contrast to the synthetic wastewater, up to 80 % of phenol was removed by potato pulp from industrial wastewater in the absence of $H_2O_2$ (Fig. 9a). This finding may suggest effective phenol sorption to pulp solids. Increasing $H_2O_2$ concentration until 1.94 mM did not affect removal efficiency. However, 2.59 mM $H_2O_2$ or higher concentrations enhanced the removal efficiency (Fig. 9b).

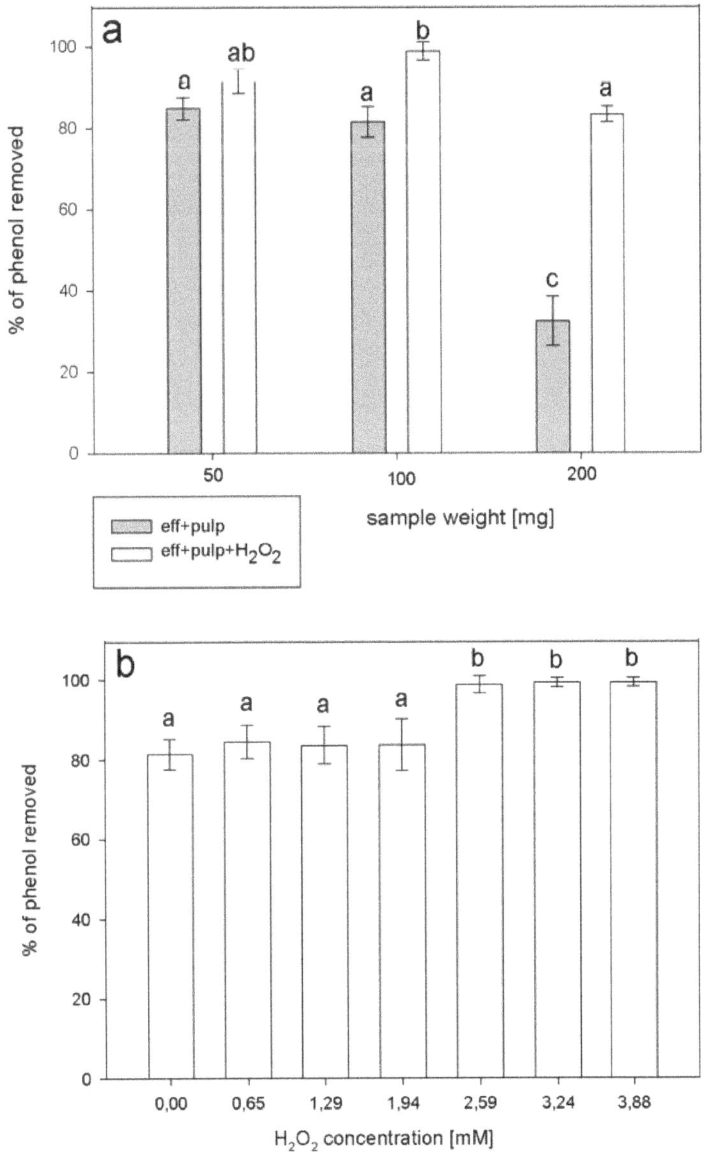

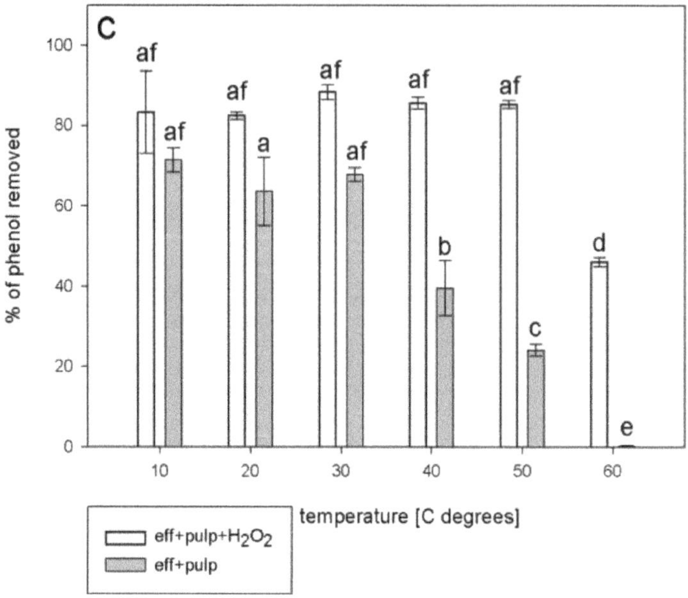


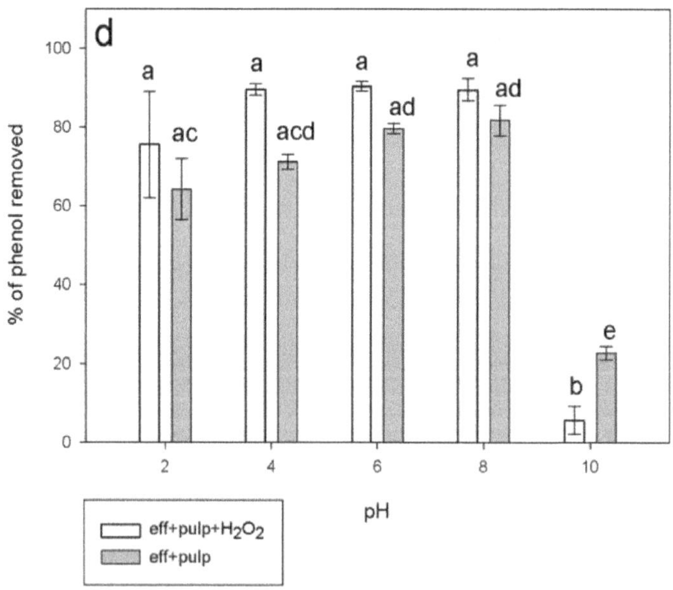


**Figure 9**: Phenol removal from industrial effluent. Different conditions for phenol removal were tested. The reactions were performed for 2 h with 50–200 mg of potato pulp and supplemented with 2.59 mM $H_2O_2$ (a). The effect of $H_2O_2$ concentration on phenol removal (b). The effect of temperature on phenol removal (c). The effect of pH on phenol removal (d). eff + pulp + $H_2O_2$ = reaction mixture composed of industrial effluent, potato pulp

and supplemented with $H_2O_2$, eff + pulp = industrial effluent incubated with potato pulp, but devoid of $H_2O_2$. Different *letters* denote significant differences at $p \leq 0.05$.

The efficiency of phenol removal in $H_2O_2$-unsupplemented medium progressively decreased under incubation temperature of 40 °C or higher. However, the removal process was still effective in $H_2O_2$-supplemented samples (Fig. 9c) suggesting that peroxidase-based phenol removal was less susceptible to high temperature than in $H_2O_2$-free controls. We also observed, that the reaction could be conducted over a wide range of the pH range. The significant decrease in phenol removal efficiency was observed only when the reaction was conducted at pH 10 (Fig. 9d).

## CONCLUSION

The bottleneck for systems using peroxidases for the decontamination of phenol-polluted wastewater is the inaccessibility of a low-cost enzyme source, which would satisfy the demands of large-scale remediation. The use of potato pulp for remediation may help to overcome this difficulty, since potato starch factories produce hundreds thousands of tons of this material every year as an industrial waste product. Potato pulp provides a good alternative to purified peroxidases, such as HRP, which are expensive and prone to inactivation, and to hairy root cultures, which—while ensuring good removal efficiencies—are also costly, due to the high cost of culture media. The present work opens the possibility of further research aimed at scaling-up the process and construction of bioreactors adapted to large-scale decontamination of phenol-polluted wastewater with potato pulp.

## ACKNOWLEDGMENTS

This work was financially supported by a grant from the National Science Centre, Poland (Grant No. 005/B/P01/2010/39) and by the Kuyavian-Pomeranian Marshal's Office PhD student scholarship program "Step into the future—scholarships for PhD students, 5$^{th}$ edition".

## REFERENCES

1. Abdallah, M. A. M. (2013). The potential of different bio adsorbents for removing phenol from its aqueous solution. *Environmental Monitoring and Assessment, 185*(8), 6495–6503.
2. Aitken, M. D., Venkatadri, R., & Irvine, R. L. (1989). Oxidation of phenolic pollutants by a lignin degrading enzyme from the white-rot fungus *Phanerochaete chrysosporium. Water Research, 23*(4), 443–450.
3. Alemzadeh, I., & Nejati, S. (2009). Phenols removal by immobilized horseradish peroxidase.*Journal of Hazardous Materials, 166*(2–3), 1082–1086.
4. Angelini, V. A., Orejas, J., Medina, M. I., & Agostini, E. (2011). Scale up of 2,4-dichlorophenol removal from aqueous solutions using *Brassica napus* hairy roots. *Journal of Hazardous Materials, 185*(1), 269–274.
5. Angelini, V. A., Agostini, E., Medina, M. I., & González, P. S. (2014). Use of

hairy roots extracts for 2,4-DCP removal and toxicity evaluation by *Lactuca sativa* test. *Environmental Science and Pollution Research, 21*(4), 2531–2539.

6. Arndt, R. E., & Wagner, E. J. (1997). The toxicity of hydrogen peroxide to rainbow trout*Oncorhynchus mykiss* and cutthroat trout *Oncorhynchus clarki* fry and fingerlings. *Journal of the World Aquaculture Society, 28*(2), 150–157.
7. Asadgol, Z., Forootanfar, H., Rezaei, S., Mahvi, A. H., & Faramarzi, M. A. (2014). Removal of phenol and bisphenol-A catalyzed by laccase in aqueous solution. *Journal of Environmental Health Science and Engineering, 12*, 93.
8. Ashraf, H., & Husain, Q. (2010). Studies on bitter gourd peroxidase catalyzed removal of p-bromophenol from wastewater. *Desalination, 262*(1), 267–272.
9. Bayramoğlu, G., & Arica, M. Y. (2008). Enzymatic removal of phenol and *p*-chlorophenol in enzyme reactor: horseradish peroxidase immobilized on magnetic beads. *Journal of Hazardous Materials, 156*(1), 148–155.
10. Bhat, N. R., & Zhang, P. (1999). Hydrogen peroxide activation of multiple mitogen-activated protein kinases in an oligodendrocyte cell line: role of extracellular signal-regulated kinase in hydrogen peroxide-induced cell death. *Journal of Neurochemistry, 72*(1), 112–119.
11. Caza, N., Bewtra, J. K., Biswas, N., & Taylor, K. E. (1999). Removal of phenolic compounds from synthetic wastewater using soybean peroxidase. *Water Research, 33*(13), 3012–3018.
12. Cooper, V. A., & Nicell, J. A. (1996). Removal of phenols from a foundry wastewater using horseradish peroxidase. *Water Research, 30*(4), 954–964.
13. Dalal, S., & Gupta, M. N. (2007). Treatment of phenol wastewater by horseradish peroxidase immobilized by bioaffinity layering. *Chemosphere, 67*(4), 741–747.
14. Diao, M., Ouédraogo, N., Baba-Moussa, L., Savadogo, P. W., N'Guessan, A. G., Bassolé, I. H., & Dicko, M. H. (2011). Biodepollution of wastewater containing phenolic compounds from leather industry by plant peroxidases. *Biodegradation, 22*(2), 389–396.
15. El Agha, A., Makris, D. P., & Kefalas, P. (2008). Peroxidase-active cell free extract from onion solid wastes: biocatalytic properties and putative pathway of ferulic acid oxidation. *Journal of Bioscience and Bioengineering, 106*(3), 279–285.
16. Feng, W., Taylor, K. E., Biswas, N., & Bewtra, J. K. (2013). Soybean peroxidase trapped in product precipitate during phenol polymerization retains activity and may be recycled. *Journal of Chemical Technology and Biotechnology, 88*(8), 1429–1435.
17. Gholizadeh, A., Kermani, M., Gholami, M., & Farzadkia, M. (2013). Kinetic and isotherm studies of adsorption and biosorption processes in the removal of phenolic compounds from aqueous solutions: comparative study. *Journal of Environmental Health Science and Engineering, 11*, 29.
18. Gómez, J. L., Bódalo, A., Gómez, E., Bastida, J., Hidalgo, A. M., & Gómez, M. (2006). Immobilization of peroxidases on glass beads: an improved alternative for phenol removal.*Enzyme and Microbial Technology, 39*(5), 1016–1022.

19. González, P. S., Capozucca, C., Tigier, H., Milrad, S., & Agostini, E. (2006). Phytoremediation of phenol from wastewater, by peroxidases of tomato hairy root cultures. *Enzyme and Microbial Technology, 39*(4), 647–653.

20. González, P. S., Agostini, E., & Milrad, S. R. (2008). Comparison of the removal of 2,4-dichlorophenol and phenol from polluted water, by peroxidases from tomato hairy roots, and protective effect of polyethylene glycol. *Chemosphere, 70*(6), 982–989.

21. González, P. S., Maglione, G. A., Giordana, M., Paisio, C. E., Talano, M. A., & Agostini, E. (2012). Evaluation of phenol detoxification by *Brassica napus* hairy roots, using *Allium cepa*test. *Environmental Science and Pollution Research International, 19*(2), 482–491.

22. Gonzálcz, P. S., Ontañon, O. M., Armcndariz, A. L., Talano, M. A., Paisio, C. E., & Agostini, E. (2013). *Brassica napus* hairy roots and rhizobacteria for phenolic compounds removal.*Environmental Science and Pollution Research International, 20*(3), 1310–1317.

23. Hamid, M., & Rehman, K. (2009). Potential applications of peroxidases. *Food Chemistry, 115*(4), 1177–1186.

24. Husain, Q., & Ulber, R. (2011). Immobilized peroxidase as a valuable tool in the remediation of aromatic pollutants and xenobiotic compounds: a review. *Critical Reviews in Environmental Science and Technology, 47*(8), 770–804.

25. Jha, P., Jobby, R., Kudale, S., Modi, N., Dhaneshwar, A., & Desai, N. (2013). Biodegradation of phenol using hairy roots of *Helianthus annuus* L. *International Biodeterioration and Biodegradation, 77*, 106–113.

26. Kalaiarasan, E., & Palvannan, T. (2014). Removal of phenols from acidic environment by horseradish peroxidase (HRP): aqueous thermostabilization of HRP by polysaccharide additives. *Journal of the Taiwan Institute of Chemical Engineers, 45*(2), 625–634.

27. Kinsley, C., & Nicell, J. A. (2000). Treatment of aqueous phenol with soybean peroxidase in the presence of polyethylene glycol. *Bioresource Technology, 73*(2), 139–146.

28. Laemmli, U. K. (1970). Cleavage of structural proteins during the assembly of the head of bacteriophage T4. *Nature, 227*(5259), 680–685.

29. Lesiecki, M., Białas, W., & Lewandowicz, G. (2012). Enzymatic hydrolysis of potato pulp. *Acta Scientiarum Polonorum. Technologia Alimentaria, 11*(1), 53–59.

30. Longoria,A.,Tinoco,R.,&Vázquez-Duhalt,R.(2008).Chloroperoxidase-mediated transformation of highly halogenated monoaromatic compounds. *Chemosphere, 72*(3), 485–490.

31. Malani, R. S., Khanna, S., & Moholkar, V. S. (2013). Sonoenzymatic decolourization of an azo dye employing immobilized horse radish peroxidase (HRP): a mechanistic study. *Journal of Hazardous Materials, 256–257*, 90–97.

32. Mao, L., Luo, S., Huang, Q., & Junhe, L. (2013). Horseradish peroxidase

inactivation: heme destruction and influence of polyethylene glycol. *Scientific Reports, 3*, 3126.

33. Mayer, F. (1998). Potato pulp: properties, physical modification and applications. *Polymer Degradation and Stability, 59*(1), 231–235.
34. Mayer, F., & Hillebrandt, J. O. (1997). Potato pulp: microbiological characterization, physical modification, and application of this agricultural waste product. *Applied Microbiology and Biotechnology, 48*(4), 435–440.
35. Michałowicz, J., & Duda, W. (2007). Phenols—sources and toxicity. *Polish Journal of Environmental Studies, 16*(3), 347–362.
36. Naghibi, F., Pourmorad, F., Honary, S., & Shamsi, M. (2003). Decontamination of water polluted with phenol using *Raphanus sativus* root. *Iranian Journal of Pharmaceutical Research, 2*(1), 29–32.
37. Paisio, C. E., González, P. S., Gerbaudo, A., Bertuzzi, M. L., & Agostini, E. (2010). Toxicity of phenol solutions treated with rapeseed and tomato hairy roots. *Desalination, 263*(1), 23–28.
38. Park, J. W., Park, B. K., & Kim, J. E. (2006). Remediation of soil contaminated with 2,4-dichlorophenol by treatment of minced shepherd's purse roots. *Archives of Environmental Contamination and Toxicology, 50*(2), 191–195.
39. Qayyum, H., Maroof, H., & Yasha, K. (2009). Remediation and treatment of organopollutants mediated by peroxidases: a review. *Critical Reviews in Biotechnology, 29*(2), 94–119.
40. Radushev, A. V., Plotnikov, A. V., & Tyryshkina, V. N. (2008). Regeneration methods of decontamination of phenol-containing waste waters. *Theoretical Foundations of Chemical Engineering, 42*(5), 781–794.
41. Rao, M. A., Scelza, R., Scotti, R., & Gianfreda, L. (2010). Role of enzymes in the remediation of polluted environments. *Journal of Soil Science and Plant Nutrition, 10*(3), 333–353.
42. Regalado, C., Garcia-Almendarez, B., & Duarte-Vasquez, M. (2004). Biotechnological applications of peroxidases. *Phytochemistry Reviews, 3*, 243–256.
43. Sakurai, A., Masuda, M., & Sakakibara, M. (2003). Effect of surfactants on phenol removal by the method of polymerization and precipitation catalysed by *Coprinus cinereus* peroxidase.*Journal of Chemical Technology and Biotechnology, 78*(9), 952–958.
44. Singh, S., Melo, J. S., Eapen, S., & D'Souza, S. F. (2006). Phenol removal using *Brassica juncea*hairy roots: role of inherent peroxidase and $H_2O_2$. *Journal of Biotechnology, 123*(1), 43–49.
45. Sosa Alderete, L. G., Ibáñez, S. G., Agostini, E., & Medina, M. I. (2012). Phytoremediation of phenol at pilot scale by tobacco hairy roots. *International Journal of Environmental Sciences, 3*(1), 398–407.
46. Suza, W., Harris, R. S., & Lorence, A. (2008). Hairy roots: from high-value metabolite production to phytoremediation. *Electronic Journal of Integrative*

*Biosciences, 3*(1), 57–65.

47. Twiner, M., Dixon, S., & Trick, C. (2001). Toxic effects of *Heterosigma akashiwo* do not appear to be mediated by hydrogen peroxide. *Limnology and Oceanography, 46*(6), 1400–1405.

48. Varsha, Y. M., Naga Deepthi, C. H., & Chenna, S. (2011). An emphasis on xenobiotic degradation in environmental clean up. *Journal of Bioremediation and Biodegradation, S11*, 001.

49. Wang, T. Y., Wu, Y. H., Jiang, C. Y., & Liu, Y. (2010). Solid state fermented potato pulp can be used as poultry feed. *British Poultry Science, 51*(2), 229–234.

50. Wu, Y., Taylor, K. E., Biswas, N., & Bewtra, J. K. (1997). Comparison of additives in the removal of phenolic compounds by peroxidase-catalyzed polymerization. *Water Research, 31*(11), 2699–2704.

# 16
CHAPTER

# BIOREMEDIATION: A GENUINE TECHNOLOGY TO REMEDIATE RADIONUCLIDES FROM THE ENVIRONMENT

Dhan Prakash[1,2], Prashant Gabani[3], Anuj K Chandel[4], Zeev Ronen[2], and Om V Singh[3]

[1]Institute of Microbial Technology (CSIR), Sector 39-A, Chandigarh -160036, India

[2]Department of Environmental Hydrology & Microbiology, Ben-Gurion University of the Negev, Sede- Boqer Campus, 84990, Israel

[3]Division of Biological and Health Sciences, University of Pittsburgh, Bradford, PA 16701, USA. 4 Department of Biotechnology, School of Engineering of Lorena, University of Sao Paulo, Lorena, Brazil.

## SUMMARY

Radionuclides in the environment are a major human and environmental health concern. Like the Chernobyl disaster of 1986, the Fukushima Daiichi nuclear disaster in 2011 is once again causing damage to the environment: a large quantity of radioactive waste is being generated and dumped into the environment, and if the general population is exposed to it, may cause serious life-threatening disorders. Bioremediation has been viewed as the ecologically responsible alternative to environmentally destructive physical remediation. Microorganisms carry endogenous genetic, biochemical and physiological properties that make them ideal agents for pollutant remediation in soil and groundwater. Attempts have been made to develop native or genetically engineered (GE) microbes for the remediation of environmental contaminants including radionuclides. Microorganism-mediated bioremediation can affect the solubility, bioavailability and mobility of radionuclides. Therefore, we aim to unveil the microbial-mediated mechanisms for biotransformation of radionuclides under various environmental conditions as developing strategies for waste management of

**Citation:** Dhan Prakash, Prashant Gabani, Anuj K. Chandel, Zeev Ronen2 and Om V. Singh, Bioremediation: a genuine technology to remediate radionuclides from the environment, doi: 10.1111/1751-7915.12059.

radionuclides. A discussion follows of '-omics'-integrated genomics and proteomics technologies, which can be used to trace the genes and proteins of interest in a given microorganism towards a cell-free bioremediation strategy.

## INTRODUCTION

In addition to the occasional disastrous accidents at nuclear facilities such as the Chernobyl disaster of 1986 and the Fukushima Daiichi nuclear disaster in 2011, the extensive use of radioactive materials at research and development, biomedical, and industrial sites has created a great accumulation of radioactive waste. Fredrickson and colleagues (2004) reported that during World War II, ~90 million gallons of high-level radioactive waste accumulated across the USA. Most radioactive wastes are generated by nuclear power plants, which contribute ~95% of the radioactivity generated from all sources (Ahier and Tracy, 1995; Tamponnet and Declerck, 2008). In the environment, even a small concentration of radionuclides can have an impact for a prolonged period of time due to their long half-life. As a result, the impact of radionuclide pollutants is growing with time. The commonly encountered radionuclides include cobalt-60 ($^{60}Co$), plutonium-239 ($^{239}Pu$), radium-226 ($^{226}Ra$), radon-222 ($^{222}Rn$), technetium-99 ($^{99}Tc$), thorium-232 ($^{232}Th$) and uranium-238 ($^{238}U$). However, the typical radionuclides produced through nuclear reactors via the splitting of elemental atoms are thallium-201 ($^{201}Tl$), iridium-192 ($^{192}Ir$), caesium-137 ($^{137}Cs$) and strontium-90 ($^{90}Sr$), which take a significantly long time to decay (Kurnaz *et al.*, 2007). Additionally, $^{238}U$ decays to form $^{226}Ra$, which has a half-life of 1600 years.

Exposure to radionuclides or radiation causes acute health effects that begin with nausea, vomiting and headaches. With increased exposure a person may also experience fatigue, weakness, fever, hair loss, dizziness disorientation, diarrhoea, blood in stool, low blood pressure and ultimately death. Foetuses are particularly vulnerable to the effects of radiation at the cellular level, which can result in smaller head or brain size, poorly formed eyes, abnormal growth and mental retardation (Nussbaum, 2007; Al-Zoughool and Krewski, 2009; Bogutskaya *et al.*, 2011). Studies have revealed that long-term exposure to radionuclides leads to an elevated risk of leukaemia, leucopenia, kidney damage and genetic damage, which can result in lethal health problems that can pass into the next generation (Mohner *et al.*, 2006).

Currently, excavation and shipping to a distant waste disposal location is the most common means of eradicating soil contaminated with radionuclides. The costs of cleaning up these sites are estimated to be in excess of a trillion dollars in the USA and 50 billion pounds sterling in the UK (Lloyd and Renshaw, 2005). Given the high costs of physiochemical approaches, there has been an unprecedented interest in microbes and plants with radionuclides for decontamination of sediments and waters impacted by nuclear waste (Lloyd*et al.*, 2003; Kumar *et al.*, 2007). Figure 1 summarizes the various biotechnological approaches for bioremediation of radionuclides. Bioremediation via microorganisms can be an attractive alternative to excavating contaminated soil. Microorganisms such as *Rhodanobacter* sp. and *Desulfuromus aferrireducens*were observed to be able to interact with these contaminants (Amachi *et al.*, 2010;

Green *et al.*, 2012). The interaction of site-specific microorganisms initiates solubility of transformed radionuclides by addition or removal of electrons, thus increasing the mobility of the contaminant and allowing it to be easily flushed from the environment (Amachi *et al.*, 2010). This microbial-mediated biotransformation presents opportunities for bioremediation of radionuclides in the environment, either to immobilize them in place or to accelerate their removal. This article aims to interpret the procedure by which microorganisms assist in eliminating radionuclides from the environment and how they influence the toxicity and transport of radionuclides.

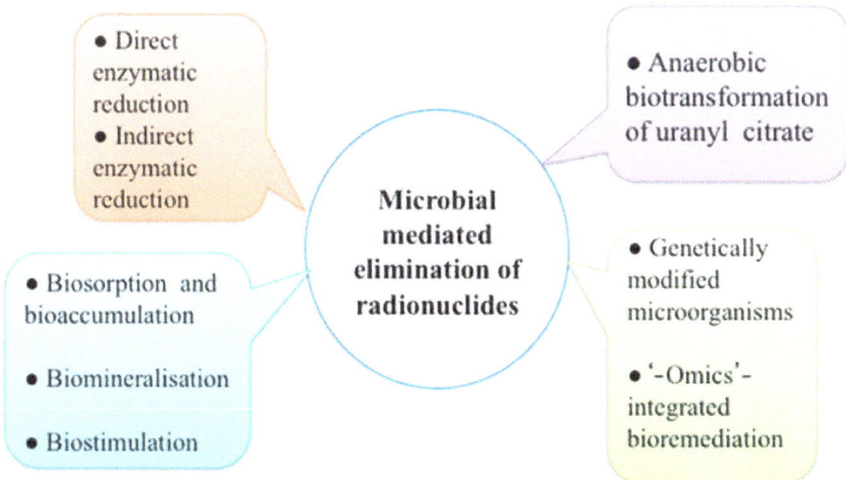


**Figure 1:** Summarization of various biotechnological approaches for bioremediation of radionuclides.

## MICROORGANISMS: AN ASSET IN ELIMINATION OF RADIONUCLIDES

Bioremediation of environmental niches such as soil, sediments and water contaminated with radionuclides can be achieved through biologically encoded changes in the oxidation state. Changes in speciation such as detoxification of mercury by methylation [$Hg(CH_3)_2$] can alter the solubility, transport properties and toxicity of radionuclides (Wang *et al.*, 2012). The bioremedial strategies for radionuclides depend on the active metabolizing capabilities of microorganisms. Radionuclides can be solubilized by direct and indirect enzymatic reduction through oxidation–reduction, change in pH and Eh (activity of electrons), biosorption by mass, biodegradation of radionuclide–organic complexes or biosorption by biomass (Holker *et al.*, 2002; Law *et al.*, 2010; Hegazy and Emam, 2011). Microbial activity during the biotransformation of radionuclides is greatly influenced by electron donors and acceptors, nutrients, and environmental factors. Figure 2 shows the possible mechanistic linkages of metals with microorganism: key interaction for bioremediation.

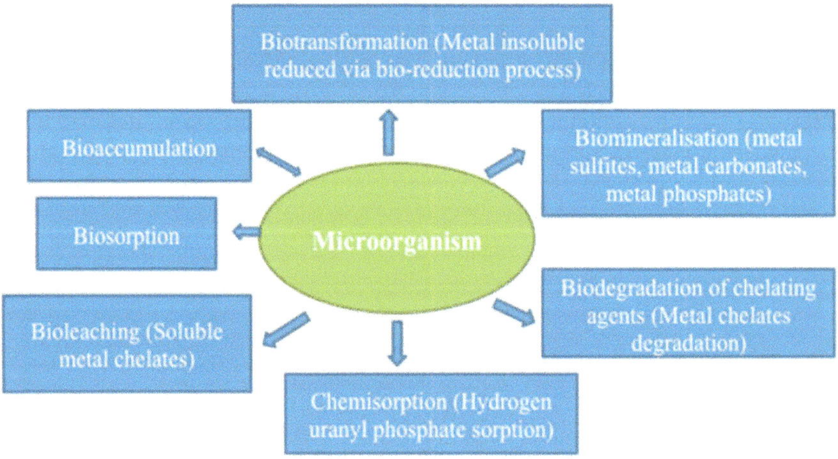


**Figure 2:** Linkage of metals with microorganism: key interaction for bioremediation.

## Direct Enzymatic Reduction of Radionuclides

The oxidized forms of radionuclides are highly soluble in aqueous medium, which makes them mobile in groundwater, whereas the reduced species are highly insoluble and often precipitate from the solution. A direct enzymatic reduction of soluble U(VI) to insoluble species is presented in Fig. 3. Wildung and colleagues (2000) reported enzymatic reduction of U(VI) on the surface of the microorganism *Shewanella putrefaciens*. A *c*-type cytochrome of mass 9.6 kDa was observed in the periplasm of *S. putrefaciens* required for U(VI) reduction. The studies showed that the mutant was unable to synthesize *c*-type cytochrome, so it did an *in vitro* reduction of U(VI). The purified tetrahaem cytochrome *c*3 protein from *Desulfovibrio vulgaris*was observed to reduce U(VI) *in vitro* using U(VI) reductase in combination with hydrogenase as its physiological electron donor (Lovley, 2003; Lovley and Phillips, 1994). *In vivo* studies showed that a homologue of *Desulfovibrio desulfuricans* (G20) confirmed the role of cytochrome *c*3 in hydrogen-dependent U(VI) reduction. Similarly, Lloyd and colleagues (2003) identified a homologous cytochrome (PpcA), a trihaem periplasmic cytochrome *c*7 of the Fe(III)-reducing bacterium *Geobacter sulfurreducens*that may also play a role in U(VI) reduction *in vitro*.

$^{99}$Tc, another long-lived radionuclide (half-life $2.13 \times 10^5$ years) that mostly occurs in nuclear wastes, has attracted considerable interest (Tamponnet and Declerck, 2008; Alliot *et al.*, 2009). Tc(VII) has feeble ligand-complexing capabilities and is difficult to remove from solution using conventional chemical methods. Metal-reducing microorganisms can reduce Tc(VII) and precipitate the radionuclide into low-valency oxide TC(IV). The studies demonstrated reduction of Tc solubility (Henrot, 1989; Pignolet *et al.*, 1989). However, the direct microbial enzymatic reduction of Tc(VII) was first observed by Lloyd and Macaskie (1996) using *S. putrefaciens* and *Geobacter metallireducens*. The reduction of Tc(VII) was quantified using a phosphorimager technique. In another study, X-ray absorption spectroscopy showed Tc(IV) as the final insoluble oxidation state after Tc(VII) was reduced by *G. sulfurreducens* and *S. putrefaciens* (Wildung *et al.*,2000). It is well reported that Tc(VII) reduction has produced fortuitous biochemical side reactions in the organisms studied to date.

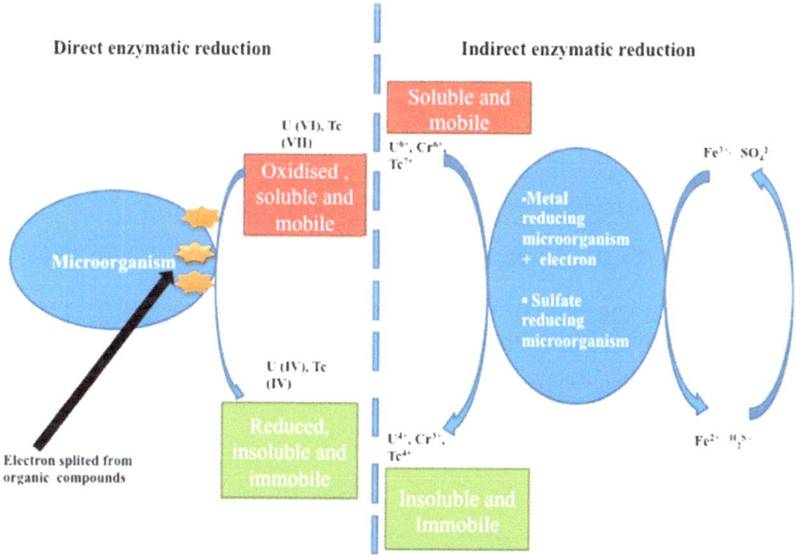


**Figure 3:** Depiction of direct enzymatic reduction and indirect mobilization of radionuclides by metal-reducing microorganisms via capturing of electrons derived by organic compounds (lactate or acetate).

Further, the biochemical basis of Tc(VII) reduction has been studied in *Escherichia coli*. Initial studies demonstrated that anaerobic cultures of *E. coli* reduced Tc(VII) with the reduced radionuclide precipitated within the cell (Lloyd *et al.*, 1997). Results obtained with wild-types and mutants defective in the synthesis of regulatory or electron transfer proteins were used to construct a model for Tc(VII) reduction by *E. coli*. The model demonstrates that three major components of formate hydrogenlyase catalyse the transfer of electrons from dihydrogen to Tc(VII). However, according to the model, the formate dehydrogenase component is required only if formate is supplied as an electron donor for Tc(VII) reduction in place of hydrogen. The model was authenticated by a mutant strain unable to synthesize all three components of hydrogenase to reduce Tc(VII) when either hydrogen or formate was supplied as an electron donor (Lloyd *et al.*, 1997). The identification of three components of formate hydrogenlyase and Tc(VII) reductase of *E. coli* opened up a new avenue to screen for organisms with naturally enhanced ability to act against Tc(VII).

Several other organisms are known to have naturally high formate hydrogenlyase activity. In addition, the uptake of hydrogenase was established to couple the oxidation of formate or hydrogen to Tc(VII) reduction (Lloyd *et al.*, 2001). *Desulfovibrio desulfuricans* and related organisms were observed to utilize formate as an efficient electron donor for Tc(VII) reduction (Lloyd *et al.*, 1999). This observation is consistent with the existence of a rudimentary formate hydrogenlyase (FHL) complex located in the periplasm of microorganisms (Peck, 1993). These studies have also been confirmed by the role of periplasmic Ni-Fe hydrogenase in Tc(VII) reduction by a relative in the N subclass of *Desulfovibrio fructosovorans* (De Luca*et al.*, 2001). Subsequent studies on the development of a bioprocess to decontaminate Tc(VII)-contaminated water have focused on the use of immobilized cells of sulfate-reducing bacteria such as *D. desulfuricans,* which are capable of treating low concentrations of Tc(VII) against a high concentration of nitrate ions, which is frequently found in nuclear waste (Lloyd *et al.*, 1999). Fujimoto and Morita (2006) reported a novel strain of Halomonas (Tc-202), isolated from a marine environment, that was capable of removing Tc(VII) from solid- and aqueous-phase material aerobically.

Although U and Tc remain the highest-priority radionuclide pollutants in most medium- and low-level radioactive wastes, other actinides including Th, Np, Pu and Am also occur at the contaminated sites (Lloyd and Macaskie, 2000; Tamponnet and Declerck, 2008). Iron-reducing bacteria such as *Geobacter* sp. and*Rhodoferax ferrireducens* have the metabolic potential to reduce these radionuclides enzymatically (Kim *et al.*, 2012). These findings are significant, as most tetravalent actinides are acquiescent to bioremediation due to their high ligand-complexing abilities. In addition, these actinides could be immobilized in sediments containing active biomass (Peretrukhin *et al.*, 1996). Although it is possible for Fe(III)-reducing bacteria to reduce and precipitate actinides in one step [e.g. soluble U(VI) to insoluble U(IV)], few studies support the direct formation of an insoluble mineral stage; rather they indicate the formation of a cation prone to bioprecipitation (Lloyd *et al.*, 2000). This phenomenon is illustrated when considering highly soluble Np(V), which was reduced to insoluble Np(IV) by *S. putrefaciens*, then removed as an insoluble phosphate biomineral by a phosphate-liberating *Citrobacter* sp. Lloyd and colleagues (2000) studies have also suggested that the reduction of Pu(IV) to Pu(III) can be achieved by Fe(III)-reducing bacteria and Pu(III) was reported to reoxidize spontaneously (Rusin *et al.*, 1994).

## Indirect Enzymatic Reduction of Radionuclides

The enzymatic bioreduction of radionuclides can be triggered through indirect reduction of soluble contaminants in sedimentary and subsurface environments by metal-reducing or sulfate-reducing microorganisms. One approach is to couple the oxidation of organic compounds or hydrogen to the reduction of iron Fe(III) or sulfur S(IV) in the form of sulfate. The iron Fe(III) can be bioreduced into Fe(II) and S(VI) into S(II) (hydrogen sulfide, $H_2S$). The product can be further reduced chemically to yield separate or multi-component insoluble species (van Hullebusch *et al.*, 2005). The reduced forms of these metals are insoluble and can precipitate as either reduced oxide or hydroxide minerals. Similarly, sulfate-reducing bacteria may be stimulated to produce hydrogen sulfide. Several microorganisms such as *Microbacterium flavescens* grown in the presence of other radionuclides (i.e. U, Th, Am and Pu) produced unidentified compounds such as organic acids, siderophores and extracellular metabolites capable of dissolving and mobilizing radionuclides into the soil. These compounds may also help to transport radionuclides within the cells (John *et al.*, 2001). Figure 3 depicts the direct enzymatic reduction and indirect mobilization of radionuclides by metal-reducing microorganisms via capturing of electrons derived by organic compounds (lactate or acetate).

Pu(IV), Th(IV), U(VI) and Fe(III) have chemical and biochemical similarities because of the ubiquity of siderophore-producing microbes. The iron-sequestering agents produced by such microorganisms are crucial to increase the solubility and bioavailability of radionuclides. Premuzic and colleagues (1985) reported the production of extracellular chelating agents in *Pseudomonas aeruginosa* that can bioaccumulate uranium. This report prompted several chelating agents for thorium and uranium when grown with these metals. Brainard and colleagues (1992) solubilized hydrous $PuO_2(s)$ using the siderophores enterobactin, desferrioxamin, carboxylate amino polycarboxylate and catecholate ligands. The report conclusively showed that enterobactin siderophores are extremely effective in solubilizing actinide oxides of plutonium, among the tested other chelators.

Several microorganisms produced extracellular complexes in the presence of Pu and Th that increased the concentration of Pu and Th in soil-column elutes compared with controls. The increased mobility of Pu and Th in soil resulted from the formation of neutral and negatively charged Pu and Th complexes. In the presence of known microbial metabolites such as synthetic ligand [ethylenediamine tetraacetic acid (EDTA)] and citric acid, Pu(VI) and Th(IV) were reduced to Pu(IV) and Th(IV) respectively before complex formation, suggesting that the latter valence state would be the dominant one associated with the organic complexes in soil (Panak and Nitsche, 2001).

## BIOSORPTION AND BIOACCUMULATION

Biosorption is the sequestration of positively charged metal ions to the negatively charged cell membranes and polysaccharides secreted on the outer surfaces of bacteria through slime and capsule formation. Sorption of metals to intact cells is directed by a multiplicity of mechanisms and interactions that are not yet fully understood. Langley and Beveridge (1999) described the role of carboxyls in the binding of metal cations to*O*-side-chains of lipopolysaccharide (LPS) and concluded that metal was most likely to bind to phosphoryl groups in the core-lipid 'A' of LPS, and the negatively charged side-chains affected cell hydrophobicity in Gram-negative bacteria. Similarly, Khani and colleagues (2005) described the effective adsorption of radionuclide U(VI) by the brown marine alga *Cystoseira indica* and observed that the pre-treatment of the alga with calcium could enhance the adsorption efficiency of several radionuclides. Several microorganisms including *Citrobacter freudii* and *Firmicutes* have been reported as radionuclide biosorbents (Haferburg *et al.*, 2007; N'Guessan *et al.*, 2008; Xie *et al.*, 2008).

Sorption of metals plays a key role in microbe-metal interactions. The precipitation of uranyl phosphate by*Citrobacter* is initiated by interaction (electrostatic) with phosphate groups in the LPS. This interaction provides nucleation sites for mineral formation and protects an outer membrane phosphatase (Macaskie *et al.*,2000). Panak and colleagues (2002) examined the sorption of Pu(VI) with bacterial resting cells and concluded that the interaction with bacteria caused changes in the oxidation state of Pu(V) due to endogenous respiration. Extended X-ray Absorption Fine Structure (EXAFS) analysis of Pu associated with the cells showed that Pu(VI) was primarily bound to the phosphate groups on the cell surface; the biosorption efficiency looked to be positively related to temperature and could occur within hours. This process is substantially faster than direct bioreduction. However, radionuclide-contaminated ground is normally poor in biomass concentration due to the higher toxicity of radionuclides. Therefore, biosorption alone may not be sufficient to bioremediate radionuclides unless the ground biomass content is increased. To improve the efficiency of biosorption, elegant strategies such as recombinant DNA technology and stimulated growth of microbes in the contaminated ground could enhance radionuclide remediation.

## Biostimulation

Biostimulation using specific communities of microorganisms is another mechanism to enhance the bioremediation of radionuclides. An *in situ* remediation method was developed to reduce the bioavailability and prevent the further spread of uranium in groundwater by promoting the activity of dissimilatory sulfate- and iron-reducing organisms (Vrionis *et al.*, 2005). In biostimulation, nitrate serves as an energetically favourable electron acceptor for metal-reducing bacteria in nitric acid co-contaminated sediments (DiChristina, 1992). A lack of microbial reduction in U(VI) has been reported if the sediment is co-contaminated with nitrate (Finneran *et al.*, 2002). Sakadevan and colleagues (1999) reported that nitrate reduction was inhibited due to the presence of several heavy metals. Therefore, an attempt was made to resolve this issue by the *ex situ* treatment and removal of heavy metals and nitrate prior to *in situ*biostimulation to reduce the U(VI) (Wu *et al.*, 2006).

At sites co-contaminated with metallic ions beyond toxic levels, the microbial resistance of endogenous microbial populations is critical for *in situ* biostimulation. A number of microbes has been shown to carry out reductive precipitation of radionuclides (e.g. *Desulfovibrio* sp., *Geobacter* sp. and *Shewanella* sp.); however, the resistance of these organisms to non-reducible heavy metals could possibly limit their *in situ* activity. Martinez and colleagues (2006) reported the presence of heavy-metal-resistance genes within the endogenous microbial community at the Oak Ridge Field Research Center (FRC) site. In the presence of heavy metals, the potential for biostimulation was improved by using ethanol for bioreduction of nitrate, followed by the successful reduction and *in situ* immobilization of U(VI) (Istok *et al.*, 2004; Nyman *et al.*, 2006). Therefore, the addition of a suitable carbon source was recommended to stimulate radionuclide bioreduction at co-contaminated sites.

Another novel method was developed to bioremediate areas contaminated with high concentrations of uranium ($>1000\,\mu m$) at low pH and high nitric acid concentrations (Wu *et al.*, 2006). The method was based on raising pH with the addition of carbon (ethanol) to stimulate the growth of denitrifying and radionuclide-reducing bacteria, using a hydraulic recirculation system composed of an outer and inner loop. An *ex situ*system was also developed to prevent clogging of the hydraulic recirculation system with nitrogen gas, aluminium, calcium precipitation and biomass due to denitrification. After 1 year at optimized conditions, the concentrations of U(VI) and nitrate had been reduced to 5 µm and 0.5 µm respectively. After 2 years of preconditioning, the level of U(VI) was reduced to 0.126 µm – below the EPA›s recommended level. These studies suggested that alternation of geochemical and hydrological parameters can effectively biostimulate radionuclide-reducing bacteria *in situ*. Remediation of high U(VI) concentrations with ethanol-biostimulated biofilm is an alternative approach that can effectively reduce 87% of U(VI) (Marsili *et al.*, 2007).

## BIOMINERALIZATION OF RADIONUCLIDES

Microorganisms can interact with metal ions and immobilize them to transformation. Some microorganisms generate biofilms to bind significant quantities of metallic ions, which can serve as a platform for the precipitation of insoluble minerals. Microorganism *Citrobacter* sp. was reported to be able to produce deposits of metal phosphate enzymatically. Polycrystalline $NaUO_2Po_4$ accumulates in and around the cell wall of *Citrobacter* by sorption to LPS and the activity of an outer membrane acid-phosphatase (Keasling *et al.*, 2000). The mineral formation is driven by two gradients in the outer membranes: an incoming $UO_4$ and an outgoing $PO_4$, resulting in total removal of U from the solution and binding of 1 mg $NaUO_2Po_4$ per mg of the cell. Keasling and colleagues (2000) prepared a recombinant version of this mechanism by cloning the gene encoding polyphosphate kinase in *P. aeruginosa,* resulting in the precipitation of a complex containing both phosphorous and uranium on the cell surface.

### Biomineralization via Microbial-Generated Ligands

Chelating agents are present in wastes because they are widely used for the decontamination of nuclear reactors and equipment, cleanup operations, and the separation of radionuclides. Several organic agents, such as citric acid, hydroxyl-acetic acid, oxalic acid, tartaric acid, EDTA, diethylenetriamine pentaacetic acid (DTPA), nitrilotriacetic acid (NTA) and *N*-hydroxyethylenediamine triacetic acid (HEDTA) have been used to complex with radionuclides. These metal chelates undergo aerobic or anaerobic biodegradation and cause the precipitation of released ions as water-insoluble hydroxides or oxides, thus retarding their migration into groundwater. Citrate has been found useful as a chelating agent in decontamination because it forms highly soluble metal–citrate complexes that can be degraded by microorganisms, resulting in subsequent re-precipitation of metals. Stable complexes such as bidentate, tridentate and polynuclear complexes can be formed with citrate and radionuclides. Francis (1998) reported that bioidentate uranium complexed with citric acid was readily biodegraded, whereas tridentate was recalcitrant.

### Anaerobic and Aerobic Biotransformation of Uranyl–Citrate

Sulfate-reducing *D. desulfuricans* and facultative iron-reducing *Schwanella alga* reduced U(VI) complexes with citrate to U(IV) anaerobically, but little uranium was precipitated (Ganesh *et al.*, 1997). Similarly,*Clostridium* sp. (ATCC 53464) metabolized glucose but not citrate and reduced U(VI)–citrate into U(IV)–citrate only in the presence of glucose. Another *Clostridium sphenoides* (ATCC 19403) metabolized bidentate Fe(III)–citrate complex. This bacterium reduced Fe(III) to Fe(II) and concomitantly metabolized citric acid. In contrast, U(VI)–citrate was reduced to U(IV)–citrate by the bacterium in the presence of electron donor glucose or uncomplex citric acid. Therefore, these results indicate that the complex of uranium with citric acid is readily available for organisms as an electron acceptor, despite their incapability to metabolize organic ligands of radionuclides.

Some organisms metabolize citric acid aerobically using the enzymes aconitase and citrate lyase. The enzyme aconitase isomerizes citrate into isocitrate in the TCA cycle. However, citrate lyase drives the anaerobic metabolism of citric acid. Francis (1998) tested the ability of *Acinetobacter, Citrobacter* and *Pseudomonas*from low-level radioactive waste sites to metabolize citric acid into a uranyl–citrate complex. These cultures failed to metabolize a stable binuclear complex of uranyl–citrate, possibly because the uranyl–citrate complex did not transport inside the cells, as revealed by $^{14}C$ labelling analysis. Furthermore, a cell-free extract of the tested organism showed that the binuclear U–citrate complex was completely degraded. Francis and colleagues (2002) tested the speciation of uranium and citric acid as a function of pH and concluded that the amount of U–citrate complex decreased rapidly above pH 6.0, whereas free citric acid was not biodegraded and transported.

Similarly, citric acid can form ternary mixed-metal complexes with a number of metals, and the complex formed affects the biodegradation and mobility of metal–citrate complexes (Dodge *et al.*, 2002). For example, the biotransformation of Fe–U–citrate complex was recalcitrant. When onefold excess citric acid was added to the 1:1:2 Fe–U–citrate complex, the excess citric acid was completely degraded. However, with twofold excess citric acid, 1:1:1 Fe–U-citric acid remained in solution after the excess citric acid was biodegraded. Therefore, this study suggested that Fe–U–mixed-metal citric acid complexes resist biodegradation and may persist in the environment, which is a limitation of current technology and a challenge to the microbial-mediated bioremediation of radionuclides (Dodge *et al.*, 2002).

## Genetically Modified Microorganisms

Genetic engineering (GE) and recombinant DNA technology have been employed to generate character-specific microorganisms for efficient removal of metal by sorption. Different protein constructs have been generated in which the bacterial cell surface is equipped with metal-binding polypeptides by fusion-binding domains to outer-membrane-anchored proteins that include metallothioneins (Valls *et al.*, 2000a), randomly generated polypeptides (Schembri *et al.*, 1999), polyhistidines and synthetic phytochelatines (Schembri *et al.*,1999; Bae *et al.*, 2000). These protein constructs proved an increase in metal binding. The metallothioneins were also tested in microcosm field study (Valls *et al.*, 2000b). Another approach attempted to enhance the metal accumulation by combining the specific metal transporter with MTs in the cytoplasm (Wolfram and Bauerfeind, 2009). A recombinant strain of *E. coli* was generated with five times the sorption ability for U(VI) radionuclides by altering the transporter genes *nixA* (*Helicobacter pylori*) and *merTP* (*Serratia marcescens*) respectively (Beckwith *et al.*, 2001). Therefore, the expression of both metal transporter proteins and metal-binding peptides may enhance a strain's ability to accumulate metal ions.

The microorganism *Deinococcus radiodurans* has been studied to detoxify Cr(VI), U(VI) and Tc(VII) from soil (Fredrickson *et al.*, 2000). A genetically engineered *D. radiodurans* strain was created by cloning the *E. coli* gene (*merA*) that provides the ability to utilize carbon and energy from catabolism of toluene and mercury (radioactive contaminants) (Brim *et al.*, 2000). Progress has been made in constructing strains of *D. radiodurans* for radionuclide remediation; however, an *in situ* bioremediation strategy has yet to be discovered. The microorganisms *Deinococcus murrayi* and *Deinococcus geothermalis* were characterized as growing at a higher temperature range (55°C), and showing remarkable resistance against chronic irradiation (50 $Gy h^{-1}$) (Brim *et al.*, 2003).

The microbial family *Geobacteriaceae* has shown potential for radioactive metal reduction (Lloyd *et al.*,2003). The gene *dcuB* from *G. sulfurreducens,* which encodes a fumarate transporter, was engineered in *G. metallireducens* to grow with fumarate as a terminal electron acceptor, revealing the ruling approach of GE with expanded respiratory capabilities (Butler *et al.*, 2006). Undoubtedly, GE microbes show promise, but their implementation for *in situ* bioremediation will require additional steps to develop safe routes of environmental cleanup.

## '-OMICS'-IMPLEMENTED RADIONUCLIDE BIOREMEDIATION

The genome of an organism comprises its entire set of hereditary information that is converted to mRNA (the transcriptome) for protein translation. The proteome of an organism is the entire set of proteins, including enzymes, that are expressed in the organism under specific environmental conditions. In order to identify genes, proteins and enzymes involved in the bioremediation of radionuclides, it is important to study the structural and functional interactions between proteins and other metabolites. Potential genes and proteins involved in the metabolism of radionuclides can be identified and studied via advanced genomics and proteomics techniques (Nagaraj and Singh, 2010; Singh *et al.*, 2011). Recent advances in next-generation sequencing, genomics and proteomics allow the expression of required proteins and enzymes of interest into radionuclide-resistant organisms for bioremediation. Further, genome-wide transcriptome analysis can provide us with a better understanding of the metabolic pathways and the physiology of the microorganisms.

The genome sequences of many microorganisms are now available and can be used for genome organization, including comparisons through microarrays (Ishii *et al.*, 2007). Among several genome-wide studies of genes and proteins involved in radionuclide reduction pathways, Methe and colleagues (2003) reported that *G. sulfurreducens* had more than 100 *c*-type cytochrome gene coding regions in its genome and many of its translated proteins were involved with radionuclide reduction pathways. DNA-microarray-mediated analysis revealed 121 genes found to be upregulated in *Shewanella oneidensis* during U(VI) reduction, compared with Cr(VI) reduction, where only 83 genes were overexpressed (Bencheikh-Latmani *et al.*, 2005). In a comparative genomics study, the organism *Thermococcus gammatolerans* was observed to be radioresistant among *Archaea*, expressing thioredoxin reductase (tgo180), a glutaredoxin-like protein (tg1302) and two peroxiredoxins (tg1253 and tg1220), which allowed the organism to cope with the stress of radionuclides (Zivanovic *et al.*, 2009). It was found that the expression of the *NiCoT* gene in *Rhodopseudomonas palustris*CGA009 and *Novosphingobium aromaticivorans* F-199 was highly increased when the organisms were grown in the presence of radioactive cobalt (Raghu *et al.*, 2008).

In addition to genomics technology, proteomics technology has been proven effective in studying proteins involved in radionuclide bioremediation. Tian and colleagues (2010) found a total of 552 differentially regulated proteins, including a cytochrome *bd* ubiquinol oxidase, on the membrane of *D. geothermalis* that were involved in the radioresistance of the organism. Many of these proteins comprised function groups including nutrient transport and metabolism, energy production and conversion, and cell wall/membrane biogenesis (Tian *et al.*, 2010). In another study, cascades of enzymatic proteins in the outer membrane of *S. oneidensis* were found to effectively reduce the levels of radionuclides such as uranium and chromium (Marshall *et al.*, 2006). In a well-known radioresistant bacterium, *D. radiodurans*, 2DE and MALDI-TOF MS revealed that 31 radiation-responsive proteins were significantly upregulated, including RecA and PprA, which are well known for DNA replication and repair (Lu *et al.*, 2009). Other proteins included those involved in the stress response, energy metabolism, transcriptional regulation, protein turnover and chaperoning (Lu *et al.*, 2009). Understanding specific genes and their protein products is essential to understand the pathways involved in the bioremediation of radionuclides. Combining '-omics'-based approaches may assist in the identification of specific microbes effective for *in situ* bioremediation of radionuclides. Table 1 summarizes the major transcripts and proteins investigated in various radionuclides remediation studies.

**Table 1:** Major transcripts and their sources for radionuclides remediation

| Radionuclides | Organisms | Gene/protein | Function | Reference |
|---|---|---|---|---|
| Uranium | *Deinococcus radiodurans* | DrPhoN | Surface-associated precipitation of uranium (5.7 g uranium $g^{-1}$ biomass) | Misra *et al.* (2012) |
| Cobalt | *Rhodopseudomonas palustris* CGA0009 | NiCoT | 85% removal of Cobalt was achieved in a two-cycle treatment with recombinant *E. coli* | Raghu *et al.* (2008) |
| Cobalt | *Novosphingobium aromaticivorans* F-199 | NiCoT | | Raghu *et al.* (2008) |
| Uranium | *Sphingomonas* sp. BSAR-1 | phoK | Recombinant *E. coli* precipitated > 90% of 0.5–5 mM of uranyl carbonate in less than 2 h | Nilgiriwala *et al.* (2008) |
| Chromate, uranyl | *Escherichia coli* | ChrR6 | Expressed chromate reductase activity in addition to convert soluble U(VI) to insoluble with U(IV) with Vmax of 8,812 ± 611 | Barak *et al.* (2006) |
| Uranium | *Salmonella enterica serovar Typhi* | phoN | Recombinant strain precipitated over 90% of the uranium from a 0.8 mM uranyl nitrate solution in 6 h | Appukuttan *et al.* (2006) |
| Mercury | *Escherichia coli* BL308 | *merA* | Biotransformation of Hg (II) to the less toxic volatile Hg (I) | Brim *et al.* (2003) |
| Cadminum, zinc, cobalt | *Ralstonia eutrophus* CH34 | *czc* | | Diels *et al.* (1995) |
| Fumarate, nitrate, dimethyl sulfoxide, trimethylamine *N*-oxide (TMAO), nitrite and insoluble iron and manganese oxides | *Shewanella oneidensis* | *fccA* | Periplasmic flavocytochrome *c* fumarate reductase | Saffarini *et al.* (2003) |
| Uranium | *Desulfovibrio vulgaris* | *ctyc3* | Reduce uranium(VI) to uranium(IV) with hydrogen as the electron donor | Payne *et al.* (2002) |
| Fe(III) | *Geobacter sulfurreducens* | PpcA | Restoration of Fe(III) reduction with acetate | Lloyd *et al.* (2003) |

## Challenges, Limitations and 5-Year View

Despite the progress that has been made in the field of radionuclide bioremediation using microorganisms, many challenges lie ahead. A basic concern is to optimize the conditions and procedures for sustained and effective bioremediation in the presence of competing anions, toxic metals, organic compounds and chelating agents. Another aspect that is of interest is the reoxidation and remobilization of reduced radionuclides by microbial metabolism and abiotic mechanisms. Careful consideration needs to be given to *in situ*biostimulation or bioaugmentation where various additives (microbes or chemical ingredients) are employed to enhance microbial activity; such additives might be disruptive to natural microbiota. It is still an issue to enhance the metabolic activity by maintaining the required growth conditions, such as pH, temperature, and levels of contaminants and nutrients, including diverse chemical parameters of selective microbial populations for *in situ* bioremediation. In addition, it is necessary to recognize the heterogeneous nature of contaminants on sites that can lead to an uneven flow of the liquid or gas containing the microbes. Due to the slow speed of achieving acceptable levels of decontamination at diverse sites, it is difficult to predict the performance of bench-scale bioremediation in field operations.

Genetically modified microorganisms have limitations in terms of liability and ethical issues from regulatory agencies such as the United States Environmental Protection Agency (USEPA). The release of GE organisms is a tedious task that requires site dependence compatibility and comparability with other organisms on site. After remediation, removal of GE microorganisms is also not an easy task. The direct immobilization of enzymes on support material could be useful for bioremediation, but it requires abundant optimization of site-specific substrates.

Recent developments in '-omics'-based technologies, i.e. genomics, transcriptomics and proteomics, are set to revolutionize many aspects of the biological sciences. A whole-genome transcription profile of *E. coli* has already uncovered potential strategies for metal detoxification (Brocklehurst and Morby, 2000). Because of the current availability of complete genome sequences of many radioresistant microorganisms, this approach will prove useful for determining the precise mechanisms of biologically relevant radionuclide–microbe interactions. However, these discoveries in environmental science are still in their infancy stage and have yet to prove their usefulness in environmental cleanup. In the future, coordinated multidisciplinary efforts need to be made to understand the overall process by which microorganisms degrade radionuclides as well as to implement the conditions that most effectively degrade the radionuclides. Focused research exploring the necessary genes and proteins of the microbial metabolic pathway towards cell-free bioremediation will pave the way to refine these bioremediation technologies before successful field practice.

## CONCLUSION

Microbial transformations of radionuclides, heavy metals and minerals are a vital part of natural biosphere processes and can have beneficial consequences for the human community. As we know, the interactions between microorganisms and radionuclides are far from simple, and it is not easy to understand the wide range of environments these organisms inhabit. Study of the molecular mechanisms behind the microbial transformation of radionuclides using '-omics'-based approaches and exploiting them in applications such as bioremediation would assist in tracking the responsible microbial metabolic products towards cell-free bioremediation and further assist in efficient removal of radionuclides from the environment.

## REFERENCES


1. Ahier BA, Tracy BL. Radionuclides in the Great Lakes basin. Environ Health Perspect.1995;103(Suppl. 9):89–101.
2. Alliot I, Alliot C, Vitorge P, Fattahi M. Speciation of technetium(IV) in bicarbonate media. Environ Sci Technol. 2009;43:9174–9182.
3. Al-Zoughool M, Krewski D. Health effects of radon: a review of the literature. Int J Radiat Biol.2009;85:57–69.
4. Amachi S, Minami K, Miyasaka I, Fukunaga S. Ability of anaerobic microorganisms to associate with iodine: 125I tracer experiments using laboratory strains and enriched microbial communities from subsurface formation water. Chemosphere. 2010;79:349–354.
5. Appukuttan D, Rao AS, Apte SK. Engineering of *Deinococcus radiodurans* R1 for bioprecipitation of uranium from dilute nuclear waste. Appl Environ Microbiol. 2006;72:7873–7878.
6. Bae W, Chen W, Mulchandani A, Mehra RK. Enhanced bioaccumulation of heavy metals by bacterial cells displaying synthetic phytochelatins. Biotechnol Bioeng. 2000;70:518–524.
7. Barak Y, Ackerley DF, Dodge CJ, Banwari L, Alex C, Francis AJ, Matin A. Analysis of novel soluble chromate and uranyl reductases and generation of an improved enzyme by directed evolution. Appl Environ Microbiol. 2006;72:7074–7082.
8. Beckwith CS, McGee DJ, Mobley HL, Riley LK. Cloning, expression, and catalytic activity of*Helicobacter hepaticus* urease. Infect Immun. 2001;69:5914–5920.
9. Bencheikh-Latmani R, Williams SM, Haucke L, Criddle CS, Wu L, Zhou J, Tebo BM. Global transcriptional profiling of *Shewanella oneidensis* MR-1 during Cr(VI) and U(VI) reduction. Appl Environ Microbiol. 2005;71:7453–7460.
10. Bogutskaya NG, Zuykov MA, Naseka AM, Anderson EB. Normal axial skeleton structure in common roach *Rutilus rutilus* (Actinopterygii: Cyprinidae) and malformations due to radiation contamination in the area of the Mayak (Chelyabinsk Province, Russia) nuclear plant. J Fish Biol.2011;79:991–1016.
11. Brainard JR, Strietelmeier BA, Smith PH, Langston-Unkefer PJ. Actinide binding and solubilization by microbial siderophores. Radiochim Acta. 1992;58–59:357–363.
12. Brim H, McFarlan SC, Fredrickson JK, Minton KW, Zhai M, Wackett LP, Daly

MJ. Engineering*Deinococcus radiodurans* for metal remediation in radioactive mixed waste environments. Nat Biotechnol. 2000;18:85–90.

13. Brim H, Venkateswaran A, Kostandarithes HM, Fredrickson JK, Daly MJ. Engineering *Deinococcus geothermalis* for bioremediation of high-temperature radioactive waste environments. Appl Environ Microbiol. 2003;69:4575–4582.
14. Brocklehurst KR, Morby AP. Metal-ion tolerance in *Escherichia coli*: analysis of transcriptional profiles by gene-array technology. Microbiology. 2000;146(Part 9):2277–2282.
15. Butler JE, Glaven RH, Esteve-Nunez A, Nunez C, Shelobolina ES, Bond DR, Lovley DR. Genetic characterization of a single bifunctional enzyme for fumarate reduction and succinate oxidation in*Geobacter sulfurreducens* and engineering of fumarate reduction in *Geobacter metallireducens*. J Bacteriol. 2006;188:450–455.
16. De Luca G, de Philip P, Dermoun Z, Rousset M, Vermeglio A. Reduction of technetium(VII) by*Desulfovibrio fructosovorans* is mediated by the nickel-iron hydrogenase. Appl Environ Microbiol.2001;67:4583–4587.
17. DiChristina TJ. Effects of nitrate and nitrite on dissimilatory iron reduction by *Shewanella putrefaciens*200. J Bacteriol. 1992;174:1891–1896.
18. Diels L, Dong Q, van der Lelie D, Baeyens W, Mergeay M. The *czc* operon of *Alcaligenes eutrophus*CH34: from resistance mechanism to the removal of heavy metals. J Ind Microbiol. 1995;14:142–153.
19. Dodge CJ, Francis AJ, Gillow JB, Halada GP, Eng C, Clayton CR. Association of uranium with iron oxides typically formed on corroding steel surfaces. Environ Sci Technol. 2002;36:3504–3511.
20. Finneran KT, Housewright ME, Lovley DR. Multiple influences of nitrate on uranium solubility during bioremediation of uranium-contaminated subsurface sediments. Environ Microbiol.2002;4:510–516.
21. Francis AJ. Biotransformation of uranium and other actinides in radioactive wastes. J Alloys Comp.1998;271–273:78–84.
22. Francis AJ, Joshi-Tope GA, Dodge CJ, Gillow JB. Biotransformation of uranium and transition metal citrate complexes by *Clostridia*. J Nucl Sci Technol. 2002;3:935–938.
23. Fredrickson JK, Kostandarithes HM, Li SW, Plymale AE, Daly MJ. Reduction of Fe(III), Cr(VI), U(VI), and Tc(VII) by *Deinococcus radiodurans* R1. Appl Environ Microbiol. 2000;66:2006–2011.
24. Fredrickson JK, Zachara JM, Balkwill DL, Kennedy D, Li SM, Kostandarithes HM, et al. Geomicrobiology of high-level nuclear waste-contaminated vadose sediments at the Hanford site, Washington state. Appl Environ Microbiol. 2004;70:4230–4241.
25. Fujimoto K, Morita T. Aerobic removal of technetium by a marine *Halomonas* strain. Appl Environ Microbiol. 2006;72:7922–7924.
26. Ganesh R, Robinson KG, Reed GD, Sayler GS. Reduction of hexavalent uranium

from organic complexes by sulfate- and iron-reducing bacteria. Appl Environ Microbiol. 1997;63:4385–4391.

27. Green SJ, Prakash O, Jasrotia P, Overholt WA, Cardenas E, Hubbard D, et al. Denitrifying bacteria from the genus *Rhodanobacter* dominate bacterial communities in the highly contaminated subsurface of a nuclear legacy waste site. Appl Environ Microbiol. 2012;78:1039–1047.

28. Haferburg G, Merten D, Buchel G, Kothe E. Biosorption of metal and salt tolerant microbial isolates from a former uranium mining area. Their impact on changes in rare earth element patterns in acid mine drainage. J Basic Microbiol. 2007;47:474–484.

29. Hegazy AK, Emam MH. Accumulation and soil-to-plant transfer of radionuclides in the Nile Delta coastal black sand habitats. Int J Phytoremediation. 2011;13:140–155.

30. Henrot J. Bioaccumulation and chemical modification of Tc by soil bacteria. Health Phys.1989;57:239–245.

31. Holker U, Schmiers H, Grosse S, Winkelhofer M, Polsakiewicz M, Ludwig S, et al. Solubilization of low-rank coal by *Trichoderma atroviride*: evidence for the involvement of hydrolytic and oxidative enzymes by using 14C-labelled lignite. J Ind Microbiol Biotechnol. 2002;28:207–212.

32. van Hullebusch ED, Peerbolte A, Zandvoort MH, Lens PN. Sorption of cobalt and nickel on anaerobic granular sludges: isotherms and sequential extraction. Chemosphere. 2005;58:493–505.

33. Ishii N, Nakahigashi K, Baba T, Robert M, Soga T, Kanai A, et al. Multiple high-throughput analyses monitor the response of *E. coli* to perturbations. Science. 2007;316:593–597.

34. Istok JD, Senko JM, Krumholz LR, Watson D, Bogle MA, Peacock A, et al. *In situ* bioreduction of technetium and uranium in a nitrate-contaminated aquifer. Environ Sci Technol. 2004;38:468–475.

35. John SG, Ruggiero CE, Hersman LE, Tung CS, Neu MP. Siderophore mediated plutonium accumulation by *Microbacterium flavescens* (JG-9) Environ Sci Technol. 2001;35:2942–2948.

36. Keasling JD, Van Dien SJ, Trelstad P, Renninger N, McMahon K. Application of polyphosphate metabolism to environmental and biotechnological problems. Biochemistry (Mosc) 2000;65:324–331.

37. Khani M, Keshtkar A, Meysami B, Zarea M, Jalali R. Biosorption of uranium from aqueous solutions by nonliving biomass of marine algae *Cystoseiraical* heterogeneity in an in situ uranium bioremediation field site. Appl Environ Microbiol. 2005;71:6308–6318.

38. Kim SJ, Koh DC, Park SJ, Cha IT, Park JW, Na JH, et al. Molecular analysis of spatial variation of iron-reducing bacteria in riverine alluvial aquifers of the Mankyeong River. J Microbiol. 2012;50:207–217.

39. Kumar R, Singh S, Singh OV. Bioremediation of radionuclides: emerging

technologies. OMICS.2007;11:295–304.

40. Kurnaz A, Kucukomeroglu B, Keser R, Okumusoglu NT, Korkmaz F, Karahan G, Cevik U. Determination of radioactivity levels and hazards of soil and sediment samples in Firtina Valley (Rize, Turkey) Appl Radiat Isot. 2007;65:1281–1289.

41. Langley S, Beveridge TJ. Effect of O-side-chain-lipopolysaccharide chemistry on metal binding. Appl Environ Microbiol. 1999;65:489–498.

42. Law GT, Geissler A, Lloyd JR, Livens FR, Boothman C, Begg JD, et al. Geomicrobiological redox cycling of the transuranic element neptunium. Environ Sci Technol. 2010;44:8924–8929.

43. Lloyd JR, Macaskie LE. A novel PhosphorImager-based technique for monitoring the microbial reduction of technetium. Appl Environ Microbiol. 1996;62:578–582.

44. Lloyd JR, Macaskie LE. Bioremediation of radionuclide-containing waste waters. In: Lovley DR, editor. Environmental Microbe–Metal Interactions. Washington, DC, USA: ASM Press; 2000. pp. 277–327.

45. Lloyd JR, Renshaw JC. Microbial transformations of radionuclides: fundamental mechanisms and biogeochemical implications. Met Ions Biol Syst. 2005;44:205–240.

46. Lloyd JR, Cole JA, Macaskie LE. Reduction and removal of heptavalent technetium from solution by*Escherichia coli*. J Bacteriol. 1997;179:2014–2021.

47. Lloyd JR, Ridley J, Khizniak T, Lyalikova NN, Macaskie LE. Reduction of technetium by*Desulfovibrio desulfuricans*: biocatalyst characterization and use in a flowthrough bioreactor. Appl Environ Microbiol. 1999;65:2691–2696.

48. Lloyd JR, Sole VA, Van Praagh CV, Lovley DR. Direct and Fe(II)-mediated reduction of technetium by Fe(III)-reducing bacteria. Appl Environ Microbiol. 2000;66:3743–3749.

49. Lloyd JR, Mabbett A, Williams DR, Macaskie LE. Metal reduction by sulfate-reducing bacteria: physiological diversity and metal specificity. Hydrometallurgy. 2001;59:327–337.

50. Lloyd JR, Leang C, Hodges Myerson AL, Coppi MV, Cuifo S, Methe B, et al. Biochemical and genetic characterization of PpcA, a periplasmic c-type cytochrome in *Geobacter sulfurreducens*.Biochem J. 369:153–161.

51. Lovley DR. Dissimilatory Fe(III) and Mn(IV) reduction 1991. Microbiol Rev. 2003;55:259–287.

52. Lovley DR, Phillips EJ. Reduction of chromate by *Desulfovibrio vulgaris* and Its c(3) cytochrome.Appl Environ Microbiol. 1994;60:726–728.

53. Lu H, Gao G, Xu G, Fan L, Yin L, Shen B, Hua Y. *Deinococcus radiodurans* PprI switches on DNA damage response and cellular survival networks after radiation damage. Mol Cell Proteomics.2009;8:481–494.

54. Macaskie LE, Bonthrone KM, Yong P, Goddard DT. Enzymically mediated bioprecipitation of uranium by a *Citrobacter* sp.: a concerted role for exocellular lipopolysaccharide and associated phosphatase in biomineral

formation. Microbiology. 2000;146(Part 8):1855–1867.

55. Marshall MJ, Beliaev AS, Dohnalkova AC, Kennedy DW, Shi L, Wang Z, et al. c-Type cytochrome-dependent formation of U(IV) nanoparticles by *Shewanella oneidensis*. PLoS Biol. 2006;4:e268.

56. Marsili E, Beyenal H, Di Palma L, Merli C, Dohnalkova A, Amonette JE, Lewandowski Z. Uranium immobilization by sulfate-reducing biofilms grown on hematite, dolomite, and calcite. Environ Sci Technol. 2007;41:8349–8354.

57. Martinez RJ, Wang Y, Raimondo MA, Coombs JM, Barkay T, Sobecky PA. Horizontal gene transfer of PIB-type ATPases among bacteria isolated from radionuclide- and metal-contaminated subsurface soils. Appl Environ Microbiol. 2006;72:3111–3118.

58. Methe BA, Nelson KE, Eisen JA, Paulsen IT, Nelson W, Heidelberg JF, et al. Genome of *Geobacter sulfurreducens*: metal reduction in subsurface environments. Science. 2003;302:1967–1969.

59. Misra CS, Appukuttan D, Kantamreddi VS, Rao AS, Apte SK. Recombinant *D. radiodurans* cells for bioremediation of heavy metals from acidic/neutral aqueous wastes. Bioeng Bugs. 2012;3:44–48.

60. Mohner M, Lindtner M, Otten H, Gille HG. Leukemia and exposure to ionizing radiation among German uranium miners. Am J Ind Med. 2006;49:238–248.

61. N›Guessan AL, Vrionis HA, Resch CT, Long PE, Lovley DR. Sustained removal of uranium from contaminated groundwater following stimulation of dissimilatory metal reduction. Environ Sci Technol. 2008;42:2999–3004.

62. Nagaraj NS, Singh OV. Using genomics to develop novel antibacterial therapeutics. Crit Rev Microbiol. 2010;36:340–348.

63. Nilgiriwala KS, Alahari A, Rao AS, Apte SK. Cloning and overexpression of alkaline phosphatase PhoK from *Sphingomonas* sp. strain BSAR-1 for bioprecipitation of uranium from alkaline solutions.Appl Environ Microbiol. 2008;74:5516–5523.

64. Nussbaum RH. The Chernobyl nuclear catastrophe: unacknowledged health detriment. Environ Health Perspect. 2007;115:A238–A239. ; author reply A239–A240.

65. Nyman JL, Marsh TL, Ginder-Vogel MA, Gentile M, Fendorf S, Criddle C. Heterogeneous response to biostimulation for U(VI) reduction in replicated sediment microcosms. Biodegradation.2006;17:303–316.

66. Panak P, Nitsche H. Interaction of aerobic soil bacteria and plutonium (VI) Radiochim Acta.2001;89:499–504.

67. Panak PJ, Booth CH, Caulder DL, Bucher JJ, Shuh DK, Nitsche H. X-ray absorption fine structure spectroscopy of plutonium complexes with *Bacillus sphaericus*. Radiochimica Acta. 2002;90:315–321.

68. Payne RB, Gentry DM, Rapp-Giles BJ, Casalot L, Wall JD. Uranium reduction by *Desulfovibrio desulfuricans* strain G20 and a cytochrome c3 mutant. Appl Environ Microbiol. 2002;68:3129–3132.

69. Peck HD., Jr . Bioenergetic strategies of the sulfate-reducing bacteria. In: Odom JM, Singleton R, editors. Sulfate-Reducing Bacteria: Contemporary Perspectives. New York, USA: Springer-Verlag; 1993. pp. 41–76.

70. Peretrukhin VF, Khizhniak NN, Lyalikova NN, German KE. Biosorption of technetium 99 and some actinides by bottom sediments of Lake Belsso Kosino of the Moscow region. Radiochem.1996;38:440–443.

71. Pignolet L, Auvray F, Fonsny K, Capot F, Moureau Z. Role of various microorganisms on Tc behavior in sediments. Health Phys. 1989;57:791–800.

72. Premuzic ET, Francis AJ, Lin M, Schubert J. Induced formation of chelating agents by *Pseudomonas aeruginosa* grown in presence of thorium and uranium. Arch Environ Contam Toxicol. 1985;14:759–768.

73. Raghu G, Balaji V, Venkateswaran G, Rodrigue A, Maruthi Mohan P. Bioremediation of trace cobalt from simulated spent decontamination solutions of nuclear power reactors using *E. coli* expressing NiCoT genes. Appl Microbiol Biotechnol. 2008;81:571–578.

74. Rusin PA, Quintana L, Brainard JR, Strietelmeier BA, Tait CD, Ekberg SA, et al. Solubilization of plutonium hydrous oxide by iron-reducing bacteria. Environ Sci Technol. 1994;28:1686–1690.

75. Saffarini DA, Schultz R, Beliaev A. Involvement of cyclic AMP (cAMP) and cAMP receptor protein in anaerobic respiration of *Shewanella oneidensis*. J Bacteriol. 2003;185:3668–3671.

76. Sakadevan K, Zheng H, Bavor HJ. Impact of heavy metals on denitrification in surface wetland sediments receiving wastewater. Water Sci Technol. 1999;40:349–355.

77. Schembri MA, Kjaergaard K, Klemm P. Bioaccumulation of heavy metals by fimbrial designer adhesins. FEMS Microbiol Lett. 1999;170:363–371.

78. Singh OV, Nagaraj NS, Gabani P. Systems biology: integrating '-omics' oriented approaches to determine foodborne microbial toxins. In: Sahu SC, Casciano DA, editors. Handbook of Systems Toxicology: From Omcis Technology to Nanotechnology. New York, USA: John Wiley & Sons; 2011. pp. 469–488.

79. Tamponnet C, Declerck S. Radionuclide (RN) pollution is a worldwide problem that arises from human activities. J Environ Radioact. 2008;99:773–774.

80. Tian B, Wang H, Ma X, Hu Y, Sun Z, Shen S, et al. Proteomic analysis of membrane proteins from a radioresistant and moderate thermophilic bacterium *Deinococcus geothermalis*. Mol Biosyst.2010;6:2068–2077.

81. Valls M, Atrian S, de Lorenzo V, Fernandez LA. Engineering a mouse metallothionein on the cell surface of *Ralstonia eutropha* CH34 for immobilization of heavy metals in soil. Nat Biotechnol.2000a;18:661–665.

82. Valls M, de Lorenzo V, Gonzalez-Duarte R, Atrian S. Engineering outer-membrane proteins in*Pseudomonas putida* for enhanced heavy-metal bioadsorption. J Inorg Biochem. 2000b;79:219–223.

83. Vrionis HA, Anderson RT, Ortiz-Bernad I, O›Neill KR, Resch CT, Peacock AD, et al. Microbiological and geochemical heterogeneity in an *in situ* uranium bioremediation field site. Appl Environ Microbiol. 2005;71:6308–6318.

84. Wang J, Feng X, Anderson CW, Xing Y, Shang L. Remediation of mercury contaminated sites – a review. J Hazard Mater. 2012;221–222:1–18.

85. Wildung RE, Gorby YA, Krupka KM, Hess NJ, Li SW, Plymale AE, et al. Effect of electron donor and solution chemistry on products of dissimilatory reduction of technetium by *Shewanella putrefaciens*. Appl Environ Microbiol. 2000;66:2451–2460.

86. Wolfram L, Bauerfeind P. Activities of urease and nickel uptake of *Helicobacter pylori* proteins are media- and host-dependent. Helicobacter. 2009;14:264–270.

87. Wu WM, Carley J, Fienen M, Mehlhorn T, Lowe K, Nyman J, et al. Pilot-scale in situ bioremediation of uranium in a highly contaminated aquifer. 1. Conditioning of a treatment zone. Environ Sci Technol.2006;40:3978–3985.

88. Xie S, Yang J, Chen C, Zhang X, Wang Q, Zhang C. Study on biosorption kinetics and thermodynamics of uranium by *Citrobacter freudii*. J Environ Radioact. 2008;99:126–133.

89. Zivanovic Y, Armengaud J, Lagorce A, Leplat C, Guerin P, Dutertre M, et al. Genome analysis and genome-wide proteomics of *Thermococcus gammatolerans*, the most radioresistant organism known amongst the Archaea. Genome Biol. 2009;10:R70.

# 17

CHAPTER

# EX-SITU BIOREMEDIATION OF U(VI) FROM CONTAMINATED MINE WATER USINGACIDITHIOBACILLUS FERROOXIDANS STRAINS

Maria Romero-González[1*], Bennett C. Nwaobi[2], Joseph M. Hufton[1] and Daniel J. Gilmour[3]

[1]Department of Geography, Kroto Research Institute, The University of Sheffield, Broad Lane, Sheffield, UK

[2]Department of Life Sciences, Imperial College London, South Kensington Campus, London, UK

[3]Department of Molecular Biology and Biotechnology, University of Sheffield, Western Bank, Sheffield, UK

## ABSTRACT

**The *ex-situ* bioremoval of U(VI) from contaminated water using *Acidithiobacillus ferrooxidans* strain 8455 and 13538 was studied under a range of pH and uranium concentrations. The effect of pH on the growth of bacteria was evaluated across the range 1.5–4.5 pH units. The respiration rate of *At. ferrooxidans* at different U(VI) concentrations was quantified as a measure of the rate of metabolic activity over time using an oxygen electrode. The biosorption process was quantified using a uranyl nitrate solution, U-spiked growth media, and U-contaminated mine water. The results showed that both strains of *At. ferrooxidans* are able to remove U(VI) from solution at pH 2.5–4.5, exhibiting a buffering capacity at pH 3.5. The respiration rate of the micro-organism was affected at U(VI) concentration of 30 mg $L^{-1}$. The kinetics of the sorption fitted a pseudo-first order equation, and depended on the concentration of U(VI). The $K_D$obtained from the biosorption experiments indicated that strain 8455 is more efficient for the removal of U(VI). A bioreactor designed to treat a solution of 100 mg U(VI) $L^{-1}$removed at least 50%**

**Citation:** Maria Romero-González, Bennett C. Nwaobi, Joseph M. Hufton and Daniel J. Gilmour, Ex-situ Bioremediation of U(VI) from Contaminated Mine Water Using Acidithiobacillus ferrooxidans Strains, doi.org/10.3389/fenvs.2016.00039.

**of the U(VI) in water. The study demonstrated that *At. ferrooxidans* can be used for the *ex-situ* bioremediation of U(VI) contaminated mine water.**

## INTRODUCTION

An increase in anthropogenic activity through uranium mining and milling for the nuclear fuel cycle, and other industrial processes, has resulted in elevated levels of uranium within the environment. Low levels of uranium are released during uranium mining and high levels are released during uranium enrichment and reprocessing (Lloyd and Renshaw, 2005). Leaching of these waste materials into nearby ground water and soil sediments, due to ineffective storage mechanisms, results in radionuclides entering the food and water supply, ultimately becoming a hazard to human health (Choy et al., 2006; Gavrilescu et al., 2009). Therefore, safe removal and detoxification of these contaminants is of utmost importance, with current bioremediation technologies showing promise.

Contamination by uranium from mining activities is extensive in surface and ground waters across the world, and the consequences of this are not only detrimental to the health of the natural environment, since they also have an impact on human health. The UNESCO estimates that groundwater provides about 50% of the global drinking water supply (WWAP, 2009). The contamination of groundwater from uranium mining activities is limiting the access to clean water in former mining communities in the US. For example, a 5-year study conducted by the US EPA in abandoned mining areas within the 27,000 square miles of the Navajo Nation showed that about 57,000 people access water from unregulated sources, and some of those sources are contaminated with uranium. From the 240 unregulated sources tested, 29 exceeded the drinking water standard for uranium (USA-EPA, 2014). In the south-west of the Punjab state of India, the levels of uranium in groundwater have been reported to be as high as 100 ppb (Alrakabi et al., 2012), potentially affecting 10 million people. A study of 11,000 public supply wells in California showed that at least 2 million people are exposed to water contaminated with uranium, amongst other trace elements (Belitz et al., 2015). These examples of ground water contamination demonstrate the urgent need to develop water-cleaning technologies for the removal of uranium.

Biologically based remediation techniques, such as bioremediation, provide a cheap and efficient alternative to current remediation technologies. These processes utilize the enzymatic reductive power of microorganisms to produce U(IV) minerals and the biosorbent, and enzymatic precipitation capabilities of bacteria to sequester U(VI) (Merroun and Selenska-Pobell, 2008; Neu et al., 2010; Newsome et al., 2014; Choudhary and Sar, 2015). The enzymatic redox transformations of U(VI) and U(IV) have been studied comprehensively (Wall and Krumholz, 2006). Numerous strains of bacteria are also capable of transforming U(VI) using a similar mechanism (Lloyd et al., 2003; Khijniak et al., 2005; Bernier-Latmani et al., 2010; Orellana et al., 2013). However, the poor stability of these reduced products within aerobic environments results in the reoxidation of U(IV) minerals, hindering the development of large scale bioremediation technologies using this technique (Wall and Krumholz, 2006; Merroun and Selenska-Pobell, 2008). Reoxidation of U(IV) minerals has been observed in the presence of oxygen (Abdelouas et al., 1999; Duff et al., 1999), nitrates/ites (Finneran et al., 2002; Senko et al., 2002, 2005b), carbonates (Choudhary and Sar, 2015), Fe(III) hydroxides (Nevin and Lovley, 2000; Senko et al., 2005a), organic ligands (Luo and Gu, 2011; Singh et al., 2014), and microbial oxidation activity (Dispirito and Tuovinen, 1982; Newsome et al., 2015). Therefore, using indigenous microorganisms' abilities to uptake and precipitate U(VI) in the presence of oxygen may be more efficient in the removal of U(VI) from surface contaminated environments.

Biosorption is a metabolism-independent biological process by which metal species, compounds, and particulate materials can be removed from solution through complexation with biomass (Gadd, 1993). The removal of U(VI) via bacterial biosorption processes has been described as metabolism-independent. Numerous bacterial strains from a diverse origin have displayed binding affinities for the radionuclide species. The process is affected by pH, ionic strength, sorbent and sorbate concentrations, cell physiological state, and the presence of other soluble polymers within solution (Dobson and Burgess, 2007; Merroun and Selenska-Pobell, 2008; Wang and Chen, 2009). Structural differences in cell wall architecture between gram-positive and gram-negative bacteria suggest differences in uranium biosorption to the cell surface. The gram-positive genus *Bacillus* has been studied extensively for its biosorption capabilities in sequestering uranium from aqueous environments. It is suggested that at very low pH, covalent bonding alongside electrostatic interactions occurs between uranium and the cell surface due to protonation of the bioavailable functional groups (Selenska-Pobell et al., 1999; Fowle et al., 2000; Kelly et al., 2002; Panak et al., 2002; Gorman-Lewis et al., 2005). As the pH increases and deprotonation of the functional groups occurs; electrostatic interactions become more favorable due to the positively charged uranyl species and negatively charged surface. This results in uranyl complexation with neutral phosphate and deprotonated carboxyl functional groups on the bacterial cell wall (Fowle et al., 2000). The composition of gram-positive cell wall suggests that these preferential surface functional groups originate from the peptidoglycan (carboxyl) and secondary teichoic and lipoteichoic acids (phosphate).

The biosorption of uranium to gram-negative bacteria has been demonstrated in a variety of strains (Haas et al., 2001; Haas and Northup, 2004; Lütke et al., 2012). *Shewanella putrificiens* was shown to biosorb uranium through forming surface complexes with carboxyl and protonated phosphate groups, with sorption being dependent on pH and ionic strength (Haas et al., 2001). A similar U(VI) biosorption mechanism has been described for other Gram negative strains including*Idiomarina loihiensis* MAH1 (Morcillo et al., 2014), *Pseudomonas fluorescens* (Lütke et al., 2012),*Cupriavidus metallidurans* CH34, and *Rhodopseudomonas palustris* (Llorens et al., 2012). Alongside the phosphate groups within the outer membrane phospholipids, lipopolysaccharides (LPS) have been suggested to be involved in uranyl binding at the gram-negative cell surface (Ferris and Beveridge, 1986; Langley and Beveridge, 1999; Barkleit et al., 2008, 2011). The experiments suggest a binding mechanism of uranyl to phosphate moieties rather than carboxyl groups. This process is governed by the concentration of LPS instead of solution pH. Based on these findings, the treatment of effluents with low to medium concentrations of U(VI) waste using bacterial biosorption may be more favorable due to the attachment of metal cations to the surface (Newsome et al., 2014).

The removal of U(VI) has been demonstrated *in situ* by underground field studies at the DOE Oak Ridge Field Research Center. Microorganisms indigenous to the site were stimulated using glycerol-3-phosphate. The precipitation of uranium was achieved using small flow-through columns under aerobic conditions and circumneutral pH-values. The precipitates formed were identified as uranium phosphate minerals using X-ray Absorption Spectroscopy (XAS) (Beazley et al., 2011). Studies conducted under similar conditions using glycerol-2-phosphate amended sediments also showed the precipitation of U(VI) phosphate bearing minerals (Salome et al., 2013). Strains isolated from same site displayed varied phosphatase capabilities. The exposure of some of these strains to uranium significantly decreased their culturability, indicating the toxic effects of uranium to bacterial growth (Beazley et al., 2007; Martinez et al., 2007). Further studies from this site suggested that, in the presence of high calcium concentrations, other bacterial strains can hydrolyze inorganic phosphate from glycerol-3-phosphate to form uranium hydroxyapatite precipitates (Shelobolina et al., 2009).

We present here the use of an extremophile microorganism, *Acidithiobacillus ferrooxidans*, for the*ex-situ* bioremediation of U(VI) contaminated acidic mine water. Our hypothesis is that the ability of microorganisms to remove U(VI) from solution can be harnessed *ex-situ* using a bioreactor. The main possible mechanisms of removal proposed are biosorption and biomineralization due to the presence of bacterial biomass.

## MATERIALS AND METHODS

### Microorganisms and Growth Conditions

Two bacterial strains of *At. ferrooxidans*, type strain NCIMB 8455 (ATCC 23270) and strain NCIMB 13538, were obtained from the National Collection of Industrial, Food, and Marine Bacteria (NCIMB), Aberdeen Scotland. *At. ferrooxidans* was originally isolated from an acid, bituminous coal mine effluent in the United States of America and has been extensively studied (Leduc and Ferroni, 1994), (Amouric et al., 2011). The NCIMB 13538 strain was isolated from an abandoned uranium mine in Dera Ghazi Khan, Pakistan by the National Institute for Biotechnology and Genetic Engineering (Faisalabad, Pakistan). No published reports confirming the identity of the bacterium are available. The prokaryote identity of this species was confirmed using 16S rRNA gene analysis, and the results can be found on Supplementary Table 1. The bacterial strains were cultured using *Acidiphilium* complex medium, modified by the omission of carbon and peptide sources, and the addition of 10 mM Fe (DSMZ, 2015). This medium consists of 2 mM $MgSO_4$ $7H_2O$, 0.8 mM $(NH_4)_2SO_4$, 0.7 mM KCl, 0.06 mM $Ca(NO_3)2$ $4H_2O$, 0.4 mM $KH_2PO_4$, 0.3 mM NaCl, 0.6 mM $K_2HPO_4$, 10 mM $FeSO_4$ $7H_2O$, 1.0 g $L^{-1}$ mannitol, and 1.7 g $L^{-1}$ bactotryptone. The effect of temperature on the bacterial growth experiments for both strains can be found in Supplementary Figures 1, 2. The optimum temperature obtained for the growth was 30°C for both strains and therefore all experiments were conducted at this temperature. A Scanning Electron Microscopy (SEM) micrograph showed that the cells are rod-shaped (Supplementary Figure 3).

## Effect of pH on the Growth of *Acidithiobacillus ferrooxidans*

The effect of variation of initial solution pH on the growth of *At. ferrooxidans* was evaluated by adjusting the starting pH of the modified acidiphilium medium to initial pH-values of 1.5, 2.5, 3.5, and 4.5. The bacteria were then incubated at 30°C for 72 h with agitation. The optical density was measured periodically at 600 nm. The value of pH was monitored constantly for the duration of the experiment using a pH meter (Orion, Thermo Scientific).

## Effect of U(VI) Concentration on the Metabolism of *Acidithiobacillus ferrooxidans*

The effect of U(VI) concentration on the metabolic activity of aerobically grown *At. ferrooxidans*strains was measured by quantifying respiration rate. An oxygen electrode (Hansatech, Ltd.) calibrated using sodium dithionite (Sigma-Aldrich) was used. A stock solution of U(VI) was prepared using uranyl nitrate hexahydrate, $UO_2(NO_3)_2$ $6H_2O$ (Sigma-Aldrich). Samples of 2 ml containing bacteria culture, incubated with a range of U(VI) concentrations ranging from 1 to 100 mg U(VI) $L^{-1}$, were analyzed for 5 min. The electrode calibration was matched for the composition of the modified acidiphilium medium by including a non-inoculated medium control. Six replicates of controls and samples were run in alternating pairs. In order to compare the results obtained between strains, cells were diluted to a concentration of $10^8$ cells $mL^{-1}$.

## Kinetics of Biosorption of U(VI) Using *Acidithiobacillus ferrooxidans*

The bacteria were grown to exponential phase using the modified acidiphilium growth medium. Cells were harvested by centrifugation at 6000 rpm for 20 min to separate the biomass from Fe(III) precipitates and hydroxides produced in the medium during growth. Afterwards, biomass pellets were washed three times with 0.05 M sulfuric acid and then two times with 0.01 M perchloric acid. A sample of 0.015 g wet mass (0.0038 g dry weight approximately) of bacteria was suspended in 10 ml sterile de-ionized water containing $1.08 \times 10^{-3}$ mmol $L^{-1}$ U(VI) and incubated at 25°C with agitation. Samples were collected at 1, 2, 4, 6, 12, and 24 h and analyzed using ICP-MS (ELAN 6000, Perkin Elmer). All experiments were performed in triplicate and controls containing uranium without bacteria were also included.

## Biosorption of U(VI) Using *Acidithiobacillus ferrooxidans*

Batch experiments were conducted to quantify the biosorption of U(VI) to *At. ferrooxidans* using simulated mine water solution at different concentrations of uranium. The bacteria samples were prepared using the same procedure as in the kinetic experiment. Samples of 0.015 g wet mass (0.0038 g dry weight approximately) of bacteria were suspended in 10 ml simulated mine water spiked with U(VI) to a solution concentration of 10, 50, 80, 100, 120, 150, and 200 mg U(VI) $L^{-1}$. Controls solutions of U(VI) with no bacteria inoculate, and bacteria with no U(VI) added, were also included. Triplicate samples were incubated at 25°C in an orbital shaker for 24 h. The supernatant was separated from the biomass pellet by centrifugation at 3000 rpm, and analyzed for its U(VI) content using ICP-MS. U(VI) biosorption was also quantified using a simulated mine water solution containing 120 mg U(VI) $L^{-1}$ and a control of acidiphilium growth medium spiked with 120 mg U(VI) $L^{-1}$. Comparison of mean values was performed using SPSS statistical software (IBM-SPSS, 2011).

## Semi-Continuous Bioreactor

A semi-continuous 5 L bioreactor containing growth media and a culture of *At. ferrooxidans* was designed. The reactor was submerged in a water bath at 30°C to ensure the micro-organism growth. The reactor was kept under continuous agitation using a magnetic stirrer and oxygen was supplied at a rate of 25 mL $min^{-1}$ using a pump connected and tubing connected to a porous stone to aid the oxygen diffusion through solution. The pH and temperature of the solution were monitored using electrode probes and samples were collected via a sampling port. The reactor was left to equilibrate for 24 h before adding an aliquot of U(VI) solution to a final volume of 100 mg U(VI) $L^{-1}$. The reactor was operated for 72 h and the concentration of U(VI) was quantified using ICP-MS. Figure 1 shows a sketch of the reactor depicting its components.

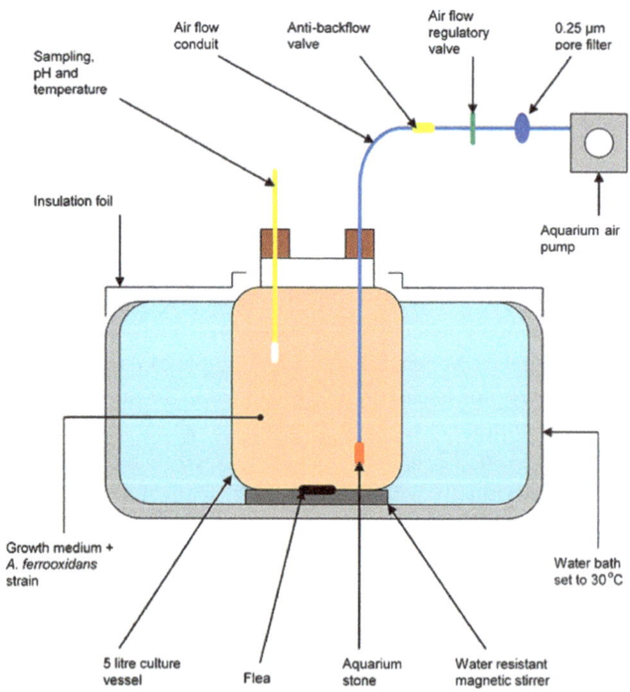


**Figure 1:** Design sketch of the bioreactor showing its components. The total volume of the vessel was 5 L of a mine water solution containing 100 mg U(VI) $L^{-1}$. Temperature and pH probes were inserted to monitor the reaction during time. Oxygen was constantly supplied via a aeration pump. The temperature was kept to a constant value of 30°C using a water bath.

# RESULTS

## Effect of Solution pH

The effect of pH on the growth of *At. ferrooxidans* was quantified at different initial growth medium pH-values. Figure 2A shows the growth curves obtained at pH 1.5, 2.5, 3.5, and 4.5 for strain 13538 at 30°C. The growth curves obtained for pH 2.5, 3.5, and 4.5 follow a typical exponential curve, with a plateau at $OD_{600}$-value 1.2. Bacteria did not grow at pH 1.5, indicating that the optimum pH-value is between 2.5 and 4.5 pH units. Figure 2B shows the profile for the pH obtained during the growth experiments. There was no variation in the pH-values during the experiment conducted at pH 1.5. Despite the initial solution pH-values being adjusted to 2.5, 3.5, and 4.5 pH units for the rest of the experiments, the pH measurements returned values of 2.5 pH units after 20 h. The pH remained stable at this value until the end of the experiment at 72 h.

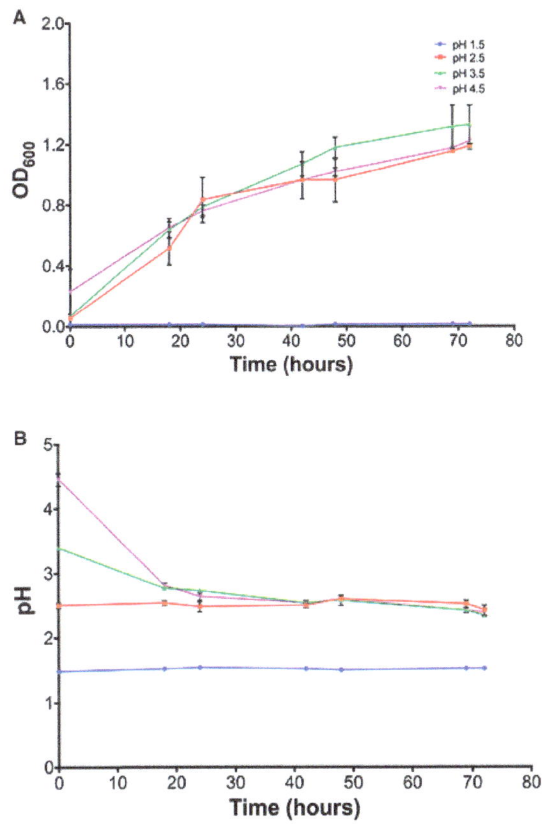


**Figure 2:** (A) Growth curves of *At. ferrooxidans* strain 13538 at different pH-values. The optical density at 600 nm was measured over a period of 72 h. Optimal bacteria growth is achieved at pH 2–4.5. (B) Profile of solution pH during growth experiments. Error bars represent 1 standard deviation of the mean.

## Effect of U(VI) Concentration on the Respiration Rate of*Acidithiobacillus ferrooxidans*

The respiration rate of *At. ferrooxidans* at different U(VI) concentrations was quantified as a measure of the rate of metabolic activity over time by monitoring oxygen consumption. Figure 3Ashows the amount of oxygen consumed by *At. ferrooxidans* strain 8455 as a function of increasing concentration of U(VI) in solution. Control experiments performed without U(VI) were used for baseline measurements of the respiration rate and a mean value of 0.9 nmoles [$10^6$ cells]$^{-1}$ h$^{-1}$ was calculated for all experiments. Four different uranium concentrations were used to test metabolic oxygen utilization in strain 8455, ranging from 1 to 30 mg L$^{-1}$. At 1 mg L$^{-1}$ uranium concentration, strain 8455 averaged 1.29 nmoles [$10^6$ cells]$^{-1}$ h$^{-1}$, a value higher than the 0.98 nmoles [$10^6$ cells]$^{-1}$h$^{-1}$ obtained from corresponding controls. Paired *t*-tests showed that there was a statistically significant difference between the means at a confidence level >99%. Six replicate experiments using strain 8455 in contact with 10 mg U(VI) L$^{-1}$ uranium solution were performed over a 2 day period to quantify errors due to differences in experimental conditions. Two controls, plus replicate experiments were run on the first day and four on the second, with consistent results. This helped negate experimental anomaly and complemented the statistical tests by confirming that perceived enhanced metabolic performance of strain 8455 at 10 mg L$^{-1}$ uranium was significant and may be due to a stress response that increases respiration rate in an attempt to detoxify the uranium. At 30 mg L$^{-1}$ uranium, the bacteria consumed oxygen at rates of 0.86 and 0.87 nmoles [$10^6$ cells]$^{-1}$h$^{-1}$ for controls, respectively. According to SPSS paired *t*-test analysis, there was no statistically significant difference between the means. Strain 8455 was hence metabolically uninhibited in terms of oxygen consumption by 30 mg L$^{-1}$ uranium. Tests at 40 mg L$^{-1}$ proved different. Average bacterial oxygen consumption was 0.4 nmoles [$10^6$ cells]$^{-1}$h$^{-1}$, lower than the corresponding control value of 0.87 nmoles [$10^6$ cells]$^{-1}$h$^{-1}$. Paired *t*-tests confirmed a statistical difference between the means at a significance level >99%. A uranium concentration of 30 mg L$^{-1}$ was hence sufficient to inhibit metabolic consumption in strain 8455.

It was found that strain 13538 was more susceptible to uranium than strain 8455 (Figure 3B). At 1 mg L$^{-1}$ uranium, strain 13538 recorded an oxygen consumption of 0.85 nmoles [$10^6$ cells]$^{-1}$ h$^{-1}$ as against 0.89 nmoles [$10^6$ cells]$^{-1}$ h$^{-1}$ registered by the corresponding controls. Paired *t*-test analysis using SPSS software showed there was no statistically significant difference between the means. Strain 13538 was metabolically uninhibited by a uranium concentration of 1 mg L$^{-1}$. Tests conducted at 5 mg L$^{-1}$ uranium showed strain 13538 consumed oxygen at a rate of 0.56 nmoles [$10^6$ cells]$^{-1}$h$^{-1}$, lower than the 0.88 nmoles [$10^6$ cells]$^{-1}$ h$^{-1}$ value recorded by the corresponding controls. The paired *t*-tests comparing the means showed a statistical difference at the 90% confidence level. As uranium concentrations of 5 mg L$^{-1}$ were clearly inhibitory to oxygen metabolic activity in strain 13538, it was expected that the results obtained at the higher metal concentrations of 10 and 30 mg L$^{-1}$ would follow a similar trend, as observed in Figure 3B.

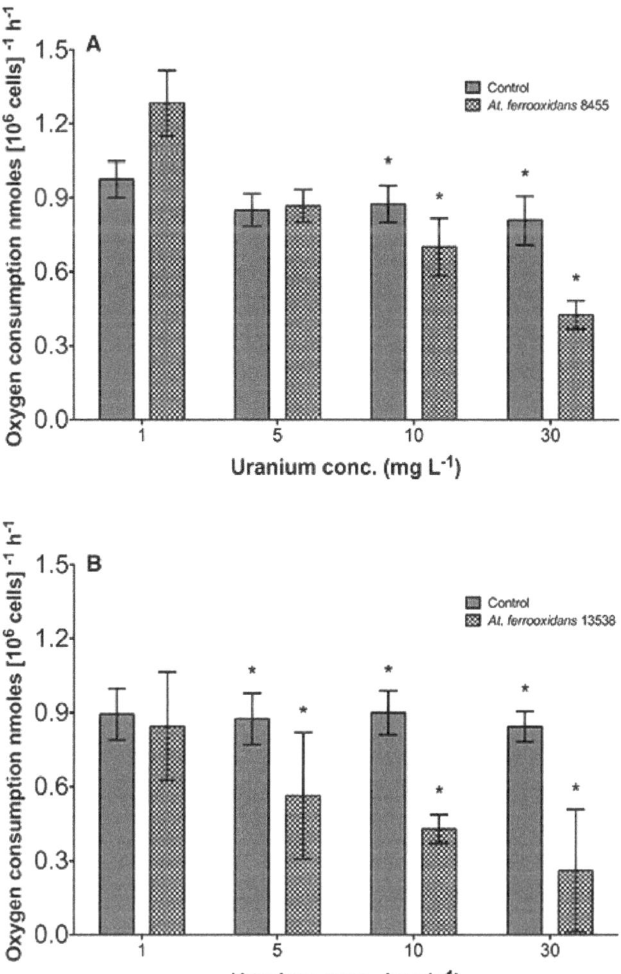


**Figure 3:** Oxygen consumption rate by *At. ferrooxidans* **8455 (A) and 13538 (B) as a function of increasing concentration of U(VI) in solution**. The respiration rate is inhibited by the increasing concentration of U(VI) in solution up to 30 mg (VI) $L^{-1}$. Data labeled with * indicate significantly different results.

## Kinetics of the Biosorption Process

Results from the kinetic experiment performed using batch type reactors with a solution of $1.08 \times 10^{-3}$ mmol U(VI) $L^{-1}$ are shown in Figure 4. A reduction of the initial concentration of U(VI) over time was observed, with a typical exponential shape in the curve for the experiments using both *At. ferrooxidans* strains. Strain 13538 is slightly more efficient in reducing the concentration of U(VI) in solution than strain 8455. The obtained data was fitted to a pseudo-first order model (Supplementary Figure 4). Fitting the data to zero order reactions was not possible since this model is not dependent on the concentrations of the reactants. Assuming that the rate of reaction depends on one reactant, in this case, the concentration of U(VI), a pseudo-first order equation was used to fit the obtained data. The observed decrease of uranium concentration in solution by strain 8455 was from $1.08 \times 10^{-3}$ to $0.95 \times 10^{-3}$ mmol $L^{-1}$ after 24 h, which equates to a total decrease of $0.13 \times 10^{-3}$ mmol $L^{-1}$ or 11.9%. Of this amount, nearly half (5.2%) was taken up by the cells after only 4 h. The amount of U(VI) from solution, when in contact with strain 13538, was determined to be from $1.08 \times 10^{-3}$ to $0.84 \times 10^{-3}$ mmol $L^{-1}$ after 24 h, resulting in a decrease of $0.25 \times 10^{-3}$ mmol $L^{-1}$ or 22.6%. A 16.5% reduction was observed after 12 h (half of the experiment's duration). The rate of reaction, *r*, was defined as (Equation 1).

$$r = -k[U(VI)] \quad (1)$$

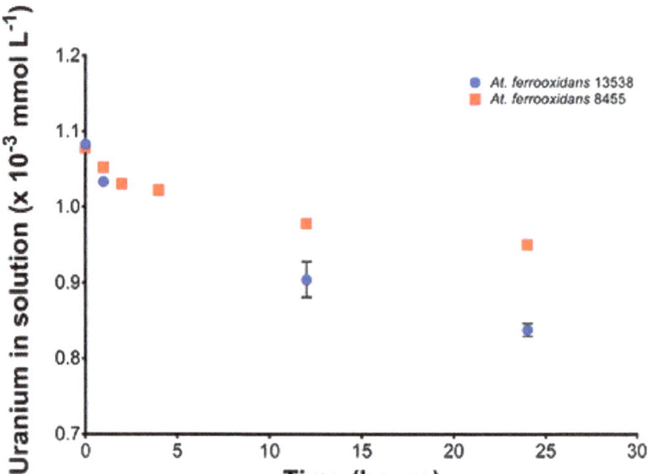


**Figure 4:** Kinetics of the biosorption removal of U(VI) by *At. ferrooxidans*strains 8455 and 13538. The data fits a pseudo-first order equation. Bacteria strains are more efficient removing U(VI) from simulated mine water solution compared to growth media or uranyl salt solution.

## Bioremediation of U(VI) from Simulated Mine Water

The effect of U(VI) concentration on the growth of *At. ferrooxidans* was evaluated using a solution of 120 mg U(VI) $L^{-1}$. Figure 5 shows changes in the concentration of uranium in solutions of U(VI), growth media, and simulated mine water. The concentration of uranium in the uranium salt solution after being in contact with *At. ferrooxidans* was 57.35 mg $L^{-1}$ for strain 8455 and 69.50 mg $L^{-1}$ for strain 13538. The removal of uranium from solution was calculated at 52.6 and 42.6% for strains 8455 and 13538, respectively. This indicated that both strains were efficient in the reduction of uranium concentration from a uranyl salt solution. The efficiency of *At. ferrooxidans* in removing uranium from a solution of modified acidiphilium media was also evaluated. The mean concentrations of uranium in solution were 106.06 and 105.77 mg $L^{-1}$ for strain 8455 and strain 13538, respectively. These values represent a removal of uranium from growth media solution of only 12% for strain 8455 and 13% for strain 13538 when in the presence of nutrients. The removal of uranium from a simulated mine water solution by strain 8455 was 48.81 mg $L^{-1}$ and 50.15 mg $L^{-1}$ for strain 13538. These represent an efficiency of removal of 59.7 and 58.5%, respectively. T-test results showed that the removal of uranium from the three solutions by each one of the strains was significantly different at 99% confidence. The statistical comparison also showed that strain 8455 was more efficient compared to strain 13538 in removing uranium from the uranyl salt solution. However, no significant differences were found when comparing the efficiency of both strains in removing uranium from the growth media or simulated mine water solution.

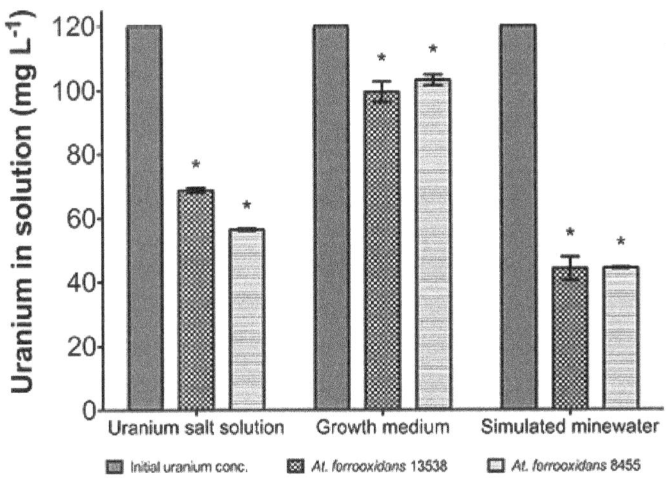


**Figure 5:** Amount of uranium removed from uranyl nitrate, growth media spiked with U(VI) and simulated mine water spiked with U(VI) solutions byAt. ferrooxidans **strains 8455 and 13538**. A higher concentration of U(VI) was removed from simulated mine water solution. Data labeled with * indicate significant different results.

The uptake of U(VI) by *At. ferrooxidans* from simulated mine water solution was quantified using batch sorption experiments. Figure 6 shows the sorption isotherm obtained for the removal of uranium from simulated mine water solution at pH 3.5 for strain 8455 and strain 13538. The sorption isotherms exhibited a typical curve shape with a plateau, indicating that the sorption capacity of uranium by *At. ferrooxidans* is limited. The maximum sorption obtained was 53% for strain 8455 and 43% for strain 13538. The isotherm data was fitted to a Langmuir equation as follows:

$$q = \frac{q_m K_L C}{1+K_L C} \qquad (2)$$

where $q$ represents a relationship between the initial and final concentration of U(VI) in solution as a function of the biomass in a given volume:

$$q = \frac{(C_i - C_f)V(L)}{biomass\ (g)} \tag{3}$$

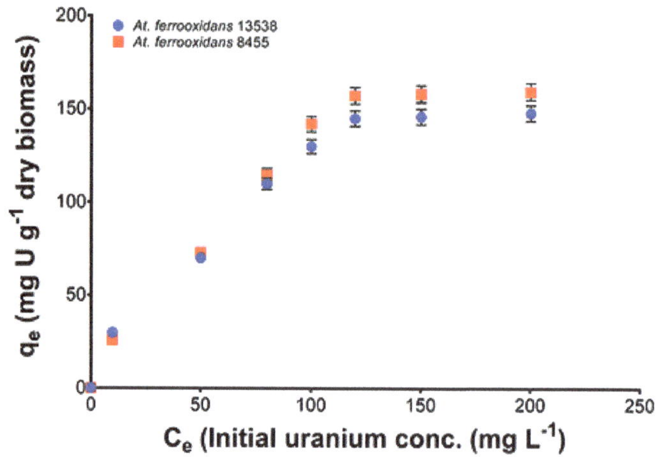


**Figure 6:** Sorption isotherms for *At. ferrooxidans* **strains 8455 and 13538 obtained at pH 3.5.** The data fitted a Langmuir equation. Error bars correspond to 1 standard deviation of the mean.

A plot of the inverse of $q$ vs. $C$ produces a straight line (Supplementary Figure 5) with a slope of $\frac{1}{K_L q_m}$ and an intercept of $\frac{1}{q_m}$, where $q_m$ is the maximum sorption capacity and $K_L$ is the Langmuir constant. The sorption parameters obtained for strain 8455 were $K_L = 0.013$ and $q_m = 227.27$ mg g dry biomass$^{-1}$. For strain 13538 the calculated sorption parameters were $K_L = 0.019$ and $q_m = 178.57$ mg g dry biomass$^{-1}$. The dissociation constant, $K_D$ was calculated by:

$$K_D = \frac{C_s}{C_w} \tag{4}$$

Using the Langmuir isotherm model, the concentration in the biomass sorbent is quantified using:

$$Cs = \frac{q_m K_L C_w}{[1 + K_L C_w]} \tag{5}$$

Assuming a concentration in solution, $C_w$ of 100 mg L$^{-1}$, the estimated $K_D$-values were 1.3 L g$^{-1}$for strain 8455 and 1.2 L g$^{-1}$ for strain 13538. The $K_D$-value obtained suggests that both strains possess similar ability to remove uranium.

The bioremoval of U(VI) using *At. ferrooxidans* was also quantified using a semi-continuous reactor. The reactor containing growth media and bacteria biomass was left to equilibrate for 24 h to half of the exponential growth phase, ensuring the availability of bacterial cells in the reactor. A spike of U(VI) solution was added to make a final concentration of 100 mg U(VI) $L^{-1}$ in the reactor. The pH throughout the bioremoval process was constant, at a value of 3.5, and no pH amendments were required during the reaction time. The oxygen concentration in solution was also found to be constant at 8.1 mg $O_2$ $L^{-1}$. The concentration of U(VI) was reduced from 100 mg U(VI) $L^{-1}$ to 51.2 mg U(VI) $L^{-1}$ in the first 24 h. The amounts of U(VI) quantified at 48 and 72 h were 22.7 mg $L^{-1}$and 15.4 mg $L^{-1}$, respectively. The removal efficiency was then calculated to be 51% at 24 h reducing to 32% at 72 h.

## DISCUSSION

### Effect of Solution pH and Oxygen on the Activity of *Acidithiobacillus ferrooxidans*

*Acidithiobacillus ferroxidans* has great potential for use in biotechnology processes as it grows well under acidic conditions typical of acid mine drainage and tolerates high concentrations of metals such as Cu(II) (Dew et al., 1999) and U(VI) (Merroun and Selenska-Pobell, 2001). These characteristics make this organism an ideal candidate for use in bioleaching or bioremediation technology. Studies conducted with solution pH in the range 1.5–4.5 show that *At. ferrooxidans*growth is best at pH 3.5 and hindered at pH 1.5. The optimum range of pH obtained (2.5–4.5) is consistent with previously reported growth conditions for *At. ferrooxidans* (Merroun and Selenska-Pobell, 2001; Mazuelos et al., 2010). Therefore, any successful biotechnology process using this organism requires an influent pH-value within this range. However, experiments conducted using*At. ferrooxidans* showed a buffering capacity around pH 2.5 (Figure 2B), indicating that the system readily equilibrates. The stabilization of the pH profile across the experiment is comparable to that obtained during an iron bio-oxidation experiment using *At. ferrooxidans* (Mazuelos et al., 2010) where pH fluctuated around a value of 2.3 under constant oxygen conditions. The observed pH profile supports the hypothesis that pH stabilization in solution is due to limitation of oxygen in solution and the formation of $Fe^{3+}$ via oxidation, as follows:

$$2Fe^{2+} + 0.5O_2 + 2H^+ \leftrightarrow 2Fe^{3+} + \qquad (6)$$

This process is limited by oxygen and hence indicates that the abiotic oxidation of $Fe^{2+}$ to $Fe^{3+}$ in solution contributes to maintenance of the buffering conditions that favor the healthy activity of the*At. ferrooxidans* culture. Since pH plays such an important role in the survival of *At. ferrooxidans*, any potential reactor needs to be conceptualized using pH as a master variable.

The metabolic activity of *At. ferrooxidans* strains 8455 and 13538 was also measured as a function of U(VI) concentration in order to evaluate the limits of resistance of the bacterial cells to the presence of U(VI). The inhibition of oxygen uptake by aerobic microorganisms may be considered a suitable method to evaluate the toxicity of chemicals. There is a marked difference in the behavior observed for strain 8455 compared to 13538 at 30 mg $L^{-1}$. Strain 13538 is more affected by the presence of U(VI) in solution, exhibiting a reduction in its respiration rate of ~70% compared to only 50% reduction for strain 8455. This behavior is comparable with results obtained for three *A. ferrooxidans* eco-types that are capable of removing 25 mg U(VI) $L^{-1}$ from solution (Merroun and Selenska-Pobell, 2001). Tolerance to concentration of U(VI) varies between strains. A study of a large number of *A. ferrooxidans* isolates found that the 16 rRNA genes have specific signatures distinguishing strains within the species (Selenska-Pobell, 2002), and these different types exhibited differences in their ability to accumulate uranium (Merroun and Selenska-Pobell, 2001). These findings have been reported for other types of micro-organisms, and lead to the conclusion that accumulation of U(VI) is type specific.

In addressing practical considerations of a bioreactor technology, strain 8455 adapted best to the conditions of the study, and so it may be more suitable for use in bioremediation. Designing a reactor will need to consider the limitations imposed by the pH of the solution, the content of oxygen supplied to the reactor and the concentration of U(VI) in solution, although this is not as critical as the first two parameters.

## Bioremoval Process of U(VI) by *Acidithiobacillus Ferrooxidans*

Immobilization of U(VI) by bacterial biomass has been reported for a wide range of micro-organisms including natural isolates from U-contaminated sites such as *At. ferrooxidans, D. radiodurans, Bacillus* sp., *Arthrobacter* sp., and *P. fluorescens* (Merroun and Selenska-Pobell, 2001; Bencheikh-Latmani and Leckie, 2003; Suzuki and Banfield, 2004). Similar behavior was observed for *At. ferrooxidans* strains 8455 and 13538, where U(VI) was efficiently removed from solution despite its nutrient composition. Interestingly, both strains seem to remove U(VI) from solution more efficiently when in the presence of other ions typical of mine waters. This result highlights the importance of the presence of $Fe^{2+}$ and the role as electron donor that this ion plays in the metabolism of *At. ferrooxidans*. Also interesting was that the initial supply of nutrients provided did not result in enhanced removal of U(VI) from solution (Figure 5). This has an impact on the cost associated with culture maintenance. *At. ferrooxidans* seems to adapt to the nutrient and electrolyte conditions supplied by mine water solution and efficiently remove U(VI) without any added growth medium components in small reactor systems. Compared to other bacteria species strains 8455 and 13538 demonstrated a good removal performance. For example, *Bacillus thuringiensis* exhibited a superior efficiency of U(VI) removal of to 420 mg $g^{-1}$ dry biomass, compared to 227 and 179 mg $g^{-1}$ dry biomass for strains 8455 and 13538, respectively. However, the efficiency of *B. thuringiensis* was quantified using a uranyl acetate solution, where acetate contributed to the maintenance of the bacterial culture (Table 1). The ability of strain 8455 for the removal of U(VI) is comparable to *Microbacterium* and Pseudomonas *aeruginosa* and exceeds the efficiency of *Stenotrophomonas* sp. and *Acinetobacter* sp. (Table 1). The behavior of strain 13538 is more similar to the one reported for *Bacillus pumilus* at pH 4. Most of the reported studies were conducted using U(VI) solutions only in contrast to the mine water solution used in this study. The obtained efficiency confirms that *At. ferrooxidans* is a good candidate organism to be used in bioremediation systems.

**Table 1:** Efficiency of different bacteria species biomass (mg $g^{-1}$ **dry biomass) for the removal of** U(VI) compared to *At. ferrooxidans* **strains**

| Bacterial species | U removal (mg $g^{-1}$ dry biomass) | References |
|---|---|---|
| *At. ferrooxidans* 8455 | 227 in mine water | This study |
| *At. ferrooxidans* 13538 | 179 in mine water | This study |
| *Acinetobacter* sp. Hg1 | 103 in U(VI) solution, pH 4 | Islam and Sar, 2016 |
| *Bacillus pumilus* | 160 in U(VI) solution, pH 4 | Suzuki and Banfield, 2004 |
| *B. thuringiensis* ZYR4 | 420 in uranyl acetate | Pan et al., 2015 |
| *Microbacterium* sp. | 230 in U(VI) solution, pH 4.5 | Nedelkova et al., 2007 |
| *P. aeruginosa* | 275 in U(VI) solution, pH 4 | Choudhary and Sar, 2011 |
| *Stenotrophomonas* sp. U18 | 80 in U(VI) solution, pH 4 | Islam and Sar, 2016 |

The biosorption process followed a typical Langmuir isotherm shape and the distribution constants ($K_D$) derived from the Langmuir model fitting indicated that *At. ferrooxidans* strain 8455 is more efficient in the removal and sorption of U(VI) than strain 1358. These results are consistent with previously observed behavior of *At. ferrooxidans* in mine water solution. For example, the difference in efficiency of U(VI) removal from solution between the different strains has been previously reported (Merroun and Selenska-Pobell, 2001), where binding and complexation of U(VI) was different between three *At. ferrooxidans* eco-types. Although, we conducted genetic characterization of the strains reported here, a comparison of the strains to determine if they belong to different eco-type groups was not conducted and hence comparison to previous studies is only speculative. The biosorption process for strains 8455 and 1358 is described as a monolayer sorption process, where $UO_2^{2+}$ ions in solution bind to the surface of *At. ferrooxidans* at specific binding sites. These binding sites are energetically similar and possess similar chemical characteristics. Since this study was focused on quantifying the efficiency of removal for a bioprocessing technology, a full characterization of the chemical nature of binding was not conducted. However, the binding and removal of $UO_2^{2+}$ ions by bacterial biomass extensively reported in the literature indicates that the main bacteria binding sites have been identified as phosphate containing sites. Merroun and Selenska-Pobell (2001) described the complexes formed by three eco-types of *At. ferrooxidans* studied by fluorescence and infrared spectroscopy as uranyl organic-phosphate compounds. Other bacterial strains have shown the ability to precipitate U(VI) from solution as a phosphate type of compound, which suggests that the mechanism of sorption and precipitation is not unique to the bacteria type but to the chemical nature of bacterial cell wall components. Beazley et al. (2007) reported the formation of organophosphate compounds by three heterotrophic bacteria isolated from soils when in the presence of U(VI). The compounds were identified and characterized using EXAFS and reported as organophosphate of the autinite/met-autinite group mineral. Choudhary and Sar (2011) observed a similar organophosphate compound formed by the interaction of *P. aeruginosa*. The precipitates formed on the cell envelope had a needle shape and were identified as uranium organophosphate complexes by X-ray diffraction analysis and infrared spectroscopy. Suzuki and Banfield (2004) reported the formation of uranyl organophosphate complexes by *D. radiodurans* and *Arthrobacter* spp. as observed from TEM and EDX analysis. More recently, studies conducted by the authors on *Idiomarina loihiensis* (Morcillo et al., 2014) showed that this bacterium forms uranyl organophosphate compounds when in the presence of U(VI) at higher salinity concentrations. These findings indicate that the preferred sorption mechanism of U(VI) by bacterial biomass is via complexation and precipitation on phosphate rich binding sites. The evidence hence suggests that *At. ferroxidans* would remove U(VI) as uranyl organophosphate as a preferred sorption and removal mechanism.

## Implications for *Ex-situ* Mine Water Treatment—toward the Design of Bioreactors

Very few technologies have been proposed for the removal of uranium from water using treatment plant technologies and most of the literature focuses on description of *in situ* remediation of uranium mill tailings sites. Therefore, there is an opportunity for advances by coupling removal of uranium with high-density sludge systems (HDS; Campbell et al., 2015). A small scale ion exchange system was developed by the US EPA as early as 1989 (Jelinek and Sorg, 1988). The removal of uranium was achieved using anion exchange resins based on chloride ion and a water softener tank. Uranium was reduced form 23.8 mg $L^{-1}$ to 1 μg $L^{-1}$. The treatment was efficient in the removal of uranium, but the re-use of the ion exchange cartridges was limited and additional uranium waste was generated.

Harnessing the power of *At. ferroooxidans* and the formation of Fe-containing sludge may provide a feasible biotechnology for the removal of uranium from contaminated waters. Using the partition parameters obtained from batch experiments, a bioreactor was designed to test the limitations of the technology. The bioreactor used in this study contained *At. ferrooxidans* culture, spiked with 100 mg $L^{-1}$ of U(VI), under constant oxygen supply to ensure the survival of the micro-organism. Care had to be taken to ensure that the aeration process did not oxidize $Fe^{2+}$ at a faster rate than*At. ferrooxidans*, otherwise the depletion of $Fe^{2+}$ would hinder the metabolism of the micro-organism. Therefore, an oxygen supply of 25 ml/min was used. The reactor was operated for 72 h and samples of the solution were analyzed for their U(VI) content. The efficiency of U(VI) removal obtained was of 51% at 24 h for strain 8455. This efficiency was reduced to 32% at 72 h of reaction. A reactor tested using strain 13538 did not operate successfully. The reduction in the efficiency of removal of U(VI) from solution indicates that there is a decay in the ability of the micro-organisms to perform the biosorption process, possibly due to the consumption of nutrient resources that sustain the microbial biomass over time. A large amount of Fe-containing sludge was formed at the bottom of the reactor, limiting the diffusion of oxygen through solution and impeding the bacteria respiration process. Therefore, removal of the sludge during the water cleaning process may be required to maximize the reactor performance. In mine water treatment, HDS are a common feature of treatment plants. Sludge formed from precipitation using lime is separated using a clarifier and recycled into a tank until density is increased and the volume is reduced by 90%. The resulting sludge is of high density producing a high quality of effluent water. The HDS formed is easier to remove from the plant and dispose of as waste. The combination of the bioreactor with a HDS system would provide an optimal solution for remediating U(VI).

The efficiency of removal achieved using the reactor was close to the batch experiments, despite the reactor limitations of nutrient and oxygen availability. Reactors packed using *At. ferrooxidans* have been previously reported for ferrous ion bio-oxidation and the removal of $H_2S$ (Mazuelos et al., 2010) and (Giro et al., 2006). However, this is the first time a bioreactor for the removal of U(VI) using *At. ferrooxidans* has been described. The results presented here demonstrate an efficient *ex-situ* treatment of mine water contaminated with U(VI) using *At. ferrooxidans*. The ability of this microorganism to operate under acidic pH and low nutrient concentration makes it ideal for use in bioreactor technology. Compared to the performance of bacteria indigenous to uranium ore (Islam and Sar, 2016), *At. ferrooxidans* ability to remove U(VI) is higher, indicating its suitability for use in bioremediation. The efficiency of a reactor such as the one described here may be improved by integration with HDS.

## AUTHOR CONTRIBUTIONS

MR designed the hypothesis and experimental work presented here and contributed to the writing of all sections of the manuscript. BN conducted the laboratory work and contributed to the writing of the manuscript. JH conducted the literature review, contributed to the writing of the literature review, and experimental section and the preparation of the manuscript. DG contributed to the design of the experimental work and the preparation of the manuscript.

## ACKNOWLEDGMENTS

The authors are grateful to the EPSRC project "Nanometre-scale observation of biodegradation in soil and groundwater" (GR/S72467/01) and the EPSRC Programme grant "Hard-soft matter interfaces: from understanding to engineering" (EP/I001514/1) for financial support. BN and JH thank EPSRC for financial support via The University of Sheffield DTA scholarships.

## REFERENCES

1. Abdelouas, A., Lutze, W., and Nuttall, H. E. (1999). Oxidative dissolution of uraninite precipitated on Navajo sandstone. *J. Contam. Hydrol.* 36, 353–375. doi: 10.1016/S0169-7722(98)00151-X
2. Alrakabi, M., Singh, G., Bhalla, A., Kumar, S., Srivastava, A., Rai, B., et al.

(2012). Study of uranium contamination of ground water in Punjab state in India using X-ray fluorescence technique. *J. Radioanal. Nucl. Chem.* 294, 221–227. doi: 10.1007/s10967-011-1585-x

3. Amouric, A., Brochier-Armanet, C., Johnson, D. B., Bonnefoy, V., and Hallberg, K. B. (2011). Phylogenetic and genetic variation among Fe(II)-oxidizing acidithiobacilli supports the view that these comprise multiple species with different ferrous iron oxidation pathways. *Microbiology* 157, 111–122. doi: 10.1099/mic.0.044537-0
4. Barkleit, A., Foerstendorf, H., Li, B., Rossberg, A., Moll, H., and Bernhard, G. (2011). Coordination of uranium(VI) with functional groups of bacterial lipopolysaccharide studied by EXAFS and FT-IR spectroscopy. *Dalton Trans.* 40, 9868–9876. doi: 10.1039/c1dt10546a
5. Barkleit, A., Moll, H., and Bernhard, G. (2008). Interaction of Uranium(VI) with lipolysaccharide. *Dalton Trans.* 21, 2879–2886. doi: 10.1039/b715669c
6. Beazley, M. J., Martinez, R. J., Sobecky, P. A., Webb, S. M., and Taillefert, M. (2007). Uranium biomineralization as a result of bacterial phosphatase activity: insights from bacterial isolates from a contaminated subsurface. *Environ. Sci. Technol.* 41, 5701–5707. doi: 10.1021/es070567g
7. Beazley, M. J., Martinez, R. J., Webb, S. M., Sobecky, P. A., and Taillefert, M. (2011). The effect of pH and natural microbial phosphatase activity on the speciation of uranium in subsurface soils. *Geochim. Cosmochim. Acta* 75, 5648–5663. doi: 10.1016/j.gca.2011.07.006
8. Belitz, K., Fram, M. S., and Johnson, T. D. (2015). Metrics for assessing the quality of groundwater used for public supply, CA, USA: equivalent-population and area. *Environ. Sci. Technol.* 49, 8330–8338. doi: 10.1021/acs.est.5b00265
9. Bencheikh-Latmani, R., and Leckie, J. O. (2003). Association of uranyl with the cell wall of *Pseudomonas fluorescens* inhibits metabolism. *Geochim. Cosmochim. Acta* 67, 4057–4066. doi: 10.1016/S0016-7037(03)00214-X
10. Bernier-Latmani, R., Vecchia, E. D., Junier, P., Lezama-Pacheco, J. S., Veeramani, H., Suvorova, E. I., et al. (2010). Non-uraninite products of microbial U(VI) reduction. *Environ. Sci. Technol.* 44, 9456–9462. doi: 10.1021/es101675a
11. Campbell, K. M., Gallegos, T. J., and Landa, E. R. (2015). Biogeochemical aspects of uranium mineralization, mining, milling, and remediation. *Appl. Geochem.* 57, 206–235. doi: 10.1016/j.apgeochem.2014.07.022
12. Choudhary, S., and Sar, P. (2011). Uranium biomineralization by a metal resistant *Pseudomonas aeruginosa* strain isolated from contaminated mine waste. *J. Hazard. Mater.* 186, 336–343. doi: 10.1016/j.jhazmat.2010.11.004
13. Choudhary, S., and Sar, P. (2015). Interaction of uranium (VI) with bacteria: potential applications in bioremediation of U contaminated oxic environments. *Rev. Environ. Sci. Biotechnol.* 14, 347–355. doi: 10.1007/s11157-015-9366-6
14. Choy, C. C., Korfiatis, G. P., and Meng, X. (2006). Removal of depleted uranium from contaminated soils. *J. Hazard. Mater.* 136, 53–60. doi: 10.1016/j.

jhazmat.2005.11.011

15. Dew, D. W., van Buuren, C., McEwan, K., and Bowker, C. (1999). "Bioleaching of base metal sulphide concentrates: a comparison of mesophile and thermophile bacterial cultures," in *Biohydrometallurgy and the Environment Toward the Mining of the 21st Century, Part A*, Vol. 9, eds A. Ballester and R. Amils (Madrid: Elsevier Science), 229–238. doi: 10.1016/s1572-4409(99)80022-4
16. Dispirito, A. A., and Tuovinen, O. H. (1982). Uranous ion oxidation and carbon-dioxide fixation by *Thiobacillus ferrooxidans.Arch. Microbiol.* 133, 28–32. doi: 10.1007/BF00943765
17. Dobson, R. S., and Burgess, J. E. (2007). Biological treatment of precious metal refinery wastewater: a review. *Miner. Eng.* 20, 519–532. doi: 10.1016/j.mineng.2006.10.011
18. DSMZ (2015). *Acidiphilium Medium*. Available online at: https://www.dsmz.de/ (Accessed 2016).
19. Duff, M. C., Hunter, D. B., Bertsch, P. M., and Amrhein, C. (1999). Factors influencing uranium reduction and solubility in evaporation pond sediments. *Biogeochemistry* 45, 95–114. doi: 10.1007/BF00992875
20. Ferris, F. G., and Beveridge, T. J. (1986). Site specificity of metallic ion binding in *Escherichia coli* k-12 lipopolysaccharide. *Can. J. Microbiol.* 32, 52–55. doi: 10.1139/m86-010
21. Finneran, K. T., Housewright, M. E., and Lovley, D. R. (2002). Multiple influences of nitrate on uranium solubility during bioremediation of uranium-contaminated subsurface sediments. *Environ. Microbiol.* 4, 510–516. doi: 10.1046/j.1462-2920.2002.00317.x
22. Fowle, D. A., Fein, J. B., and Martin, A. M. (2000). Experimental study of uranyl adsorption onto *Bacillus subtilis*. *Environ. Sci. Technol.* 34, 3737–3741. doi: 10.1021/es991356h
23. Gadd, G. M. (1993). Interactions of fungi with toxic metals. *New Phytol.* 124, 25–60. doi: 10.1111/j.1469-8137.1993.tb03796.x
24. Gavrilescu, M., Pavel, L. V., and Cretescu, I. (2009). Characterization and remediation of soils contaminated with uranium. *J. Hazard. Mater.* 163, 475–510. doi: 10.1016/j.jhazmat.2008.07.103
25. Giro, M. E. A., Garcia, O., and Zaiat, M. (2006). Immobilized cells of *Acidithiobacillus ferrooxidans* in PVC strands and sultite removal in a pilot-scale bioreactor. *Biochem. Eng. J.* 28, 201–207. doi: 10.1016/j.bej.2005.11.005
26. Gorman-Lewis, D., Elias, P. E., and Fein, J. B. (2005). Adsorption of aqueous uranyl complexes onto *Bacillus subtilis* cells.*Environ. Sci. Technol.* 39, 4906–4912. doi: 10.1021/es047957c
27. Haas, J. R., Dichristina, T. J., and Wade, R. (2001). Thermodynamics of U(VI) sorption onto *Shewanella putrefaciens*. *Chem. Geol.* 180, 33–54. doi: 10.1016/S0009-2541(01)00304-7

28. Haas, J. R., and Northup, A. (2004). Effects of aqueous complexation on reductive precipitation of uranium by *Shewanella putrefaciens*. *Geochem. Trans.* 5, 41–48. doi: 10.1186/1467-4866-5-41

29. IBM-SPSS (2011). *IBM SPSS Statistics for Windows, Version 20.0, 20th Edn.* Armonk, NY: IBM Corp.

30. Islam, E., and Sar, P. (2016). Diversity, metal resistance and uranium sequestration abilities of bacteria from uranium ore deposit in deep earth stratum. *Ecotoxicol. Environ. Saf.* 127, 12–21. doi: 10.1016/j.ecoenv.2016.01.001

31. Jelinek, R. T., and Sorg, T. J. (1988). Operating a small full-scale ion exchange system for uranium removal. *Am. Water Works Assoc. J.* 80, 79–83.

32. Kelly, S. D., Kemner, K. M., Fein, J. B., Fowle, D. A., Boyanov, M. I., Bunker, B. A., et al. (2002). X-ray absorption fine structure determination of pH-dependent U-bacterial cell wall interactions. *Geochim. Cosmochim. Acta* 66, 3855–3871. doi: 10.1016/S0016-7037(02)00947-X

33. Khijniak, T. V., Slobodkin, A. I., Coker, V., Renshaw, J. C., Livens, F. R. Bonch-Osmolovskaya, E. A., et al. (2005). Reduction of Uranium (VI) phosphate during growth of thermophilic bacterium *Thermoterrabacterium ferrireducens*. *Appl. Environ. Microbiol.* 71, 6423–6426. doi: 10.1128/AEM.71.10.6423-6426.2005

34. Langley, S., and Beveridge, T. J. (1999). Effect of O-side-chain-lipopolysaccharide chemistry on metal binding. *Appl. Environ. Microbiol.* 65, 489–498.

35. Google Scholar

36. Leduc, L. G., and Ferroni, G. D. (1994). The chemolithotrophic bacterium *Thiobacillus ferrooxidans*. *FEMS Microbiol. Rev.* 14, 103–119. doi: 10.1111/j.1574-6976.1994.tb00082.x

37. Llorens, I., Untereiner, G., Jaillard, D., Gouget, B., Chapon, V., and Carriere, M. (2012). Uranium interaction with two multi-resistant environmental bacteria: *Cupriavidus metallidurans* CH34 and *Rhodopseudomonas palustris*. *PLoS One* 7:e51783. doi: 10.1371/journal.pone.0051783

38. Lloyd, J. R., Leang, C., Myerson, A. L. H., Coppi, M. V., Cuifo, S., Methe, B., et al. (2003). Biochemical and genetic characterization of PpcA, a periplasmic c-type cytochrome in *Geobacter sulfurreducens*. *Biochem. J.* 369, 153–161. doi: 10.1042/bj20020597

39. Lloyd, J. R., and Renshaw, J. C. (2005). Microbial transformations of radionuclides: fundamental mechanisms and biogeochemical implications. *Metal Ions Biol. Syst.* 44, 205–240.

40. Lütke, L., Moll, H., and Bernhard, G. (2012). Insights into the uranium(VI) speciation with *Pseudomonas fluorescens* on a molecular level. *Dalton Trans.* 41, 13370–13378. doi: 10.1039/c2dt31080e

41. Luo, W., and Gu, B. (2011). Dissolution of uranium-bearing minerals and mobilization of uranium by organic ligands in a biologically reduced sediment. *Environ. Sci. Technol.* 45, 2994–2999. doi: 10.1021/es103073u

42. Martinez, R. J., Beazley, M. J., Taillefert, M., Arakaki, A. K., Skolnick, J., and Sobecky, P. A. (2007). Aerobic uranium (VI) bioprecipitation by metal-resistant bacteria isolated from radionuclide- and metal-contaminated subsurface soils. *Environ. Microbiol.* 12:9. doi: 10.1111/j.1462-2920.2007.01422.x

43. Mazuelos, A., Carranza, F., Romero, R., Iglesias-Gonzalez, N., and Villalobo, E. (2010). Operational pH in packed-bed reactors for ferrous ion bio-oxidation. *Hydrometallurgy* 104, 186–192. doi: 10.1016/j.hydromet.2010.06.002

44. Merroun, M. L., and Selenska-Pobell, S. (2001). Interactions of three eco-types of *Acidithiobacillus ferrooxidans* with U(VI).*Biometals* 14, 171–179. doi: 10.1023/A:1016658209397

45. Merroun, M. L., and Selenska-Pobell, S. (2008). Bacterial interactions with uranium: an environmental perspective. *J. Contam. Hydrol.* 102, 285–295. doi: 10.1016/j.jconhyd.2008.09.019

46. Morcillo, F., González-Muñoz, M. T., Reitz, T., Romero-González, M. E., Arias, J. M., and Merroun, M. L. (2014). Biosorption and biomineralization of U(VI) by the marine bacterium *Idiomarina loihiensis* MAH1: effect of background electrolyte and pH.*PLoS ONE* 9:e91305. doi: 10.1371/journal.pone.0091305

47. Nedelkova, M., Merroun, M. L., Rossberg, A., Hennig, C., and Selenska-Pobell, S. (2007). *Microbacterium* isolates from the vicinity of a radioactive waste depository and their interactions with uranium. *FEMS Microbiol. Ecol.* 59, 694–705. doi: 10.1111/j.1574-6941.2006.00261.x

48. Neu, M. P., Boukhalfa, H., and Merroun, M. L. (2010). Biomineralization and biotransformations of actinide materials. *MRS Bull.* 35, 849–857. doi: 10.1557/mrs2010.711

49. Nevin, K. P., and Lovley, D. R. (2000). Potential for nonenzymatic reduction of Fe(III) via electron shuttling in subsurface sediments. *Environ. Sci. Technol.* 34, 2472–2478. doi: 10.1021/es991181b

50. Newsome, L., Morris, K., and Lloyd, J. R. (2014). The biogeochemistry and bioremediation of uranium and other priority radionuclides. *Chem. Geol.* 363, 164–184. doi: 10.1016/j.chemgeo.2013.10.034

51. Newsome, L., Morris, K., Shaw, S., Trivedi, D., and Lloyd, J. R. (2015). The stability of microbially reduced U(IV); impact of residual electron donor and sediment ageing. *Chem. Geol.* 409, 125–135. doi: 10.1016/j.chemgeo.2015.05.016

52. Orellana, R., Leavitt, J. J., Comolli, L. R., Csencsits, R., Janot, N., Flanagan, K. A., et al. (2013). U(VI) reduction by diverse outer surface c-type cytochromes of *Geobacter sulfurreducens*. *Appl. Environ. Microbiol.* 79, 6369–6374. doi: 10.1128/AEM.02551-13

53. Pan, X. H., Chen, Z., Chen, F. B., Cheng, Y. J., Lin, Z., and Guan, X. (2015). The mechanism of uranium transformation from U(VI) into nano-uramphite by two indigenous *Bacillus thuringiensis* strains. *J. Hazard. Mater.* 297, 313–319. doi: 10.1016/j.jhazmat.2015.05.019

54. Panak, P. J., Knopp, R., Booth, C. H., and Nitsche, H. (2002). Spectroscopic studies on the interaction of U(VI) with *Bacillus sphaericus. Radiochim. Acta* 90, 779–783. doi: 10.1524/ract.2002.90.9-11_2002.779

55. Salome, K. R., Green, S. J., Beazley, M. J., Webb, S. M., Kostka, J. E., and Taillefert, M. (2013). The role of anaerobic respiration in the immobilization of uranium through biomineralization of phosphate minerals. *Geochim. Cosmochim. Acta* 106, 344–363. doi: 10.1016/j.gca.2012.12.037

56. Selenska-Pobell, S. (2002). "Diversity and activity of bacteria in uranium waste piles," in *Interactions of Microorganisms with Radionuclides*, Vol. 2, eds M. J. Keith-Roach and F. R. Livens (Amsterdam: Elsevier Science), 225–254.

57. Selenska-Pobell, S., Panak, P., Miteva, V., Boudakov, I., Bernhard, G., and Nitsche, H. (1999). Selective accumulation of heavy metals by three indigenous *Bacillus strains, B. cereus, B. megaterium* and *B. sphaericus*, from drain waters of a uranium waste pile. *FEMS Microbiol. Ecol.* 29, 59–67. doi: 10.1111/j.1574-6941.1999.tb00598.x

58. Senko, J. M., Istok, J. D., Suflita, J. M., and Krumholz, L. R. (2002). *In-situ* evidence for uranium immobilization and remobilization. *Environ. Sci. Technol.* 36, 1491–1496. doi: 10.1021/es011240x

59. Senko, J. M., Mohamed, Y., Dewers, T. A., and Krumholz, L. R. (2005a). Role for Fe(III) minerals in nitrate-dependent microbial U(IV) oxidation. *Environ. Sci. Technol.* 39, 2529–2536. doi: 10.1021/es048906i

60. Senko, J. M., Suflita, J. M., and Krumholz, L. R. (2005b). Geochemical controls on microbial nitrate-dependent U(IV) oxidation.*Geomicrobiol. J.* 22, 371–378. doi: 10.1080/01490450500248911

61. Shelobolina, E. S., Konishi, H., Xu, H., and Roden, E. E. (2009). U(VI) sequestration in hydroxyapatite produced by microbial glycerol 3-phosphate metabolism. *Appl. Environ. Microbiol.* 75, 5773–5778. doi: 10.1128/AEM.00628-09

62. Singh, G., Sengoer, S. S., Bhalla, A., Kumar, S., De, J., Stewart, B., et al. (2014). Reoxidation of biogenic reduced uranium: a challenge toward bioremediation. *Crit. Rev. Environ. Sci. Technol.* 44, 391–415. doi: 10.1080/10643389.2012.728522

63. Suzuki, Y., and Banfield, J. F. (2004). Resistance to, and accumulation of, uranium by bacteria from a uranium-contaminated site. *Geomicrobiol. J.* 21, 113–121. doi: 10.1080/01490450490266361

64. USA-EPA (2014). *Federal Actions to Address the Impact of Uranium Contamination in the Navajo Nation.* San Francisco, CA.

65. Wall, J. D., and Krumholz, L. R. (2006). Uranium reduction. *Annu. Rev. Microbiol.* 60, 149–166. doi: 10.1146/annurev.micro.59.030804.121357

66. Wang, J., and Chen, C. (2009). Biosorbents for heavy metals removal and their future. *Biotechnol. Adv.* 27, 195–226. doi: 10.1016/j.biotechadv.2008.11.002

67. WWAP (2009). *The United Nations World Water Development Report 3: Water in a Changing World.* Paris: UNESCO; London: Earthscan.

# 18
CHAPTER

# OVER-THE-COUNTER MONOCYCLIC NON-STEROIDAL ANTI-INFLAMMATORY DRUGS IN ENVIRONMENT—SOURCES, RISKS, BIODEGRADATION

Ariel Marchlewicz, Urszula Guzik, and Danuta Wojcieszyńska

Department of Biochemistry, Faculty of Biology and Environmental Protection, University of Silesia in Katowice, Jagiellonska 28, 40-032 Katowice, Poland

## ABSTRACT

**Recently, the increased use of monocyclic non-steroidal anti-inflammatory drugs has resulted in their presence in the environment. This may have potential negative effects on living organisms. The biotransformation mechanisms of monocyclic non-steroidal anti-inflammatory drugs in the human body and in other mammals occur by hydroxylation and conjugation with glycine or glucuronic acid. Biotransformation/biodegradation of monocyclic non-steroidal anti-inflammatory drugs in the environment may be caused by fungal or bacterial microorganisms. Salicylic acid derivatives are degraded by catechol or gentisate as intermediates which are cleaved by dioxygenases. The key intermediate of the paracetamol degradation pathways is hydroquinone. Sometimes, after hydrolysis of this drug, 4-aminophenol is formed, which is a dead-end metabolite. Ibuprofen is metabolized by hydroxylation or activation with CoA, resulting in the formation of isobutylocatechol. The aim of this work is to attempt to summarize the knowledge about environmental risk connected with the presence of over-the-counter anti-inflammatory drugs, their sources and the biotransformation and/or biodegradation pathways of these drugs.**

**Citation:** Ariel Marchlewicz, Urszula Guzik, and Danuta Wojcieszyńska, Over-the-Counter Monocyclic Non-Steroidal Anti-Inflammatory Drugs in Environment—Sources, Risks, Biodegradation, DOI 10.1007/s11270-015-2622-0.

## INTRODUCTION

In an age of high level care of human health, many pharmaceuticals are commonly used to cure or prevent diseases and other ailments, such as headache, muscle pain, or inflammatory conditions. Presently, over-the-counter drugs are very popular, especially over-the-counter monocyclic and polycyclic non-steroidal anti-inflammatory drugs (NSAIDs). Among these drugs, the most popular and the most often used are monocyclic NSAIDs, such as ibuprofen, acetaminophen, and salicylic acid (and its derivatives, like mesalazine), due to their availability (Ziylan and Ince 2011). For example, the yearly intake of ibuprofen is up to 300 t in Germany, 162 t in England, and 58 t in Poland (Sosnowska et al. 2009; Guzik et al. 2013a). In the USA during 2001–2005, about 29 billion doses of paracetamol in all forms were sold (Li et al. 2014). The high intake of these widely available drugs may lead to their or their metabolites' presence in the environment. In connection with the presence of NSAIDs in the environment, there is a risk of long-term exposure, causing chronic toxic effect in organisms living there. This may cause negative effects for living creatures and the accumulation of drugs or their metabolites in the food chain (Carlsson et al. 2006; Sosnowska et al. 2009). Current knowledge about the microbial metabolism of non-steroidal anti-inflammatory drugs is still very little, and the fact that we can find them in the environment suggests that sewage treatment plants are not currently adapted to completely remove these drugs before they reach the environment. These drugs and their metabolites are found in wastewater influent and effluent from wastewater treatment plants. For example, in Germany, acetylsalicylic acid was detected in the sewage effluents at 0.22 μg/L (Heberer 2002) and paracetamol was detected in groundwater used as a source of public drinking water in California at 1.89 μg/L (Li et al. 2014). Observed metabolites are formed as a result of the metabolism by activated sludge microorganisms or enter the treatment plants as the human body biotransformation products with municipal wastewater (Buser et al. 1999; Zwiener et al. 2002; Marco-Urrea et al. 2009; Ziylan and Ince 2011). The fate of medicines, including monocyclic NSAIDs, in the natural environmental is still less known. The main aim of this work is a compilation of the actual knowledge about sources and risks connected with the presence of monocyclic non-steroidal anti-inflammatory drugs in the environment. Moreover, authors describe microbiological degradation of the three most widespread painkillers, antipyretic and anti-inflammatory and over-the-counter drugs—acetylsalicylic acid, paracetamol, and ibuprofen.

## SOURCES OF PHARMACEUTICALS IN THE ENVIRONMENT

The development of modern analytical methods makes it possible to detect NSAIDs in the environment. Considering the high intake of drugs, it may be assumed that this has an impact on the presence of pharmaceuticals in wastewaters and surface waters. The main sources of drugs that reach the environment are excreted in non-metabolized form or slightly modified, i.e., hydroxylated, conjugated, and disposed of through the toilet (Buser et al. 1999; Heberer2002; Khan and Ongerth 2002; Zwiener et al. 2002; Metcalfe et al. 2004). Nearly half of respondents declared the disposal of medications in the household trash. That is why the presence of pharmaceuticals is expected in the landfill leachate or leachate-contaminated groundwater (Kuspis and Krenzelok 1996; Musson and Townsend 2009). Hospitals' wastewaters and discharges from pharmaceutical production also constitute a significant source of pharmaceuticals in the environment. In many countries, hospitals and pharmaceutical factories do not have separate wastewater treatment plants (WWTP); therefore, these contaminants pass into the general wastewater treatment system (Metcalfe et al. 2004). The burdened sewage flows into WWTPs, but not all NSAIDs are removed in biological sewage treatment with activated sludge. Drug detection in WWTP effluent confirmed inadequacy of wastewater treatment plants to completely removing these pollutants from sewage (Ternes1998; Buser et al. 1999; Tixier et al. 2003; Lee et al. 2005; Gómez et al. 2007; Salgado et al.2010). Consequently, this leads to the detection of drugs even in surface waters such as lakes or rivers (Winkler et al. 2001; Dębska et al. 2005; Roberts and Thomas 2006; Vieno et al. 2006; Togola and Budzinski 2008; Pailler et al. 2009). Additionally, Kolpin et al. (2004) observed an increased concentration of drugs in downstream, above places of outflows from WWTP.

## ENVIRONMENTAL RISK OF MONOCYCLIC NON-STEROIDAL ANTI-INFLAMMATORY DRUGS

Although NSAIDs are observed in the environment in low concentrations, there is little known about the long-term effects of low concentrations of these drugs on living organisms. The most data about toxicity of ibuprofen, paracetamol, and acetylsalicylic acid are based on acute toxicity and short-term chronic toxicity tests (Table 1). In acute toxicity tests, high concentrations of substances that may result in unrealistic effects are usually used. In many cases, metabolites of drugs are not taken into account in toxicity tests; therefore, it is difficult to evaluate the real risk of NSAIDs and their metabolites on the environment (Webb 2004). Marco-Urrea et al. (2009), using Microtox toxicity test with *Photobacterium phosphoreum* as a tested organism, proved that hydroxylated derivatives of ibuprofen (which was also found in sewage and surface waters) are more toxic than the original compound. Pomati et al. (2004) showed that even microgram per liter concentration of ibuprofen can influence the growth of aquatic phototrophs. For example, *Lemna minor* exhibited inhibition of growth after 7-day exposure to low concentration of ibuprofen. Under these conditions, the little effect on abscisic acid production was also observed (Pomati et al. 2004; Brausch et al. 2012; Murdoch and Hay 2013). High sensitivity to ibuprofen was also found for phytoplankton. Depending on tested organisms, $EC_{50}$ value was between 1 and 315 mg/L after 72–120-h exposition to this drug (Brausch et al. 2012).

**Table 1**: Toxicity of selected monocyclic NSAIDs

| Organism | Drug | Duration | Concentration (mg/L) | References |
|---|---|---|---|---|
| *Daphnia magna* | ASA | 24 h $EC_{50}$ | 1468 | Lilius et al. (1994) |
| *Artemia salina* | ASA | 24 h $EC_{50}$ | 382 | Calleja et al. (1994) |

| | | | | |
|---|---|---|---|---|
| *Streptocephalus proboscideus* | ASA | 24 h $EC_{50}$ | 178 | Calleja et al. (1994) |
| *Daphnia magna* | ASA | 24 h $EC_{50}$ | 168 | Calleja et al. (1994) |
| *Brachionus calyciflorus* | ASA | 24 h $EC_{50}$ | 141 | Calleja et al. (1994) |
| *Desmodesmus subspicatus* | ASA | 24 h $EC_{50}$ | 106.7 | Cleuvers (2004) |
| *Daphnia magna* | ASA | 24 h $EC_{50}$ | 88.1 | Cleuvers (2004) |
| *Daphnia magna* | Ibuprofen | 48 h $EC_{50}$ | 51.4 | Han et al. (2010) |
| *Moina macrocopa* | Ibuprofen | 48 h $EC_{50}$ | 72.6 | Han et al. (2010) |
| *Keletonema cost-stum* | Ibuprofen | 96 h $EC_{50}$ | 7.1 | Halling-Sorensen et al. (1998) |
| *Daphnia magna* | Ibuprofen | 48 h $EC_{50}$ | 9.06 | Halling-Sorensen et al. (1998) |
| *Lepomismacro-chirus* | Ibuprofen | 96 h $EC_{50}$ | 173 | Halling-Sorensen et al. (1998) |
| *Thamnocephalus platyurus* | Ibuprofen | 24 h $EC_{50}$ | 19.59 | Kim et al. (2009) |
| *Oryzias latipes* | Ibuprofen | 96 h $EC_{50}$ | >100 | Kim et al. (2009) |
| *Hydra attenuata* | Ibuprofen | 96 h $LC_{50}$ | 22.36 | Quinn et al.(2008) |
| *Hydra attenuata* | Ibuprofen | 96 h $EC_{50}$ | 1.65 | Quinn et al. (2008) |
| *Daphnia magna* | Ibuprofen | 24 h $EC_{50}$ | 101.2 | Cleuvers (2004) |
| *Desmodesmus subspicatus* | Ibuprofen | 24 h $EC_{50}$ | 343.2 | Cleuvers (2004) |
| *Artemia salina* | Paracetamol | 24 h $EC_{50}$ | 577 | Calleja et al.(1994) |
| *Daphnia magna* | Paracetamol | 24 h $EC_{50}$ | 55.5 | Calleja et al. (1994) |
| *Brachionus calyciflorus* | Paracetamol | 24 h $EC_{50}$ | 5306 | Calleja et al. (1994) |
| *Daphnia magna* | Paracetamol | 24 h $EC_{50}$ | 13 | Kühn et al. (1989)) |
| *Daphnia magna* | Paracetamol | 48 h $EC_{50}$ | 9.2 | Kühn et al. (1989)) |
| *Daphnia magna* | Paracetamol | 24 h $EC_{50}$ | 293 | Henschel et al. (1997) |
| *Daphnia magna* | Paracetamol | 48 h $EC_{50}$ | 50 | Henschel et al. (1997) |
| *Brachydanio rerio* | Paracetamol | 48 h $EC_{50}$ | 378 | Henschel et al. (1997) |
| *Scenedesmus subspicatus* | Paracetamol | 24 h $EC_{50}$ | 134 | Henschel et al. (1997) |
| *Daphnia magna* | SA | 24 h $EC_{50}$ | 230 | Wang and Lay (1989) |
| *Daphnia magna* | SA | $EC_{50}$ | 118 | Henschel et al. (1997) |

| | | | | |
|---|---|---|---|---|
| *Brachydanio rerio* | SA | 48 h $EC_{50}$ | 37 | Henschel et al. (1997) |
| *Scenedesmus subspicatus* | SA | 72 h $EC_{50}$ | >100 | Henschel et al. (1997) |

*SA* salicylic acid, *ASA* acetylsalicylic acid (aspirin), *ND* no data

After 14 days of exposure of *Daphnia magna* to ibuprofen at concentration of 20, 40, and 80 mg/L, significant effects in the total reproduction of daphnias were observed. Reproduction decreased with the increase of the drug concentration and totally stopped at 80 mg/L. Additionally, the time of first reproduction was delayed in a concentration of 40 mg/L. A low concentration of ibuprofen within the range of 1–100 ng/L caused a decrease in the activity of*Gammarus pulex*. This is very important information, because this concentration range of ibuprofen corresponds more to the concentration in the environment (de Lange et al. 2006). Pounds et al. (2008) showed some chronic effects of ibuprofen on mollusc *Planorbis carinatus*in the drug concentrations 0.41, 1.02, 2.43, and 5.36 mg/L. In the lowest dose, the authors did not observe snails laying eggs. A dose of ibuprofen at 5.36 mg/L caused an inhibition of egg hatching (Pounds et al. 2008). Han et al. (2010) also observed a delay in *Oryzias latipes* egg hatching after exposure to 0.1 µg/L of ibuprofen. After 120 days of ibuprofen exposure, the survival of fish was also significantly lower when compared to the control population (Han et al.2010). The obtained results are significant, because the drug concentration used in the experiment is observed in the environment (Dębska et al. 2005; Pailler et al. 2009).

Wu et al. (2012) described *p*-aminophenol as a major metabolite of paracetamol metabolism in microbes. 4-Aminophenol is one of the most toxic phenols, which causes the kidney or the liver damage (Newton et al. 1982; Song and Chen 2001). Li et al. (2014) observed the appearance of N-acetyl-*p*-benzoquinone imine (NAPQI) during biodegradation of acetaminophen in soil. Additionally, the increased sorption of paracetamol in soil biosolids was observed, which may cause a decrease of acetaminophen mineralization. Simultaneously, it may affect the half-life of drugs in the environment (Li et al. 2014).

N-acetyl-*p*-benzoquinone imine is also one of the first phase metabolites of paracetamol detoxification in humans, excreted as a glutathione conjugate with urine (Tsikas et al. 2011; Li et al. 2014). It is defined as highly hepatotoxic (Bender and MacCrehan 2006; Hinson et al.2010; Tsikas et al. 2011). Bender et al. (2004) suggest that NAPQI may be a potent inhibitor of human topoisomerase IIα. Toxicological research conducted in the presence of high concentration of NSAIDs does not provide information about the influence of long-term exposure to low concentration of drugs. The answer to this question may be provided only by long-term research of many generations of aquatic organisms living in the presence of low drug concentration (Rzepa 2009).

## ACETYLSALICYLIC ACID BIODEGRADATION BY MICROORGANISMS

In humans and animals, acetylsalicylic acid is immediately hydrolyzed to salicylic acid which can be removed from the body unchanged or in the form of conjugates with glycine (as a salicyluric acid) or with glucuronic acid, or hydroxylated at the C-5 position of the ring to gentisate (de Gaetano et al. 1985; Ingelman-Sundberg et al. 1991; Paterson et al. 2008). Apart from excretion from human and animal organisms, salicylic acid is a widespread molecule in plants. It plays a role in several physiological processes, like stomatal closure, flower induction, heat production and, most of all, its main function is defense against pathogen attack (Verberne et al. 2000). If we take this into consideration, it is not strange that in nature, there are effective salicylate degradation mechanisms. Many bacterial strains, like *Micrococcus*, *Sphingomonas*,*Amycolatopsis*, *Streptomyces*, *Pseudomonas*, *Alcaligenes*, *Pseudoramibacter*, *Rhodococcus*(Chakrabarty 1972; Shamsuzzaman and Barnsley 1974; Haribabu et al. 1984; Grund et al. 1990; Grund et al. 1992; Civilini et al. 1999; Hintner et al. 2001; Ishiyama et al. 2004; Deveryshetty et al. 2007; Jouanneau et al. 2007; Silva et al. 2007; Lanfranconi et al. 2009) and fungi, like*Sclerotinia*, *Trichosporon*, *Aspergillus*, *Fusarium*, *Rhodotorula*, *Cryptococcus* (Anderson and Dagley 1980; Kuswandi and Roberts 1992; Middelhoven 1993; Iwasaki et al. 2009; Qi et al.2012; Penn and Daniel 2013) are capable of degrading salicylate (Table 2) via a few catabolic pathways.

**Table 2**: Microorganisms degrading selected monocyclic NSAIDs

| **Strain** | **Drug** | **Concentration** | **References** |
|---|---|---|---|
| *Pseudomonas putida* R1 | SA | 10 mM | Chakrabarty (1972) |
| *Acinetobacter lwoffii* | ASA | 2.77 mM | Grant (1971) |
| *Amycolatopsis rugosa* DSM 43387 | SA | ND | Grund et al. (1990) |
| *Amycolatopsis rugosa* DSM 43388 | SA | ND | Grund et al. (1990) |
| *Streptomyces niger* DSM 40302 | SA | ND | Grund et al. (1990) |
| *Streptomyces olivaceiscloticus*-DSM 40595 | SA | ND | Grund et al. (1990) |
| *Rhodococcus* sp. B4 | SA | ND | Grund et al. (1992) |
| *Pseudaminobacter salicylatoxidans* B12 | SA | 5 mM | Hintner et al. (2001) |
| *Pseudomonas* sp. | SA | 3.5 mM | Shamsuzzaman and Barnsley (1974) |
| *Pseudomonas aeruginosa* 2NR | SA | ND | Civilini et al. (1999) |
| *Streptomyces* sp. WA46 | SA | 7.24 mM (on agar plate) | Ishiyama et al. (2004) |
| *Rhodococcus ruber* IEGM 77 | ASA | 1.38–2.77 mM | Ivshina et al. (2006) |
| *Pseudomonas* sp. PPD | SA | ND | Deveryshetty et al. (2007) |
| *Alcaligenes* sp. PPH | SA | ND | Deveryshetty et al. (2007) |
| *Sphingomonas* sp. CHY-1 | SA | 0.5–1 mM | Jouanneau et al. (2007) |
| *Pseudomonas fluorescens* HK44 | SA | 0.18–1.45 mM | Silva et al. (2007) |

| | | | |
|---|---|---|---|
| *Trichosporon cutaneum* | SA | ND | Anderson and Dagley (1980) |
| *Aspergillus nidulans* | SA | 10 mM | Kuswandi and Roberts (1992) |
| *Fusarium graminearum* | SA | 0.1–20 mM | Qi et al. (2012) |
| *Sclerotinia sclerotiorum* | SA | 1–10 mM | Penn and Daniel (2013) |
| *Trichosporon moniliiforme* WU-0401 | SA | 70 mM | Iwasaki et al. (2009) |
| Rhodococcus spp. | Paracetamol | 1.65–3.31 mM | Ivshina et al. (2006) |
| *Delftia tsuruhatensis* | Paracetamol | 0.007 mM | de Gusseme et al. (2011) |
| *Pseudomonas aeruginosa* | Paracetamol | 0.007 mM | de Gusseme et al. (2011) |
| *Stenotrophomonas* sp. f1 | Paracetamol | 2.64 mM | Zhang et al. (2013) |
| *Pseudomonas* sp. f2 | Paracetamol | 16.54 mM | Zhang et al. (2013) |
| *Pseudomonas* sp. fg-2 | Paracetamol | 13.23 mM | Zhang et al. (2013) |
| *Penicillium* sp. | Paracetamol | 0.66 mM | Hart and Orr (1975) |
| Filamentous fungi | Paracetamol | 0.99 mM | Huang et al. (2006) |
| *Nocardia* sp. NRRL 5646 | Ibuprofen | 4.85 mM | Chen and Rosazza (1994) |
| *Sphingomonas* Ibu-2 | Ibuprofen | 2.42 mM | Murdoch and Hay (2005) |
| *Patulinobacter* sp. I11 | Ibuprofen | 0.002 mM | Almeida et al. (2013) |
| *Bjerkandera* sp. R1 | Ibuprofen | 0.0049–0.0097 mM | Rodarte-Morales et al. (2011) |
| *Bjerkandera adusta* | Ibuprofen | 0.0049-0.0097 mM | Rodarte-Morales et al. (2011) |
| *Phanerochaete chrysosporium* | Ibuprofen | 0.0049–0.0097 mM | Rodarte-Morales et al. (2011) |
| *Trametes versicolor* | Ibuprofen | 0.097 mM | Marco-Urrea et al. (2009) |
| *Irpex lacteus* | Ibuprofen | 0.097 mM | Marco-Urrea et al. (2009) |
| *Ganoderma lucidum* | Ibuprofen | 0.097 mM | Marco-Urrea et al. (2009) |
| *Phanerochaete chrysosporium* | Ibuprofen | 0.097 mM | Marco-Urrea et al. (2009) |

*SA* salicylic acid, *ASA* acetylsalicylic acid (aspirin), *ND* no data

The strategy for degradation of aromatic structure comprises hydroxylation and cleavage of the aromatic ring. Hydroxylation into the dihydroxylated intermediates, the first step in the oxidative degradation of aromatic compounds, is catalyzed by oxygenases belonging to three groups: hydroxylating dioxygenases, activated-ring monooxygenases, or nonactivated-ring monooxygenases. As a result of hydroxylation, the key intermediates such as catechol, protocatechuic acid, hydroxyquinol, or gentisic acid are formed. These products are substrates for ring-cleaving dioxygenases. Salicylates are mainly transformed to catechol and gentisate, which are cleaved in the next step by dioxygenases from two groups—intradiol or extradiol (Guzik et al. 2013b; Guzik et al. 2014).

Two of the most important enzymes involved in salicylate decomposition are salicylate 1-hydroxylase and salicylate 5-hydroxylase (monooxygenases). Salicylate monooxygenases belong to one of the three groups of hydroxylating oxygenases—

activated-ring monooxygenases (Wojcieszyńska et al. 2011). The general structure of these groups includes a three-component protein with separate functional units: reductase with a flavin cofactor, ferrodoxin with a Rieskie-type iron-sulfur cluster [2Fe-2S] and hexamer-built $\alpha_3\beta_3$terminal oxygenase with [2Fe-2S] a cluster and one nonheme iron(II) per $\alpha$ subunit (Mason and Cammack 1992; Bertini et al.1996). These catalytic proteins are able to insert one atom of molecular oxygen into the structure of the aromatic ring, simultaneously reducing the second oxygen atom to water. All salicylate hydroxylases need NADH to remain active. The oxidation of NADH is directly connected with FAD reduction (Katagiri et al. 1965; Sze and Dagley 1984; Grund et al. 1992; Suzuki et al. 1996; Fuenmayor et al. 1998; Balashova et al. 2001; Zhou et al. 2002; Jouanneau et al. 2007).

Chakrabarty (1972) and Deveryshetty et al. (2007) examined the ability of *Pseudomonas putida*R1 and *Alcaligenes* sp. PPH, respectively, to degrade salicylate. It was decomposed to catechol, the key intermediate, and further to 2-hydroxymuconic semialdehyde as a product of ring fission. Because of the significant activity of catechol 2,3-dioxygenase the authors concluded that *P. putida* R1 was capable of *meta* cleavage. On the other hand, decomposition of salicylate via catechol may also run via *ortho* cleavage with *cis,cis*-muconic acid as an aromatic ring cleavage product. Not only bacteria, such as *Amycolatopsis*, *Streptomyces*, or *Pseudomonas*, but also fungi, like *Fusarium*, *Rhodotorula*, *Trichosporon*, and *Sclerotinia* show that kind of catechol ring fission (Shamsuzzaman and Barnsley 1974; Anderson and Dagley 1980; Sze and Dagley 1984; Grund et al. 1990; Fuenmayor et al. 1998; Civilini et al. 1999; Ishiyama et al.2004; Iwasaki et al. 2009; Qi et al. 2012; Penn and Daniel 2013).

Salicylate degradation may also lead via hydroxylation to gentisate. For example, this pathway was reported for *Rhodococcus* sp. B4 or *Streptomyces* sp. WA46 strains (Grund et al. 1992; Ishiyama et al. 2004). Enzymes engaged in this pathway need for their activity NADH, CoA and ATP as cofactors. Salicylate is converted to gentisate via salicylyl-CoA and gentisyl-CoA. In the first step, salicylyl-AMP ligase and probably salicylyl-CoA synthetase catalyze the formation of the thioester bond between salicylate and CoA and create salicylyl-CoA. The next step is the hydroxylation by salicylyl-CoA 5-hydroxylase with the formation of gentisyl-CoA. Hydrolysis to gentisate opens the way to ring cleavage by gentisate 1,2-dioxygenase to maleylpyruvate which leads to central metabolism. Civilini et al. (1999) showed the ability of *Pseudomonas aeruginosa* 2NR to convert salicylate into both intermediates, calechol, and gentisate.

Iwasaki et al. (2009) described a different model of decomposition of salicylate via catechol by yeast *Trichosporon moniliiforme* WU-0401. In their study, a non-oxidative way with phenol as an intermediate of salicylate degradation to catechol was presented. Salicylate was immediately transformed to phenol with, simultaneously, decarboxylation, but without hydroxylation. Before that, phenol was hydroxylated to catechol and further cleaved in the *ortho* position.

Most bacteria degrade salicylate via oxidative decarboxylation to catechol or via hydroxylation to gentisate. *Pseudaminobacter salicylatoxidans* B12 is capable of direct ring cleavage using NADH-independent salicylate 1,2-dioxygenase forming 2-oxohepta-3,5-dienedioic acid as an aliphatic product (Hintner et al. 2001).

## MICROBIAL DEGRADATION AND TRANSFORMATION OF PARACETAMOL

Paracetamol (acetaminophen) is an analgesic and antipyretic drug, and is one of the most popular over-the-counter drugs (Chandrasekharan et al. 2002). Acetaminophen is in most cases metabolized in the liver via three metabolic pathways. Most of this drug is secreted as conjugates with glucuronic acid (60 %) or sulfate (30 %) (Herd et al. 1991). A small amount of the medicine (8 %) may be dehydrogenated by cytochrome P450 to a toxic derivative N-acetyl-*p*-benzoquinone imine (Bock et al. 1987; Herd et al. 1991; Bessems and Vermeulen 2001). Nevertheless, knowledge about the further fate of acetaminophen in the environment is still limited. Hart and Orr (1975) obtained *Penicillium* sp. able to transform paracetamol to 4-aminophenol and acetate, probably with the use of aryl acylamidase. 4-Aminophenol is a dead-end metabolite (Fig. 1). Ivshina et al. (2006) reported the ability of *Rhodococcus* strains to degrade paracetamol with three detectable metabolites: 4-aminophenol, catechol, and hydroquinone. 4-Aminophenol may undergo oxidative deamination to hydroquinone (de Gusseme et al. 2011; Wei et al. 2011; Wu et al. 2012; Zhang et al. 2013). Further degradation of 1,4-hydroxybenzene could proceed in two ways. Hydroquinone may be directly cleaved by hydroquinone 1,2-dioxygenase with 4-hydroxymuconic semialdehyde as an aliphatic product (Jain et al. 1994; Rieble et al. 1994; Daubaras et al. 1996). This pathway was also suggested by de Gusseme et al. (2011), which observed degradation of hydroquinone by *Delftia tsuruhatensis*and *Pseudomonas aeruginosa*. The second way was described by Takenaka et al. (2003). These authors showed that *Burkholderia* sp. AK-4 converted 4-aminophenol to 1,4-hydroxybenzene and further to 1,2,4-trihydroxybenzene. Then 1,2,4-trihydroxybenzene was cleaved by hydroxyhydroquinone 1,2-dioxygenase to maleylacetic acid, which is introduced to the basic metabolism (Mason and Cammack 1992; Chauhan et al. 2000; Miyauchi et al. 1999; Moonen et al. 2008; Kolvenbach et al. 2011) (Fig. 1). Zhang et al. (2013) described the conversion of acetaminophen to hydroquinone, which was next transformed to an aliphatic product hexa-3-enedioic acid. Unfortunately, the authors did not determine the enzyme engaged in ring cleavage. However, it seems that hexa-3-enedioic acid was a product of aromatic ring fission or, if not, it that means some intermediate metabolites between aromatic and aliphatic compounds were passed over. Hexa-3-enedioic acid is similar to muconic acid—a product of *ortho ring*cleavage of catechol, but the authors did not find catechol in the studied samples. Based on reported intermediates a primary pathway of acetaminophen degradation could be proposed. The mechanism may be based on cutting off two carbon atoms in the form of formic acid (Fig. 1) (Zhang et al. 2013).

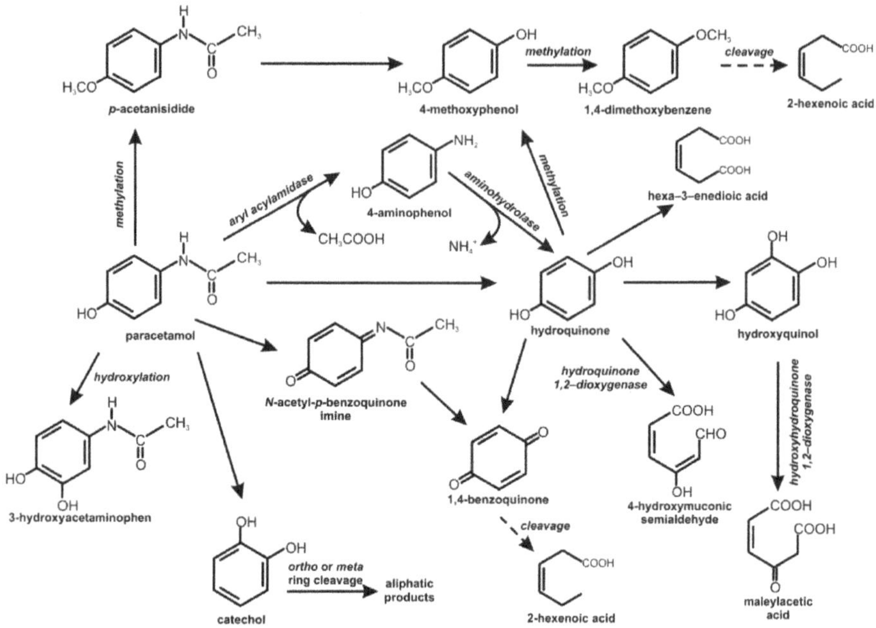


**Figure 1**: Biotransformation of paracetamol (Hart and Orr 1975; Takenaka et al. 2003; Kolvenbach et al. 2011; Wu et al. 2012; Zhang et al. 2013; Guzik et al. 2014; Li et al. 2014).

Furthermore, Huang et al. (2006) described the formation of glucoside conjugates with paracetamol by soil filamentous fungi via O- and N-linkages. This is a similar way to the human detoxication routes of xenobiotics in phase II of detoxication (Halling-Sorensen et al. 1998).

In 2014, Li et al. described degradation pathway of paracetamol in soil microorganisms. It was shown that in the first step, aromatic ring of paracetamol is hydroxylated to 3-hydroxyacetaminophen, oxygenated to N-acetyl-*p*-benzoquinone imine, or methylated to *p*-acetanisidide. It is suggested that cytochrome P-450 may be engaged in these processes. N-acetyl-*p*-benzoquinone imine is then metabolized to 1,4-benzoquinone which is more stable and critical toxic metabolite. *p*-Acetanisidide is transformed to 4-methoxyphenol and in the next step to the 1,4-dimethoxybenzene. The presence of 2-hexenoic acid in the soil extract suggests the cleavage of the aromatic ring of paracetamol (Li et al. 2014).

## BIODEGRADATION/BIOTRANSFORMATION OF IBUPROFEN

Ibuprofen (2-(4-(2-methylpropyl)phenyl)propanoic acid) is one of the most popular and commonly used non-steroidal anti-inflammatory drugs. This makes it the third most popular drug in the world. Ibuprofen is also one of the dominating medicines present in sewage because of its relatively high therapeutic dose (600–1200 mg per day), and significant levels are excreted from the human body (even 70–80 %). This drug may be secreted as an unchanged molecule or as an unchanged molecule in conjugation with glucuronide (product of the second phase of detoxication that may be hydrolyzed in the environment) or as a few metabolites: hydroxyibuprofen (two isomers), carboxyibuprofen, and carboxyhydratropic acid (Halling-Sorensen et al. 1998; Buser et al. 1999; Zwiener et al. 2002). Nonetheless, little is still known about the environmental metabolism of ibuprofen, whose concentration in the environment ranges from nanograms per liter to micrograms per liter (Calamari et al. 2003; Bendz et al.2005; Tauxe-Wuersch et al. 2005; Nakada et al. 2006; Roberts and Thomas 2006; Gómez et al.2007; Lin et al. 2009; Pailler et al. 2009).

Many reports describe only the initial steps of ibuprofen transformation. Rodarte-Morales et al. (2011) used three species of ligninolytic fungi: *Bjerkandera* sp. R1, *Bjerkandera adusta*, and*Phanerochaete chrysosporium* to check their ability to degrade pharmaceuticals, including ibuprofen (Table 2). They reported a rapid decrease of ibuprofen in growth medium, explaining that fact by degradation of the drug. However, these authors did not search for intermediates occurring in the degradation process, and they noted only the loss of the parent compound. It may be suggested that ibuprofen was not completely mineralized. *Trametes versicolor*, *Irpex lacteus*, *Ganoderma lucidum*, and *P. chrysosporium* are fungi capable of degrading ibuprofen (Marco-Urrea et al. 2009). All of them, excluding *P. chrysosporium*, degraded 10 mg/L ibuprofen to non-detectable levels. *P. chrysosporium* showed the lowest degradation level, between 70 and 88 %. Simultaneously, it should be noted that the authors tested the in vitro activity of laccase (also with laccase mediators) and manganese peroxidase. Moreover, they used inhibitors of the cytochrome P-450 complex to analyze the participation of these enzymes in ibuprofen degradation. In all cases, the researchers did not observe the contribution of the examined enzymes in the metabolism of ibuprofen. It may be suggested that the metabolism of ibuprofen could run through another pathway. The major metabolites which were found were hydroxylated in the isopropyl chain from ibuprofen—1- and 2-hydroxyibuprofen after the first hours of the experiment, and 1,2-dihydroxyibuprofen as a final metabolite after 7 days of cultivation (Fig. 2). Hydroxylated and carboxylated derivatives are frequent in the microbial metabolism (Zwiener et al. 2002; Quintana et al. 2005). It is noteworthy that ibuprofen derivatives are more toxic than the parent compound and may accumulate in the environment (Marco-Urrea et al. 2009). Despite low concentrations of these compounds in the ecosystem, they may be hazardous. However, the long-term effects of the organism exposure to this drug cannot be defined (Perry and Zylstra 2007). On the basis of European Union law, Cleuvers (2004) did not classify ibuprofen as toxic to aquatic organisms. However, a study with *Daphnia magna* and green algae showed that ibuprofen may be a toxic factor, especially in the presence of other drugs. This study informs about the effect of high-dose toxicity in short time exposure. In the environment, pharmaceuticals are not in high concentrations, so greater emphasis should be put on studies on the chronic toxicity of drugs (Cleuvers 2004).

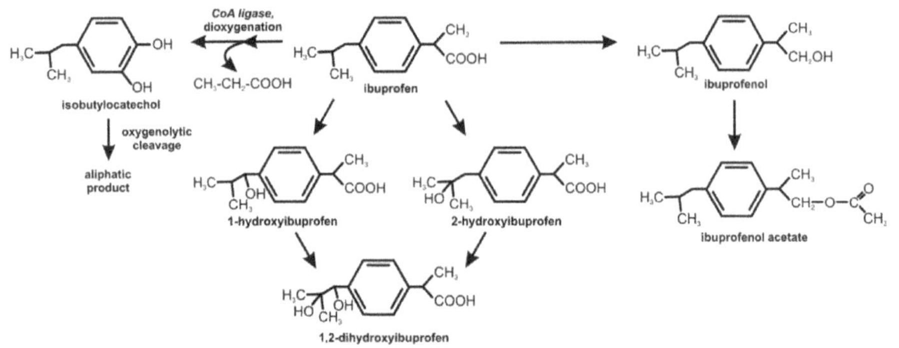


**Figure 2**: Microbiological transformation of ibuprofen (Chen and Rosazza 1994; Murdoch and Hay2005; Quintana et al. 2005; Guzik et al. 2014).

During the degradation of ibuprofen by the lignolytic bacterium *Nocardia* sp. NRRL 5646, two metabolites, ibuprofenol and ibuprofenol acetate, were observed (Fig. 2). These products underwent further mineralization (Chen and Rosazza 1994).

Murdoch and Hay (2005, 2013) characterized one of the most completed ibuprofen degradation pathways in *Sphingomonas* Ibu-2 bacteria, capable of using ibuprofen as a source of carbon and energy. Based on genetic analyses, they proposed five-gene cluster (*ipfABDEF*), coding enzymes putatively involved in ibuprofen catabolism. Two of these genes (*ipfA*, *ipfB*) were very similar to genes coding two subunits of dioxygenases; the third gene was identified as a gene coding enzyme for the removal/addition of acyl groups—acyl-CoA synthetase (IpfD); the fourth one (*ipfF*) was described as a coenzyme A ligase gene; and for the gene *ipfE*, no function was found. Two additional genes *ipfH* and *ipfI* encode ferredoxin reductase and ferredoxin components of an aromatic dioxygenase system, respectively (Murdoch and Hay 2013). As the first step in the decomposition of ibuprofen by strain Ibu-2, thioesterification with coenzyme A with the participation of coenzyme A ligase was suggested. Removal of the propionic acid chain and dioxygenation reaction led to isobutylocatechol formation. This compound undergoes oxygenolytic cleavage (Murdoch and Hay 2005; Murdoch and Hay 2013) (Fig. 2). Ibuprofen biotransformation by *Variovorax* Ibu-1 occurs with trihydroxyibuprofen as a metabolite. This compound may be a dead-end metabolite or is substrate to the *meta*-ring fission reaction (Murdoch and Hay 2015).

*Patulinobacter* sp. I11 did not grow with ibuprofen as the only source of carbon and energy. However, degradation of ibuprofen was observed in the presence of yeast extract and tryptone. This suggests that ibuprofen could not induce the expression of enzymes responsible for its decomposition. In the bacterial genome, homologous genes were found coding enzymes potentially involved in ibuprofen decomposition, such as acyl-CoA synthetase, a protein containing a Rieske-like (2Fe-2S) iron-sulfur cluster (dioxygenase-like protein), and enoyl-CoA hydratase/isomerase (Almeida et al. 2013).

## CONCLUSION

The occurrence of micropollutants in the environment, such as non-steroidal anti-inflammatory drugs, is a relatively new problem. The presence of these drugs in the environment poses a risk of long-term exposure, causing chronic toxic effects for organisms. Although the salicylic acid pathway is very well described, little is known about the biotransformation/biodegradation of other non-steroidal anti-inflammatory drugs such as ibuprofen or paracetamol. Paracetamol degradation pathways lead through hydroquinone as a key intermediate, whereas ibuprofen is metabolized by hydroxylation or activation with CoA. However, sometimes biotransformation of monocyclic NSAIDs leads to the accumulation of intermediates more toxic than the parent compounds. An understanding the drug mineralization processes is key to creating commercially available solutions for this increasing problem.

## REFERENCES

1. Almeida, B., Kjeldal, H., Lolas, I., Knudsen, A. D., Carvalho, G., Nielsen, K. L., Barreto Crespo, M. T., Stensballe, A., & Nielsen, J. L. (2013). Quantitative proteomic analysis of ibuprofen-degrading *Patulibacter* sp. strain I11. *Biodegradation, 24*, 615–630.
2. Anderson, J. J., & Dagley, S. (1980). Catabolism of aromatic acids in *Trichosporon cutaneum. Journal of Bacteriology, 141*(2), 534–543.
3. Balashova, N. V., Stolz, A., Knackmuss, H. J., Kosheleva, I. A., Naumov, A. V., & Boronin, A. M. (2001). Purification and characterization of a salicylate hydroxylase involved in 1-hydroxy-2-naphthoic acid hydroxylation from the naphthalene and phenanthrene-degrading bacterial strain *Pseudomonas putida* BS202-P1. *Biodegradation, 12*, 179–188.
4. Bender, M., & MacCrehan, W. A. (2006). Transformation of acetaminophen by chlorination produces the toxicants 1,4-benzoquinone and N-acetyl-p-benzoquinone imine. *Environmental Science and Technology, 40*, 516–522.
5. Bender, R. P., Lindsey, R. H., Burden, D. A., & Osheroff, N. (2004). N-Acetyl-p-benzoquinone imine, the toxic metabolite of acetaminophen, is a topoisomerase II poison. *Biochemistry, 43*, 3731–3739.
6. Bendz, D., Paxéus, N. A., Ginn, T. R., & Loge, F. J. (2005). Occurrence and fate of pharmaccutically active compounds in the environment, a case study: Höje river in Sweden.*Journal of Hazardous Materials, 122*, 195–204.
7. Bertini, I., Cremonini, M. A., Ferretti, S., Lozzi, I., Luchinat, C., & Viezzoli, M. S. (1996). Arene hydroxylases: metalloenzymes catalysing dioxygenation of aromatic compounds. *Coordination Chemistry Review, 151*, 145–160.
8. Bessems, J. G. M., & Vermeulen, N. P. E. (2001). Paracetamol (acetaminophen)-induced toxicity: molecular and biochemical mechanisms, analogues and protective approaches. *Critical Review in Toxicology, 31*(1), 55–138.
9. Bock, K. W., Wiltfang, J., Blume, R., Ullrich, D., & Bircher, J. (1987). Paracetamol as a test drug to determine glucuronide formation in man. Effects of inducers and

of smoking. *European Journal of Clinical Pharmacology, 31*, 677–683.

10. Brausch, J. M., Connors, K. A., Brooks, B. W., & Rand, G. M. (2012). Human pharmaceuticals in the aquatic environment: a review of recent toxicological studies and consideration for toxicity testing. *Reviews of Environmental Contamination and Toxicology, 218*, 1–99.
11. Buser, H. R., Poiger, T., & Muller, M. D. (1999). Occurrence and environmental behavior of the chiral pharmaceutical drug ibuprofen in surface waters and in wastewater. *Environmental Science and Technology, 33*, 2529–2535.
12. Calamari, D., Zuccato, E., Castiglioni, S., Bagnati, R., & Fanelli, R. (2003). Strategic survey of therapeutic drugs in the Rivers Po and Lambro in Northern Italy. *Environmental Science and Technology, 37*, 1241–1248.
13. Calleja, M. C., Personne, G., & Geladi, P. (1994). Comparative acute toxicity of the first 50 multicentre evaluation of in vitro cytotoxicity chemicals to aquatic non vertebrates. *Archives of Environmental Contamination and Toxicology, 26*, 69–78.
14. Carlsson, C., Johansson, A. K., Alvan, G., Bergman, K., & Kühler, T. (2006). Are pharmaceuticals potent environmental pollutants? Part I: Environmental risk assessments of selected active pharmaceutical ingredients. *Science of the Total Environment, 364*, 67–87.
15. Chakrabarty, A. M. (1972). Genetic basis of the biodegradation of salicylate in *Pseudomonas. Journal of Bacteriology, 112*(2), 815–822.
16. Chandrasekharan, N. V., Dai, H., Roos, K. L. T., Evanson, N. K., Tomsik, J., Elton, T. S., & Simmons, D. L. (2002). COX-3, a cyclooxygenase-1 variant inhibited by acetaminophen and other analgesic/antipyretic drugs: cloning, structure, and expression. *Proceedings of the National Academy of Sciences, 99*(21), 13926–13931.
17. Chauhan, A., Samanta, S. K., & Jain, R. K. (2000). Degradation of 4-nitrocatechol by*Burkholderia cepacia:* a plasmid-encoded novel pathway. *Journal of Applied Microbiology, 88*(5), 764–772.
18. Chen, Y., & Rosazza, J. P. N. (1994). Microbial transformation of ibuprofen by a *Nocardia*species. *Applied and Environmental Microbiology, 60*, 1292–1296.
19. Civilini, M., de Bertoldi, M., & Tell, G. (1999). Molecular characterization of *Pseudomonas aeruginosa* 2NR degrading naphthalene. *Letters in Applied Microbiology, 29*, 181–186.
20. Cleuvers, M. (2004). Mixture toxicity of the anti-inflammatory drugs diclofenac, ibuprofen, naproxen, and acetylsalicylic acid. *Ecotoxicology and Environmental Safety, 59*, 309–315.
21. Daubaras, D. L., Saido, K., & Chakrabarty, A. M. (1996). Purification of hydroxyquinol 1,2-dioxygenase and maleylacetate reductase: the lower pathway of 2,4,5-trichlorophenoxyacetic acid metabolism by *Burkholderia cepacia*

AC1100. *Applied and Environmental Microbiology, 62*(11), 4276–4279.

22. de Gaetano, G., Cerletti, C., Dejana, E., & Latini, R. (1985). Pharmacology of platelet inhibition in humans: implications of the salicylate–aspirin interaction. *Circulation, 72*(6), 1185–1193.

23. de Gusseme, B., Vanhaecke, L., Verstraete, W., & Boon, N. (2011). Degradation of acetaminophen by *Delftia tsuruhatensis* and *Pseudomonas aeruginosa* in a membrane bioreactor. *Water Research, 45*, 1829–1837.

24. de Lange, H. J., Noordoven, W., Murk, A. J., Lurling, M., & Peeters, E. T. H. M. (2006). Behavioural responses of *Gammarus pulex* (Crustacea, Amphipoda) to low concentrations of pharmaceuticals. *Aquatic Toxicology, 78*, 209–216.

25. Dębska, J., Kot-Wasik, A., & Namiesnik, J. (2005). Determination of nonsteroidal anti-inflammatory drugs in water samples using liquid chromatography coupled with diode–array detector and mass spectrometry. *Journal of Separation Science, 28*, 2419–2426.

26. Deveryshetty, J., Suvekbala, V., Varadamshetty, G., & Phale, P. S. (2007). Metabolismof 2-, 3- and 4-hydroxybenzoates by soil isolates *Alcaligenes* sp. strain PPH and *Pseudomonas* sp. strain PPD. *FEMS Microbiology Letters, 268*, 59–66.

27. Fuenmayor, S. L., Wild, M., Boyes, A. L., & Williams, P. A. (1998). A gene cluster encoding steps in conversion of naphthalene to gentisate in *Pseudomonas* sp. strain U2. *Journal of Bacteriology, 180*(9), 2522–2530.

28. Gómez, M. J., Martínez Bueno, M. J., Lacorte, S., Fernández-Alba, A. R., & Agüera, A. (2007). Pilot survey monitoring pharmaceuticals and related compounds in a sewage treatment plant located on the Mediterranean coast. *Chemosphere, 66*, 993–1002.

29. Grant, D. J. W. (1971). Degradation of acetylsalicylic acid by a strain of *Acinetobacter lwoffii*. *Journal of Applied Bacteriology, 34*, 689–698.

30. Grund, E., Denecke, B., & Eichenlaub, R. (1992). Naphthalene degradation *via* salicylate and gentisate by *Rhodococcus* sp. strain B4. *Applied and Environmental Microbiology, 58*(6), 1874–1877.

31. Grund, E., Knorr, C., & Eichenlaub, R. (1990). Catabolism of benzoate and monohydroxylated benzoates by *Amycolatopsis* and *Streptomyces* spp. *Applied and Environmental Microbiology, 56*(5), 1459–1464.

32. Guzik, U., Hupert-Kocurek, K., Mazur, A., & Wojcieszyńska, D. (2013a). Biotransformacja wybranych niesteroidowych leków przeciwzapalnych w środowisku. *Bromatologia i Chemia Toksykologiczna, 1*, 105–112 (in Polish).

33. Guzik, U., Hupert-Kocurek, K., & Wojcieszyńska, D. (2013b). Intradiol dioxygenases—the key enzymes in xenobiotics degradation. In: Chamy, R., & Rosenkranz, F. (Ed.), Biodegradation of Hazardous and Special Products (pp. 129–153). Rijeka

34. Guzik, U., Hupert-Kourek, K., & Wojcieszyńska, D. (2014). Microbial degradation non-steroidal anti-inflammatory drugs. *Postepy Mikrobiologii, 53*,

61–69 (in Polish).

35. Halling-Sorensen, B., Nors Nielsen, S., Lanzky, P. F., Ingerslev, F., Holten Lützhof, H. C., & Jorgensen, S. E. (1998). Occurrence, fate, and effects of pharmaceutical substances in the environment—a review. *Chemosphere, 36*, 357–393.

36. Han, S., Choi, K., Kim, J., Ji, K., Kim, S., Aho, B., Yun, J., Choi, K., Khim, J. S., Zhang, X., & Giesy, J. P. (2010). Endocrine disruption and consequences of chronic exposure to ibuprofen in Japanese medaka *Oryzias latipes* and freshwater cladocerans *Daphnia magna* and *Moina macrocopa. Aquatic Toxicology, 98*, 256–264.

37. Haribabu, B., Kamath, A. V., & Vaidyanathan, C. S. (1984). Degradation of substituted benzoic acids by a *Micrococcus* species. *FEMS Microbiology Letters, 21*(2), 197–200.

38. Hart, A., & Orr, D. L. J. (1975). The degradation of paracetamol (4-hydroxyacetanilide) and other substituted acetanilides by a *Penicillium* species. *Antonie van Leeuwenhoek, 41*, 239–247.

39. Heberer, T. (2002). Occurrence, fate, and removal of pharmaceuticals residues in the aquatic environment: a review of recent research data. *Toxicology Letters, 131*, 5–17.

40. Henschel, K. P., Wenzel, A., Diederich, M., & Fliedner, A. (1997). Environmental hazard assessment of pharmaceuticals. *Regulatory Toxicology and Pharmacology, 25*, 220–225.

41. Herd, B., Wynne, H., Wright, P., James, O., & Woodhouse, K. (1991). The effect of age on glucuronidation and sulphation of paracetamol by human liver fractions. *British Journal of Clinical Pharmacology, 32*, 768–770.

42. Hinson, A. H., Roberts, D. W., & James, L. P. (2010). Mechanisms of acetaminophen-induced liver necrosis. *Handbook of Experimental Pharmacology, 196*, 369–405.

43. Hintner, J. P., Lechner, C., Riegert, U., Kuhm, A. E., Storm, T., Reemtsma, T., & Stolz, A. (2001). Direct ring fission of salicylate by a salicylate 1,2-dioxygenase activity from*Pseudaminobacter salicylatoxidans. Journal of Bacteriology, 183*(23), 6936–6942.

44. Huang, H. H., Lin, L. H., Zhang, P., Qi, X. L., & Zhong, D. F. (2006). Formation of glucoside conjugate of acetaminophen by fungi separated from soil. *European Journal of Drug Metabolism and Pharmacokinetics, 31*(2), 103–108.

45. Ingelman-Sundberg, M., Kaur, H., Terelius, Y., Persson, J. O., & Halliwell, B. (1991). Hydroxylation of salicylate by microsomal fractions and cytochrome P-450. *Biochemical Journal, 276*, 753–757.

46. Ishiyama, D., Vujaklija, D., & Davies, J. (2004). Novel pathway of salicylate degradation by*Streptomyces* sp. strain WA46. *Applied and Environmental Microbiology, 70*(3), 1297–1306.

47. Ivshina, I. B., Rychkova, M. I., Vikhareva, E. V., Chekryshkina, L. A., & Mishenina, I. I. (2006). Catalysis of the biodegradation of unusable medicines by

alkanotrophic *Rhodococci*. *Applied Biochemistry and Microbiology, 42*, 392–395.

48. Iwasaki, Y., Gunji, H., Kino, K., Hattori, T., Ishii, Y., & Kirimura, K. (2009). Novel metabolic pathway for salicylate biodegradation *via* phenol in yeast *Trichosporon moniliiforme*. *Biodegradation, 21*, 557–564.
49. Jain, R. K., Dreisbach, J. H., & Spain, J. C. (1994). Biodegradation of*p*-nitrophenol *via* 1-, 2-, 4-benzenetriol by an *Arthrobacter* sp. *Applied and Environmental Microbiology, 60*(8), 3030–3032.
50. Jouanneau, Y., Micoud, J., & Meyer, C. (2007). Purification and characterization of a from*Sphingomonas* sp. strain CHY-1 three-component salicylate 1-hydroxylase. *Applied and Environmental Microbiology, 73*(23), 7515–7521.
51. Katagiri, M., Maeno, H., Yamamoto, S., Hayaishi, O., Kitao, T., & Oae, S. (1965). Salicylate hydroxylase, a monooxygenase requiring flavin adenine dinucleotide II. The mechanism of salicylate hydroxylation to catechol. *The Journal of Biological Chemistry, 240*, 3414–3417.
52. Khan, S. J., & Ongerth, J. E. (2002). Estimation of pharmaceutical residues in primary and secondary sewage sludge based on quantities of use and fugacity modeling. *Water Science and Technology, 46*, 105–113.
53. Kim, J. W., Ishibashi, H., Yamauchi, R., Ichikawa, N., Takao, Y., Hirano, M., Koga, M., & Arizono, K. (2009). Acute toxicity of pharmaceutical and personal care products on freshwater crustacean *Thamnocephalus platyurus* and fish *Oryzias latipes*. *Journal of Toxicological Sciences, 342*, 227–232.
54. Kolpin, D. W., Skopiec, M., Meyer, M. T., Furlong, E. T., & Zaugg, S. D. (2004). Urban contribution of pharmaceuticals and other organic wastewater contaminants to streams during differing flow conditions. *Science of the Total Environment, 328*, 119–130.
55. Kolvenbach, B. A., Lenz, M., Benndorf, D., Rapp, E., Fousek, J., Vlcek, C., Schäffer, A., Gabriel, F. L. P., Kohler, H. P. E., & Corvini, P. F. X. (2011). Purification and characterization of hydroquinone dioxygenase from *Sphingomonas* sp. strain TTNP3. *AMB Express, 1*(8), 2–11.
56. Kühn, R., Pattard, M., Pernak, K. D., & Winter, A. (1989). Results of the harmful effects of selected water pollutants (anilines, phenols, aliphatic compounds) to *Daphnia magna*. *Water Research, 23*(4), 495–499.
57. Kuspis, D. A., & Krenzelok, E. P. (1996). What happens to expired medications? A survey of community medication disposal. *Veterinary and Human Toxicology, 38*(1), 48–49.
58. Kuswandi, K., & Roberts, C. F. (1992). Genetic control of the protocatechuic acid pathway in*Aspergillus nidulans*. *Journal of General Microbiology, 138*, 817–823.
59. Lanfranconi, M. P., Christie-Oleza, J. A., Martin-Cardona, C., Suárez-Suárez, L. Y., Lalucat, J., Nogales, B., & Bosch, R. (2009). Physiological role of NahW, the additional salicylate hydroxylase found in *Pseudomonas stutzeri* AN10. *FEMS Microbiology Letters, 300*, 265–272.
60. Lee, H. B., Peart, T. E., & Svoboda, M. L. (2005). Determination of endocrine-

disrupting phenols, acidic pharmaceuticals, and personal-care products in sewage by solid-phase extraction and gas chromatography–mass spectrometry. *Journal of Chromatography. A, 1094*, 122–129.

61. Li, J., Ye, Q., & Gan, J. (2014). Degradation and transformation products of acetaminophen in soil. *Water Research, 49*, 44–52.
62. Lilius, H., Isomaa, B., & Holmstrom, T. (1994). A comparison of the toxicity of 50 reference chemicals to freshly isolated rainbow trout hepatocytes and *Daphnia magna. Aquatic Toxicology, 30*, 47–60.
63. Lin, A. Y. C., Yu, T. H., & Lateef, S. K. (2009). Removal of pharmaceuticals in secondary wastewater treatment processes in Taiwan. *Journal of Hazardous Materials, 167*, 1163–1169.
64. Marco-Urrea, E., Pérez-Trujillo, M., Vicent, T., & Caminal, G. (2009). Ability of white-rot fungi to remove selected pharmaceuticals and identification of degradation products of ibuprofen by*Trametes versicolor. Chemosphere, 74*, 765–772.
65. Mason, J. R., & Cammack, R. (1992). The electron transport proteins oh hydroxylating bacterial dioxygenases. *Annual Review of Microbiology, 46*, 277–305.
66. Metcalfe, C., Miao, X. S., Hua, W., Letcher, R., & Servos, M. (2004). Pharmaceuticals in the Canadian environment. In K. Kümmerer (Ed.), *Pharmaceuticals in the environment. Sources, fate, effects and risks* (pp. 67–90). Berlin: Springer-Verlag.
67. Middelhoven, W. J. (1993). Catabolism of benzene compounds by ascomycetous and basidiomycetous yeasts and yeast-like fungi. *Antonie van Leeuwenhoek, 63*, 125–144.
68. Miyauchi, K., Adachi, Y., Nagata, Y., & Takagi, M. (1999). Cloning and sequencing of a novel*meta*-cleavage dioxygenase gene whose product is involved in degradation of gamma-hexachlorocyclohexane in *Sphingomonas paucimobilis. Journal of Bacteriology, 181*(21), 6712–6719.
69. Moonen, M. J. H., Synowsky, S. A., van den Berg, W. A. M., Westphal, A. H., Heck, A. J. R., van den Heuvel, R. H. H., Fraaije, M. W., & van Berkel, W. J. H. (2008). Hydroquinone dioxygenase from *Pseudomonas fluorescens* ACB: a novel member of the family of nonheme-iron(II)-dependent dioxygenases. *Journal of Bacteriology, 190*(15), 5199–5209.
70. Murdoch, R. W., & Hay, A. G. (2005). Formation of catechols *via* removal of acid side chains from ibuprofen and related aromatic acids. *Applied and Environmental Microbiology, 71*(10), 6121–6125.
71. Murdoch, R. W., & Hay, A. G. (2013). Genetic and chemical characterization of ibuprofen degradation by *Sphingomonas* Ibu-2. *Microbiology, 159*, 621–632.
72. Murdoch, R. W., & Hay, A. G. (2015). The biotransformation of ibuprofen to trihydroxyibuprofen in activated sludge and by *Variovorax* Ibu-1. *Biodegradation, 26*, 105–113.

73. Musson, S. E., & Townsend, T. G. (2009). Pharmaceutical compound content of municipal solid waste. *Journal of Hazardous Materials, 162*, 730–735.

74. Nakada, N., Tanishima, T., Shinohara, H., Kiri, K., & Takada, H. (2006). Pharmaceutical chemicals and endocrine disrupters in municipal wastewater in Tokyo and their removal during activated sludge treatment. *Water Research, 40*, 3297–3303.

75. Newton, J. F., Kuo, C. H., Gemborys, M. W., Mudge, G. H., & Hook, J. B. (1982). Nephrotoxicity of *p*-aminophenol, a metabolite of acetaminophen, in the Fischer 344 rat. *Toxicolology and Applied Pharmacology, 65*, 336–344.

76. Pailler, J. Y., Pfister, K. L., Hoffmann, L., & Guignard, C. (2009). Solid phase extraction coupled to liquid chromatography–tandem mass spectrometry analysis of sulfonamides, tetracyclines, analgesics and hormones in surface water and wastewater in Luxembourg. *Science of the Total Environment, 40*, 4736–4743.

77. Paterson, J. R., Baxter, G., Dreyer, J. S., Halket, J. M., Flynn, R., & Lawrence, J. R. (2008). Salicylic acid sans aspirin in animals and man: persistence in fasting and biosynthesis from benzoic acid. *Journal of Agricultural and Food Chemistry, 56*, 11648–11652.

78. Penn, C. D., & Daniel, S. L. (2013). Salicylate degradation by the fungal plant pathogen*Sclerotinia sclerotiorum. Current Microbiology, 67*(2), 218–225.

79. Perry, L. L., & Zylstra, J. (2007). Cloning of a gene cluster involved in the catabolism of*p*-nitrophenol by *Arthrobacter* sp. strain JS443 and characterization of the *p*-nitrophenol monooxygenase. *Journal of Bacteriology, 189*(21), 7563–7572.

80. Pomati, F., Netting, A. G., Calamari, D., & Neilan, B. A. (2004). Effects of erythromycin, tetracyclin and ibuprofen on the growth of *Synechocystis* sp. and *Lemna minor. Aquatic Toxicology, 67*, 387–396.

81. Pounds, N., Maclean, S., Webley, M., Pascoe, D., & Hutchinson, T. (2008). Acute and chronic effects of ibuprofen in the mollusc *Planorbis carinatus* (Gastropoda: Planorbidae).*Ecotoxicology and Environmental Safety, 70*, 47–52.

82. Qi, P. F., Johnston, A., Balcerzak, M., Rocheleau, H., Harris, L. J., Long, X. Y., Wei, Y. M., Zheng, Y. L., & Ouellet, T. (2012). Effect of salicylic acid on *Fusarium graminearum*, the major casual agent of fusarium head blight in wheat. *Fungal Biology, 116*, 413–426.

83. Quinn, B., Gagne, F., & Blaise, C. (2008). An investigation into the acute and chronic toxicity of eleven pharmaceuticals (and their solvents) found in wastewater effluent on the cnidarians,*Hydra attenuate. Science of the Total Environment, 389*, 306–314.

84. Quintana, J. B., Weiss, S., & Reemtsma, T. (2005). Pathways and metabolites of microbial degradation of selected acidic pharmaceutical and their occurrence in municipal wastewater treated by a membrane bioreactor. *Water Research, 39*, 2654–2664.

85. Rieble, S., Joshi, D. K., & Gold, M. H. (1994). Purification and characterization

of a 1-, 2-, 4-trihydroxybenzene 1,2-dioxygenase from the basidiomycete *Phanerochaete chrysosporium*. *Journal of Bacteriology, 176*(16), 4838–4844.

86. Roberts, P. H., & Thomas, K. V. (2006). The occurrence of selected pharmaceuticals in wastewater effluent and surface waters of the lower Tyne catchment. *Science of the Total Environment, 356*, 143–153.

87. Rodarte-Morales, A. I., Feijoo, G., Moreira, M. T., & Lema, J. M. (2011). Degradation of selected pharmaceutical and personal care products (PPCPs) by white-rot fungi. *World Journal of Microbiology and Biotechnology, 27*, 1839–1849.

88. Rzepa, J. (2009). Oznaczanie leków i pestycydów w wodach powierzchniowych. In B. K. Głoda (Ed.), *Post py Chromatografii* (pp. 67–77). Siedlce: Wydawnictwo Akademii Podlaskiej (in Polish).

89. Salgado, R., Noronha, J. P., Oehmen, A., Carvalho, G., & Reis, M. A. M. (2010). Analysis of 65 pharmaceuticals and personal care products in 5 wastewater treatment plants in Portugal using a simplified analytical methodology. *Water Science and Technology, 6212*, 2862–2871.

90. Shamsuzzaman, K., & Barnsley, E. A. (1974). The regulation of naphthalene oxygenase in Pseudomonads. *Biochemical and Biophysical Research Communications, 60*, 582–589.

91. Silva, T. R., Valdman, E., Valdman, B., & Leite, S. G. F. (2007). Salicylic acid degradation from aqueous solutions using *Pseudomonas fluorescens* HK44: parameters studies and application tools. *Brazilian Journal of Microbiology, 38*, 39–44.

92. Song, H., & Chen, T. S. (2001). *p*-Aminophenol-induced liver toxicity: tentative evidence of a role for acetaminophen. *Journal of Biochemical and Molecular Toxicology, 15*, 34–40.

93. Sosnowska, K., Styszko-Grochowiak, K., & Gołaś, J. (2009). Leki w środowisku – źródła, przemiany, zagrożenia. IV Krakowska Konferencja Młodych Uczonych (pp. 395–404) (in Polish).

94. Suzuki, K., Mizuguchi, M., Ohnishi, K., & Itagaki, E. (1996). Structure of chromosomal DNA coding for *Pseudomonas putida* S-1 salicylate hydroxylase. *Biochimica et Biophysica Acta, 1275*, 154–156.

95. Sze, I. S., & Dagley, S. (1984). Properties of salicylate hydroxylase and hydroxyquinol 1,2-dioxygenase purified from *Trichosporon cutaneum*. *Journal of Bacteriology, 159*(1), 353–359.

96. Takenaka, S., Okugawa, S., Kadowaki, M., Murakami, S., & Aoki, K. (2003). Metabolic pathway of 4-aminophenol in *Burkholderia* sp. strain AK-5 differs from that of aniline and aniline with C-4 substituents. *Applied and Environmental Microbiology, 69*(9), 5410–5413.

97. Tauxe-Wuersch, A., de Alencastro, L. F., Grandjean, D., & Tarradellas, J. (2005). Occurrence of several acidic drugs in sewage treatment plants in Switzerland and risk assessment. *Water Research, 39*, 1761–1772.

98. Ternes, T. A. (1998). Occurrence of drugs in German sewage treatment plants and rivers. *Water Research, 32*, 3245–3260.

99. Tixier, C., Singer, H. P., Oellers, S., & Müller, S. R. (2003). Occurrence and fate of carbamazepan, clofibric acid, diclofenac, ibuprofen, ketoprofen and naproxen in surface water.*Environmental Science and Technology, 37*, 1061–1068.

100. Togola, A., & Budzinski, H. (2008). Multi-residue analysis of pharmaceutical compounds in aqueous samples. *Journal of Chromatography A, 1177*, 150–158.

101. Tsikas, D., Trettin, A., Zorner, A. A., & Gutzki, F. M. (2011). In-source formation of N-acetyl-*p*-benzoquinone imine (NAPQI), the putatively toxic acetaminophen (paracetamol) metabolite, after derivatization with pentafluorobenzyl bromide and GC–ECNICI-MS analysis. *Journal of Chromatography B, 879*, 1476–1484.

102. Verberne, M. C., Verpoorte, R., Bol, J. F., Marcado-Blanco, J., & Linthorst, H. J. M. (2000). Overproduction of salicylic acid in plants by bacterial transgenes enhances pathogen resistance.*Nature Biotechnology, 18*, 779–783.

103. Vieno, N. M., Tuhkanen, T., & Kronberg, L. (2006). Analysis of neutral and basic pharmaceuticals in sewage treatment plants and in recipient rivers using solid phase extraction and liquid chromatography–tandem mass spectrometry detection. *Journal of Chromatography A, 1134*, 101–111.

104. Wang, W. H., & Lay, J. P. (1989). Fate and effects of salicylic acid compounds in freshwater systems. *Ecotoxicology and Environmental Safety, 17*(3), 308–316.

105. Wei, F., Zhou, Q. W., Leng, S. Q., Zhang, L. L., & Chen, J. M. (2011). Isolation, identification and biodegradation characteristics of a new bacterial strain degrading paracetamol. *Chinese Journal of Environmental Science, 32*, 1812–1819.

106. Winkler, M., Lawrence, J. R., & Neu, T. R. (2001). Selective degradation of ibuprofen and clofibric acid in two model river biofilm systems. *Water Research, 3513*, 3197–3205.

107. Wojcieszyńska, D., Greń, I., Hupert-Kocurek, K., & Guzik, U. (2011). Modulation of FAD-dependent monooxygenase activity from aromatic compounds-degrading *Stenotrophomonas maltophilia* strain KB2. *Acta Biochimica Polonica, 58*, 421–426.

108. Wu, S., Zhang, L., & Chen, J. (2012). Paracetamol in the environment and its degradation by microorganism. *Applied Microbiology and Biotechnology, 96*, 875–884.

109. Zhang, L., Hu, J., Zhu, R., Zhou, Q., & Chen, J. (2013). Degradation of paracetamol by pure bacterial cultures and their microbial consortium. *Applied Microbiology and Biotechnology, 97*, 3687–3698.

110. Zhou, N. Y., Al–Dulayymi, J., Baird, M. S., & Williams, P. A. (2002). Salicylate 5-hydroxylase from *Ralstonia* sp. strain U2: a monooxygenase with close relationships to and shared electron transport proteins with naphthalene dioxygenase. *Journal of Bacteriology, 184*(6), 1547–1555.

111. Ziylan, A., & Ince, N. H. (2011). The occurence and fate of anti-inflammatory and

analgesic pharmaceuticals in sewage and fresh water: treatability by conventional and non-conventional processes. *Journal of Hazardous Materials, 187*, 24–36.

112. Zwiener, C., Seeger, S., Glauner, T., & Frimmel, F. H. (2002). Metabolites from the biodegradation of pharmaceutical residues of ibuprofen in biofilm reactors an batch experiments. *Analytical and Bioanalytical Chemistry, 372*, 569–575.

19

CHAPTER

# *BACILLUS THURINGIENSIS* B1(2015B) IS A GRAM-POSITIVE BACTERIA ABLE TO DEGRADE NAPROXEN AND IBUPROFEN

Ariel Marchlewicz, Dorota Domaradzka, Urszula Guzik, and Danuta Wojcieszyńska

Department of Biochemistry, Faculty of Biology and Environmental Protection, University of Silesia in Katowice, Jagiellonska 28, 40-032 Katowice, Poland

## ABSTRACT

**A Gram-positive bacterium, designated as strain B1(2015b), was isolated from the soil of the chemical factory "Organika-Azot" in Jaworzno, Poland. On the basis of 16S rRNA gene sequence analysis, the isolated strain was classified as a *Bacillus thuringiensis* species. Strain B1(2015b) is able to degrade ibuprofen and naproxen, however, these compounds are not sufficient carbon sources for this strain. In the presence of glucose, *Bacillus thuringiensis* B1(2015b) degrades ibuprofen and naproxen with higher efficiency. Twenty milligrams per liter of ibuprofen was degraded within 6 days and 6 mg $l^{-1}$ of naproxen was removed within 35 days. Simultaneously, the growth of the bacterial culture was observed. The obtained results suggest that *Bacillus thuringiensis* B1(2015b) appears to be a powerful and useful tool in the bioremediation of non-steroidal anti-inflammatory drugs-contaminated environment.**

## INTRODUCTION

Non-steroidal anti-inflammatory drugs (NSAIDs) enter the environment as a result of pharmaceutical industry activity and the improper disposal of unused or expired drugs, waste generated in hospitals and stock-raising farms (Wu et al. 2012). In recent years, an increasing intake of the over-the-counter drugs such as ibuprofen,

**Citation:** Ariel Marchlewicz, Dorota Domaradzka, Urszula Guzik, and Danuta Wojcieszyńska, Bacillus thuringiensis B1(2015b) is a Gram-Positive Bacteria Able to Degrade Naproxen and Ibuprofen, DOI 10.1007/s11270-016-2893-0.

naproxen, paracetamol, ketoprofen, diclofenac, and acetylsalicylate has been observed (Wojcieszyńska et al. 2014; Domaradzka et al.2015). Ibuprofen is the third most highly consumed pharmaceutical in the world (Murdoch and Hay 2015). For example, sale of ibuprofen in Poland and Germany reached 58 tonnes in 2000 and 345 tonnes in 2001, respectively (Nikolaou et al. 2007; Styszko et al. 2010). In 2000, 35 tonnes of naproxen were consumed in England (Nikolaou et al. 2007). This drug belongs to the most frequently detected pharmaceutical in the aquatic environment (Grenni et al. 2013). Although ibuprofen and naproxen are detected in surface water, ground water or even drinking water at a concentration within the ng $l^{-1}$ to μg $l^{-1}$ range, they may accumulate in aquatic organisms (Li et al. 2015a, b; Jeffries et al. 2015). The concentration of ibuprofen determined in the wild fish plasma and bile samples was 100 to 1000-fold higher than in water (Jeffries et al. 2015). It was also shown that ibuprofen, as a non-selective cyclooxygenase inhibitor, exhibits ecotoxic effects in fish mainly through the endocrine disruption (Yu et al. 2006). In contrast, photoderivatives of naproxen tends to be ecotoxic in acute and chronic conditions (Marotta et al. 2013).

In the environment, naproxen and ibuprofen may undergo sorption, desorption, biotransformation, and abiotic transformation (Lahti and Oikari 2011). The physico-chemical transformations lead to the formation of more toxic intermediates (Marotta et al. 2013; Li et al.2015a). Therefore, bioremediation processes are an attractive alternative to these methods. Bioremediation strategies are cost-effective and enable the mineralization of NSAIDs into the safer products (Ahmed et al. 2001). However, these pharmaceuticals are difficult to biodegrade, and the efficiency of these processes is not satisfactory (Rodriguez-Rodriguez et al. 2010; Li et al. 2015b). The activated sludge process is used to treat the wastewaters containing non-steroidal anti-inflammatory drugs, but the removal of these drugs has been found to be incomplete (Rodarte-Morales et al. 2011). There is a need to isolate microorganisms with a high capacity for non-steroidal anti-inflammatory drug degradation. However, until now, only a few pure bacterial strains have been described to be able to degrade ibuprofen or naproxen. Murdoch and Hay (2005, 2015) described *Variovorax* Ibu-1 and *Sphingomonas* Ibu-2 engaged in the degradation of ibuprofen. Probably, degradation of ibuprofen by these strains occurs through catechol derivatives and the *meta* ring-fission (Murdoch and Hay 2005, 2015). During the degradation of ibuprofen by lignolytic bacteria *Nocardia* sp. NRRL 5646 formation of two metabolites, ibuprofenol and ibuprofenol acetate was observed (Chen and Rosazza 1994). Only one bacterial strain able to degrade naproxen was described. *Stenotrophomonas maltophilia*KB2 degrades this drug through the hydroxylation of the derivative of naproxen to hydroxyquinol, which is then cleaved by hydroxyquinol 1,2-dioxygenase (Wojcieszyńska et al.2014).

Due to poor knowledge about the metabolism of the non-steroidal anti-inflammatory drugs in the environment, it is necessary to search for new pure bacterial strains that are able to degrade these compounds. In this study, the isolation and characterization of a Gram-positive *Bacillus thuringiensis* B1(2015b), which exhibits the ability to degrade naproxen and ibuprofen, has been reported for the first time.

## MATERIALS AND METHODS

### Isolation of Pharmaceuticals Degrading Bacterium

Non-steroidal anti-inflammatory drugs-degrading strain was isolated from the soil of the chemical factory "Organika-Azot" in Jaworzno, Poland, using the classical enrichment technique with naproxen as a selection factor. The mixed microbial population from the soil was introduced to 0.85 % NaCl solution and shook at 30 °C in an aeration chamber. After 3 h, 1 ml samples were serially diluted from $10^{-1}$ to $10^{-3}$ with saline and spread onto the agar plates containing mineral salts medium ($Na_2HPO_4$ $12H_2O$ 3.78 g; $KH_2PO_4$ 0.5 g; $NH_4Cl$ 5 g; $MgSO_4{\cdot}7H_2O$; per liter of distilled water) with 6 mg $l^{-1}$ naproxen to obtain pure cultures. The agar plates were incubated at 30 °C for 24 h and single colonies were isolated and transferred to the nutrient agar plates to test their purity. Single colonies showing different morphological characteristics were proliferated in a nutrient broth medium (at 30 °C on a rotary shaker at 130 rpm), harvested by centrifugation (5,000×*g* at 4 °C for 15 min) and washed with fresh sterile mineral salts medium. In order to verify, which strain is able to degrade naproxen and ibuprofen, cultures in a 250-ml flask containing 100 ml of a sterile mineral salts medium supplemented with 6 mg $l^{-1}$ of naproxen or 20 mg $l^{-1}$ of ibuprofen were inoculated with previously prepared cells to the final optical density of about 0.8–1.0 in absorbance scale at $\lambda = 600$ nm. The cultures were incubated with shaking at 130 rpm at 30 °C, and samples were drawn at regular intervals to monitor growth and naproxen or ibuprofen degradation. The strain able to degrade both naproxen and ibuprofen was kept as a freezing bacterial stock.

### Morphological, Physiological, and Biochemical Characterization of the Isolated Strain

The isolated strain was phenotypically and biochemically characterized using standard techniques (Gram staining, colony shape, size, and color on nutrient agar plate, etc.), according to Bergey's Manual of Determinative Bacteriology (Holt et al. 1994). Additional biochemical and physiological characteristics were determined using the API Coryne system (BioMerieux, Lyon, France). Isolation of fatty acids was performed according to Sasser (1990). Analysis of FAMEs was performed using an HP 5890 gas chromatograph (Hewlett Packard, Rolling Meadows, IL, USA) equipped with an HP 25 m × 0.2 mm cross-linked methyl-silicone capillary column. The initial oven temperature was 170 °C, increased 5 °C min$^{-1}$ to 260 °C, the increased 40 °C min$^{-1}$and held constant at 320 °C for 1.5 min. Helium was used as the carrier gas. Fatty acid methyl esters (FAMEs) were identified with Sherlock software (TSBA library, version 3.9. Microbial ID, Newark, NJ, USA) based on the actual calibration retention times run prior to sample analysis.

### Analytical Methods

The concentration of non-steroidal anti-inflammatory drugs: ibuprofen, naproxen, paracetamol, and diclofenac (introduced as pharmaceutical substances) were determined with the HPLC technique using Merck Hitachi HPLC reversed-phase chromatograph equipped with a column Ascentis Express ® C18 HPLC Column (100 × 4.6 mm), pre-column Opti-Solw ® EXP, and UV/VIS DAD detector. The mobile phase consisted of acetonitrile and 1 % acetic acid (50:50 *v/v*for naproxen, vanillic acid, protocatechuic acid, benzoic acid, and 4-hydroxybenzoic acid assay and 5:95 *v/v* for ibuprofen assay) at a flow rate of 1 ml/min. The mobile phase consisted of acetonitrile, 1 % acetic acid, and methanol (50:30:20 *v/v/v* for diclofenac assay and 20:60:20 *v/v/v* for phenol assay) or methanol and 1 % acetic acid (5:95 *v/v*) for paracetamol assay at a flow rate of 1 ml/min. The detection wavelength was set at 260 nm (naproxen, vanillic acid, protocatechuic acid, benzoic acid, and 4-hydroxybenzoic acid), 240 nm (ibuprofen and paracetamol), 272 nm (phenol), and 276 nm (diclofenac) (Wojcieszyńska et al. 2014). Ibuprofen, naproxen, paracetamol, diclofenac, phenol, vanillic acid, protocatechuic acid, benzoic acid, and 4-hydroxybenzoic acid were identified by comparing the HPLC retention times and UV-visible spectra with those of the external standards. The concentration of glucose in the culture supernatant was determined using colorimetric anthrone method (Gerhardt et al.1994). The concentration of salicylic acid was determined by the method with iron(III) chloride (Poljudek-Fabini and Bejrih 1981).

## Naproxen and Ibuprofen Degradation Experiments

Strain B1(2015b) was routinely cultivated in the nutrient broth at 30 °C and 130 rpm for 24 h. After this, cells were harvested by centrifugation (5,000×*g* at 4 °C for 15 min), washed with a fresh sterile medium, and used as inoculum.

Degradation of naproxen or ibuprofen in monosubstrate systems were performed in 500-ml Erlenmeyer flasks containing 250 ml of the mineral salts medium (Greń et al. 2010) inoculated with cells to a final optical density of about 0.8 at $\lambda = 600$ nm (OD600). Naproxen was added to obtain a final concentration of 6 mg $l^{-1}$, and all cultures were incubated with shaking at 30 °C for 35 days. The chromatographic analyses of the culture fluid and measurements of the cultures growth were carried out every 7 days. For the studies on ibuprofen degradation, strain B1(2015b) was grown in a mineral salts medium supplemented with ibuprofen (concentration range 1–25 mg $l^{-1}$). The residual ibuprofen concentration in the culture filtrates was assayed by liquid chromatography every 24 h.

For studies on the cometabolic degradation of naproxen or ibuprofen, 1 mg $l^{-1}$ glucose was added. Cultures in 250 ml of sterile mineral salt medium supplemented with glucoses and 6 mg $l^{-1}$ naproxen or 1–25 mg $l^{-1}$ ibuprofen were inoculated with cells to a final optical density of about 0.1 at $\lambda = 600$ nm (OD600) and incubated at 30 °C with shaking at 130 rpm. If the complete degradation of the suitable growth substrate was observed, a successive dose of glucose was introduced and the culture was left for incubation until it reached $OD_{600} = 1.0$. All cultures were grown in triplicates.

Additionally, control cultures (250 ml) for each drug were prepared: an uninoculated control consisted of the mineral salts medium only (abiotic degradation control), and a heat-killed control consisted of bacterial cells destroyed by autoclaving (adsorption onto biomass control). The optical density of the heat-killed control was the same as for the examined cultures.

## Phylogenetic Analysis

Bacterial DNA was isolated from the pure culture using the DNA commercial kit (GenElute Bacterial Genomic DNA Kit, Sigma-Aldrich). For 16S rRNA gene amplification, the bacteria-specific primers: 8 F 5′AGTTTGATCATCGCTCAG 3′ and 1492R 5′GGTTACCTTGTTACGACTT3′ were used. Amplification was carried out through a program consisting of initial denaturation at 94 °C for 300 s, three cycles at 94 °C for 45 s, 57 °C for 30 s, and 72 °C for 120 s; three cycles at 94 °C for 45 s, 56 °C for 30 s, and 72 °C for 120 s; three cycles at 94 °C for 45 s, 55 °C for 30 s, and 72 °C for 120 s; 26 cycles at 94 °C for 45 s, 53 °C for 30 s, and 72 °C for 120 s; and a final elongation step at 72 °C for 300 s. The nucleotide sequencing of the gene was done by using the Big Dye$^{R}$ Terminator Cycle Sequencing Kit (Applied Biosystem) and AbiPrism®3100 Genetic Analyzer. The MegaBLAST program was used for homology searches with the standard default program. Multiple sequence alignments were performed and the neighbor-joining phylogenetic tree was constructed using CLC Sequence Viewer 7.0.2 program. The 16S rRNA gene sequence determined in this study has been deposited in the GeneBank database of NCBI under the accession number KP895873.1.

# RESULTS AND DISCUSSION

## Isolation and Identification of Strain that Degrades Aromatic Compounds

Until now, only a few strains have been isolated and characterized as naproxen or ibuprofen degraders. The ability to degrade ibuprofen was shown for *Sphingomonas* sp. Ibu-2 and *Variovorax* Ibu-1 (Murdoch and Hay 2005, 2015) as well as for *Patulibacter* sp. strain I11 (Almeida et al. 2013). The only strain able to degrade naproxen is *Stenotrophomonas maltophilia* KB2, which degrades this compound under cometabolic conditions (Wojcieszyńska et al. 2014). However, a bacterial strain that would have shown the ability to degrade both ibuprofen and naproxen has not yet been described.

The strain marked as B1(2015b), which was isolated from the soil of the chemical factory “Organika-Azot” in Jaworzno, Poland, is a Gram-positive rod-shaped bacterium able to utilize two of the five most commonly used non-steroidal anti-inflammatory drugs: ibuprofen and naproxen. In contrast, the strain is not capable of degrading salicylic acid, acetaminophen, and diclofenac. Moreover, this strain is able to use various aromatic compounds as a carbon and energy source: phenol, vanillic acid, protocatechuic acid, benzoic acid, and 4-hydroxybenzoic acid. Microbiological and biochemical characterization of the strain revealed that it is aerobic, oxidase, and catalase positive (Table 1). Colonies of strain B1(2015b) were found to be circular, smooth, convex, and cream-colored. The biochemical and physiological characteristics of strain B1(2015b) are summarized in Table 1. The analysis of the fatty acid profile showed significant contents of 18:0 *anteiso* and 16:0 and 18:0 fatty acids (Table 2). It is known that straight and branched chain fatty acids are biomarkers of Gram-positive bacteria (Piotrowska-Seget and Mrozik 2003). However, we also observed an unusual amount of 18:1 ω9c fatty acids. These results coincide with the results obtained by Li et al. (2010) who observed 30.96 % of 18:1 fatty acids in *Bacillus subtilis*. Partial sequence analysis of the 16S rRNA gene allows classifying the isolate with 98 % similarity as a member of the genus *Bacillus*. Comparison of the 16S rDNA gene sequence of the isolate with the 16S rDNA gene sequences of bacteria identified to a species level showed that selected strain belongs to the species *Bacillus thuringiensis* (Fig. 1). In accordance with these data, the isolate B1(2015b) was included in the genus *Bacillus* and named as *Bacillus thuringiensis* sp. B1(2015b).

**Table 1**: Differential phenotypic characteristics of strain B1(2015b)

| **Characteristic** | **Results** |
|---|---|
| Growth in the absence of NaCl | + |
| Growth in the presence of 1.5 % (*w/v*) NaCl | + |
| Growth in the presence of 3 % (*w/v*) NaCl | + |
| Growth at 4 °C | + |
| Growth at 20 °C | + |
| Growth at 30 °C | + |
| Growth at 42 °C | – |
| Oxidase | + |
| Catalase | + |
| Hydrolysis of esculin | + |
| Hydrolysis of gelatin | – |
| Arginine dihydrolase | – |
| Urease | – |
| Indol production | – |
| Nitrate reduction | – |
| Pyrazinamidase | – |
| Pyrrolidonyl arylamidase | – |
| Alkaline phosphatase | + |
| β-Glucuronidase | – |
| β-Galactosidase | – |
| α-Glucosidase | + |
| N-acetyl-β-glucosaminidase | – |
| Assimilation of: | |

| | |
|---|---|
| Glucose | + |
| Arabinose | − |
| Mannose | + |
| Mannitol | + |
| Maltose | + |
| Gluconate | + |
| Caprate | − |
| Adipate | − |
| Malate | − |
| Citrate | − |
| Phenylacetate | + |
| N-acetyl-glucosamine | + |
| Fermentation of: | |
| Glucose | + |
| Ribose | − |
| Xylose | − |
| Mannitol | − |
| Maltose | − |
| Lactose | − |
| Saccharose | − |
| Glycogen | − |

+ positive reaction, − negative reaction

**Table 2**: Percentage of total fatty acid from *Bacillus thuringiensis* B1(2015b)

| **Fatty acids** | **% of total fatty acids** |
|---|---|
| Saturated | |
| 14:0 | 0.21±0.011 |
| 16:0 | 9.09±0.218 |
| 17:0 | 0.28±0.025 |
| 18:0 | 4.58±0.24 |
| 18:0 2OH | 0.34±0.02 |
| 18:0 *anteiso* | 47.42±0.26 |
| 19:0 *iso* | 0.31±0.05 |

| | |
|---|---|
| 20:0 | 0.19±0.00 |
| Unsaturated | |
| 16:1 ω7c | 0.19±0.03 |
| 17:1 ω8c | 0.16±0.00 |
| 18:1 ω9c | 37.46±0.13 |
| Sat./unsat. ratio | 0.18 |

–OH indicates the position of hydroxyl group from the acid end

*ω* methyl end of fatty acid, *c cis* configuration of the double bound, *iso anteiso*-branched fatty acids

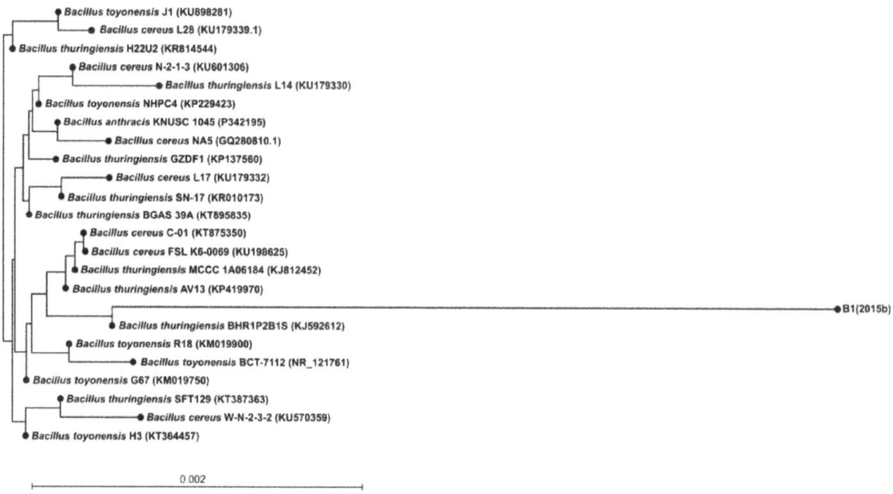


**Figure 1**: Neighbor-joining tree showing the phylogenetic position of the non-steroidal anti-inflammatory-degrading bacterium strain B1(2015b) and selected and related species of the genus *Bacillus* based on partial 16S rRNA gene sequences. The GenBank accession number for each microorganism used in the analysis is shown in parentheses after the species name.

## Degradation of Ibuprofen and Naproxen by Strain B1(2015b)

It is the first report on Gram-positive bacterium that is able to degrade both naproxen and ibuprofen. The described strain was isolated from a post-industrial landfill site belonging to a chemical factory which has produced veterinary drugs, hygiene products, and pesticides since 1928. It caused contamination of the surrounding area by cyanides, heavy metals, and pesticides such as dieldrin, endrin, α-, β-, and γ-hexachlorocyclohexane, phenols, hexachlorobenzene, dichlorodiohenyltrichloroethane (DDT), dichlorodiphenyldichloroethylene (DDE), and dichlorodiphenyldichloroethane (DDD).

These contaminations probably influence the adaptation of microflora in this area. That is why we could assume that it is a good source of microorganisms able to degrade different aromatic compounds. The bacterium isolated during this study was classified as a *Bacillus*. It is known that Gram-positive bacteria, including the*Bacillus* species, are tolerant to many toxic compounds such as phenols, policyclic aromatic hydrocarbons, heavy metals, and different organic solvents. The high tolerance to different factors is connected with the structure of cellular membranes, synthesis of the surface active agents, and specific enzymes (Satchanska et al. 2006; Trivedi et al. 2011; Solyanikova et al.2014; Swaathy et al. 2014).

The isolated strain, *Bacillus thuringiensis* B1(2015b), exhibits the ability to degrade 5 mg $l^{-1}$ of ibuprofen within 2 days (Fig. 2a) and small amounts of naproxen (Fig. 3a). However, we observed a decrease in the optical density of bacterial cultures (Figs. 2a and 3a). It indicates that these compounds are not a sufficient carbon source. It is known that an additional carbon source may enhance the degradation capability of the strain by increasing the biomass (Zhong et al. 2007; Zhang et al. 2009). Quintana et al. (2005) observed the intensification of naproxen and ibuprofen degradation by active sludge in the presence of powdered milk used as a growth substrate. In this study, glucose was used as a carbon and energy source. It resulted in the significant improvement of NSAIDs' degradation activity of the examined strain. Simultaneously, the decrease of the biomass was not observed. Under this condition, strain B1(2015b) was adapted to the degradation of 20 mg $l^{-1}$ of ibuprofen within 6 days. However, 46.56 % of 25 mg/l of ibuprofen was degraded by this strain during 20 days (Fig. 2b).

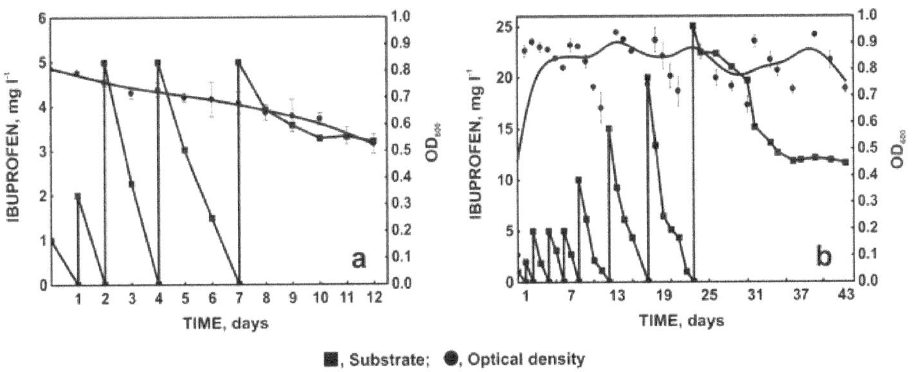


**Figure 2**: The adaptation of *Bacillus thuringiensis* B1(2015b) to increasing concentration of ibuprofen and changes of microbial biomass monitored as optical density at 600 (**a**without additional carbon source; **b** with 1 ml mg $l^{-1}$ glucose as a simple carbon source).

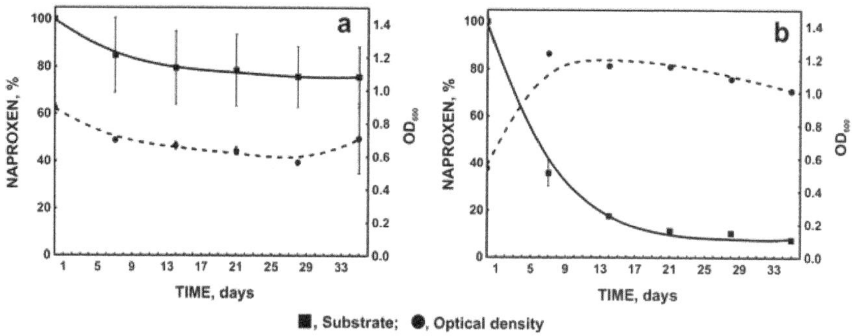


**Figure 3**: Degradation of 6 mg $l^{-1}$ naproxen by strain B1(2015b) and changes of microbial biomass monitored as optical density at 600 nm (a without additional carbon source; b with 1 ml mg $l^{-1}$ glucose as a simple carbon source).

Murdoch and Hay (2005, 2015) also described bacterial strains *Sphingomonas* Ibu-2 and *Variovorax* Ibu-1 able to degrade ibuprofen at high concentration. However, *Bacillus thuringiensis* B1(2015b) was additionally able to totally degrade 6 mg $l^{-1}$ of naproxen within 35 days (Fig. 3b). Complete degradation of naproxen by a pure bacterial strain has not yet been observed. The described *Stenotrophomonas maltophilia* strain KB2 degraded 78 % of 6 mg $l^{-1}$of naproxen in the presence of glucose (Wojcieszyńska et al. 2014).

To conclude, bacterial strain isolated from the soil of the chemical factory "Organika-Azot" in Jaworzno, Poland, was identified as *Bacillus thuringiensis* B1(2015b). This strain showed the ability to degrade 20 mg $l^{-1}$ of ibuprofen during 6 days and 6 mg $l^{-1}$ of naproxen during 5 weeks under cometabolic condition. Because of these facts, *Bacillus thuringiensis* B1(2015b) may play a key role in the bioremediation of NSAIDs-contaminated environments.

## ACKNOWLEDGMENTS

This work was financed by the National Science Centre (Poland), granted on the basis of decision DEC-2013/09/B/NZ9/00244. We thank Dr Renata Zub (Department of Molecular Biology, Institute of Oncology, Warsaw, Poland) for DNA sequencing.

## REFERENCES

1. Ahmed, S., Javed, M. A., Tanvir, S., & Hameed, A. (2001). Isolation and characterization of a*Pseudomonas* strain that degrades 4-acetamidophenol and 4-aminophenol. *Biodegradation, 12*, 303–309.
2. Almeida, B., Kjeldal, H., Lolas, I., Knudsen, A. D., Carvalho, G., Nielsen, K. L., et al. (2013). Quantitative proteomic analysis of ibuprofen-degrading *Patulibacter* sp. strain I11.*Biodegradation, 24*, 615–630.
3. Chen, Y., & Rosazza, J. P. N. (1994). Microbial transformation of ibuprofen by a *Nocardia*species. *Applied and Environmental Microbiology, 60*, 1292–1296.

4. Domaradzka, D., Guzik, U., & Wojcieszyńska, D. (2015). Biodegradation and biotransformation of polycyclic non-steroidal anti-inflammatory drugs. *Reviews in Environmental Science and Biotechnology, 14*, 229–239.

5. Gerhardt, P., Murray, R. G. E., Wood, W. A., & Krieg, N. R. (1994). *Methods for general and molecular bacteriology*. Washington DC: ASM. 1-55581-048-9.

6. Greń, I., Wojcieszyńska, D., Guzik, U., Perkosz, M., & Hupert-Kocurek, K. (2010). Enhanced biotransformation of mononitrophenols by *Stenotrophomonas maltophilia* KB2 in the presence of aromatic compounds of plant origin. *World Journal of Microbiology and Biotechnology, 26*, 289–295.

7. Grenni, P., Patrolecco, L., Ademollo, N., Tolomei, A., & Caracciolo, A. B. (2013). Degradation of gemfibrozil and naproxen in a river water ecosystem. *Microchemical Journal, 107*, 158–164.

8. Holt, J. G., Krieg, N. R., Sneath, P. H. A., Staley, J. T., & Williams, S. T. (1994). *Bergey's manual of determinative bacteriology*. USA: Williams and Wilkins.

9. Jeffries, K. M., Brander, S. M., Britton, M. T., Fangue, N. A., & Connon, R. E. (2015). Chronic exposure to low and high concentration of ibuprofen elicit different gene response patterns in a euryhaline fish. *Environmental Science and Pollution Research, 22*, 17397–17413.

10. Lahti, M., & Oikari, A. (2011). Microbial transformation of pharmaceuticals naproxen, bisprolol, and diclofenac in aerobic and anaerobic environments. *Archives of Environmental Contamination and Toxicology, 61*, 202–210.

11. Li, Y., Wu, S., Wang, L., Li, Y., Shi, F., & Wang, X. (2010). Differentiation of bacteria using fatty acid profiles from gas chromatography—tandem mass spectrometry. *Journal of the Science of Food and Agriculture, 90*, 1380–1383.

12. Li, F. H., Yao, K., Lv, W. Y., Liu, G. G., Chen, P., Huang, H. P., & Kang, Y. P. (2015a). Photodegradation of ibuprofen under UV–vis irradiation: mechanism and toxicity of photolysis products. *Bulletin of Environmental Contamination and Toxicology, 94*, 479–483.

13. Li, X., Toledo, R. A., Wang, S., & Shim, H. (2015b). Removal of carbamazepine and naproxen by immobilized *Phanerochaete chrysosporium* under non-sterile condition. *New Biotechnology, 32*, 282–289.

14. Marotta, R., Spasiano, D., Di Somma, I., & Andreozzi, A. (2013). Photodegradation of naproxen and its photoproducts in aqueous solution at 254 nm: a kinetic investigation. *Water Research, 47*, 373–383.

15. Murdoch, R. W., & Hay, A. G. (2005). Formation of catechols *via* removal of acid side chains from ibuprofen and related aromatic acid. *Applied and Environmental Microbiology, 71*, 6121–6125.

16. Murdoch, R. W., & Hay, A. G. (2015). The biotransformation of ibuprofen to trihydroxyibuprofen in activated sludge and by *Variovorax* Ibu-1. *Biodegradation, 26*, 105–113.

17. Nikolaou, A., Meric, S., & Fatta, D. (2007). Occurrence patterns of pharmaceuticals

in water and wastewater environments. *Analytical and Bioanalytical Chemistry, 387*, 1225–1234.

18. Piotrowska-Seget, Z., & Mrozik, A. (2003). Signature lipid biomarker (SLB) analysis in determining changes in community structure of soil microorganisms. *Polish Journal of Environmental Studies, 12*, 669–675.
19. Poljudek-Fabini, R., & Bejrih, T. (1981). *Organicheskij analiz* (p. 622). Leningrad: Himija (in Russian).
20. Quintana, J. B., Weiss, S., & Reemtsma, T. (2005). Pathways and metabolites of microbial degradation of selected acidic pharmaceutical and their occurrence in municipal wastewater treated by a membrane bioreactor. *Water Research, 39*, 2654–2664.
21. Rodarte-Morales, A. I., Feijoo, G., Moreira, M. T., & Lema, J. M. (2011). Degradation of selected pharmaceutical and personal care products (PPCPs) by white-rot fungi. *World Journal of Microbiology and Biotechnology, 27*, 1839–1849.
22. Rodriguez-Rodriguez, C. E., Marco-Urrea, E., & Caminal, G. (2010). Degradation of naproxen and carbamazepine in spiked sludge by slurry and solid-phase *Trametes versicolor* systems.*Bioresource Technology, 101*, 2259–2266.
23. Sasser, M. (1990). *Technical note #*, 101–110.
24. Satchanska, G., Topalova, Y., Ivanov, I., & Golovinsky, E. (2006). Xenobiotic biotransformation potential of *Pseudomonas rhodesiae* KCM-$R_5$ and *Bacillus subtilis* KCM-$RG_5$, tolernat to heavy metals and phenol derivatives. *Biotechnology & Biotechnological Equipment, 20*, 97–102.
25. Solyanikova, I. P., Robota, I. V., Mazur, D. M., Lebedev, A. Y., & Golovleva, L. A. (2014). Application of *Bacillus* sp. strain VT-8 for decontamination of TNT-polluted sites.*Microbiology, 83*, 577–584.
26. Styszko, K., Sosnowska, K., Wojtanowicz, P., Gołaś, J., Górecki, J., & Macherzyński, M. (2010). Sorption of ibuprofen on sediments from the Dobczyce (southern Poland) drinking water reservoir. *Archives of Environmental Protection, 36*, 81–91.
27. Swaathy, S., Kavitha, V., Pravin, A. S., Mandal, A. B., & Gnanamani, A. (2014). Microbial surfactant mediated degradation of anthracene in aqueous phase by marine *Bacillus licheniformis* MTCC5514. *Biotechnology Reports, 4*, 161–170.
28. Trivedi, N., Gupta, V., Kumar, M., Kumari, P., Reddy, C. R. K., & Jha, B. (2011). Solvent tolerant marine bacterium *Bacillus aquimaris* secreting organic solvent stable alkaline cellulase.*Chemosphere, 83*, 706–712.
29. Wojcieszyńska, D., Domaradzka, D., Hupert-Kocurek, K., & Guzik, U. (2014). Bacterial degradation of naproxen—undisclosed pollutant in the environment. *Journal of Environmental Management, 145*, 157–161.
30. Wu, S., Zhang, L., & Chen, J. (2012). Paracetamol in the environment and its degradation by microorganism. *Applied Microbiology and Biotechnology, 96*, 875–884.

31. Yu, J. T., Bouwer, E. J., & Coelhan, M. (2006). Occurrence and biodegradability studies of selected pharmaceuticals and personal care products in sewage effluent. *Agricultural Water Management, 86*, 72–80.

32. Zhang, G., Yang, X., Xie, F., Chao, Y., & Qian, S. (2009). Cometabolic degradation of mono-chloro benzoic acids by *Rhodococcus* sp. R04 grown on organic carbon sources. *World Journal of Microbiology and Biotechnology, 25*, 1169–1174.

33. Zhong, Y., Luan, T., Wang, X., Lan, C. H., & Tam, N. F. Y. (2007). Influence of growth medium on cometabolic degradation of polycyclic aromatic hydrocarbons by *Sphingomonas* sp. strain PheB4. *Applied Microbiology and Biotechnology, 75*, 175–186.

# 20
CHAPTER

# COMETABOLIC DEGRADATION OF NAPROXEN BY PLANOCOCCUS SP. STRAIN S5

Dorota Domaradzka, Urszula Guzik, Katarzyna Hupert-Kocurek, and Danuta Wojcieszyńska

Department of Biochemistry, Faculty of Biology and Environmental Protection, University of Silesia in Katowice, Jagiellońska 28, 40-032 Katowice, Poland

## ABSTRACT

**Naproxen is a non-steroidal anti-inflammatory drug frequently detected in the influent and effluent of sewage treatment plants. The Gram-positive strain Planococcus sp. S5 was able to remove approximately 30 % of naproxen after 35 days of incubation in monosubstrate culture. Under cometabolic conditions, with glucose or phenol as a growth substrate, the degradation efficiency of S5 increased. During 35 days of incubation, 75.14±1.71 % and 86.27±2.09 % of naproxen was degraded in the presence of glucose and phenol, respectively. The highest rate of naproxen degradation observed in the presence of phenol may be connected with the fact that phenol is known to induce enzymes responsible for aromatic ring cleavage. The activity of phenol monooxygenase, naphthalene monooxygenase, and hydroxyquinol 1,2-dioxygenase was indicated in Planococcus sp. S5 culture with glucose or phenol as a growth substrate. It is suggested that these enzymes may be engaged in naproxen degradation.**

## INTRODUCTION

Increasing interest in non-steroidal anti-inflammatory drugs (NSAIDs) as pollutants of the aquatic environment has been observed during the last decade. After the publication of an article by Ternes in 1998, where the presence of 32 pharmaceuticals and their

**Citation:** Dorota Domaradzka, Urszula Guzik, Katarzyna Hupert-Kocurek, and Danuta Wojcieszyńska, Cometabolic Degradation of Naproxen by Planococcus sp. Strain S5, DOI 10.1007/s11270-015-2564-6.

metabolites in an aquatic environment was described, a significant increase in the number of studies in which the authors indicated pharmaceuticals in soil samples, sediments, stream and river waters, wastewater treatment plants, and even drinking water has been carried out. NSAIDs have been detected in the environment in concentrations ranging from nanogram per liter to microgram per liter (Ternes 1998; Heberer 2002; Grenni et al. 2013; Grossberger et al. 2014). Many of them are slowly metabolized in the environment because of their persistence. They are excreted as demethylated and hydroxylated derivatives, acyl glucuronide conjugates, or parent drugs which accumulate in the environment (Vree et al. 1993; Christen et al. 2010). Non-steroidal anti-inflammatory drugs are the main inhibitors of cyclooxygenases. These enzymes characterize high homology regardless of organism ancestry. Due to this homology, NSAIDs may cause adverse effects in non-target organisms. Generally, environmental risk assessments are based on acute toxicity tests, which are carried out for higher concentrations of NSAIDs than those observed in the environment. However, there is little information about the chronic toxicity of environmentally relevant concentrations of these drugs. Accumulation of diclofenac in the liver, kidney, gills, and muscle of rainbow trout was reported by Schwaiger et al. (2004), Straub and Steward (2007), and Christen et al. (2010), while Melvin et al. (2014) suggested a minor influence of the exposure to low concentrations of NSAID mixtures on amphibian development. Due to the contamination of the environment by pharmaceuticals, it is necessary to find a way for their efficient degradation.

It is known that naproxen may be removed from the environment by physicochemical processes. However, during these processes, radical intermediates and/or secondary pollutants may be formed (Zhang et al. 2013). For example, Marotta et al. (2013) observed photolysis of naproxen to 1-(6-metoxy-2-naphthyl)ethanol and 2-acetyl-6-methoxynaphthalene. Moreover, physicochemical processes are very expensive and require harsh reaction conditions (Zhang et al. 2013). Bioremediation processes are a good alternative to above methods. However, effective degradation of naproxen in biological systems was not observed. Until now, only Wojcieszyńska et al. (2014) observed transformation of naproxen with about 28 % efficiency in monosubstrate culture. During removal of toxic and stable compounds, more effective are cometabolic systems. Cometabolism is defined as transformation of difficult to degrade and very stable xenobiotic compounds (cometabolite) simultaneously with degradation of easy to assimilate organic substrates (growth substrate) (Rieger et al. 2002; Greń et al. 2010). Introduction of growth substrate into the culture leads to cofactor synthesis, which are necessary for cometabolite degradation, and biomass production. Moreover, if the structure of growth substrate is similar to cometabolite, enzymes required for cometabolite transformation will be induced (Cornelissen and Sijm 1996). Schmidt et al. (1987) showed that degradation of resistant nitrophenols was more efficient in the presence of readily degraded organic substrates. It is known that naproxen exist in the environment together with either toxic or non-toxic substances which can provide sources of carbon, nitrogen, and energy. Therefore, the drugs can be more easily utilized by microorganisms colonizing polluted areas. Naproxen, as a polycyclic NSAID and naphthalene derivative, is more difficult to biodegrade by microorganisms like fungi and bacteria than monocyclic NSAIDs (e.g., paracetamol or ibuprofen). Quintana et al. (2005) observed the transformation of naproxen only in a cometabolic system. The utilization of this drug without an additional source of carbon did not occur. Moreover, the cometabolic degradation of naproxen in the presence of additional carbon sources was observed in our earlier work on the degradation of this drug by *Stenotrophomonas maltophilia* KB2 (Wojcieszyńska et al. 2014).

It was shown that non-steroidal anti-inflammatory drugs including naproxen were detected in the effluents of sewage treatment plants (Nakada et al. 2005; Brun et al. 2006; Brozinski et al.2013). For example, in Canadian Sewage Treatment Plants effluents, naproxen was detected at concentration of 0.2–14 μg/L (Brun et al. 2006). Moreover, naproxen was detected in bream and roach bile caught downstream of a wastewater treatment plant (Brozinski et al. 2013).

Due to the increasing consumption and inefficient removal of naproxen in sewage treatment plants as well as accumulation of this drug and its metabolites in organisms, it is in urgent need of searching for microorganisms able to mineralize naproxen. The present work is a continuation of studies on the identification of bacterial strains able to efficiently utilize NSAIDs. This is the first study on the Gram-positive strain *Planococcus* sp. S5 engaged in the degradation of naproxen under cometabolic conditions. The identification of enzymes involved in naproxen degradation is also presented.

## MATERIAL AND METHODS

### Bacterial Strain and Chemicals

The *Planococcus* sp. strain S5 (GenBank AY028621) used in this study was isolated from activated sludge from the sewage treatment plant in Bytom Miechowice, Poland (Łabużek et al.2003). It was cultivated in nutrient broth at 30 °C with agitation at 130 rpm for 24 h. After this time, 6 mg/L of naproxen was added to the culture. After 48 h, bacterial cells were harvested by centrifugation (5000×*g* at 4 °C for 15 min), washed with a sterile mineral salt medium, and used as an inoculum in experiments on the degradation of naproxen.

Naproxen was obtained from Sigma-Aldrich (USA). Acetonitrile and 1 % acetic acid were HPLC grade. Phenol was obtained from Merck (Germany), and glucose was purchased from POCH (Poland). All other chemicals used were of analytical grade.

### In Vivo Degradation Studies

Biological degradation of naproxen in monosubstrate and cometabolic systems was carried in 500-mL Erlenmeyer flasks containing 250 mL of a sterile mineral salt medium, which consisted of (g/L): $Na_2HPO_4 \cdot 12H_2O$ 3.78, $KH_2PO_4$ 0.5, $NH_4Cl$ 5.0, $MgSO_4 \cdot 7H_2O$ 0.2, and 0.01 yeast extract. The inoculum of bacterial cells prepared as described above was added to the medium to an initial optical density of about 1.4 and 0.05 at $\lambda = 600$ nm for the monosubstrate and cometabolic systems, respectively. Additionally, two control cultures (250 mL) were prepared: an uninoculated control consisted of the mineral salt medium only, and a heat-killed control consisted of bacterial cells destroyed by autoclaving. The optical density of the heat-killed control was the same as for the examined cultures.

Naproxen was added to each flask to the final concentration of 6 mg/L. In the monosubstrate system, naproxen was the sole carbon and energy source, while for the cometabolic transformation of naproxen, 282.33 mg/L of phenol or 1 mg/L of glucose was added as growth substrates. When the complete degradation of growth substrate and $OD_{600}<0.8$ was observed, the suitable growth substrate was added to the culture. All cultures were grown in triplicate and incubated with shaking at 130 rpm at 30 °C for 35 days.

## Determination of Substrate Concentration

Naproxen concentration was verified using HPLC (Merck HITACHI) equipped with a LiChromospher® RP-18 column (4 × 250 mm), liChroCART® 250-4 Nucleosil 5 C18, and a DAD detector. The mobile phase consisted of acetonitrile and 1 % acetic acid (50:50 *v*/*v*) at a flow rate of 1 mL/min. The detection wavelength was set at 260 nm (Wojcieszyńska et al. 2014). One-milliliter samples were taken every 7 days and centrifuged. Naproxen in supernatant was identified by comparison of HPLC retention time and UV-visible spectra with those of the external standards.

The concentration of phenol and glucose in supernatants was determined by the colorimetric method, with *p*-nitroaniline (Łurie and Rybnikova 1974) and 3,5-dinitrosalicylic acid, respectively (Miller 1959).

## Enzyme Assays

After 35 days of incubation, cells of *Planococcus* sp. strain S5 were harvested by centrifugation (4500×*g* for 15 min at 4 °C) and the obtained pellet was washed with 50 mM phosphate buffer, pH 7.0. Subsequently, the cells were disrupted by sonication (6 times for 15 s) and centrifuged at 9000×*g* for 30 min at 4 °C. The obtained crude extract was used for the enzyme assays. The activity of monooxygenase was determined spectrophotometrically by measuring NADH oxidation ($\lambda$=340 nm; $\varepsilon$=6220/M cm) (Divari et al. 2003). The naphthalene 1,2-dioxygenase, hydroxyquinol 1,2-dioxygenase, catechol 1,2-dioxygenase, and catechol 2,3-dioxygenase activity was measured spectrophotometrically by the formation of *cis*, *cis*-dihydrodiol ($\lambda$=262 nm; $\varepsilon$=8230/M cm), maleylacetate ($\lambda$=243 nm; $\varepsilon$=44,520/M cm), *cis*, *cis*-muconic acid ($\lambda$=260 nm;$\varepsilon$=16,800/M cm), and 2-hydroxymuconic semialdehyde ($\lambda$=375 nm; $\varepsilon$=3600/M cm), respectively (Cidaria et al. 1994; Wei et al. 2010; Wojcieszyńska et al. 2011). In order to determine gentisate dioxygenase, the formation of maleylpyruvate ($\lambda$=330 nm; $\varepsilon$=1080/M cm) was indicated (Feng et al. 1999). Protein concentration was measured by the method of Bradford (1976). One unit of enzyme activity was defined as the amount of enzyme required to generate 1 μmol of product per minute.

## Statistical Analysis

All experiments were performed in triplicate. The values of enzyme activities were analyzed by one-way ANOVA ($p<0.05$) using STATISTICA 10.0 PL software package.

# RESULTS AND DISCUSSION

### Biodegradation of Naproxen by *Planococcus* sp. S5

In recent years, biological methods have been increasingly used for the removal of various xenobiotics from the environment. Microorganisms possessing the efficient enzymatic systems are capable of degrading non-steroidal inflammatory drugs such as naproxen. To date, the transformation of this drug has mainly been observed in white-rot fungi (WRF).

Naproxen at a concentration of 10 mg/L and 55 μg/L was removed by *Trametes versicolor* after 6 and 5 h of incubation, respectively, while *Phanerochaete chrysosporium* was able to utilize 1 mg/L of the drug during 4 days of incubation (Marco-Urrea et al. 2010; Rodarte-Morales et al.2011). Degradation of this pharmaceutical by activated sludge, biosolids, and mixed cultures of soil and river water has also been described (Quintana et al. 2005; Monteiro and Boxal 2009; Grenni et al. 2013). Lin et al. (2015) showed that *Pseudomonas* sp. CE21, a cefalexin-degrading bacterium isolated from the activated sludge, was also capable of eliminating sulfamethoxazole and naproxen to some extent.

Gram-negative bacteria are more tolerant to xenobiotics than Gram-positive ones, mainly because of differences in the bacterial cell wall structure (Isken and Bont 1998; Greń et al.2012). Although Gram-positive bacteria do not have additional permeability barriers, especially the outer membrane, some Gram-positive strains able to degrade aliphatic and aromatic hydrocarbons, antibiotics, and chlorophenols derivatives have been reported (Park et al. 2001; Rusnak et al. 2001; Lăzăroaie 2010; Shen et al. 2012). *Planococcus* sp. S5 is a Gram-positive strain known to degrade the aromatic structure of salicylate, benzoate, hydroxybenzoate, dihydroxybenzoate, and phenol (Łabużek et al. 2003; Hupert-Kocurek et al. 2012). Due to the degradation potential of this strain and the high activity of its catechol 2,3-dioxygenase strain, S5 was used for the degradation of naproxen (Hupert-Kocurek et al. 2012). In the first set of experiments, naproxen was used as a single substrate at a concentration of 6 mg/L. After 35 days of incubation, *Planococcus* sp. strain S5 removed approximately 29.93 ± 7.94 % of this substrate (Fig. 1a). However, the decrease in the optical density of the culture (from 1.4 to 0.5) indicates that naproxen was not a sufficient carbon and energy source for the strain. There were no changes in naproxen concentration in the controls. The results obtained for the uninoculated as well as the heat-killed control indicated that the pharmaceutical was eliminated from the culture by the biological way only. Low levels of naproxen degradation were also observed by Lin et al. (2015) and Wojcieszyńska et al. (2014). *Pseudomonas* sp. strain CE21 was able to degrade 78 % of 0.01 mg/L and 100 % of 0.1 mg/L of naproxen during 3 days (Lin et al. 2015).*S. maltophilia* strain KB2 degraded 28 % of 6 mg/L of naproxen during 35 days (Wojcieszyńska et al. 2014).

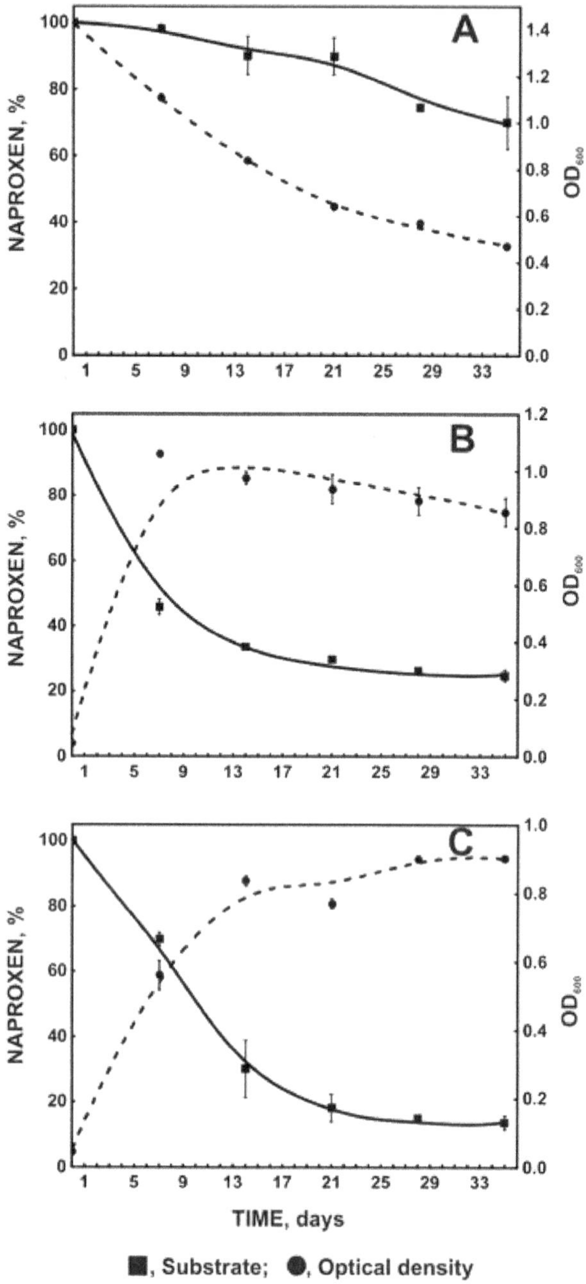


**Figure 1**: Degradation of 6 mg/L naproxen by *Planococcus* sp. strain S5 and changes of microbial biomass monitored as optical density at 600 nm (a without additional carbon source, bwith 1 mg/L glucose as a simple carbon source, c with 282.33 mg/L phenol as a carbon source).

### Cometabolic Degradation of Naproxen by *Planococcus* sp. S5

Some of the microorganisms are able to degrade xenobiotics only under cometabolic conditions (Annweiller et al. 2001; Greń et al. 2010; Arulazhagan et al. 2014). The addition of a cosubstrate which is a readily available source of carbon and energy increases biomass production. The presence of the growth substrate may also induce enzymes of xenobiotic decomposition pathways and in this way increase the xenobiotic degradation rate (Schmidt et al. 1987; Quintana et al. 2005; Greń et al. 2012). For example, the addition of powdered milk as an external carbon source allowed the degradation of three acidic pharmaceuticals by activated sludge (Quintana et al. 2005). Under cometabolic conditions, about 30 % removal of bezafibrate, 49 % removal of naproxen, and 96 % removal of ibuprofen were observed after 26 days of the experiment, while in monosubstrate systems these drugs were not degraded (Quintana et al. 2005).

In the cometabolic cultures with glucose or phenol as an additional carbon source, the growth of the biomass of *Planococcus* sp. S5 was observed (from 0.05 to 0.8 or from 0.05 to 0.9 in the presence of glucose or phenol, respectively) (Fig. 1b, c). Moreover, under this condition, significant improvement of naproxen degradation efficiency was observed. 75.14 ± 1.71 % and 86.27 ± 2.09 % of naproxen was removed after 35 days of incubation in cultures with glucose or phenol as the carbon source, respectively (Fig. 1b, c). The calculated specific naproxen degradation rates for strain S5 grown in the presence of glucose or phenol as growth substrates were 19 and 11 μg/h, respectively, whereas the specific naproxen degradation rate in monosubstrate culture was 0.5 μg/h. These results indicate that easily assimilable carbon source significantly accelerates initial degradation velocity. Glucose used as growth substrate stimulated the degradation of non-steroidal anti-inflammatory drugs by *P. chrysosporium*. Eighty-three percent of naproxen was transformed after 23 h (Rodarte-Morales et al. 2012). The other cosubstrate we used for the degradation study of naproxen was phenol. In the presence of this aromatic compound, the largest decrease in naproxen concentration was determined (Fig. 1c). Conversely, in our studies on naproxen degradation by *S. maltophilia*, degradation of this drug was most effective in the presence of glucose as a growth substrate (Wojcieszyńska et al. 2014). The highest percentage of naproxen degradation observed in the present study could be explained by the fact that phenol is known to induce enzymes responsible for aromatic ring cleavage (Krastanow et al. 2013). In the natural environment, xenobiotic compounds are present together with natural growth substrates. Therefore, cometabolic systems better reflect the degradation processes that take place in nature.

## Activation of Enzymes During Cometabolic Degradation of Naproxen

In the degradation of naproxen by white rot fungi laccases, and peroxidases like lignin peroxidases, manganese-dependent peroxidases and versatile peroxidases are engaged (Pointing 2001; Lloret et al. 2010; Marco-Urrea et al. 2010; Rodriguez-Rodriguez et al. 2010a,b; Tran et al. 2010). The formation of naproxen metabolite-1-(6-metoxynaphthalen-2-yl) ethanone as a result of laccase activity was observed by Marco-Urrea et al. (2010), whereas manganese peroxidase activity during this drug degradation was shown in *Bjerkandera* and *P. chrysosporium* (Rodarte-Morales et al. 2011). An alternative mechanism of non-steroidal anti-inflammatory drug degradation involves intracellular cytochrome P-450. It plays a key role in 6-desmethylnaproxen formation in cultures of *T. versicolor* and *P. chrysosporium*. Cytochrome P-450 and detoxification enzymes analogous to mammal enzymes of phase II were engaged in the formation of 6-desmethylnaproxen and desmethylnaproxen-6-*o*-sulfate during the transformation of naproxen by *Cunninghamella* sp. (Zhong et al. 2003; Marco-Urrea et al.2010; Rodarte-Morales et al. 2011; Rodarte-Morales et al. 2012). Conversely, there is little information about enzymes engaged in the degradation of naproxen by bacteria. In our earlier work, we indicated the contribution of naphthalene dioxygenase and phenol monooxygenase as hydroxylating enzymes, and hydroxyquinol 1,2-dioxygenase and gentisate 1,2-dioxygenase as aromatic ring cleaving enzymes in the degradation of naproxen by *S. maltophilia* strain KB2 (Wojcieszyńska et al. 2014).

Naproxen is one of the naphthalene derivatives. Therefore, during its degradation, the induction of enzymes involved in the degradation of polycyclic aromatic compounds was expected. Because of low biomass concentration in the culture with naproxen as the sole carbon and energy source, enzymes were isolated from cultures supplemented with glucose or phenol as cosubstrates. The activity of phenol monooxygenase, naphthalene dioxygenase, and hydroxyquinol 1,2-dioxygenase observed in *Planococcus* sp. S5 cultures either with glucose or phenol (Tab. 1) suggested the participation of these enzymes in naproxen degradation. These results may also indicate that hydroxylation is the first step in naproxen degradation. Generally, naphthalene dioxygenase hydroxylates C1 and C2 of naphthalene (Jeffrey et al. 1975). However, due to the propionic chain at C2 of naproxen, naphthalene dioxygenase in *Planococcus* sp. S5 may catalyze hydroxylation of this compound at C7 and C8 positions. The activity of phenol monooxygenase indicates the possibility of a third hydroxylation of the aromatic ring. The intradiol cleavage of the ring between C7 and C8 catalyzed by hydroxyquinol 1,2-dioxygenase is suggested to be the second step of naproxen degradation by Gram-positive strain S5. These two steps (hydroxylation and aromatic ring fission) are typical for xenobiotic degradation strategy in bacteria (Latus et al. 1995; Perpetuo et al. 2009; Guzik et al. 2013).

**Table 1**: Specific activity of enzymes engaged in naproxen degradation under cometabolic conditions

| **Enzyme** | **Specific enzyme activity (U/mg protein)** | |
|---|---|---|
| | **Naproxen + glucose** | **Naproxen + phenol** |
| Phenol monooxygenase | 28.32 ± 7.5 | 34.70 ± 5.8 |
| Naphthalene dioxygenase | 1.36 ± 1.0* | 4.74 ± 1.0* |
| Hydroxyquinol 1,2-dioxygen-ase | 190.80 ± 79.4 | 250.67 ± 18.8 |
| Catechol 1,2-dioxygenase | 0.0 ± 0.0 | 0.0 ± 0.0 |
| Catechol 2,3-dioxygenase | 0.0 ± 0.0* | 1070.62 ± 5.7* |
| Gentisate 1,2-dioxygenase | 517.98 ± 18.3* | 0.0 ± 0.0* |

*Values of enzyme activities which differ significantly ($p<0.05$) in dependence on the growth substrate

The key intermediate in naphthalene metabolism is salicylate, which can be cleaved by the*ortho* or *meta* pathway enzymes (Seo et al. 2009). Because in the presence of sodium salicylate, in *Planococcus* sp. S5 activity of catechol 2,3-dioxygenase as well as catechol 1,2-dioxygenase was observed (Łabużek et al. 2003), these enzymes were expected to be involved in naproxen degradation. Isolation and determination of catechol dioxygenase activity revealed the activity of catechol 2,3- dioxygenase in the culture with phenol as a growth substrate (Tab. 1), while in the culture with naproxen and glucose as a cosubstrate no activity of this enzyme was observed (Tab. 1). There was no activity of catechol 1,2-dioxygenase in cometabolic systems with phenol or glucose as growth substrates (Tab. 1). Catechol 2,3- dioxygenase of *Planococcus* sp. S5 is known to be induced in the presence of phenol (Hupert-Kocurek et al. 2012). Therefore, the activity of this enzyme only in the system with naproxen and phenol indicates its engagement in the cleavage of dihydroxylated derivative of phenol rather than in naproxen degradation. In the presence of glucose as a cosubstrate, the activity of gentisate 1,2-dioxygenase was observed (Tab. 1). It is known that salicylate can undergo transformation to gentisate, which is then cleaved by gentisate 1,2-dioxygenase to maleylpyruvic acid (Seo et al. 2009). The activity of gentisate 1,2-dioxygenase in the *Planococcus* sp. S5 culture with naproxen and glucose indicates that naproxen degradation is a more complicated process than was thought.

## CONCLUSION

Apart from white-rot fungi, bacteria are potent microorganisms with which to degrade and remove non-steroidal anti-inflammatory drugs from the environment. Under cometabolic conditions, the Gram-positive bacterium *Planococcus* sp. S5 degraded naproxen with degradation rates 19 and 11 μg/h in the presence of glucose and phenol, respectively. In the presence of easily assimilable source of carbon, faster degradation of naproxen was observed. For that reason, selection of proper conditions (e.g., kind and concentration of the carbon source) may reduce the treatment time. The activity of hydroxylating and aromatic-ring-cleaving enzymes confirms their key role in pathways of microbial degradation of naproxen. Obtained results suggest the possible application of cometabolic systems for the treatment of naproxen burdened wastewater. It should be noticed that the occurrence of naproxen in environmental waters is in the range of nanogram per liter to microgram per liter. Therefore, further naproxen biodegradation tests by *Planococcus* sp. S5 at the level of microgram per liter of naproxen would be necessary.

## REFERENCES

1. Annweiller, E., Michaelis, W., & Meckenstock, R. U. (2001). Anaerobic cometabolic conversion of benzothiophen by a sulfate-reducing enrichment culture and in a tar-oil-contaminated aquifer. *Appl Environ Microbiol, 67*, 5077–5083.

2. Arulazhagan, P., Sivaraman, C., Kumar, S. A., Slam, M., & Banu, J. R. (2014). Co-metabolic degradation of benzo(e)pyrene by halophilic bacterial consortium at different saline conditions.*J Environ Biol, 35*, 445–452.

3. Bradford, M. (1976). A rapid and sensitive method for the quantitation of microgram quantities of protein utilizing the principle of protein-dye binding. *Anal Biochem, 7*, 248–258.

4. Brozinski, J. M., Lahti, M., Meierjohann, A., Oikari, A., & Kronberg, L. (2013). The anti-inflammatory drugs diclofenac, naproxen and ibuprofen are found in the bile of wild fish caught downstream of a wastewater treatment plant. *Environ Sci Tech, 47*, 342–348.

5. Brun, G. L., Bernier, M., Losier, R., Doe, K., Jackman, P., & Lee, H. B. (2006). Pharmaceutically active compounds in Atlantic Canadian Sewage Treatment plant effluents and receiving waters, and potential for environmental effects as measured by acute and chronic aquatic toxicity.*Environ Toxicol Chem, 25*, 2163–2176.

6. Christen, V., Hickmann, S., Rechenberg, B., & Fent, K. (2010). Highly active human pharmaceuticals in aquatic systems: a concept for their identification based on their mode of action. *Aquat Toxicol, 96*, 167–181.

7. Cidaria, D., Deidda, F., & Bosetti, A. (1994). A rapid method for naphthalene dioxygenase assay in whole cell of naphthalene *cis*-dihydrodiol dehydrogenase blocked *Pseudomonas fluorescens*: screening of potential inducers of dioxygenase activity. *Appl Microbiol Biotechnol, 41*, 689–693.

8. Cornelissen, G., & Sijm, D. (1996). An energy budget model for the biodegradation and cometabolism of organic substances. *Chemosphere, 33*, 817–830.

9. Divari, S., Valetti, F., Caposito, P., Pessione, E., Cavaletto, M., Griva, E., et al. (2003). The oxygenase component of phenol hydroxylase from *Acinetobacter radioresistens* S13. *J Biochem, 270*, 2244–2253.

10. Feng, Y., Khoo, H. E., & Poh, C. L. (1999). Purification and characterization of gentisate 1,2-dioxygenases from *Pseudomonas alcaligenes* NCIB 9867 and *Pseudomonas putida* NCIB 9869.*Appl Environ Microbiol, 65*, 946–950.

11. Grenni, P., Patrolecco, L., Ademollo, N., Tolomei, A., & Caracciolo, A. B. (2013). Degradation of gemfibrozil and naproxen in a river water ecosystem. *Microchem J, 107*, 158–164.

12. Greń, I., Hupert-Kocurek, K., Osiecka, M., Guzik, U., & Wojcieszyńska, D. (2012). Toxicity of 4-chlorophenol under cometabolic conditions depending on the bacterial cell wall structure?*Architecture Civil Engineering Environment ACEE, 3*, 101–107.

13. Greń, I., Wojcieszyńska, D., Guzik, U., Perkosz, M., & Hupert-Kocurek, K. (2010). Enhanced biotransformation of mononitrophenols by *Stenotrophomonas maltophilia* KB2 in the presence of aromatic compounds of plant origin. *World J Microbiol Biotechnol, 26*, 289–295.

14. Grossberger, A., Hadar, Y., Borch, T., & Chefetz, B. (2014). Biodegradability

of pharmaceutical compounds in agricultural soils irrigated with treated wastewater. *Environ Pollut, 185*, 168–177.

15. Guzik, U., Hupert-Kocurek, K., & Wojcieszyńska, D. (2013). Intradiol dioxygenases—the key enzymes in xenobiotics degradation. In R. Chamy & F. Rosenkranz (Eds.), *Biodegradation of hazardous and special products* (pp. 129–153). Rijeka: InTech.
16. Heberer, T. (2002). Tracking persistent pharmaceutical residues from municipal sewage to drinking water. *J Hydrol, 266*, 175–189.
17. Hupert-Kocurek, K., Guzik, U., & Wojcieszyńska, D. (2012). Characterisation of catechol 2,3-dioxygenase from *Planococcus* sp. strain S5 induced by high phenol concentration. *Acta Biochim Pol, 59*, 345–351.
18. Isken, S., & Bont, J. A. M. (1998). Bacteria tolerant to organic solvents. *Extremophiles, 2*, 229–238.
19. Jeffrey, A. M., Yeh, H. J., Jerina, D. M., Patel, T. R., Davey, J. F., & Gibson, T. D. (1975). Initial reactions in the oxidation of naphthalene by *Pseudomonas putida. Biochemistry, 11*, 575–584.
20. Krastanow, A., Alexieva, Z., & Yemendzhiev, H. (2013). Microbial degradation of phenol and phenolic derivatives. *Eng Life Sci, 13*, 76–87.
21. Latus, M., Seitz, H., Eberspacher, J., & Lingens, F. (1995). Purification and characterization of hydroquinol 1,2-dioxygenase from *Azotobacter* sp. strain GP1. *Appl Environ Microbiol, 61*, 2453–2460.
22. Lăzăroaie, M. M. (2010). Multiple responses of gram-positive and gram-negative bacteria to mixture of hydrocarbons. *Braz J Microbiol, 41*, 649–667.
23. Lin, B., Lyu, J., Lyu, X. J., Yu, H. Q., Hu, Z., Lam, J. C., et al. (2015). Characterisation of cefalexin degradation capabilities of two *Pseudomonas* strains isolated from activated sludge. *J Hazard Mater, 282*, 158–164.
24. Lloret, L., Eibes, G., Lu-Chau, T. A., Moreira, M. T., Feijoo, G., & Lema, J. M. (2010). Laccase-catalyzeg degradation of anti-inflammatories and estrogens. *Biochem Eng J, 51*, 124–131.
25. Łabużek, S., Hupert-Kocurek, K., & Skurnik, M. (2003). Isolation and characterisation of new*Planococcus* sp. strain able for aromatic hydrocarbons degradation. *Acta Microbiol Pol, 52*, 395–404.
26. Łurie, J., & Rybnikova, I. (1974). *Chemical analysis of industrial sewages.* Moscow: Gaschmizdat.
27. Marco-Urrea, E., Perez-Trujillo, M., Blanquez, P., Vincent, T., & Caminal, G. (2010). Biodegradation of the analgesic naproxen by *Trametes versicolor* and identification of intermediates using HPLC-DAD-MS and NMR. *Bioresour Technol, 101*, 2159–2166.
28. Marotta, R., Spasiano, D., Di Somma, I., & Andreozzi, R. (2013). Photodegradation of naproxen and its photoproducts in aqueous solution at 254 nm: a kinetic investigation. *Water Res, 47*, 373–383.

29. Melvin, S. D., Cameron, M. C., & Lanctot, C. M. (2014). Individual and mixture toxicity of pharmaceuticals naproxen, carbamazepine and sulfamethoxazole to Australian striped marsh frog tadpoles (*Limnodynastes peronii*). *J Toxicol Environ Health A, 77*, 337–345.

30. Miller, G. L. (1959). Use of dinitrosalicylid acid reagent for determination of reducing sugar.*Anal Chem, 31*, 426–428.

31. Monteiro, S. C., & Boxal, A. B. A. (2009). Factors affecting the degradation of pharmaceuticals in agricultural soils. *Environ Toxicol Chem, 28*, 2546–2554.

32. Nakada, N., Komori, K., & Suzuki, Y. (2005). Occurrence and fate of anti-inflammatory drugs in wastewater treatment plants in Japan. *Environ Sci, 12*, 359–369.

33. Park, S. H., Ko, Y. J., & Kim, C. H. K. (2001). Toxic effects of catechol and 4-chlorobenzoate stresses on bacterial cells. *J Microbiol, 39*, 206–212.

34. Perpetuo, E. A., Marquez, R. C. P., Mendes, M. A., Lima, W. C., Menck, C. F. M., & Oller do Nescimento, C. A. (2009). Characterization of phenol monooxygenase gene from*Chromobacterium violaceum*: potential use for phenol biodegradation. *Biotechnol Bioprocess Eng, 14*, 694–701.

35. Pointing, S. B. (2001). Feasibility of bioremediation by white-rot fungi. *Appl Microbiol Biotechnol, 57*, 20–33.

36. Quintana, J. B., Weiss, S., & Reemtsma, T. (2005). Pathways and metabolites of microbial degradation of selected acidic pharmaceutical and their occurrence in municipal wastewater treated by membrane bioreactor. *Water Res, 39*, 2654–2664.

37. Rieger, P. G., Meier, H. M., Gerle, M., Voget, U., Groth, T., & Knackmuss, H. J. (2002). Xenobiotics in the environment: present and future strategies to obviate the problem of biological resistance. *J Biotechnol, 94*, 101–123.

38. Rodarte-Morales, A. I., Feijoo, G., Moreira, M. T., & Lema, M. J. (2011). Degradation of selected pharmaceutical and personal care products (PPCPs) by white-rot fungi. *World J Microbiol Biotechnol, 27*, 1839–1846.

39. Rodarte-Morales, A. I., Feijoo, G., Moreira, M. T., & Lema, M. J. (2012). Biotransformation of three pharmaceutical active compounds by the fungus *Phanerochaete chrysosporium* in a fed batch stirred reactor under air and oxygen supply. *Biodegradation, 23*, 145–156.

40. Rodriguez-Rodriguez, C. E., Marco-Urrea, E., & Caminal, G. (2010a). Naproxen degradation test to monitor *Trametes versicolor* activity in solid-state bioremediation processes. *J Hazard Mater, 179*, 1152–1155.

41. Rodriguez-Rodriguez, C. E., Marco-Urrea, E., & Caminal, G. (2010b). Degradation of naproxen and carbamazepine in spiked sludge by slurry and solid-phase *Trametes versicolor* systems.*Bioresour Technol, 101*, 2259–2266.

42. Rusnak, K., Troyanovich, J., Mierzwa, R., Chu, M., Patel, M., & Weinstein, M. (2001). An antibiotic with activity against gram-positive bacteria from the

gentamicin-producing strain of*Micromonospora purpurea. Appl Microbiol Biotechnol, 56*, 502–503.

43. Schmidt, S., Scow, K., & Alexander, M. (1987). Kinetics of *p*-nitrophenol mineralization by a*Pseudomonas* sp. effects of second substrates. *Appl Environ Microbiol, 53*, 2617–2623.

44. Schwaiger, J., Ferling, H., Malow, U., Wintermayr, H., & Negele, R. D. (2004). Toxic effect of the non-steroidal anti-inflammatory drug diclofenac. Part I: histopathological alterations and bioaccumulation in rainbow trout. *Aquat Toxicol, 68*, 141–150.

45. Seo, J. S., Keum, Y. S., & Li, Q. X. (2009). Bacterial degradation of aromatic compounds. *Int J Environ Res Public Health, 6*, 278–309.

46. Shen, X. H., Zhou, N. Y., & Liu, S. J. (2012). Degradation and assimilation of aromatic compounds by *Corynebacterium glutamicum*: another potential for applications for this bacterium? *Appl Microbiol Biotechnol, 95*, 77–89.

47. Straub, J. O., & Steward, K. M. (2007). Deterministic and probabilistic acute-based environmental risk assessment for naproxen for Western Europe. *Environ Toxicol Chem, 26*, 795–806.

48. Ternes, T. A. (1998). Occurrence of drugs in German sewage treatment plants and rivers. *Water Res, 32*, 3245–3260.

49. Tran, N. H., Urase, T., & Kusakabe, O. (2010). Biodegradation characteristics of pharmaceutical substances by whole fungal culture *Trametes versicolor* and its laccase. *Journal of Water and Environment Technology, 8*, 125–140.

50. Vree, T. B., Biggelaar-Martea, M., Verwey-van Wissen, C. P. W. G. M., Vree, M. L., & Guelen, P. J. M. (1993). The pharmacokinetics of naproxen, its metabolite *O*-desmethylnaproxen, and their acyl glucuronides in humans. Effect of cimetidine. *Br J Clin Pharmacol, 35*, 467–472.

51. Wei, M., Zhang, J. J., Liu, H., & Zhou, N. Y. (2010). *para*-Nitrophenol 4-monooxygenase and hydroxyquinol 1,2-dioxygenase catalyze sequential transformation of 4-nitrocatechol in*Pseudomonas* sp. strain WBC-3. *Biodegradation, 21*, 915–921.

52. Wojcieszyńska, D., Domaradzka, D., Hupert-Kocurek, K., & Guzik, U. (2014). Bacterial degradation of naproxen—undisclosed pollutant in the environment. *J Environ Manage, 145*, 157–161.

53. Wojcieszyńska, D., Guzik, U., Greń, I., Perkosz, M., & Hupert-Kocurek, K. (2011). Induction of aromatic ring cleavage dioxygenases in *Stenotrophomonas maltophilia* strain KB2 in cometabolic systems. *World J Microbiol Biotechnol, 27*, 805–811.

54. Zhang, L., Hu, J., Zhu, R., Zhou, Q., & Chen, J. (2013). Degradation of paracetamol by pure bacterial cultures and their microbial consortium. *Appl Microbiol Biotechnol, 97*, 3687–3698.

55. Zhong, D. F., Sun, L. L., & Huang, H. H. (2003). Microbial transformation of naproxen by*Cunninghamella* species. *Acta Pharmacol Sin, 24*, 442–447.

# INDEX